In this series original contributions of physical research in all fields are contained, such as

— research monographs
— collections of papers to a single topic
— reports on congresses of exceptional interest for physical research.

This series is aimed at promoting quick information and communication between physicists of the various special branches.

In diese Reihe werden Originalbeiträge zu allen Gebieten der physikalischen Forschung aufgenommen wie

— Forschungsmonographien
— Sammlungen von Arbeiten zu einem speziellen Thema
— Berichte von Tagungen, die für die physikalische Forschung besonders aktuell sind.

Die Reihe soll die schnelle Information und gute Kommunikation zwischen den Physikern der verschiedenen Fachgebiete fördern.

Manuscripts
in English comprising at least 100 pages and not more than 500 pages can be admitted to this series. With respect to a quick publication the manuscripts are reproduced photomechanically. Authors who are interested in this series please turn directly to the Akademie-Verlag Berlin. Here you will get more detailed information about the form of the manuscripts and the modalities of publication.

Manuskripte
in englischer Sprache, die mindestens 100 Seiten und nicht mehr als 500 Seiten umfassen, können in diese Reihe aufgenommen werden. Im Interesse einer schnellen Publikation werden die Manuskripte auf fotomechanischem Wege reproduziert. Autoren, die an der Veröffentlichung entsprechender Arbeiten in dieser Reihe interessiert sind, wenden sich bitte direkt an den Akademie-Verlag Berlin. Sie erhalten dort genauere Informationen über die Gestaltung der Manuskripte und die Modalitäten der Veröffentlichung.

Electron Microscopy in Plasticity and Fracture Research of Materials

International Symposium
Holzhau, October 1989

Physical Research

Volume 14

Electron Microscopy in Plasticity and Fracture Research of Materials

edited by
U. Messerschmidt
F. Appel
J. Heydenreich
V. Schmidt

Electron Microscopy in Plasticity and Fracture Research of Materials

edited by
U. Messerschmidt, F. Appel, J. Heydenreich, and V. Schmidt

With 345 Figures and 27 Tables

Proceedings of the International Symposium held at
Holzhau near Dresden, 8 to 13 October 1989

organized by the
Institute of Solid State Physics and Electron Microscopy of the
Academy of Sciences of the German Democratic Republic (G.D.R.)
and the
Section of Electron Microscopy of the Physical Society of the G.D.R.
under the auspices of the
International Federation of Societies for Electron Microscopy
and the
European Committee of Societies for Electron Microscopy

Akademie-Verlag Berlin

Editors:

Prof. Dr. sc. nat. Ulrich Messerschmidt
Dr. sc. nat. Fritz Appel
Prof. Dr. sc. nat. Johannes Heydenreich
Prof. Dr. sc. nat. Volker Schmidt

Institut für Festkörperphysik und Elektronenmikroskopie
Halle/Saale
Akademie der Wissenschaften der DDR

Reproduction of the original authors' manuscripts

ISBN 3-05-500711-5

Erschienen im Akademie-Verlag Berlin, Leipziger Str. 3—4, Berlin, DDR-1086

Lizenznummer: 202 · 100/452/90
Printed in the German Democratic Republic
Gesamtherstellung: VEB Druckerei „Thomas Müntzer", Bad Langensalza, 5820
Lektorin: Heike Höpcke
LSV 1185
Bestellnummer: 764 026 7 (9253)

PREFACE

The present volume contains the proceedings of the International Symposium on Electron Microscopy in Plasticity and Fracture Research of Materials held at Holzhau near Dresden, October 8 to 13, 1989. On this occasion, the editors would like to commemorate the 70th birthday of Professor Heinz Bethge.

The mechanical properties of materials depend on their crystal structure as well as on the defect structure from the macroscopic level down to the atomic level. According to general rules in the physics of plasticity and fracture, the structural elements controlling the macroscopic properties of materials are the smaller the higher the strength is. Electron microscopy plays a key role in analysing the microstructures of materials down to the nanometer level, necessary to improve the properties of up-to-date structural materials. It was the aim of the symposium to review the progress in the understanding of the microprocesses of plasticity and fracture attained by applying the broad spectrum of methods of electron microscopy.

The papers of the proceedings have been arranged in the following topics:

- Deformation of non-ferrous metals and alloys
- Microstructure and deformation of steels
- Deformation of ionic crystals and semiconductors
- Microstructure and mechanical behaviour of ceramics
- High-temperature deformation and recovery
- Fatigue
- Fracture
- Properties and interaction of lattice defects

The present volume comprises 16 keynote papers and 46 contributed papers presented at the symposium. The keynote papers are labelled by in the list of contents. The editors are grateful to all authors for submitting their manuscripts in due time.

The papers printed here were reproduced from the offset-ready manuscripts of the authors. In some cases, the editors felt it necessary to retype or revise the manuscripts. They did, however, not touch the contents and hope that the changes improved the quality of the papers.

The editors acknowledge substantial help in organizing the symposium by many coworkers of the Institute of Solid State Physics and Electron Microscopy, especially Johanna and Hartwig Baraniak, and in revising and retyping some papers by Heike Messerschmidt and Christine Scholz. They are also indebted to the Akademie-Verlag Berlin for rapid publication.

U. Messerschmidt
F. Appel
J. Heydenreich
V. Schmidt

Halle, December 1989

CONTENTS

---

* Keynote paper

1. DEFORMATION OF NON-FERROUS METALS AND ALLOYS

AN IN SITU STUDY OF STRENGTH ANOMALIES - COMPARISON BETWEEN PRISMATIC GLIDE IN BERYLLIUM AND CUBE GLIDE IN THE γ' PHASE OF A SUPERALLOY

D. Caillard, N. Clément, A. Couret

CEMES / LOE du CNRS, 29 rue Jeanne Marvig, BP 4347, 31055 TOULOUSE Cedex, France

ABSTRACT

The dislocation mechanisms at the origin of the stress anomalies of beryllium (prismatic glide) and single crystals of the γ' phase of a $CMSX_2$ superalloy (cube glide) are studied in situ and compared. A common explanation is that the nonplanar dissociation increases as the temperature is increased.

1) INTRODUCTION

The mechanical strength of most materials with a normal behaviour more or less regularly decreases at increasing temperatures. Under such conditions, the behaviour is said to be anomalous when the strength varies in the opposite sense, i.e. increases in some temperature range.

This property has very important applications in the elaboration of materials with high mechanical strength at high temperatures, and sufficient ductility at low temperatures. The main examples of this are $Ni_3(Al,X)$ and nickel-based superalloys, extensively used in the aircraft industry.

Strength anomalies are however not restricted to the above mentioned ordered alloys with the $L1_2$ structure. On the contrary, they appear in various structures, for instance ordered alloys with the $L1_2$, $L1_0$ and B_2 structures, and metals and alloys with the HCP structure.

Numerous models have been proposed to account for this anomalous behaviour. They are all based on the increase of the efficiency of obstacles opposing dislocation movements, as the temperature is increased. Although they are generally specific to the corresponding crystalline structures they can be classified in two groups, according to the origin of this increase in the efficiency of the obstacles.

i) Thermally activated mechanism. This corresponds to an increase of the frequency of dislocation obstacle interaction, by any mechanism which obeys the well-known rules of thermal activation.

ii) Non thermally activated mechanism. On the contrary, this generally corresponds to an increase of the obstacle strength by a continuous evolution of one or several parameters involved.

On the other hand, the elastic limit corresponds to the initiation of extensive dislocation movements across these obstacles. These movements can be a priori controlled either by athermal or thermally activated mechanisms. In the latter case, the crossing of obstacles tends to be easier at increasing temperatures, which tends to annihilate the strength anomaly. One important condition for obtaining a strength anomaly is thus that the increase of obstacle efficiency overcome the above classical effect of temperature, when obstacle crossing is thermally activated.

Most models, derived for the different structures mentioned above, belong to group i). They are based on the thermally activated locking of dislocations. The corresponding obstacles can be intrinsic, which corresponds to an interaction between dislocations and the crystal lattice |1-4|. They can also be extrinsic, i.e. correspond to the thermal activation of different slip systems. This includes the case of a forest mechanism with an increasing density of trees |5,6|, and the case of the trapping of mobile dislocations by cross slip and thermally activated glide on cross slip planes |7|.

The other models which are based on a non-thermally activated increase of the obstacle strength (group ii), have also been proposed long ago for $L1_2$ structures |8-10|, and recently for beryllium |11-12|, and for the $L1_2$ γ' phase of a superalloy (cube glide) |13,14|.

Most of these models are classified and described in |12|, and in other reviews |15,16|. When they are all considered in detail, however they are seen to be very different, mainly because they are based on specific properties of the corresponding crystalline structure. This sometimes makes it difficult to transpose them from one structure to another. This point being not very satisfying, it seems reasonable to assume that, on the contrary, strength anomalies in various materials result from only one, or very few, mechanisms that may be present in every crystalline structure. Starting from this assumption, we have performed a complete study in the simple case of prismatic glide in beryllium, and applied the results obtained to the more complex case of cube glide in the $L1_2$ structure. The main results are presented in parallel and compared in this paper. The study of octahedral glide in the $L1_2$ structure is in progress, and will be published later.

2) EXPERIMENTAL

It is always difficult to determine deformation processes unambiguously, on the basis of classical macroscopic experiments and images of the substructure after deformation. Several dislocation mechanisms are usually compatible with the measured activation parameters. It is also obviously difficult to deduce the type of dislocation movements from fixed pictures.

On the contrary, in situ experiments combine several advantages of both techniques, without however pretending to replace them. For instance, it is now possible to perform quantitative measurements on physical events which are perfectly characterized. This includes the measurement of local stresses, dislocation velocities, waiting times, activation energies and activation areas, on dislocations, the movement of which is unambiguously attributed to one given mechanism.

This obviously requires high accuracy in all in situ measurements, and total control of all possible artefacts. Since the latter point is often very poorly understood, we briefly explain how it is achieved.

First of all, experiments at low voltages allow irradiation effects to be eliminated, since these may sometimes be undesirable in high voltage electron microscopy.

The remaining problem is thus the representativity of deformation experiments in thin-foils. The main thin foil effects could be the following :

- Image forces. Very simple calculations show that these become non-negligible with regard to the applied stress σ at distances from the surface smaller than $\tau / \sigma = R$, where τ is the dislocation line tension and R their radius of curvature.

- Line tension effects. Dislocations ending at the surfaces tend to adopt an orientation which shortens their length, and does not too much increase their energy per unit length.

- Movements out of the foil. Some dislocation families tend to disappear at the surfaces, when they move under the applied stress.

- Easier initiation of cross slip or kinks at the surfaces. This effect can occur by a combination of image forces and line tension effects.

All these effects cannot be always avoided, but their importance can be very easily estimated, and very simple rules can be used, in order to be sure that they play a negligible role :

- The sample thickness has to be much larger than $\tau / \sigma = R$. This corresponds to the image force criterion, and has been discussed long ago by Martin and Kubin |17|. This condition is easily fulfilled when the stress is high (case of Peierls' stresses, for instance).

- It must be checked that dislocations not ending at free surfaces (distance larger than R) behave like dislocations ending at free surfaces.

- Keeping the same tensile axis and varying the foil plane must lead to consistent results.

Several checks are also very useful. For instance :

- In the case of Peierls' stresses, kink nucleation at free surfaces is negligible when the velocity of dislocations is proportional to their length (nucleation all along the dislocation).

- Measuring microscopic parameters compatible with macroscopic ones ensures similar behaviour. This is particularly convincing when the σ (T) curves are the same (see para. 4).

- A substructure similar to that observed after bulk deformation is also a good indication of similar behaviour.

In short, we consider that even if thin foils present some differences from the bulk material, there is always some degree of similar behaviour, which can be the more easily estimated as the sample thickness is larger than $5 \tau / \sigma$. This allows us to build a model, which must imperatively be compatible with results from other techniques.

All the results presented below satisfy the above criteria.

In situ studies have been performed in a Jeol 200 CX electron microscope, using two straining devices : one at high temperatures (300-1300 K) and one at low temperatures (80-300 K).

Microsamples have been cut out of single crystals of beryllium, and single crystals of the phase of the $CMSX_2$ superalloy. Their orientations have been chosen in such a way as to inhibit basal glide and to promote prismatic glide in beryllium, and to activate both octahedral and cube glide in γ'. Details concerning sample preparation can be found in |11,13,19|.

3) MICROSCOPIC DEFORMATION MECHANISMS

3.1) Observations

A complete description and analysis of the dislocation behaviour can be found in |11-14|. The most striking point is that the same dislocation behaviour is observed in both materials, and in the whole temperature range investigated. In the prismatic planes of beryllium, between 80 and 370 K, and in the cube plane of γ', between 140 and 1150 K, it is characterized by very rectilinear screw portions, moving in a jerky manner. This behaviour is illustrated on Figs. 1 to 5. Rectilinear screws remain immobile during 1/25 to a few seconds, and suddenly jump to a neighbouring position, distant 100 to a few 1000 Å. The jumps are so rapid that they cannot be recorded. Only starting and final positions can be observed simultaneously on the corresponding frames, because of the remanence of the fluorescent screen (examples : Fig. 1 at t = 0.42 s, or Fig. 3 at t = 0.42 s). In the whole temperature range investigated, dislocations multiply on sources, and on open and closed loops after cross slip as previously observed in magnesium |20|.

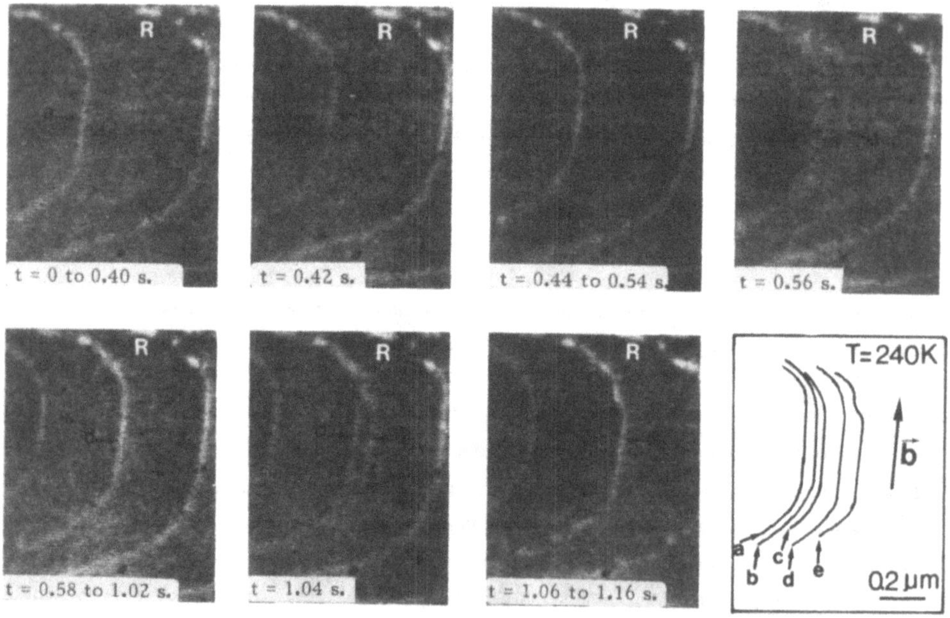

Fig. 1 - Jerky movement of screw dislocations (locking-unlocking) in Be, at T = 240 K.
a - e : successive positions of one dislocation - b : Burgers vector - R : fixed point.

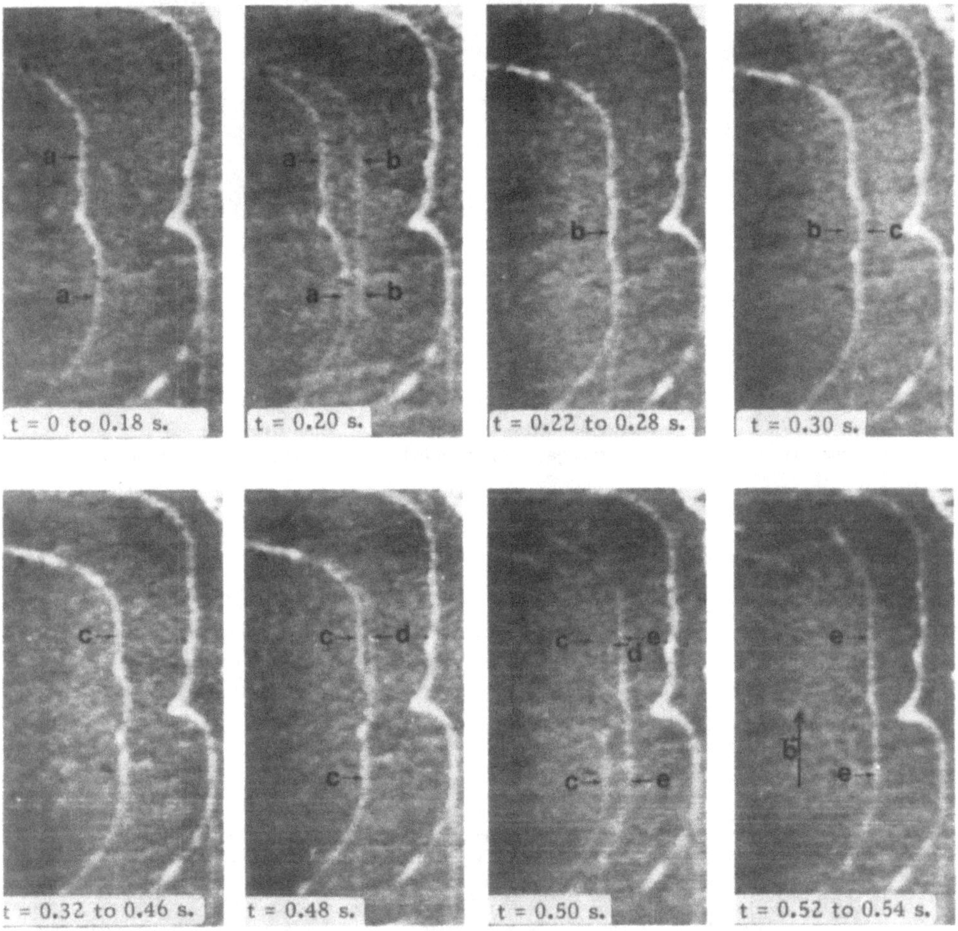

Fig. 2 - Jerky movement of screw dislocations (locking-unlocking) in Be, at T = 300 K.
a - e : successive positions of one dislocation - b : Burgers vector. Note the movement of one segment far from surfaces at t = 0.30 s.

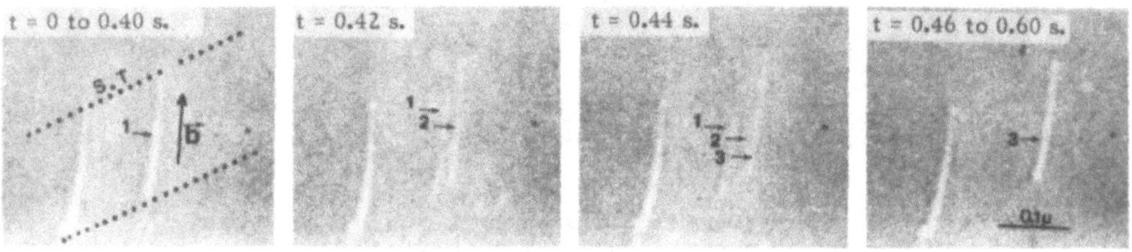

Fig. 3 - Jerky movement of screw dislocations (locking-unlocking) in γ ', at T = 570 K.
1 - 3 : successive positions of one dislocation. ST : slip trace (cube plane). Note a double jump between t = 0.42 and t = 0.44 s.

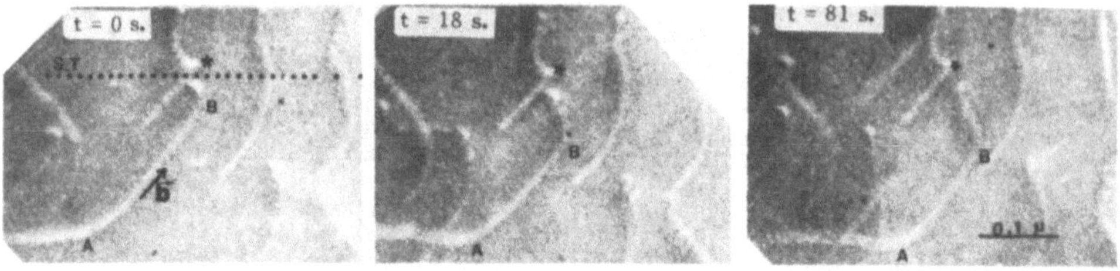

Fig. 4 - Jerky movement of screw segment AB in γ ' at T = 300 K (low strain rate).
b : Burgers vector. ST : slip traces (cube plane). Note that segment AB has no connexion with the surfaces.

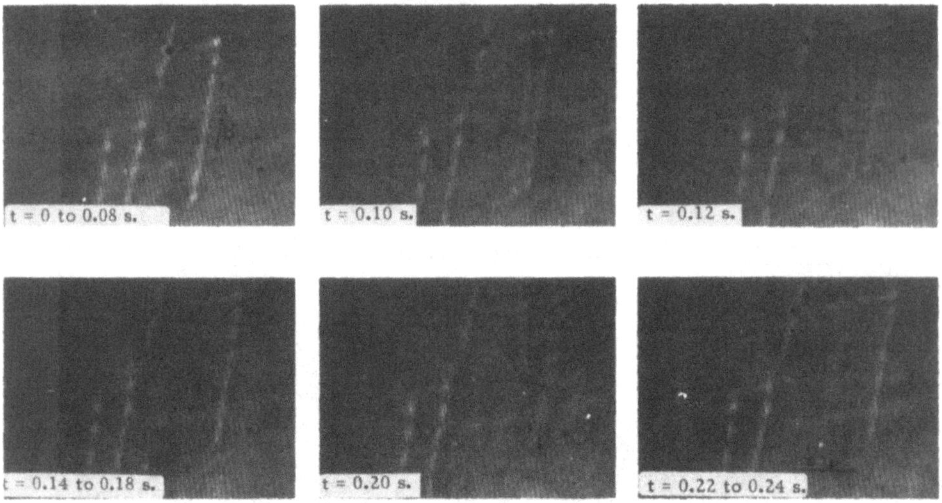

Fig. 5 - Jerky movement of screw dislocations (locking-unlocking) in γ ' at T = 220 K.
1 - 4 : successive positions of one dislocation - X is a fixed point.

3.2) A common interpretation

This interpretation has been derived for the case of beryllium |11,18|, and then transposed to γ' |13,19|.

From their rectilinear shape, it is clear that locked screw dislocations strongly interact with the crystal lattice, by spreading out of their glide plane. This situation is generally associated with a Peierls' mechanism. In the present case, however, dislocations move rapidly over hundred interatomic distances (instead of one in the case of Peierls) before being locked again. This implies that a metastable glissile configuration also exists and that the movement can be described by a series of cross slip events between a stable sessile configuration (long lifetime) and a metastable glissile configuration (short lifetime). The corresponding configurations are schematized on Fig. 6. Additional proofs of this mechanism can be found in a more detailed study |11,13|, based on slip trace observations and statistics of waiting times and jump length at different stresses and temperatures.

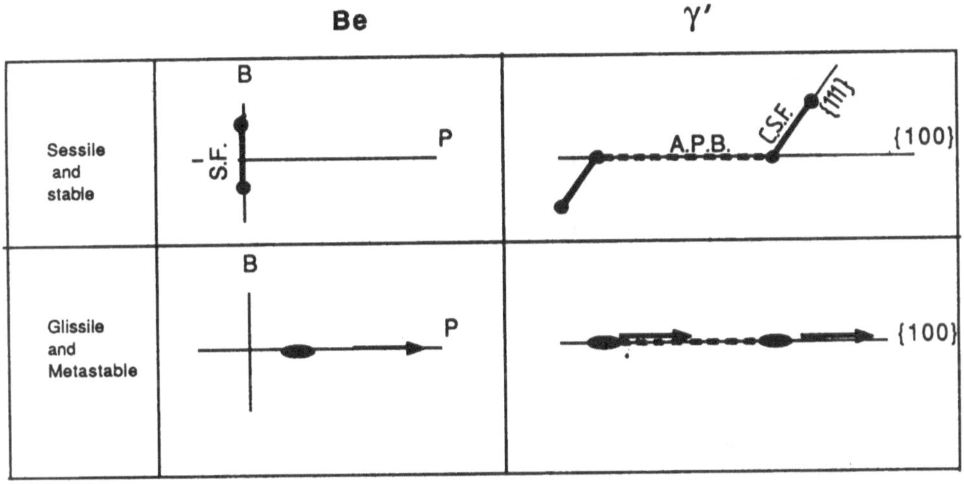

Fig 6 - Scheme of dislocation core structures in Be and γ '.

Under such conditions, the dislocation velocity can be written as : $V = V_o (P_{UL}/P_L)$, where V_o is the instantaneous dislocation velocity (glissile configuration), P_{UL} and P_L the respective probabilities of unlocking and locking per unit time.

The latter probabilities are :

$$P_{UL} \propto \exp - \frac{\Delta G_{UL}}{kT}$$

and $P_L \propto \exp - \frac{\Delta G_L}{kT}$, where ΔG_{UL} and ΔG_L are the corresponding cross slip activation energies. It can be demonstrated that $\Delta G_{UL} >> \Delta G_L$, in agreement with the corresponding models and a larger lifetime in the sessile configuration.

The resulting dislocation velocity is thus :

$$V \propto \exp - \frac{(\Delta G_{UL} - \Delta G_L)}{kT} \quad \# \exp - \frac{\Delta G_{UL}}{kT}$$

4) CORRELATION WITH STRENGTH ANOMALIES

The local stress acting on individual dislocations has been deduced from classical radii of curvature measurements |12,14|, and compared with macroscopic results.

In the case of beryllium, in situ measurements reproduce very well the macroscopic curve of Régnier, including the remarkable stress anomaly between 200 and 330 K (Fig. 7).

In the case of cube slip in γ', in situ measurements also give indications of a stress anomaly between 140 and 600 K (Fig. 8). This unexpected result is corroborated by macroscopic data obtained in Ni_3 (Al,Nb) by Lall, Chin and Pope |21|. These results clearly exhibit a slight stress anomaly, but no emphasis has been placed on this fact, and no explanation has been proposed. A similar behaviour can also be deduced from the analysis of results of Umakoshi et al. in Ni_3 (Al, Ta) |22|, and Saburi et al. in Ni_3 (Al, W) |23| (see discussion in |14|).

The presence of a stress anomaly relating to cube glide is consistent with the observation of a mixture of cube and octahedral glide at rather low temperatures, since normal behaviour would otherwise imply a very high CRSS value below about 600 K. Cube glide is however extensively present only in ternary Ni_3(Al,X) alloys, particularly those with a high Ta content |15|.

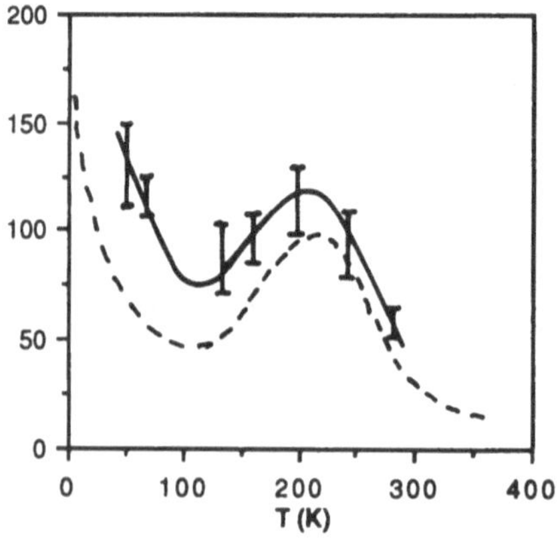

Fig. 7 - Temperature variation of the CRSS of prismatic glide in Be. Full line : local in situ measurements. Dotted line : macroscopic results of Régnier |1|.

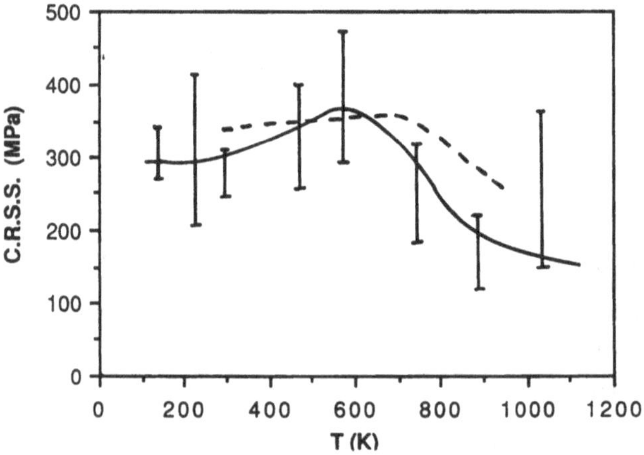

Fig. 8 - Temperature variation of the CRSS of cube glide in γ'. Full lines : local in situ measurements. Dotted lines : macroscopic results of Lall, Chin and Pope in Ni_3(Al,Nb) |21|.

5) DISCUSSION AND MODEL OF STRENGTH ANOMALY

The above results show that the behaviour of cube glide in γ' is exactly similar to that of prismatic glide in beryllium. Both glide systems present a stress anomaly, which is evidenced by in situ and macroscopic measurements, and the corresponding dislocation movements can be described by the same "locking-unlocking" mechanism. The measurement of similar stress anomalies in "in situ" and macroscopic experiments proves that the behaviour of the bulk material is well reproduced in our experiments (cf. § 1), i.e. the physical process at the origin of the stress anomalies is present in the thin foils.

It thus becomes possible to give a common explanation, on the basis of the above set of results.

From an examination of the different possible models in § 1, it could seem that the strength anomalies come from the thermally activated increase of the frequency of dislocation locking (case i)). As a matter of fact, the corresponding probability P_L increases at increasing temperatures ($P_L \alpha \exp - (\Delta G_L/kT)$). However, the unlocking process is also thermally activated, with a probability $P_{UL} \alpha \exp - (\Delta G_{UL}/kT)$. Since $\Delta G_{UL} >> \Delta G_L$ (§ 3), the temperature variation of P_{UL} (probability of crossing obstacles) is necessarily stronger than that of P_L (probability of forming obstacles), which conflicts with the conditions mentioned in § 1 for obtaining a stress anomaly. This becomes obvious when we consider the expression for the dislocation velocity

$$V \alpha \exp - \left(\frac{\Delta G_{UL} - \Delta G_L}{kT} \right)$$

which has a positive total activation energy ($\Delta G_{UL} - \Delta G_L \# \Delta G_{UL}$), in agreement with normal behaviour.

Since no additional thermally activated locking mechanism can be evidenced, the true origin of the stress anomaly is necessarily in the second kind of hypothesis (case ii) of § 1), i.e. a non thermally activated increase of the obstacle strength. In the present case, the obstacles consist in the non planar dissociation of screw dislocations, which is thus expected to increase at increasing temperatures. This conclusion is supported by the very unusual behaviour of the probability of unlocking by cross slip, P_{UL}, which remains constant at increasing temperatures in the domain of the strength anomaly, in spite of a simultaneous increase of stress and temperature |12,14|.

If we assume that dislocations consist of Shockley partials separated by well defined stacking-fault ribbons (elastic approximation), the origin of the strength anomaly thus corresponds to a decrease of the stacking fault energy in the case of beryllium, and to a decrease of the complex stacking fault (CSF) energy in γ' at increasing temperatures, in agreement with the scheme of figure 6.

Since this behaviour is a priori transposable to every glide system for which dislocations are dissociated in a non planar manner, it should be considered as a possible explanation for strength anomalies in different glide systems (octahedral in γ') and different materials. In particular, it readily explains the positive stress dependence of the strain rate, which is often observed in the domain of stress anomalies |15,24|.

6) CONCLUSIONS

In situ deformation studies in beryllium (prismatic glide) and the γ ' phase of a nickel based superalloy (cube glide) lead to very similar dislocation properties :

- The movement of screw dislocations is controlled by a double cross slip mechanism ("locking-unlocking") in the whole temperature range investigated.

- The stress versus temperature curves, obtained in situ and by conventional macroscopic experiments, exhibit similar anomalies.

The analysis of these results leads to a common origin for both stress anomalies, namely a continuous increase of the obstacle strength (non-planar dissociation), as the temperature is increased.

REFERENCES

1 P. Régnier and J.M. Dupouy, Phys. Stat. Sol., 39 (1970) 179.

2 S. Takeuchi and E. Kuramoto, Acta Met., 21 (1973) 415.

3 V. Paĭdar, D.P. Pope and V. Vitek, Acta Met., 32 (1984) 435.

4 Y.Q. Sun and P.M. Hazzledine, Phil. Mag. A 58 (1988) 603.

5 P.H. Thornton, R.G. Davies and T.L. Johnston, Metall. Trans., 1 (1970) 207.

6 A. Korner, Phil. Mag. A 58 (1988) 507.

7 P. Veyssière, MRS Conf., Mat. Res. Soc. Symp. Proc., 133 (1989) 175.

8 R.G. Davies and N.S. Stoloff, Trans. Am. Inst. Min. Metall. Petrol. Engrs. 223 (1965) 714.

9 T.L. Johnston, A.S. Tetelman and A.J. Mc Evily, High Strength Materials, ed. V.F. Zacklay (Wiley, New York) (1965) p. 365.

10 S.M. Copley and B.H. Kear, Trans. Am. Inst. Min. Metall. Petrol. Engrs. 239 (1967) 977.

11 A. Couret and D. Caillard, Phil. Mag. 59 (1989) 783.

12 A. Couret and D. Caillard, Phil. Mag. 59 (1989) 801.

13 N. Clément, A. Couret and D. Caillard, to be published.

14 N. Clément, G. Molénat and D. Caillard, to be published.

15 D.P. Pope and S.S. Ezz, Int. Metals Rev. 29 (1984) 136.

16 T. Suzuki, Y. Mishima and S. Miura, ISIJ International, 29 (1989) 1.

17 J.L. Martin and L.P. Kubin, Phys. Stat. Sol. (a) 56 (1979) 487.

18 A. Couret and D. Caillard, Proc. ICSMA 8, Tampere, Finland, Ed. by P.O. Kettunen, T.K. Lepisto and M.E. Lehtonen (1988) p. 211.

19 N. Clément, D. Caillard, P. Lours and A. Coujou, Ibidem (1988) p. 205.

20 A. Couret and D. Caillard, Acta Met. 33 (1987) 1447 and 1455.

21 C. Lall, S. Chin and D.P. Pope, Metall. Trans. 10A (1979) 1323.

22 Y. Umakoshi, D.P. Pope and V. Vitek, Acta Met. 32 (1984) 449.

23 T. Saburi, T. Hamana, S. Nenno and H. Pak, Jap. J. of Appl. Phys. 16 (1977) 267.

24 N. Baluc, J. Stoiber, J. Bonneville and J.L. Martin, Proc. 4th Israël Mat. Eng. Conf., Israël Journal of Technology, 24 (1988) 269.

THE FORMATION OF ORDERED DISLOCATION MICROSTRUCTURES

L.P. Kubin*, G. Canova**

* L.E.M. (CNRS-ONERA, OM), 29, Av. de la Division Leclerc, BP 72, 92322 Châtillon Cedex, France.
** L.P.M.M. (UA CNRS 1215), Faculté des Sciences, Ile du Saulcy, 57045 Metz Cedex, France.

ABSTRACT

During plastic flow both mobile and forest dislocations tend to form regular arrangements. The basic mechanisms responsible for these patterning effects are the long range and the local dislocation-dislocation interactions. The modelization of long range elastic interactions is, however, made difficult by their non local character. For this reason numerical simulations have been developed. Preliminary results of a 3D simulation of collective dynamic dislocation behaviour are presented. They deal with the development of dislocation loops during the early stages of easy glide in a model copper crystal.

INTRODUCTION

During plastic flow, organized dislocation patterns emerge from initially random configurations, as has been shown by many T.E.M. observations. Beyond a certain value of the plastic strain, immobile or forest dislocations form two- or three-dimensional wall, vein or cell structures which delimitate dislocation-rich and dislocation-poor regions. The perfection of these arrangements depend on experimental conditions as well as on material properties. In addition, well-defined scaling laws relating the preferred wavelengths present in the microstructure, e.g. cell diameters or the width of the dislocation-poor channels in cyclic deformation, to the inverse of the applied stress seem to be followed by many materials. Such type of patterning, which is essentially that of immobile dislocations, is referred to as "forest patterning" and for more detail the reader is referred to the review articles [1-4]. As an example, Fig. 1 shows a very regular wall and channel structure obtained in a polycrystalline ferritic steel after cycling at room temperature.

Mobile dislocations also exhibit patterning, as revealed by the examination of etch markings or of slip traces at the surface of deformed single or polycrystals. Slip lines and bands are often arranged in form of static patterns or propagate along the gauge length of the specimen in an orderly (Lüder-like) manner with a constant propagation velocity. These aspects have been experimentally examined and discussed by Neuhäuser and coworkers [5, 6]. They are illustrated by Fig. 2 in the case of a Cu-5 at. % Ni crystal.

These observations indicate that plastic flow is inherently nonuniform at microscopic scale as well as at mesoscopic scale, a feature which is presently not well understood. Indeed, most of the theoretical elaborations based on dislocation theory either assume that dislocation densities are uniformly distributed in space or make ad hoc hypothesis about the geometry of their arrangements. Understanding macroscopic plasticity in terms of dislocation theory requires, therefore, a better knowledge of the collective behaviour of dislocations and of their tendency to self-organization. This is undoubtedly a very challenging and very difficult task.

Early studies [7] treated this problem by assuming that dislocation patterns can be described through energy minimization principles, i .e. by making the hypothesis that the microstructure is somehow in thermodynamic equilibrium. More recent elaborations stress, on the contrary, the fact that plastic flow is basically a dissipative process far from thermodynamic equilibrium. They make use of a body of knowledge already developed in other domains such as hydrodynamic instabilities, chemical instabilities or dendritic growth. A summary of recent elaborations on the non linear aspects of plasticity can be found in the Proceedings of three recent conferences [8-10].

The aim of such studies is to define the critical conditions for the emergence of patterns from initially random dislocation configurations or for the transition between different types of organized microstructures and to predict the geometry of these arrangements in terms of materials properties.

In what follows, the physical mechanisms leading to dislocation patterning, i.e. the long range and local dislocation-dislocation interactions are briefly discussed. The main difficulty encountered in all elaborations resides in the description of the long range elastic interactions between dislocations. This was the main motivation leading to the development of numerical simulations which are thought to involve less approximations than analytical models and which may serve as a basis for further theoretical elaboration and as checks for some simplifying hypothesis. Here, the main features of a 3D numerical simulation are presented and a few preliminary results are reported.

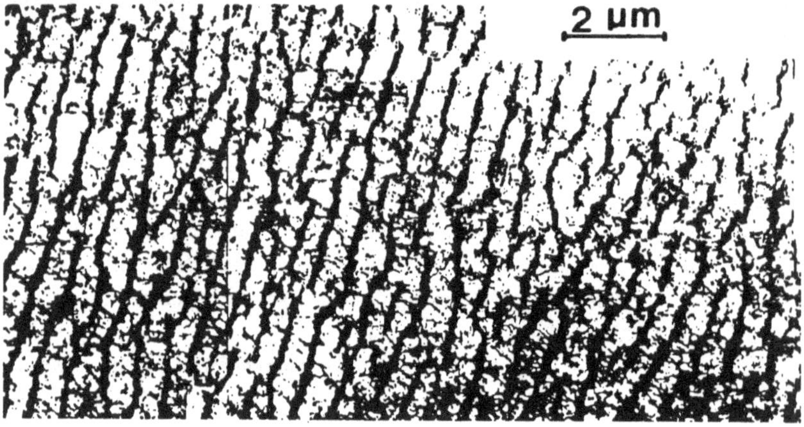

Fig. 1 Wall and channel dislocation structure in a polycrystalline stainless steel after 50 fatigue cycles at room temperature with a plastic strain amplitude of 4×10^{-3} (after [4]).

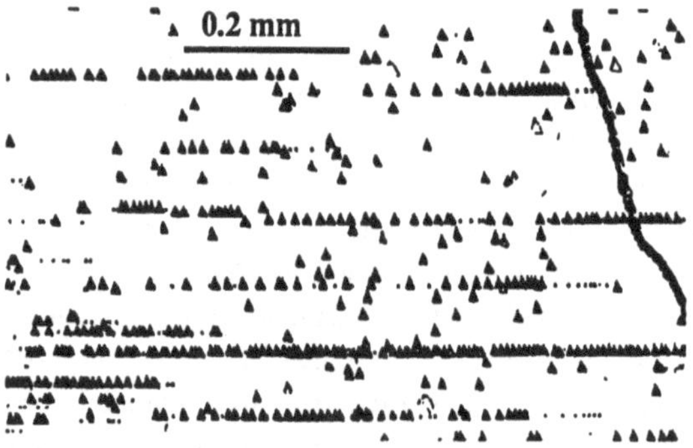

Fig. 2 Slip patterning in a Cu-5 at.% Ni specimen after a stress pulse of 4.5 MPa at room temperature. The dislocations were revealed by etching before and after the pulse. After Arkan and Neuhäuser [11], by Courtesy of Phys. Stat. Solidi.

Slip Patterning

At the beginning of plastic flow, the production of mobile dislocation by multiplication mechanisms results in a negative value of the local strain hardening rate and in strain localization [12, 13]. In addition to multiplication, several mechanisms are thought to govern slip line patterning: the annihilation by cross-slip of screw dislocations gliding in parallel planes separated by a distance smaller than a critical value y_s, the mutual trapping of dislocations; particularly of edge dislocations, into dipolar and multipolar walls perpendicular to the slip plane, the passing stresses between dislocations of the same sign. Fig. 3 shows the result of a 2D simulation of the motion of screw dislocations during creep in a model crystal of size 31x13 μm^2. In this simulation screw dislocations, which are seen end-on, move to the right or to the left inside a cellular space with cell parameter $2y_s$. Therefore, the slip pattern of Fig. 3 is scaled by the critical annihilation distance, y_s, which is inversely proportional to the applied stress. Actually, these 2D simulations have shown that other types of configurations can be obtained: wavy slip patterns and propagating patterns. These results are discussed in more detail in [4, 14, 15].

The same type of behaviour is obtained every time the strain hardening rate becomes negative, either because of a high dislocation multiplication rate, or because the microstructure undergoes strain softening. This is the case, in particular for channelling effects due to precipitate shearing [16] or for the destruction of short-range order by mobile dislocations in dilute solutions [17]. This last case is actually the one illustrated by Fig. 2, which explains its resemblence with Fig. 3.

Fig. 3 Slip patterning in a 2D simulation. Screw dislocations, seen end-on, are represented by triangles which point towards the right or the left according to their sign. The traces of the slip planes are horizontal. The initial configuration consisted of randomly distributed dislocations which have evolved under an applied shear stress of 35 MPa (J. Lépinoux, unpublished work).

Forest Patterning

With increasing strain, an increasing density of mobile dislocations gets trapped into immobile configurations. The processes leading to the formation of forest patterning are poorly known but they undoubtedly result from dislocation-dislocation interactions since they are enhanced by a decrease in the mean dislocation spacing. The formation of an organized microstructure coincides with the occurrence of a significant internal stress and the two features are clearly related. Then, the dislocation-dislocation interaction must prevail over the interaction between isolated dislocations and localized obstacles such as the Peierls relief. This explains why, during low temperature straining, cell structures are formed at the beginning of plastic flow in F.C.C. metals and at some larger strains (e.g. 0.1) in B.C.C. metals.

There are two types of dislocation-dislocation interactions. Local or contact interactions involve such mechanisms as direct annihilation, the annihilation of screw dislocations by cross-slip, the trapping of edge dislocations into dipolar configurations, the annihilation of closely spaced dipoles under the effect of their self-stresses, the formation of dislocation junctions, multiplication by Frank-Read sources. All of these mechanisms can conveniently be described by rate equations. For instance, the annihilation of screw dislocations by cross-slip contributes to the mobile density ρ_m by a quadratic term of the form [4]:

$$d\rho_m/dt = -2y_s v \rho_m^2 \tag{1}$$

where v is the average velocity of mobile dislocations. These mechanisms as well as the couplings between mobile and forest densities introduce non linearities in the balance equations which describe the evolution of the dislocation densities with time.

The long range interactions define the internal stresses. For each differential element of a dislocation line at the position **r** in the crystal, the internal stress $\sigma_i(\mathbf{r})$ is the sum of the long range elastic interactions between this segment and all the other dislocation lines or loops present in the crystal. This is written in the form:

$$\sigma_i(\mathbf{r}) = \propto \int \rho(\mathbf{r}') \, F(\mathbf{r}) \, \mathbf{I}(|\mathbf{r}'-\mathbf{r}|) \, d\mathbf{r}' \tag{2}$$

where F is a correlation function which takes into account the possibility for the dislocations to have positive or negative signs and **I** is the pair interaction function which is inversely proportional to the distance $|\mathbf{r}'-\mathbf{r}|$ (it involves, in addition, an angular dependence for edge dislocations). This non local expression is the main difficulty encountered in analytical models of dislocation patterning. The dynamic balance equation for the evolution of a density of mobile or forest dislocations, ρ, in time and space can be written as:

$$\partial\rho/\partial t + \mathrm{div}\mathbf{J} = g \tag{3}$$

where $\mathbf{J} = \rho\mathbf{v}$ is the flux associated with the dislocation species considered and g contains all the reaction terms describing the local interactions. The velocity **v** is determined by the local stress, i.e. it depends on the applied stress minus the local value of the internal stress, as given by Eq. (2). Unfortunately, this internal stress depends on the spatial distribution of the dislocation densities i.e. on the solution of the problem. A simplification may arise if the microstructure is organized in such a way that the long range internal stresses are effectively screened. An expansion around **r** of the integral in eq. (2) is then possible and the internal stress, then, becomes a local quantity.

The reaction-diffusion models [18-21] make use of this simplifying hypothesis and, within a few approximations, it is shown that the flux term in Eq. (3) reduces to a diffusion-like term of the form $D\Delta\rho$, D being an effective diffusion coefficient. These reaction-diffusion forms are thought to apply to fatigue patterning where the organization of the forest is typically multipolar, involving effectively a short cut-off radius for the long range internal stresses.

This is, however, not the case in many other situations as has been shown by 2D simulations [4]. The result illustrated by Fig (4) was obtained as follows. An initial random distribution of straight, parallel screw dislocations with density $\rho_0 = 10^{12}\ m^{-2}$ was left to rearrange under the effect of its internal stresses and in the absence of any applied stress. Once a stable configuration was obtained, the interaction energy, E, was measured for each dislocation within a cylinder centered on the dislocation and of radius R. The average value of E over the whole configuration, $E_{av}(R)$, is plotted in Fig. 4 as a function of the reduced truncation distance $R/\rho_0^{-1/2}$. We see that there is, indeed, a cut-off radius, since beyond a certain distance the average interaction energy is a constant, but its value is quite large, typically 5 µm in the present case. Although this value is probably larger than what would be expected e.g. in creep conditions, typically 1 µm, the use of reaction-diffusion forms to describe patterning in monotonic deformation appears quite problematic.

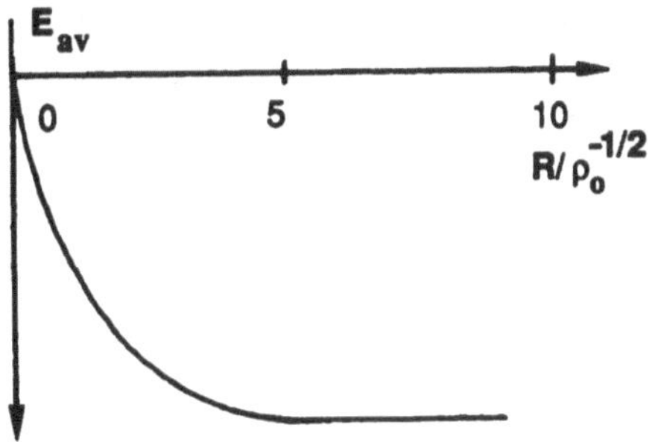

Fig. 4 Average elastic interaction energy, E_{av} (arbitrary units), as a function of the reduced truncation distance, $R/\rho_0^{-1/2}$. This measurement was performed on a two-dimensional simulated dislocation configuration (see text for detail).

THREE-DIMENSIONAL SIMULATIONS

Simulation Technique

The method used involves a discretization in time and space. It describes the motion and interaction of dislocation loops in a single crystal under any type of external loading.

A lattice which has the symmetries of the crystal considered is defined. In the present case we deal with an FCC lattice and the <110> segments joining a corner to the middle of an adjacent face define the possible sites for screw segments, while <112> segments which join a corner to the middle of an opposite face define the possible sites for edge segments. Dislocation loops or segments are decomposed as a set of elementary segments of edge or screw character and these elementary segments are distributed along the sites of the underlying lattice. It follows that the elementary translation of an edge segment is a <110> translation and that of a screw segment a <112> translation.

The basis for the discretization in space is the maximum distance around a dislocation segment within which it is not possible to find another segment of same Burgers vector. This distance is assumed to be twice the critical annihilation distance for the annihilation of an edge dipole under the effect of its self-stresses, y_e. The value of y_e has been estimated to about 1.6 nm in copper at room temperature by Essmann and Mughrabi [23]. In an FCC cell of lattice parameter a, the elementary climb distance is $d_{111} = a/\sqrt{3}$. Therefore, we have $a/\sqrt{3} = 2y_e = 3.2$ nm and a = 5.54 nm. This defines the length of the elementary dislocation segments: $a/\sqrt{2}$ = 3.92 nm for screw segments and $a\sqrt{3}/\sqrt{2}$ = 6.78 nm for edge segments. In the simulation, the reference frame for the nodes of this FCC lattice is a cubic lattice of parameter a/2 = 2.77 nm and the simulated crystal contains about 2 10^{11} such cubic cells. Its size is $(15\ \mu m)^3$, with periodic boundary conditions or image forces at the outer surfaces. (Numerical values slightly different from those indicated here were given in [3, 22]; they correspond to an earlier version of the simulation.)

The interactions between the elementary dislocation segments are computed within rather large truncation distances, up to 5 µm, using the stress field of finite dislocation segments. The local internal stresses are defined through the Peach-Köhler formula. They are detracted from the applied stresses, which defines the local effective stresses. According to the local value of the effective stress, τ^*, a dislocation segment can, then, glide, cross-slip or remain sessile. During its motion, new segments are created to maintain the continuity of the dislocation lines. The relation between stress and velocity is of the form $\tau^* b = Bv$, where b is the Burgers vector and B a viscous drag coefficient (B = 5 10^{-5} Pa/s).

Local mechanisms which involve the interaction between the stress fields of two dislocations such as annihilation or dipole trapping naturally occur within the simulation. The contact interactions between two dislocations of different Burgers vectors may lead to either crossing, with the formation of two jogs, or to a reaction with the formation of a dislocation junction. Jog formation was neglected on the grounds that it does not influence the mechanical properties of FCC crystals at room temperature. The formation of attractive junctions was taken into account by attributing a zero velocity to every elementary dislocation segment entering a rectangular area centered around the meeting point of the interacting segments.

The initial configurations consist of a random array of segments pinned at their ends with a density ρ = 10^{11} m^{-2} and with lengths randomly distributed around an average value of $\rho^{-1/2}$. This configuration, which represents the Frank net of annealed crystals, together with attractive junctions and edge segments left by double cross-slip events, provides the anchoring points around which dislocation sources can operate.

Any type of monotonic or cyclic, uniaxial or multiaxial deformation conditions can be prescribed and the activity on the possible slip systems follow naturally from the orientation of the external constraints with respect to the simulated crystal.

Only a brief account of the technique has been given here and more detail can be found in [22, 3], in particular about the numerical outputs of the simulation. In the present stage, the maximum strain values which have been obtained are of the order of 10^{-4} and only easy glide has been investigated. With further implementation of the program it is hoped that strain values sufficiently large to allow for a study of dislocation patterning will be reached.

Preliminary Results

Two finite dislocation segments of same Burgers vector and of edge and screw character have non-zero interaction stresses, while this is not the case for infinite segments. A consequence is that the simulated loops have a line tension, as illustrated by Fig. 5. On Fig. 5, the value of the shear stress needed to maintain a dislocation loop in equilibrium is plotted as a function of the loop diameter. According to dislocation theory, we have $\tau = 2\Gamma/D$, with $\Gamma = (\mu b/4\pi)\ \mathrm{Ln}(D/2b)$, where Γ is the line energy and μ the shear modulus (μ = 4.2 10^{10} Pa for copper). The relation between τ and D measured on simulated loops is in good agreement with this theoretical value.

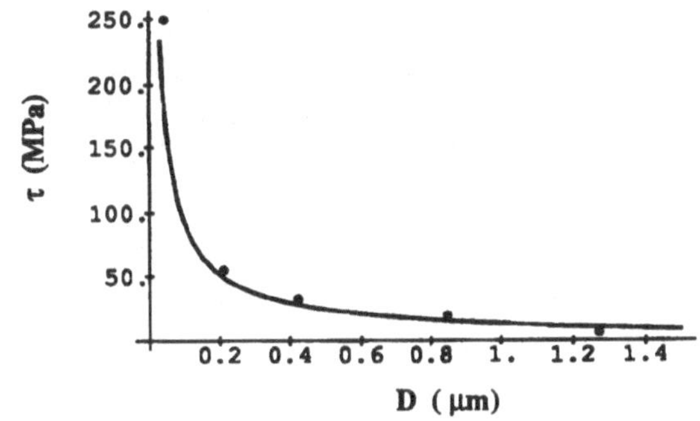

Fig. 5 The shear stress τ needed to maintain a loop of diameter D in equilibrium vs. D. The full curve represents the relation given by dislocation theory and the solid circles values measured on simulated loops.

Figure 6 shows several steps of the operation of a Frank-Read source under a constant stress of 100 MPa, from an initial segment pinned at its ends. The last step (Fig. 6 d) was obtained after a time interval of $1.2\ 10^{-11}$ s and a strain of $5\ 10^{-6}$. These values are indicative of the very short elementary time intervals needed for an accurate description of multiplication mechanisms, particularly at small strains.

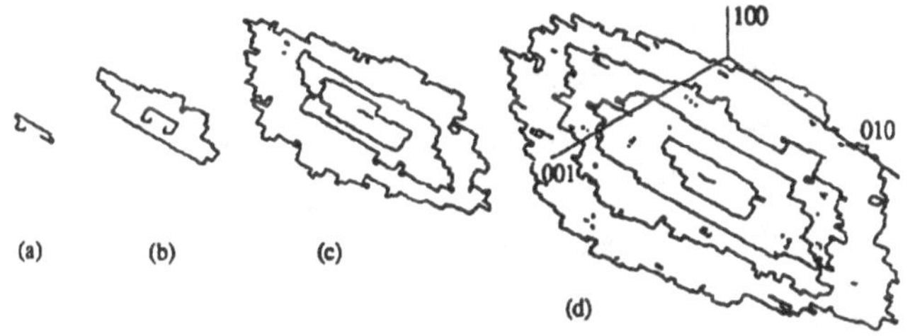

Fig. 6 Several stages of operation of a Frank-Read source from an initial segment pinned at its ends. Number of time steps: (a): 1, (b):3, (c):7, (d):11.

Figure 7 shows an initial configuration which consists of a Frank net containing only one Burgers vector, (1/2)[$\bar{1}$01]. Fig. 8 illustrates the development of easy glide from this initial configuration under a constant tensile stress of 50 MPa parallel to [$\bar{1}$12]. The configuration of Fig. 8 corresponds to a total strain of $1.4\ 10^{-5}$, obtained after a (simulated) time interval of $1.75\ 10^{-8}$ s. The dislocation density has increased by one order of magnitude with respect to its initial value and a few dislocation sources have started operating. At this stage, the average internal stress is practically negligible and the average dislocation velocity is of the order of 200 m/s. Figures 9 (a) and (b) show the same configuration as Fig. 8, but viewed along directions parallel to the screw and edge segments respectively.

In further developments of this 3 D simulation, solute atoms and precipitates will be introduced and the possibility of surmounting localized obstacles by thermal activation will be taken into account.

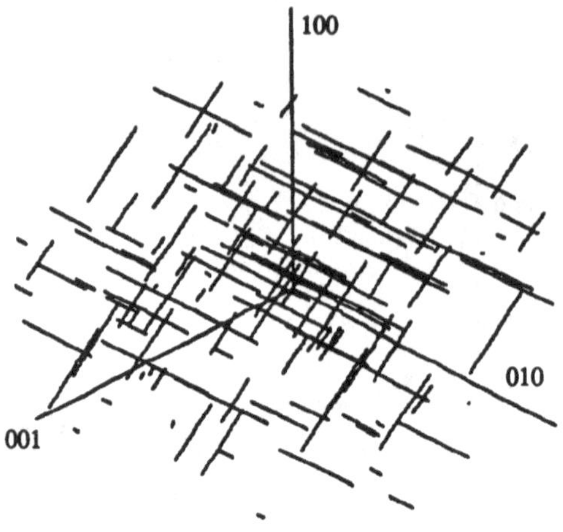

Fig. 7 An initial configuration consisting of dislocation segments of same Burgers vector pinned at their ends (density $\rho = 10^{11}$ m^{-2}, average length $\rho^{-1/2} = 3.16$ µm). The simulated crystal has a size of (15 µm)^3.

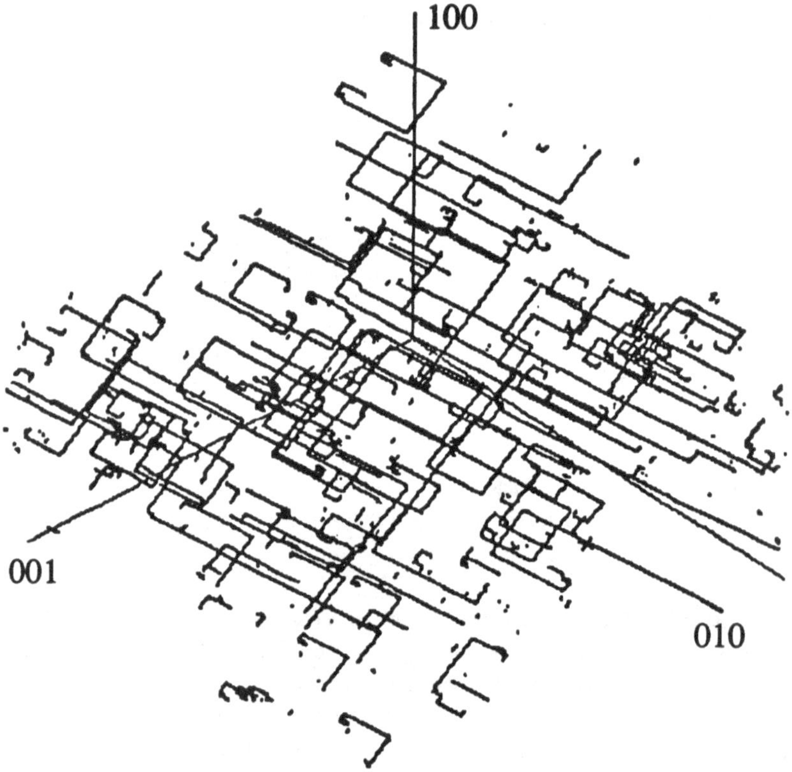

Fig. 8 An early stage of easy glide in the (111) plane of a (copper) FCC crystal, seen along a direction parallel to [111]. The initial configuration is that of Fig. 7.

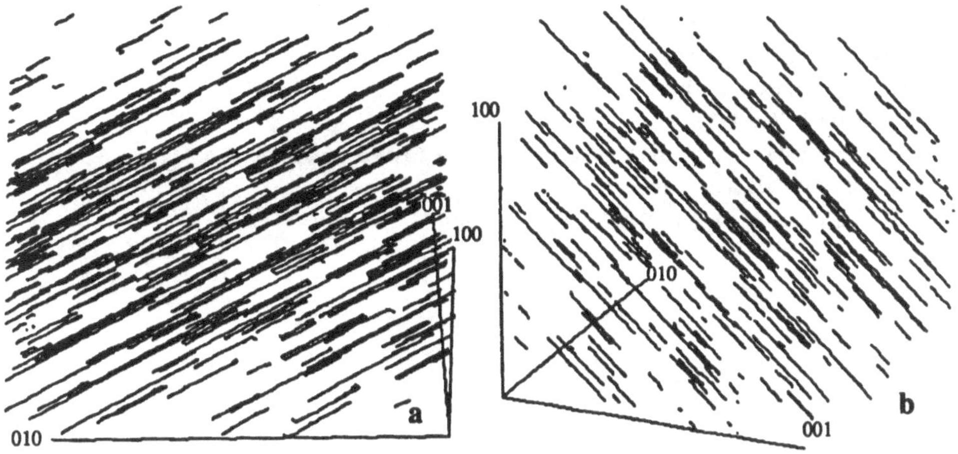

Fig. 9 Same configurations as in Fig. 8, but viewed along a direction (a): close to [101], (b) close to [1$\bar{2}$1].

REFERENCES

1 J. D. Embury, In: Strengthening methods in crystals, A. Kelly and R.B. Nicholson (Eds.), Applied Science Publishers, London 1971, p. 331

2 F. Louchet and Y. Bréchet, In: Non Linear Phenomena in Materials Science, L.P. Kubin and G. Martin (Eds.), Trans Tech Publications, Aedermannsdorf, Switzerland 1988, p.473

3 L.P. Kubin, Y. Estrin and G. Canova, In: Patterns, Defects and Materials Instabilities, D. Walgraef and N. Ghoniem (Eds.), Kluwer Acad. Publications, in Press

4 L.P. Kubin and J. Lépinoux, In: Strength of Metals and Alloys (ICSMA 8), Vol. 1, P.O. Kettunen, T.K. Lepistö and M.E. Lehtonen (Eds.), Pergamon Press, Oxford 1988, p. 35

5 H. Neuhäuser, In: Dislocations in Solids, Vol. 6, F.R.N. Nabarro (Ed.), North-Holland, Amsterdam, 1983, p. 321

6 H. Neuhäuser, In: Mechanical Properties and Behaviour of Solids: Mechanical Properties, V. Balakrishnan and C.E. Bottani (Eds.), World Scientific, Singapore, 1986, p.209

7 D.L. Holt, J. Appl. Phys. 41 (1970) 3197

8 "Patterns, Defects and Microstructures in Nonequilibrium systems", D. Walgraef (Ed.), Martinus Nijhoff, Dordrecht, 1987

9 "Nonlinear Phenomena in Materials Science", L.P. Kubin and G. Martin (Eds.), Trans Tech Publications, Aedermannsdorf, Switzerland, 1988

10 " Patterns, Defects and Materials Instabilities", D. Walgraef and N. Ghoniem (Eds.), Kluwer Acad. Publications, in Press

11 O.B. Arkan and H. Neuhäuser, Phys. Stat. Sol (a) 99 (1987) 385

12 Y. Estrin and L.P. Kubin, Acta Met. 34 (1986) 2455

13 L.P. Kubin and Y. Estrin, Rev. Phys. Appliquée 23 (1988) 573

14 J. Lépinoux and L.P. Kubin, Scripta Met. 21 (1987) 833

15 J. Lépinoux, Doctorate Thesis, University of Poitiers, 1987

16 J. Olfe and H. Neuhäuser, Phys. Stat. Sol. (a) 109 (1988) 149

17 A. Luft and Ch. Ritschel, Phys. Stat. Sol. (a) 72 (1982) 225

18 D. Walgraef and E.C. Aifantis, J. Appl. Phys. 58 (1985) 688
19 E.C. Aifantis, Mat. Sci. Eng. 81 (1986) 563
20 E.C. Aifantis, Int. J. of Plasticity 3 (1987) 211
21 J. Kratochvil, Rev. Phys. Appliquée 23 (1988) 419
22 G. Canova and L.P. Kubin, In: Continuum Models of Discrete Systems 6, G.A. Maugin (Ed.), Longman Scientific and Technical, in press
23 U. Essmann and H. Mughrabi, Phil. Mag A 40 (1979) 731

STUDY OF LARGE DEFORMATION OF POLYCRYSTALLINE MATERIALS BY SCANNING ELECTRON MICROSCOPY

K. Luecke

Institut fuer Allgemeine Metallkunde und Metallphysik
RWTH Aachen, 5100 Aachen, F.R.G.

INTRODUCTION

The main process of plastic deformation is slip on the various slip systems. This implies that for predicting the properties of the deformed state, the shear strains on the different systems have to be known. Mostly, these are calculated by applying models based on the assumption of "ideal slip". This means (i) that under a homogeneous external stress also slip is homogeneous within a crystal or, in the polycrystalline case, within a grain, and (ii) that under the same external stress equally oriented crystals or grains experience the same slip. The method most frequently used to obtain information on the various shear strains consists in measuring the deformation-induced crystallographic orientation change of the crystal which is very sensitive to the distribution of slip on the various slip systems.

There are mainly two models for calculating the shear strains occurring on the various systems from the observed orientation changes both being based on ideal slip behaviour:

<u>The Sachs model</u> [1, 3]. Here the systems with the highest external resolved shear stresses are assumed to become active but this assumption is practically always violated, e.g. in single crystals by forming slip inhomogeneities like deformation bands etc. or in polycrystals because of the incompatibility of the deformations of neighbouring grains.

<u>The Taylor model</u> [2, 3]. In order to avoid these incompatibilities Taylor assumed that all grains carry out identical deformations equal to the macroscopic deformation which is to be achieved by polyslip. Since the agreement between the observed orientation changes and those being calculated by the Taylor theory is reasonably good, the Taylor theory together with some of its modifications ("relaxed constraints models") is generally considered to be a good description of polycrystalline deformation. It is now the purpose of the present paper to test this theory more thoroughly by directly checking the underlying assumptions (i) and (ii) of ideal slip by applying an etching method.

Due to the strong interaction between the deformation of a grain and that of its neighbours both assumptions (i) and (ii) are strongly violated.

METHODICAL

The present way of checking the homogeneity of slip is measuring the crystallographic orientation changes induced by the deformation continuously over the sample surface. There are several methods of local orientation determination available, e.g. Kikuchi lines or channeling methods, but for the present purpose they have the disadvantage that the orientation is obtained only point-wise so that it is very difficult to gain a continuous survey of local orientation changes required for studying the inhomogeneity of slip.

In the present paper a continuous and very simple method recently developed by Koehlhoff and Luecke [4, 5] for copper and copper alloys is applied. By etching with the proper solution and under proper conditions it is possible to reveal the close-packed {111} faces. Fig. 1 gives an example for copper. The resolution of the etching reaches down to 1 μm.

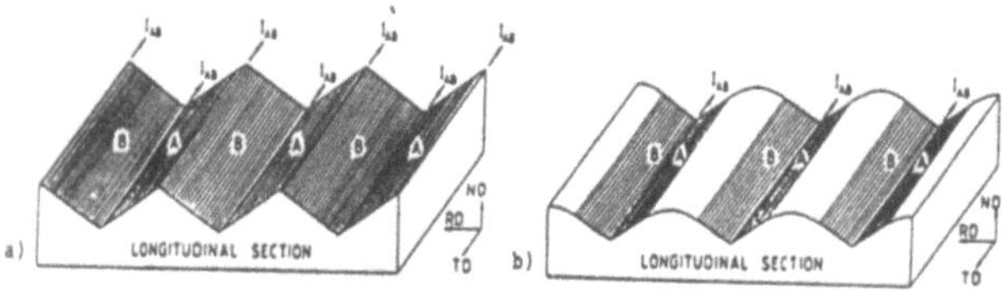

Fig. 2 Schematic representation of surface relief corresponding to Fig. 3a a) ideal b) real

Magnifications between 200 and 5000 times achieved by an optical or a scanning electron microscope allow one to obtain either a large survey of the orientations over an area covering many grains or a highly local orientation determination. Since in a relief of {111} faces the ridges are always in <110> directions the micrograph of such an etching pattern consists of projection lines of the <110> directions. From such a pattern, on principle it is possible to determine the orientation of the crystal. The most accurate determination is attained by comparing the observed pattern to standard patterns being displayed for the various orientations in an atlas of etching patterns [6]. Fig. 3 gives some examples with the figures exhibiting, besides the usual {111} poles, the projection lines of the <110> direction which must run parallel to the lines in the micrographs.

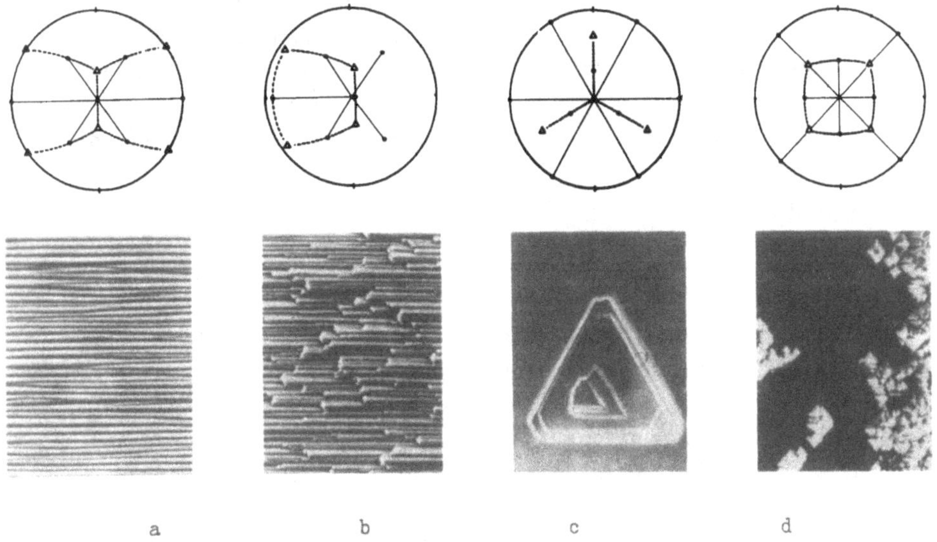

RD

10 μm

Fig. 3 Stereographic projection of {111} poles and <110> directions (——) and etched pattern in rolling plane a) {011}<100>. b) {011}<100> rotated 5° around the rolling direction. c) {111}<112>. d) {001}<100>

For the case that one <110> direction is situated in the surface, one obtains a roof-like etching relief as schematically indicated in Fig. 2. A corresponding etching is presented in Fig. 3a. It consists of straight lines parallel to the <110> direction in the surface. A tilt of the orientation by 5° around the rolling direction reveals additional <110> lines as can be seen in the schematic drawing in Fig. 4 as well as in the etching pattern in Fig. 3b.

Fig. 3c represents the case that three <110> directions are being in the surface, i.e. that the surface is the {111} plane. Then only this plane is revealed and no further etching features occur (except for etch pits at dislocations). Finally, in the case of two <110> directions in the surface, i.e. of a {001} surface, quadratic etch patterns are

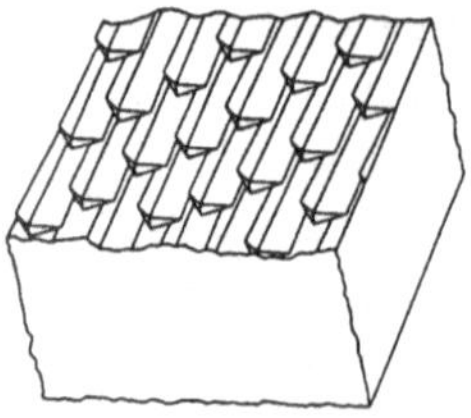

Fig. 4 Schematic representation of surface relief corresponding to Fig. 3b

obtained (Fig. 3d). These figures clearly demonstrate that it is possible to distinguish between the different orientations by means of etch patterns. The accuracy of this method of orientation determination is quite good for some of the orientations, e.g. about 3° in cases like those in Figs. 3a and 3c. In other cases, e.g. the slightly tilted cube orientation, an accuracy of 7° to 10° can be achieved.

EXAMPLES OF NON-TAYLORIAN POLYCRYSTALLINE DEFORMATION

By this simple method important information on the deformed state can be obtained, information, which is difficult to get by any other method. In the following some examples are given. Samples are copper and brass sheets rolled to various extents. In most of the figures the specimens are viewed from the transverse direction.

<u>Abrupt</u> <u>and</u> <u>gradual</u> <u>orientation</u> <u>transitions.</u> During deformation in many grains large orientation gradients are formed. These can consist in sharp boundaries or in gradual orientation transitions. An example of the first case is displayed in Fig. 5 for a {111}<211> single crystal.

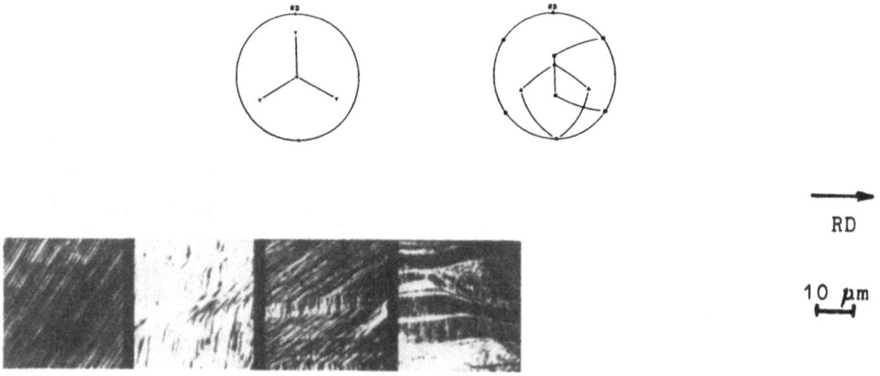

Fig. 5 Orientation splitting in cold-rolled {111}<112> copper

Rolling first yields gradual orientation gradients, which finally lead to a splitting of the crystal into bands of {112}<111> and {011}<100>

orientations. They are separated by sharp boundaries which comprise an orientation difference of 55°. An example of the second case is given in Fig. 6. There, in one grain orientations are observed ranging from {112}<111> to {011}<100>, i.e. continuously changing over an angle of 55°.

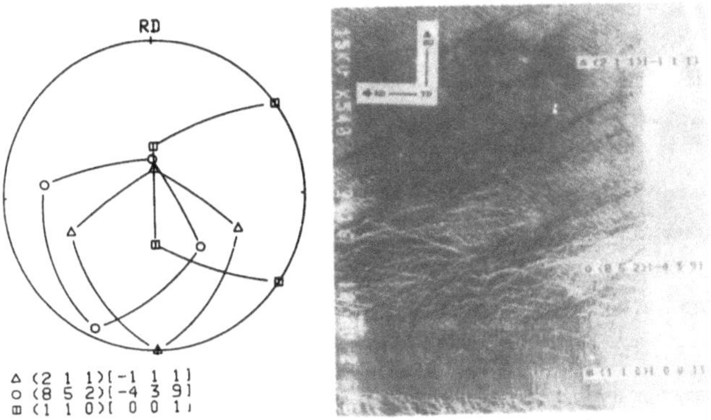

Fig. 6 Orientation change of one grain in 65% cold-rolled copper

<u>Interaction between grains.</u> While the orientation splitting occurring in the first case is due to a basic instability of the {111}<211> orientation and thus takes place even in single crystals, the continuous orientation change in the second case must be due to different neighbours at different sides of the grain. A more transparent example of such an interaction is given in Fig. 7. Two variants of the {112}<111> orientation form a grain boundary first parallel then perpendicular to the rolling direction.

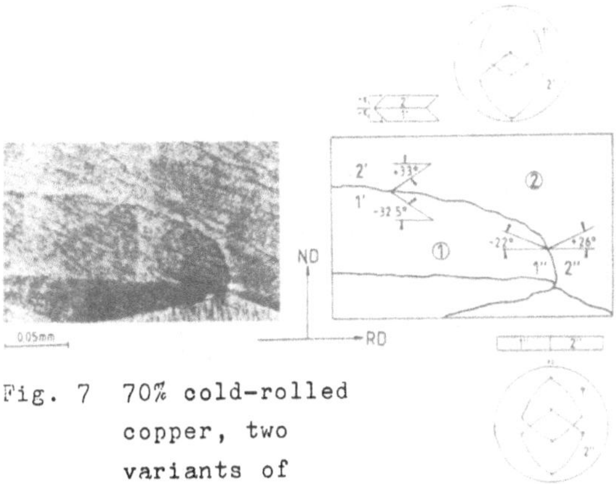

Fig. 7 70% cold-rolled copper, two variants of {112}<111>

At the grain boundary parallel to the rolling direction (1', 2') the shear components + ε_c and - ε_c compensate each other and both orientations remain stable. At the grain boundary parallel to the normal direction (1", 2") the two shear components are incompatible. This implies a rotation of the two C orientations around the transverse direction towards an orientation which exhibits no shearing.

Twinning. Fig. 8 gives an example of orientation splitting by deformation twinning in a {011}<211> single crystal. The two small deformation twins on both sides of the matrix cause a rotation of the matrix regions close to the twin boundaries from {011}<211> to {011}<100> whereas in the middle of the matrix the orientation remains stable. This rotation is also due to the necessary compatible deformation at the twin boundary. Another example of twinning is given in Fig. 9. Originally C-oriented grains are twinning during rolling. The resulting twin-matrix bands rotate towards {111}<112>. Then shear bands develop.

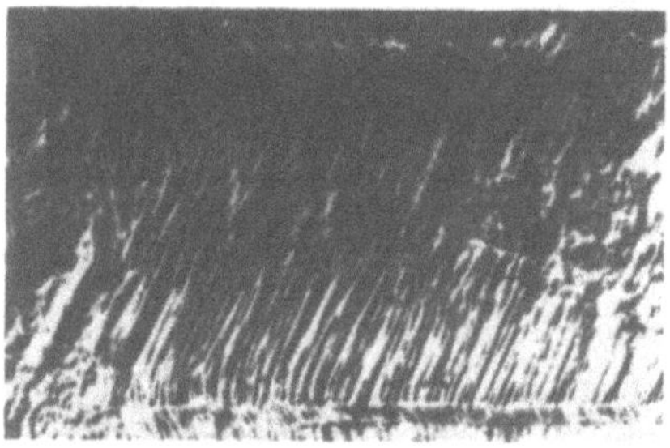

RD

Fig. 8 Twinning in {110}<112> Cu30%Zn 90% cold-rolled

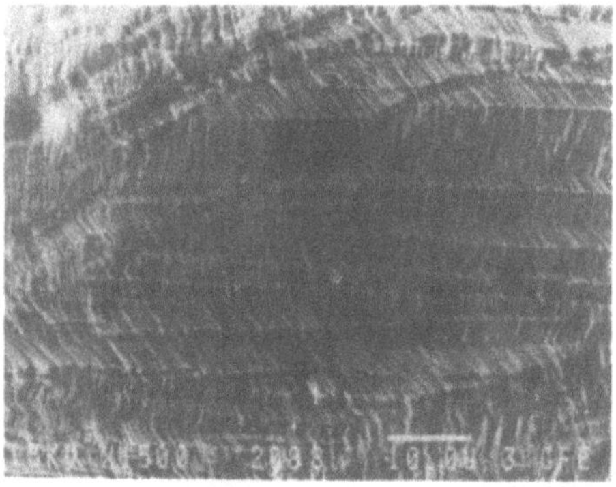

Fig. 9 Twinning in 78% cold-rolled Cu9%Zn

Extremely inhomogeneous deformation. Shear bands are an example of an extreme deviation of the assumption of ideal slip. Fig. 10 shows the abrupt orientation change from matrix to shear band. In Fig. 11 a single grain has been sheared off so strongly that it has been divided into separate parts.

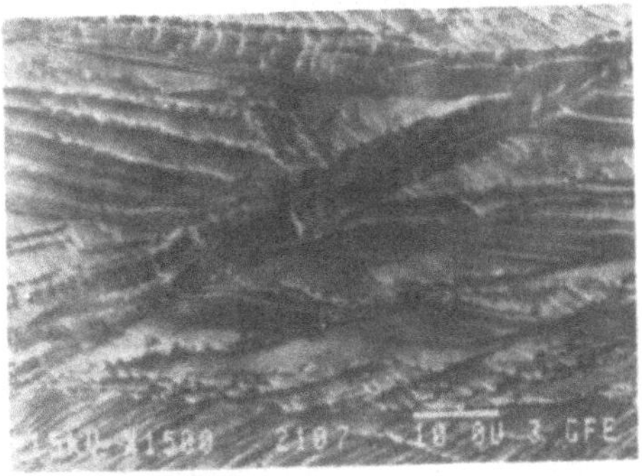

Fig. 10 Shear band in 78% cold-rolled Cu9%Zn

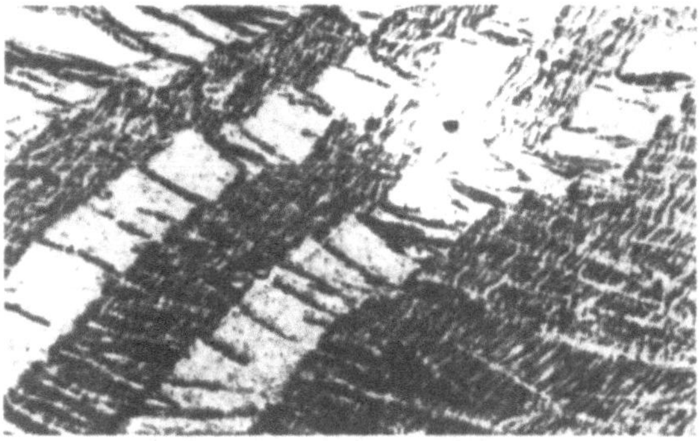

RD

20 µm

Fig. 11 65% cold-rolled Cu9%Zn

CONCLUDING REMARKS

The above examples show that due to the difference of the deformation of a central grain and its neighbours, different parts of this grain may experience different orientation changes. This means strong orientation gradients may occur within a grain and because of the different neighbourhoods equally oriented grains may deform differently. Both is in contrast to the above assumptions (i) and (ii) of homogeneous slip and thus in contrast to the suppositions of the Taylor type theories. The deviations are so drastic that it is less a question why the theory in some cases fails to predict the experimental results than a question why it leads to a relatively good agreement at all. One should further consider that the present method allows one to investigate the deviation from the Taylor type theories in detail so that the present results represent a first step into the post-Taylorian age.

Of course, the present method can be applied to investigate also other phenomena than plastic deformation, e.g. the mechanisms of recrystallization. Fig. 12 gives an example of growth inhibition by the Goss orientation. There a nucleus grows in a C-oriented region whereas the neighboured Goss grain is not consumed. Fig. 13 shows recrystallized grains in a C/S matrix. The recrystallized orientations are the cube orientation and its twins with a higher amount of the latter.

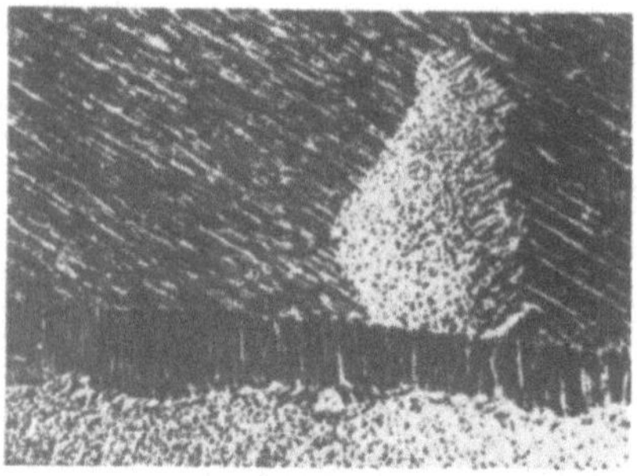

Fig. 12 {111}<112> copper 80% cold-rolled and partially recrystallized

20 µm

Fig. 13 93% cold-rolled and partially recrystallized copper

ACKNOWLEDGEMENTS

The author is grateful to G.D. Koehlhoff, M.G. Sindel, X. Sun and W. Wang for their helpful collaboration. He acknowledges the financial support by the Deutsche Forschungsgemeinschaft.

REFERENCES

1 G. Sachs, Z. Ver. Deutsch. Ing. 72 (1928) 734
2 G.I. Taylor, J. Inst. Met. 62 (1938) 307
3 J. Hirsch and K. Luecke, Acta met. 36 (1988) 2883
4 G.D. Koehlhoff, J. Hirsch, U.v. Schlippenbach, and K. Luecke, Proc. ICOTOM 6, Tokyo (1981), p. 489
5 G.D. Koehlhoff, X. Sun and K. Luecke, Proc. ICOTOM 8, Santa Fe (1987), p. 183
6 W. Wang et al., to be published

TEM-INVESTIGATIONS ON MICROWIRES A CONTRIBUTION TO THE ELUCIDATION OF THEIR DEFORMATION BEHAVIOUR

H. Berek, S. Gärtner, H.H. Daut, R. Juknat, D. Häußler

VEB Mansfeld Kombinat Wilhelm Pieck
Forschungsinstitut für Nichteisenmetalle
Lessingstraße 41, Freiberg, DDR 9200

ABSTRACT

The subject of the paper is the interdependency between the microstructure investigated in an analytical transmission electron microscope and the drawability of microwire. Results are demonstrated for two different Al-alloys with two different structures but good drawability. It is shown that for each alloy the optimal structure has to be found anew.

INTRODUCTION

The application of gold for the manufacturing of microwire is commonly known. Our task was the development and small scale production of microwire from alternative materials like Al-alloys. In the course of the work two different problems had to be solved, at first to find a material state with good drawability and second to guarantee a sufficient bondability of the manufactured wires under industrial conditions. For this purpose the wire drawing process itself had to be mastered, that means all questions of control, drawing die quality or suitable lubricant should be solved. Furthermore, coarse inhomogeneities as cavities, voids or big precipitations had to be avoided. Anyway, drawing of microwire in the range of some 30 μm means working at boundary case conditions regarding formability of the material. That's why good drawability can only be realized after optimization of the structure. Some general conclusions regarding the drawability of materials based on the theory of plasticity can be found in the literature. This is how the microstructure should look like :

- small subgrains with a narrow distribution curve and irregular orientations
- dissolved atoms with a big lattice distortion but small diffusion coefficients and small precipitation rates

- incoherent, medium sized and globularly formed precipitations which are homogeneously distributed
- low dislocation density and a great number of active glide systems

ALCu4Mg1 MICROWIRE

The precipitation behaviour of Al alloys containing Cu and Mg is well known. There exist two kinds of intermetallics, first Al2Cu (the ϑ phase) and second Al2CuMg (the S phase). After homogenization and during aging at room temperature first GP zones are formed. They are transformed to the coherent ϑ'' and S'' phases, respectively, at higher temperature. At an aging temperature above 100°C ϑ' and S' are formed especially at dislocations. The stable and incoherent ϑ and S phases grow at temperatures above 200°C. Possible nucleation sites are ϑ' and S' particles or grain boundaries. During this process the former ϑ' and S' particles are dissolved. We used DC cast material, which was then hydrostatically extruded. Starting points for the wire drawing process were homogenized rods with a diameter of 4 mm.
The precipitation behaviour is illustrated in Fig. 1. The precipitation rate, the yield strength and the elongation were plotted against the aging temperature.

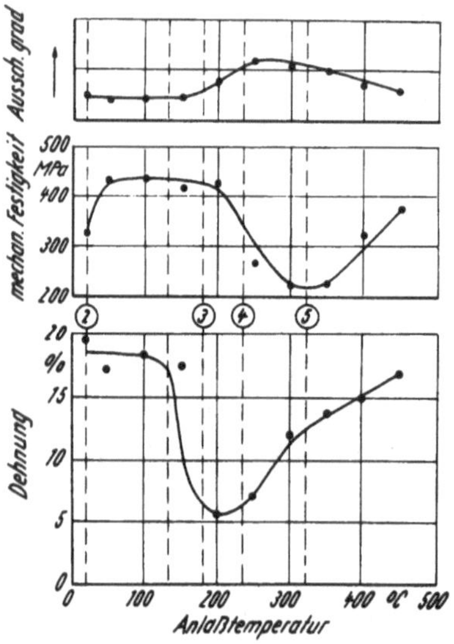

Fig. 1 Precipitation behaviour of homogenized AlCu4Mg1

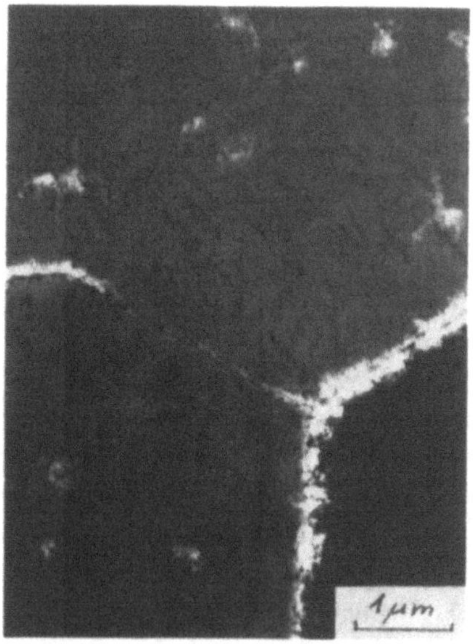

Fig. 2 TEM image of the structure after aging at 235 C/6.5 h

During our study it was noted that neither the homogenized state nor the state of maximum strength showed sufficient drawability. In order to find out the optimal aging temperature before drawing we carried out investigations at the marked temperatures (see Fig. 1). Fig. 2 is a TEM image of the structure after aging the sample 6.5 h at 235°C. In the volume of the grains the semicoherent θ' and S' precipitates can be recognized. The precipitation of the stable and incoherent θ and S particles started mainly at grain boundaries. The semicoherent θ' and S' particles strongly resist deformation. But a state where all of these particles are dissolved and where the bigger stable phases dominate showed also a rather poor deformation behaviour. Our drawing experiments proved succesful with material aged at temperatures between 250°C and about 300°C. In this range the precipitation degree reaches its maximum while yield strength and elongation have medium values.

AlSi1 MICROWIRE

Al with 1 % Si is an age hardenable alloy, too. But in this case other problems occur. The material also was DC cast. After casting it was homogenized at a temperature of 560°C for 16 h. Due to quenching in water we realized that Si stayed in solution. According to positron annihilation experiments carried out at the university in Halle a few years ago there is a great number of vacancy clusters stabilized by Si atoms in the quenched material. These clusters are able to work as nucleation sites for Si particles. Other possible nucleation sites are dislocations. It is well known that Si particles nucleate at temperatures above 150°C because of the enhanced mobility of vacancies.
Our TEM investigations on the quenched material showed that the assumption of a homogenized state is valid.
Starting points for the wire drawing process were again hydrostatically extruded rods with a diameter of 4mm. Fig. 3 shows the structure of the extruded state of different samples. It is a partly recovered structure. It is important to note that the samples were extruded at room temperature. However, the subgrain structure and the existence of precipitations indicate that the material was heated due to the deformation at a temperature of at least 150 °C. During this dynamical recovery dislocations are rearranged by the well known processes of gliding, sliding or annihilation. This clearly can be recognized in Fig. 4. Here we have some dislocation free areas in the middle of the subgrains and ordered dislocation arrays especially on subgrain boundaries.
The subgrain size of the sample in Fig. 4 is inhomogeneously distributed. Some precipitations occur which are also inhomogeneously

distributed. And, indeed, the corresponding sample showed a rather poor drawability. This is in accord with the general remarks about the interdependency between the structure and the drawability mentioneo before. The precipitations are mainly located at subgrain boundaries. X-ray maps demonstrate that most of the particles contain only Si (see Fig. 5). Sometimes we found Cu in the samples after extrusion. We believe that Cu-particles may have their origin from tools of the power press. The extrusion temperature is high enough to partly dissolve Cu. After that a small scale segregation on subgrain boundaries is possible.

Fig. 3 Microstructures of two different samples of AlSi1 in the extruded state

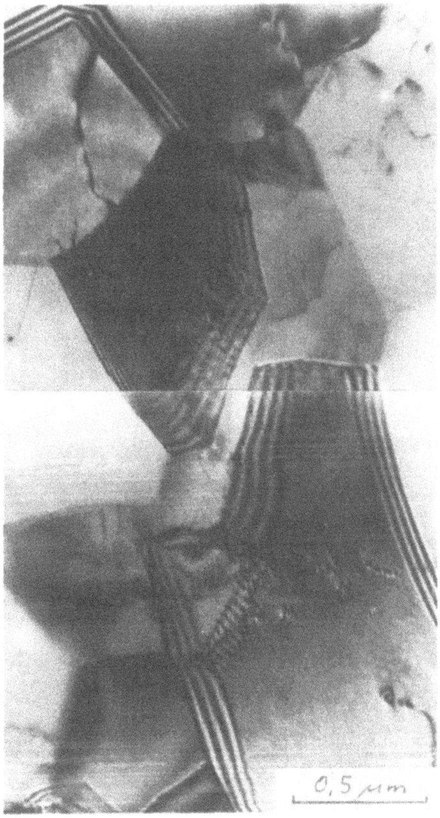

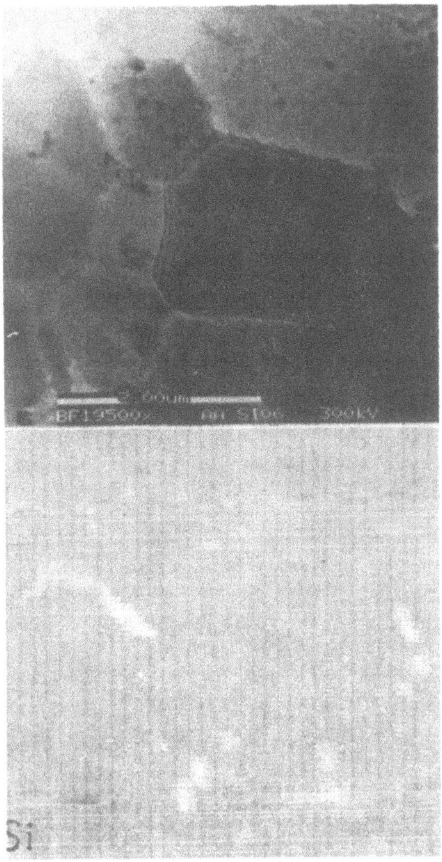

Fig. 4
TEM-image of the recovered structure in the extruded state of AlSi1

Fig. 5
X-ray map of the Silicon distribution and corresponding STEM-image of the extruded state of AlSi1

We found out that good drawability is given if the structure of the material shows a rather small subgrain size with a narrow distribution curve. This structure is a result of the processes of dynamic recovery. The occurence of Si-particles mainly on subgrain boundaries indicates that the temperature during extrusion is higher than 150 The nucleation sites for this particles are obviously dislocations. In this case one has to conclude that the precipitation occurs during or after recovery and that the subgrain size is not determined by the Si-distribution. The kinetics of these processes are a topic of further investigations.

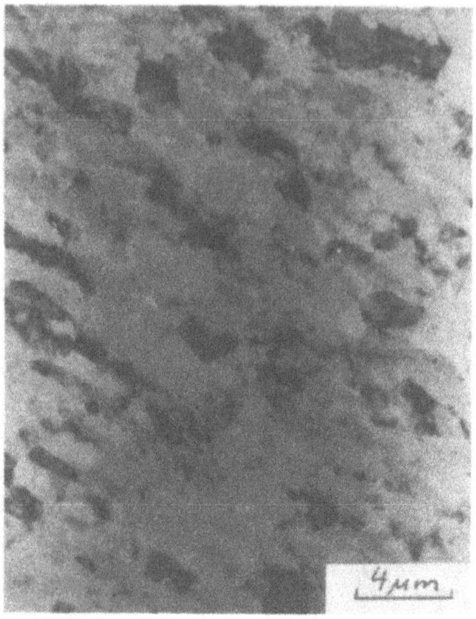

Fig. 6 Ultramicrotome section of an AlSi1 microwire in the final state

Fig. 6 shows the final state of the AlSi1 microwire. It was very difficult to find a suitable preparation technique for the TEM investigations. Jet polishing and ion thinning proved not succesful.
The image in Fig. 6 was taken after slicing the sample directly with an ultramicrotome by using a diamond knife. The deformed structure could be demonstrated clearly and without disturbing scratches.

SUMMARY

Two kinds of materials were demonstrated showing two different structures but good drawability. In the first case it was possible to find a suitable precipitation distribution before drawing. In the second case an optimal subgrain size seemed to be the most important fact. So one can point out some general considerations. But the optimal structure has to be found for each alloy anew.

REFERENCES

1 G.E.R. Schulze, Metallphysik, Akademie Verlag, Berlin 1967
2 E. Hornbogen, Aluminium 43 (1967) 166
3 L.F.Mondolfo, Aluminium Alloys : Structure, Properties, Butterworths, London/Boston 1976
4 H.J. McQueen, Metallurgical Transactions A 8A (1977) 807
5 S. Gärtner, Laborbericht Nr. 3/89, FNE Freiberg
6 P. Furrer, P.M.B. Rodrigues, Z. Werkstofftechnik 16 (1985) 291
7 P.A. Hollinshead, T. Sheppard, Mat. Sci. Technol. 3 (1987) 1019
8 R.K. Wyss, R.E. Sanders, Metallurgical Transactions A 19A (1988) 2523
9 I.R. Dover, J.D. Embury, Materials Forum 9 (1986) 188

PLASTIC DEFORMATION NEAR GRAIN BOUNDARIES IN MOLYBDENUM BICRYSTALS

M. M. Kantor*, L. M. Olshansky*, U. Messerschmidt**

*Baikov Institute of Metallurgy Acad.Sci.of U.S.S.R.
Leninsky Prosp., 49, Moscow 117911 U.S.S.R.
**Institute of Solid State Physics and Electron Microscopy,
GDR-Academy of Sciences, 402 Halle, Weinberg 2, GDR.

ABSTRACT

By means of the in situ straining SEM and HVEM, processes of deformation and fracture have been studied in molybdenum bicrystals containing high-angle grain boundaries of controllable misorientation with boundary planes perpendicular to the deformation axis. It has been found that in situ experiments in SEM permit direct measurements of intergranular sliding by the displacement of slip lines along grain boundaries during deformation of bicrystals. The rate of intergranular sliding is different at various stages of deformation of bicrystals. The mechanism of total deformation can be characterized by two main stages. At the initial stage of loading with the stresses close to the yield limit, plastic deformation takes place mainly due to intragranular slip and fragmentation. At stresses exceeding the yield limit, grain boundary sliding has the decisive effect on plastic deformation of bicrystals. Grain boundary sliding is based on a dislocation mechanism. Intergranular sliding can only be analysed in terms of motion of grain boundary dislocations.

The cold brittleness of molybdenum and tungsten, which is characteristic of even their single crystals, can be enhanced substantially by the tendency of polycrystals of these metals to intergranular fracture. Earlier, the intergranular fracture at low temperatures is due primarily to segregates and precipitates formed by interstitial impurities in regions near grain boundaries /1, 2/. The studies of the strength of grain boundaries in molybdenum and tungsten bicrystals, however, suggest that intergranular fracture at low temperatures can be associated with the intrinsic structure of grain boundaries /3-10/. The actual cause of intergranular fracture is however still unknown. For clarification of this problem, it is essential to investigate the role of grain boundaries as obstacles to slip and how they disturb the homogeneous plastic flow when the latter propagates through grain boundaries.

The present report contains some results of a direct investigation of the role of grain boundaries in the processes of deformation and fracture, which have been obtained in in situ experiments in a scanning and a high-voltage transmission electron microscope.

The samples of the study were molybdenum bicrystals containing high-angle boundaries of a quite definite misorientation. The molybdenum bicrystals were grown as in /11/. The misorientation of the adjacent crystallites was additionally controlled by Laue back-reflection photography; the accuracy of determination of orientations amounted to 2°. The investigated bicrystals contained twist boundaries with twist axes ⟨100⟩ and ⟨110⟩. The specimens for experiments in the scanning microscope were bars with the dimensions 3 mm x 3 mm x 20 mm. The tensile axis of specimens was coincident with the crystallographic direction ⟨100⟩ and a boundary was normal to this axis. The observation surface coincided with a {100} plane of one of the crystallites in a bicrystal.

In specimens selected for the experiments in HVEM the grain boundary was approximately perpendicular to the tension axis. Plates of 0.8 mm thick were cut out so that boundary was at about midway from the upper and lower face of the plate. The work-hardened layer of the metal was removed by electrolytic polishing in a mixture of 95% H_2SO_4 and 5% CH_3OH. The final thinning of plates was carried out by electrolytic polishing in the same electrolyte.

Deformation in the scanning electron microscope was performed by three-point bending with the spacing between the supports being .17 mm and the deformation rate 0.05 mm/min. The stress-strain curve was recorded simultaneously. Deformation in the high-voltage transmission electron microscope was carried out by using an attachment developed by U.Messerschmidt and F.Appel /12/.

During deformation in the scanning microscope, slip lines were observed to appear when attaining the yield stress. The results of the analysis of slip traces suggest that at very beginning slip occurs in a single system of the type {110}⟨111⟩ which is the most heavily stressed system in the given scheme of loading. With a further increase of the stress applied, the density of slip lines in the indicated system increases and at least two other systems are activated (Fig.1). It should also be noted that, as the yield limit is exceeded in the region of tension, the crystallites disintegrate into individual fragments (Fig.2).

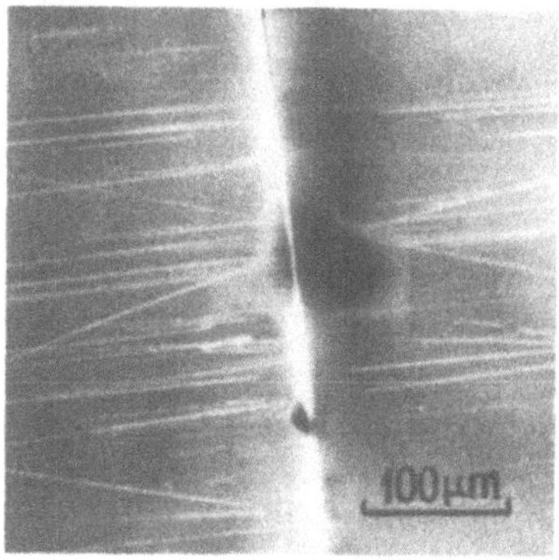

Fig.1. Dislocation slip near a grain boundary in deformation of a molybdenum bicrystal

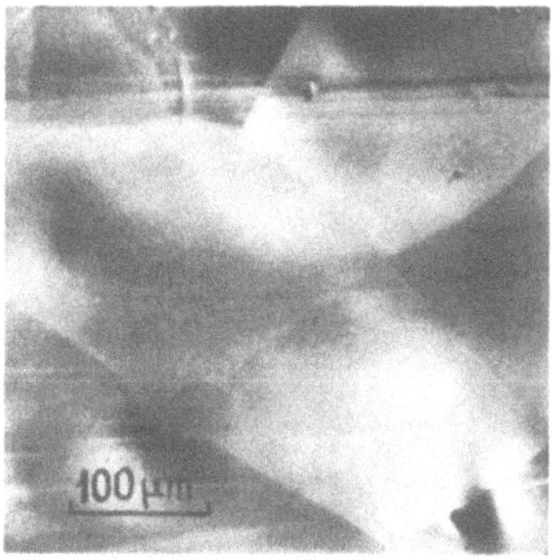

Fig.2. Deformation fragmentation of crystallites under load

Similarly, as the yield limit is exceeded due to concentrating shear stresses at a grain boundary, the process of grain-boundary slip takes place. Since no boundary migration takes place in molybdenum at the room temperature, it is possible, in in situ experiments, to measure the displacement of one crystallite relative to the other along their common interface without using reference scratch marks, i.e. by directly measuring the displacement of slip lines along a grain boundary in the initially polished surface of the specimen (Fig.3). The results of measurements of the amount of sliding in boundary planes depending on the stress applied are shown in Fig.4(a). Within the range of deformations corresponding to stresses close to the yield limit, the displacement increases slowly, but as the stresses approach the fracture strength, the rate of displacement becomes enhanced. This is the evidence that plastic deformation at the first loading stage takes place due to

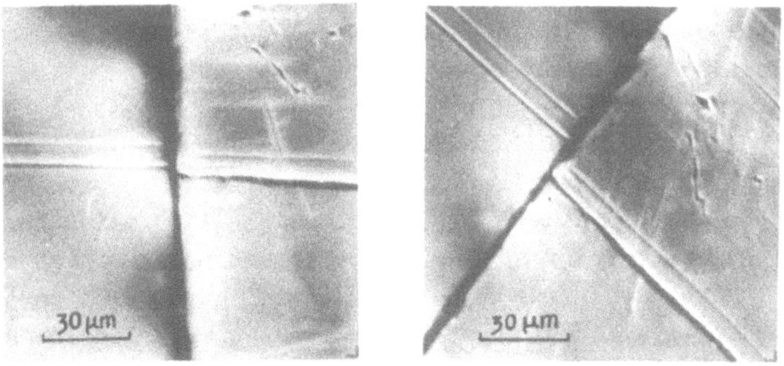

Fig.3. Displacement of slip lines along a grain boundary in deformation of molybdenum bicrystal: (a)-after the first loading stage; (b)-after the second loading stage

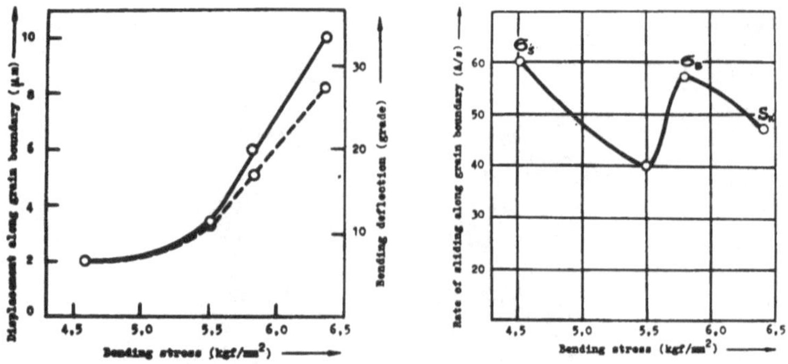

Fig.4. Dependence of the amount (a) and rate of grain-boundary sliding (b) on applied stress in tests of molybdenum bicrystals by the method of three-point bending

two processes: grain-boundary sliding and intragranular deformation, whereas at the second stage it occurs predominantly owing to intergranular sliding. This agrees well with the dependence shown in Fig.4(b). At the deformation near the yield limit, the initial rate of sliding is approximately proportional to the shear stress and follows a decreasing exponent (which indicates that the second process gives the major contribution to plastic deformation), whereas the rate of displacement at the grain boundary increases drastically near the ultimate strength and reaches its maximum at σ_t, i.e. plastic deformation in this part of the diagram occurs only due to grain-boundary sliding. In the next part of the deformation curve, the rate of slip decreases again, which can be explained by the appearance and propagation of numerous microcracks in the boundary plane, whose nucleation takes place in the sites of intersection of slip lines of several systems with the interface of the crystallites.

It has been found from the results of crystallographic analyses that slip takes place predominantly in {110} planes of both grains, but, since the grains are misoriented relative to each other, the slip planes are not matched in the interface. This mismatch is probably compensated by the slip of structural dislocations.

This assumption is confirmed by the results of in situ HVEM experiments with straining of bicrystals with ⟨110⟩ twist boundaries. In all cases the fracture of specimens took place along a high-angle boundary, though boundaries of a low misorientation also participated in the process of plastic deformation. Already at very low loads (of an order of 10 g), it was possible to observe motion of lattice dislocations near low-angle boundaries and their interaction with the wall dislocations. The process proceeded as had been described earlier /13/. Consequently, deformation at the initial stage of loading takes place only due to intragranular slip.

At the next stage of loading, with the stresses close to the yield limit, plastic deformation in the interface of high-angle boundaries begins: the dislocation density increases linearly and reaches $10^9 cm^{-2}$ at a deformation of 6% (Fig.5). It is likely that the structure of these boundaries is completely rearranged at this stage, in particular, dislocations of orientational mismatching appear in the grain-boundary face. The Burgers vectors of these dislocations differ from those of

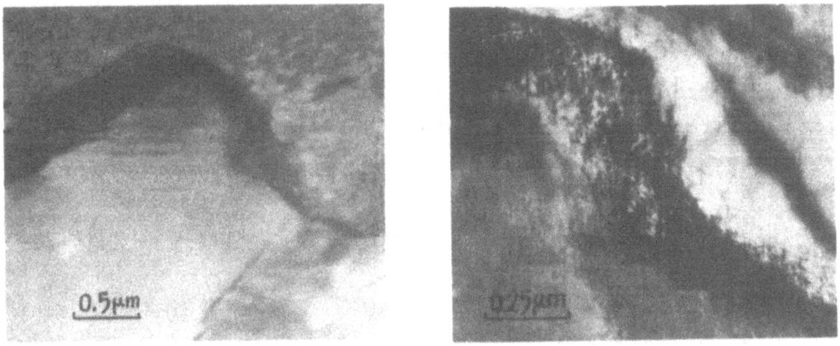

Fig.5. Dislocation structure of a high-angle boundary before and after deformation: (a)-before the loading; (b)-after the loading

lattice dislocations and are random relative to the boundary plane. Dislocations cease to accumulate at the interface as soon as the internal stresses in the boundary plane reach the theoretical fracture strength. In that case, the dislocations merge together and form a brittle micro-crack which opens either into a grain or into the boundary plane depending on the amount of misorientation of the adjacent grains.

The results of investigation of deformation of bicrystals make it possible to draw the following conclusions:

1. In situ experiments inside a scanning electron microscope permit direct measurements of intergranular sliding by the displacement of slip lines along grain boundaries during deformation of bicrystals.

2. The rate of intergranular sliding is different at various stages of deformation of bicrystals and has a cyclic nature depending on the stress applied.

3. The mechanism of total deformation can be characterized by two main stages. At the initial stage of loading with the stresses close to

the yield limit, plastic deformation takes place mainly due to intragranular slip and fragmentation. At stresses exceeding the yield limit, grain-boundary sliding has the decisive effect on plastic deformation.

4. Grain-boundary sliding is based on a dislocation mechanism. Intergranular sliding can only be analysed in terms of the motion of structural dislocations and the associated elementary rearrangements in the structure of boundaries.

REFERENCES

1 D.McLean, Grain Boundaries in metals, Claredon Press, Oxford 1957

2 V.I.Trefilov, Yu.V.Milman, S.A.Firstov, Physical Foundations of Strength and Plasticity of Refractory Metals, Naukova Dumka, Kiev 1975 (in Russian)

3 A.Kobylansky, G.Goux, Compt.Rendus, serie C, 242, 1937 (1971)

4 Ch.V.Kopetskii and A.I.Pashkovskii, Dokl.Acad.Nauk SSSR, 210, 72 (1973)

5 A.A.Yastrebkov, Yu.P.Ivakin, Fiz.Metallov i Metallovedenie, 36, 135 (1973)

6 A.A.Yastrebkov, V.M.Lakeenkov, Problemy Prochnosti, No.1, 75 (1975)

7 J.B.Brosse, R.Fillit, M.Biscondi, Scripta Metall., 15, 619 (1981)

8 H.Kurishita, S.Kuba, H.Kubo, H.Yoshinaga, Trans.Japan Inst.of Metals, 26, 332 (1985)

9 H.Kurishita, A.Oishi, H.Kubo, H.Yoshinaga, Trans.Japan Inst.of Metals, 26, 341 (1985)

10 V.G.Suraeva, V.G.Glebovsky, I.M.Shulga, L.S.Shindlerman, Scripta Metall., 19, 411 (1985)

11 A.A.Yastrebkov, V.I.Lakeenkov, Pribori i Technika eksperimenta, No.4, 215, (1973)

12 U.Messerschmidt, F.Appel, Ultramicroscopy, 1, 223, (1976)

13 I.I.Novikov, M.M.Kantor, V.M.Plastinin, V.N.Timofeev, V.A.Yermishkin, Kristall und Technik, 14, 1283 (1979)

THE EFFECT OF L1₂ ORDERED δ' PRECIPITATES ON THE MECHANICAL PROPERTIES OF Al-Li ALLOYS

J. Lendvai*, H.-J. Gudladt+, J. Schneider+, W. Wunderlich+ and V. Gerold+

*Institute for General Physics, Eötvös University, H-1088 Budapest Múzeum krt 6-8

+ Max-Planck-Institut für Metallforschung, Institut für Werkstoffwissenschaft, Stuttgart and Institut für Metallkunde der Universitaet Stuttgart, Seestraße 92, D-7000 Stuttgart 1

ABSTRACT

The deformation characteristics of high purity binary Al-Li alloys have been studied under monotonic as well as under cyclic loading on single- and polycrystalline specimens. The $L1_2$ ordered δ' (Al_3Li) precipitates are sheared by paired dislocations in underaged specimens and are bypassed by Orowan looping when the particle diameter becomes larger than about 50 nm. In peak-aged and in moderately overaged specimens particle shearing and bypassing occurs simultaneously. Extended planar dislocation arrays are observed on isolated slip planes in underaged single crystals in consequence of local glide plane softening. The strengthening of the alloy can be satisfactorily described by the theory of order strengthening in the underaged and by the theoretical estimation of the Orowan stress in the overaged state.

INTRODUCTION

The increase in strength during ageing of binary Al-Li alloys is due to the formation and growth of $L1_2$ ordered δ' (Al_3Li) particles. The δ' precipitates remain spherical and fully coherent with the matrix up to extremely large ($r > 100$ nm) particle sizes. Binary Al-Li alloys provide an excellent possibility to compare theoretical models for strengthening by ordered particles with experimental results [1-6].

In the present work the dislocation microstructure of single- and polycrystalline Al-Li specimens deformed in monotonic compression as well as in low cycle fatigue (LCF) was studied by TEM. The experimentally determined strengthening is correlated with the size and volume fraction of precipitates.

EXPERIMENTALS

The investigations were carried out on high purity binary Al-Li samples prepared from 6N purity Al and 4N purity Li. Most of the results were obtained on single slip oriented single crystals prepared by a modified Bridgman technique but in a few cases results of experiments carried out on coarse grained (average grain size about 0.5 mm) polycrystalline specimens were also included. The $\tau = \sigma/M$ conversion with $M = 2.24$ has been used in the case of polycrystals as discussed in an earlier work [6].

The samples were solution heat treated in a salt bath at 838 K for 30 min, water quenched and naturally aged at room temperature for about 4 days prior to the artificial ageing. Symmetrical push-pull fatigue tests ($R\varepsilon = -1$) at 0.3 Hz under plastic strain control were carried out and the values of the plastic strain amplitude, γ_{pl} ranged from 10^{-4} to 10^{-2}. Disk-shaped samples with a thickness of 0.2 mm were spark-cut for TEM investigations. Thin foils were prepared by the ordinary electrolytic jet polishing with a mixture of methanol and HNO_3 (2:1) at 253 K, 10 V.

Monotonic testing was made in compression on cylindrical specimens of 6 mm diameter and 9 mm length at an initial normal strain rate of $9\times10^{-5}s^{-1}$ The CRSS values were obtained as the average of the results of measurements on two specimens.

DISLOCATION MICROSTRUCTURES

Fig. 1 shows the dislocation structure of an underaged (433 K / 10 h) single crystal specimen fatigued at $\gamma_{pl} = 2\times10^{-4}$. Dislocation pairs are clearly visible demonstrating that during deformation the ordered particles are cut by dislocations. Fig. 2 shows the dislocation arrangement in a (110) oriented foil prepared of an underaged single crystal which has been fatigued up to fracture at $\gamma_{pl} = 2\times10^{-4}$. The dislocation arrangement is characteristic of extremely localized deformation where large amount of shear is accumulated on isolated slip planes. In the activated slip planes very high numbers of nearly parallel dislocations can be observed. The distance of the activated slip planes (1 to 2 μm) correlates well with the distance of slip steps which could be observed on the polished surface of the fatigued specimens [7,8]. On the activated slip planes the particles are completely sheared off since the shear in one direction attains higher values than the particle diameter.

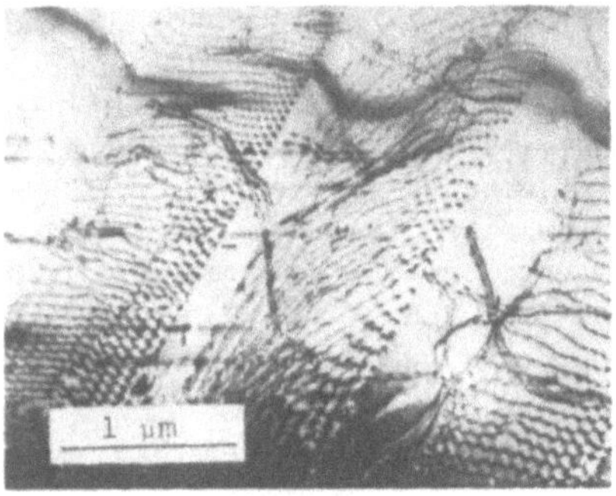

Fig.1 Dislocation structure of an underaged specimen fatigued at $\gamma_{pl} = 2\times10^{-4}$

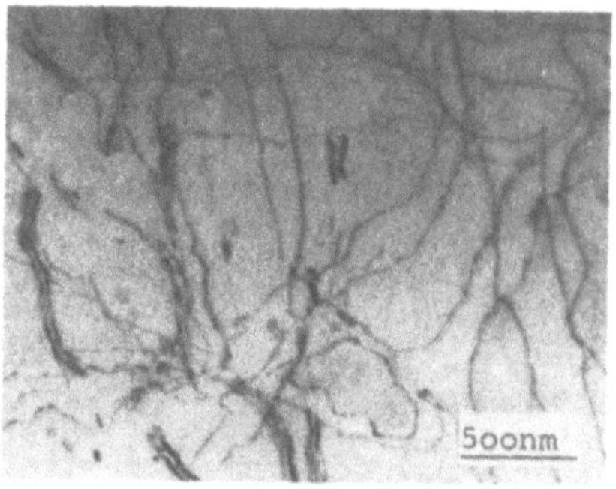

Fig.2 Planar dislocation arrays on the primary slip planes of an underaged specimen fatigued at $\gamma_{pl} = 2 \times 10^{-4}$ up to fracture (110) foil; $g = (\bar{1}11)$

Fig. 3 shows the micrograph of a specimen deformed up to 2% in compression. The specimen was aged for 1000 min at 473 K which corresponds to a nearly peak-aged state of the alloy. Paired dislocations are observed together with Orowan loops. Evidently cutting and bypassing occur simultaneously. The same characteristic features can be inferred from the dislocation structure of a slightly overaged (473 K / 3000 min) specimen which has been fatigued at $\gamma_{pl} = 7.8\times10^{-4}$ (Fig. 4). A few paired dislocations besides many Orowan loops have been observed even in specimens aged for 10000 min at 473 K, whereas in strongly overaged samples (493 K / 10000 min) only Orowan loops were found.

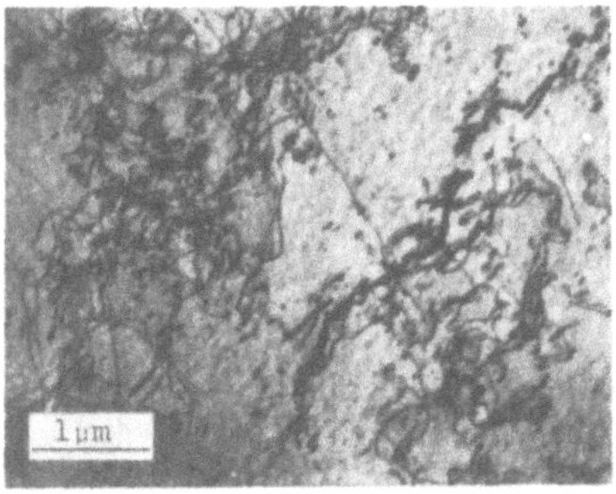

Fig.3 TEM micrograph of a nearly peak-aged single crystal specimen deformed up to 2% in compression

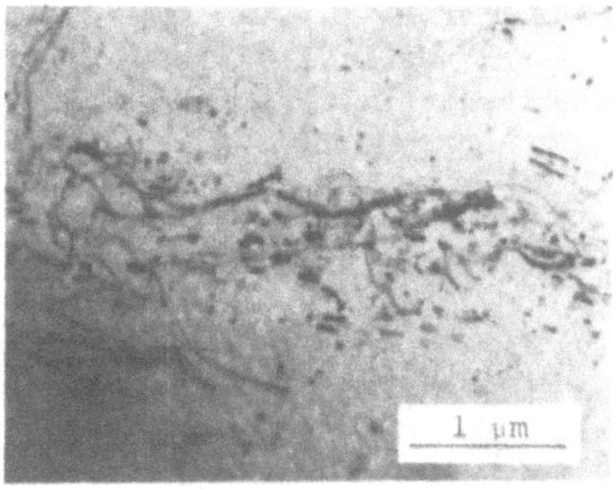

Fig.4 Dislocation arrangement in a slightly overaged sample fatigued at $\gamma_{pl} = 7.8 \times 10^{-4}$

THE STRENGTHENING EFFECT OF δ' PARTICLES

The stress response of the Al-Li alloys investigated here showed a constant value, τ_c, for the first 10 to 100 cycles in the plastic strain controlled LCF tests. For underaged and peak-aged specimens the τ_c stress was found to be practically independent of γ_{pl} in the range from 10^{-4} to 10^{-2} [6-8]. On the other hand τ_c depends strongly on the ageing condition and, therefore, on the precipitate microstructure.

Similarly, in the case of monotonic tests the CRSS was found to be strongly dependent on the ageing conditions. In order to determine the strengthening introduced by the δ' particles the contribution of the solid solution matrix to the strength of the alloy has to be also determined. From the results of compression tests carried out on a series of single crystals of different compositions (7 to 11 at% Li) as well as on the basis of the work of Miura et al. [9] the strength of the solid solution matrix, τ_m, after ageing at 473 K (c_{Li} = 6.6 at%) and at 493 K (c_{Li} = 7.07 at%) was determined as 7.6 and 8.6 MPa, respectively. The strengthening due to the δ' particles was obtained by linearly substracting these τ_m values from the measured CRSS and τ_c values.

It is well established that the strengthening introduced by the δ' particles is mainly due to order strengthening [2,4-6,10]. Other possible contributions like the increase in the particle/matrix interfacial area or the effect of coherency strains are negligibly small since the particle/matrix interface energy is of the order of 10 mJ/m^2 [11,12] and the misfit strain is smaller than 10^{-3} [13,14]. Therefore in the following only the effect of order strengthening will be considered.

According to the theory of order strengthening [2,15] in underaged alloys, where particles become sheared by paired dislocations the contribution of the particles to the strength, $\Delta\tau_o$, can be given as

$$\Delta\tau_o = \frac{\gamma}{2b}\left[\left(\frac{3\pi^2\ \gamma\ f\ r}{32\ T}\right)^{1/2} - f\right], \qquad (1)$$

where γ is the APB energy on the (111) slip planes, b the Burgers vector, r the particle radius, f the volume fraction and T the line tension of dislocations. In a previous investigation γ was determined as 165 mJ/m^2 [6]. If the particles become larger than $r_1 = 4T/\gamma\pi$ then the theory predicts that $\Delta\tau_o$ becomes constant i.e. independent of the particle size. The value of r_1 in the present case is about 20 nm [6]. With further increasing particle size the Orowan bypassing sets in [7,16]. In this regime the strengthening effect can be given as [17]

$$\Delta\tau_o = 0.84 \frac{Gb}{2\pi(1-\nu)^{1/2} L_2} \ln \frac{1.6\, r}{r_o} \qquad (2)$$

where G (= 30 GPa) is the shear modulus of the alloy, L_2 the mean planar particle distance, r_o (= b) the inner cut-off radius.

In order to compare the experimentally determined $\Delta\tau$ and the theoretically calculated $\Delta\tau_o$ values the average particle radius and the volume fraction has to be determined. The particle radius, r, has been measured on the basis of TEM micrographs and for all ageing treatments between 433 and 493 K an $r \propto t^{1/3}$ kinetics has been found [6]. The volume fraction was determined from measuring calorimetrically the heat of dissolution of the precipitates or from the data of the metastable miscibility gap [18]. A correction has been applied accounting for the increased solid solubility of the matrix in equilibrium with small particles [12]. The experimentally obtained $\Delta\tau$ strengthening values as well as the particle size and volume fraction data are listed in Tab. 1.

Table 1 Experimentally determined strengthening data and microstructural parameters

Concentration at%	Heat treatment T(K) / t (min)	r (nm)	f	$\Delta\tau$ (MPa)	Remark
9.1	473 / 30	6.1	0.13	43	C; S
9.1	473 / 100	9.1	0.15	52	C; S
9.1	473 / 300	13.1	0.16	63	C; S
9.1	473 / 600	16.5	0.16	73	C; S
9.1	473 / 1000	19.5	0.16	77	C; S
9.1	493 / 1440	25.4	0.14	69	C; S
9.1	493 /10000	48.1	0.14	41	C; S
8.5	473 / 600	16.5	0.12	76	LCF; P
8.5	473 / 1000	18.0	0.12	77	LCF; S
8.5	493 / 3600	34.0	0.10	50	LCF; S

C: compression test; LCF: low cycle fatigue; S: single crystal; P: polycrystal

In Fig. 5 the $\Delta\tau$ values from Tab. 1 are plotted as a function of the average particle radius. The curves in the figure represent the results of the theoretical calculations based on Eqs. (1) and (2). In the calculations for the underaged regime γ = 165 mJ/m^2 was used [6].

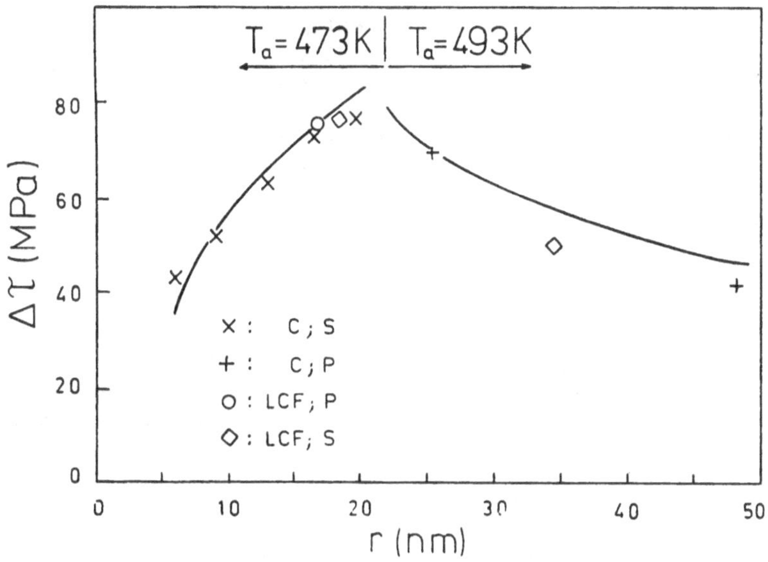

Fig. 5 Experimentally determined strengthening contribution of the δ' particles as a function of particle size. Solid lines represent theoretical curves calculated from Eqs. (1) and (2). (C: compression; LCF: low cycle fatigue; S: single crystal; P: polycrystal)

CONCLUSIONS

The results described in the present work show that the $L1_2$ ordered δ' particles are cut by paired dislocations moving on (111) matrix planes in underaged Al-Li samples. The leading dislocation of a pair forms an APB in the particles while the trailing one restores the order. By the interaction through the APB energy between the two dislocations of a pair they are bound to one slip plane and consequently slip becomes planar. Strong strain localization on isolated (111) slip planes is associated with planar slip in the course of both monotonic and cyclic deformation. Slip becomes localized in consequence of glide plane softening : the following dislocation pairs encounter a reduced resistance of the particles sheared by the preceding pairs in the same slip plane. As pointed out in a recent review on planar slip [19], if the particles with an atomic order are sheared obstacle damage is restricted to the shearing plane only, leading to planar slip with isolated active slip planes.

In the dislocation microstructure of peak-aged and even of overaged specimens dislocation pairs and Orowan loops are observed simultaneously indicating that both particle shearing by paired dislocations and bypassing take place. Similar observations were reported in the case of Ni-base superalloys containing ordered γ' particles [20]. With the occurrence of bypassing slip localization becomes less pronounced. The dislocation structure in specimens deformed to a few per cent deformation in monotonic compression and up to fracture in plastic strain controlled LCF tests was found to be similar.

It was found that in the underaged regime the theory of order strengthening correctly predicts the strengthening contribution of the δ' precipitates under both monotonic and cyclic conditions if γ the specific APB energy for the (111) slip planes in Eq. (1) is taken as 165 mJ/m^2 [6]. It has to be mentioned, however, that the constant (size independent) strength regime predicted by the theory [2,20] in the peak-aged condition was not observed in our experiments. The experimentally determined particle strengthening in overaged specimens is also in reasonable agreement with the Orowan stress calculated from Eq. (2).

REFERENCES

1 M Tamura, T. Mori and T. Nakamura, Trans JIM, 14 (1973) 355
2 A.J. Ardell, Metall. Trans. 16A (1985) 2131
3 Y. Miura, A Matsui, M. Furukawa and M. Nemoto, In: Aluminium-Lithium Alloys III, C. Baker, P.J. Gregson, S.J. Harris and C.J. Peel (Eds.), The Inst. of Metals, London 1986, p. 427
4 J.C. Huang and A.J. Ardell, Acta metall. 36 (1988) 2995
5 J.C. Huang and A.J. Ardell, Mater. Sci. Engng. A104 (1988) 149
6 H.-J. Gudladt, J. Lendvai and J.Schneider, Acta metall. 37 (1989) in press
7 J. Schneider, H.-J. Gudladt and V. Gerold, J. de Physique, Colloque C3, suppl. au n° 9, 48 (1987) C3-745
8 H.-J. Gudladt, J. Lendvai, J. Schneider, W. Wunderlich and V. Gerold, In: Aluminium-Lithium 5, to be published
9 Y. Miura, S. Nishitani, M. Furukawa and M.Nemoto, In: Strength of Metals and Alloys (ICSMA 8), P.O. Kettunen, T.K. Lepistoe and M.E. Lehtonen (Eds.), Pergamon Press, Oxford 1988, p. 995
10 M. Furukawa, Y. Miura and M.Nemoto, Trans. JIM, 26 (1985) 230
11 S.F. Baumann and D.B. Williams, Scripta metall. 18 (1984) 611
12 J. Lendvai, H.-J. Gudladt, W. Wunderlich and V. Gerold, Z. Metallkde, 80 (1988) 310
13 B. Noble and G.E. Thompson, Metals. Sci. 5 (1971) 114
14 D.B. Williams and J.W. Edington, ibid, 9 (1975) 529
15 D. Raynor and J.M. Silcock, ibid, 4 (1970) 121
16 M. Furukawa, Y. Miura and M. Nemoto, Trans. JIM, 26 (1985) 226
17 V. Gerold, In: Dislocations in Solids, Vol. 4, F.R.N. Nabarro (Ed.), North Holland, Amsterdam 1979, p.219
18 G. Cocco, G. Fagherazzi and L. Schiffini, J. Appl. Cryst., 10 (1977) 325
19 V. Gerold and H.P. Karnthaler, Acta metall. 37 (1989) in press
20 E. Nembach and G. Neite, Prog. Mat. Sci. 29 (1985) 177

THE STRUCTURE OF F.C.C. METALS AFTER HIGH STRAINS

V.I. Levit

Institute of Metal Physics, Ural Branch of the
Academy of Sciences of the USSR, Sverdlovsk

ABSTRACT

Single crystals of nickel, copper, and of the nickel-based alloy nimonic 80A were deformed by torsion at quasi-hydrostatic pressure to a true logarithmic degree of deformation of about 7. The parameters describing the structure and texture of the ultra-fine grained polycrystalline aggregates correlate with the dislocation mobility in the materials.

RESULTS AND DISCUSSION

There is no agreement in the literature concerning the behaviour of materials at high plastic deformations over the temperature range below 0.2 T_{melt}. As a rule, fracture processes stop the deformation before the steady state with a work hardening coefficient $d\sigma/de \rightarrow 0$ is reached. As theoretically shown [1], the occurrence of this state is governed by the dynamic softening rather than by the thermal one.

The present paper describes the deformation behaviour of f.c.c. single crystals highly deformed by torsion, with the fracture being suppressed by high pressure.

By applying the sparking-method single crystals of nickel (99.97 %), copper (99.96 %), and of the nickel-based alloy nimonic 80A were cut into discs of 0.5 mm thickness and 5 mm in diameter. Then they were reduced to a thickness of 0.2 mm by mechanical and electrolytical polishing. The planes of the discs were parallel to (001) and (111) planes, respectively. The torsion deformation was realized by using a Bridgman apparatus [2]. The degree of deformation e was approximately calculated by $e=\ln(\varphi r/l)$, where φ is the angle of twisting, r and l are the radius and the thickness of the specimen, respectively [3]. The deformation temperature was not higher than 295 K. Some nickel specimens were deformed at 77 K in the same apparatus equipped with a cooling device. In this case the samples for TEM were electrolytically polished at 240 K and investigated in a JEM 200 CX microscope at T=100 K.

The measurements of the micro-hardness and the investigation of the samples were performed at a distance of $r=2\pm0.2$ mm from the disc centre. X-ray investigations of the distribution of the reflection intensity were performed on the whole disc plane.

For medium deformations ($e\approx3$) the behaviour of nickel and copper is similar and agrees well with other results [3, 4]. The cell structure is

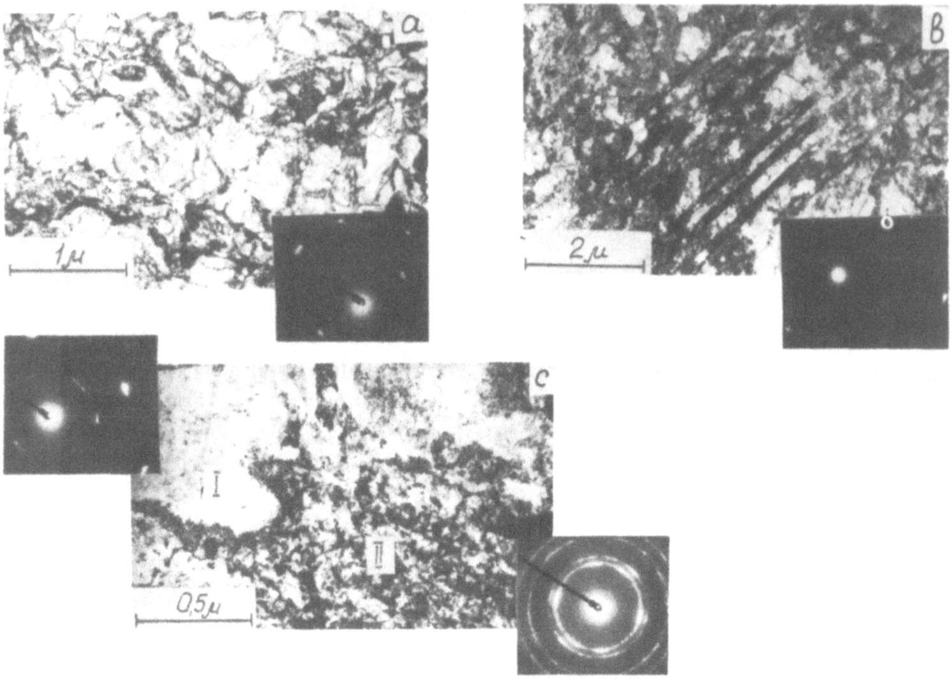

Fig. 1: Structure of single crystals of Ni (a, b) and nimonic 80A (c) strained for $\varphi=0.3$; (a, c) $T_{str}=293$ K, (b) $T_{str}=77$ K

formed and with advancing deformation the cell size decreases whereas the misorientation increases. The typical structure is shown in Fig. 1a.

If the deformation is performed at 77 K, microtwins occur and a cell structure develops, Fig.1b. In nimonic 80A, which has a low stacking fault energy, no cell structure is formed. If the density of the dislocations reaches the value of about 10^{11} cm^{-2}, shear bands are generated in the alloy structure, which have a much higher local deformation and misorientation than the average sample. Fig. 1c shows the edge of a shear band with a local misorientation of 50°. In the area marked by I the dislocations are homogeneously distributed as in the main part of the sample. The azimuthal misorientation on an area of 4 μm^2 does not exceed 3 to 4°, which indicates the single crystalline character of the imaged part of the foil.

The electron diffraction pattern shows that the misorientation is not due to the discrete turns of neighbouring areas but due to the large gradient of the elastic distorsions of the lattice.

After a deformation of $\varphi=2\pi$ the structures of nickel and copper reveal that the separate microcrystalline areas lack in dislocations and are discretely misoriented by large angles (20° and more, Figs. 2a, b). In the structure of copper recrystallized grains occur, which are larger than the neighbouring microcrystallites and are mostly joined by 120° triple nodes. The electron diffraction pattern shows distinct uniaxial reflexes arising from the separate microcrystallites indicating the

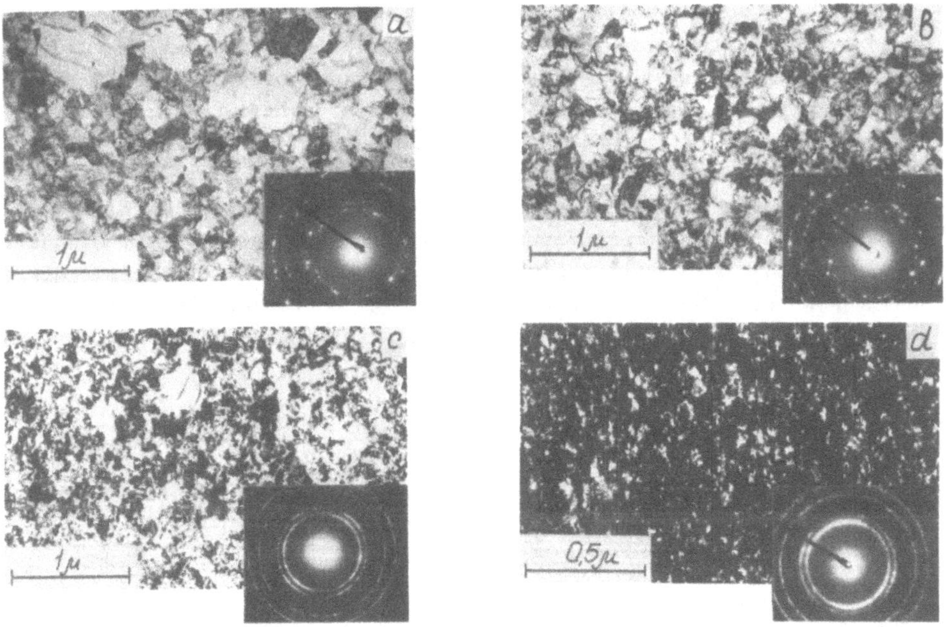

Fig 2: Structure of single crystals of Cu (a), Ni (b, c), and nimonic 80A (d) after a high strain of $\varphi=20\pi$, ($e \geq 7$). (a, b, d) $T_{str}=293$ K, (c) $T_{str}=77$ K

perfection of the inner structure. The size of the microcrystallites in nickel is smaller and there are no regular triple nodes. Deformation at 77 K causes the size of the microcrystallites to further decrease and separate cells to form having a size of 0.2 μm and a high dislocation density in their interior (Fig. 2c). In nimonic 80A the microcrystallites could not be resolved in the bright-field image. Thus, dark-field imaging was applied yielding sizes of 11 to 15 nm (Fig. 2d).

An increase of the twisting angle from 2π to 20π corresponds to a relative elongation of about 10 in a usual tensile test. It should be pointed out that this high deformation does almost not (copper and nickel) change the structures and the numerical parameters describing them, or only little (nimonic 80A), in comparison to the structure changes occurring at the earlier stages of deformation. This is also confirmed by the data of X-ray investigations and micro-hardness measurements (Fig. 3).

According to a parabolic law of work hardening with a decreasing work hardening coefficient, the hardness increases with the twisting angle showing a sharp jump in the interval $\pi \leq \varphi \leq 2\pi$.

The increase of the strength due to extremely high deformations has been observed for nickel, silver and some other alloys [4, 5], further trends of the behaviour, however, are reported here for the first time. At $\varphi > 2\pi$, in nimonic 80A, the micro-hardness increases slowly, whereas it is almost constant in nickel, slightly decreasing in copper. It is interesting to note that at high strains copper and nickel form ultrafine grained polycrystalline aggregates, the intensity distributions of which are entirely equal, as it is observed for textureless structures. The result is independent of the initial orientation, with <001> and <111> being the main components of the torsion texture in f.c.c. Nimonic 80 shows a weak <111> texture; the tendency to form such structures is also noticeable in nickel deformed at liquid nitrogen temperature.

The comparison of X-ray and TEM data gives the following result. High macroscopic misorientations were detected by the X-ray technique at $\varphi \geq 0.3$, whereas TEM revealed highly misoriented neighbouring microcrystallites at $\varphi \geq \pi$. In the single crystals high macroscopic misorientations can be produced by a set of homogeneously distributed dislocations in

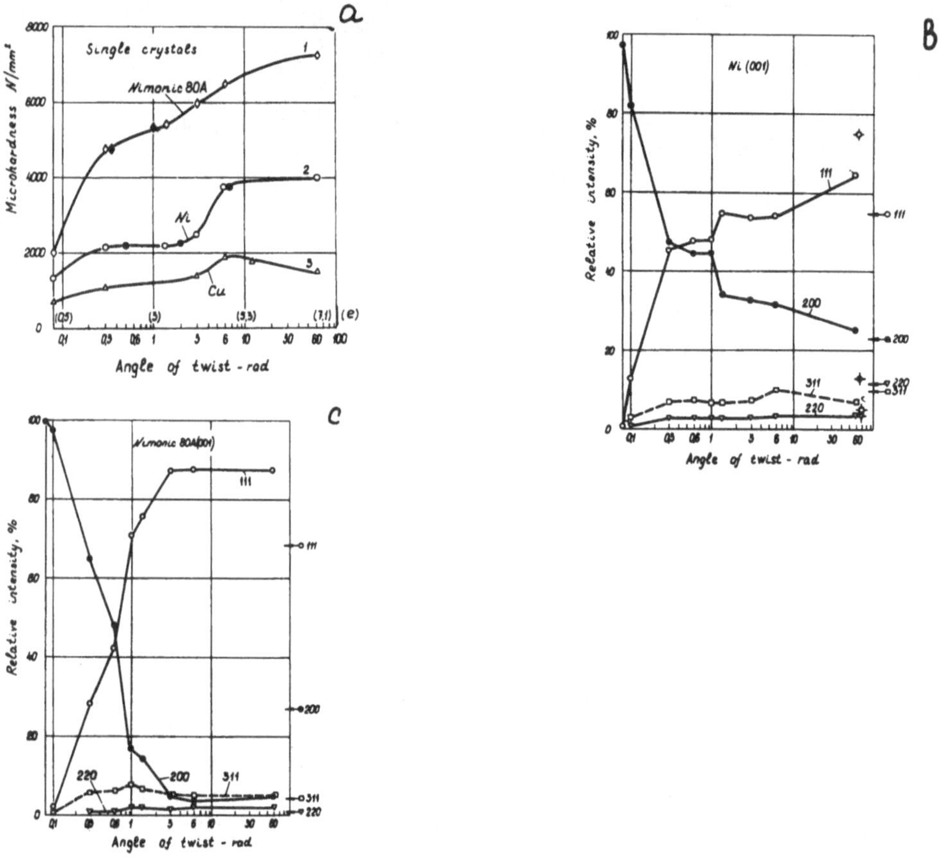

Fig. 3: Micro-hardness (a) and X-ray intensities (b, c) of single crystals after straining. The arrows on the right side show the textureless level of intensity: o, ● T_{str}=77 K

neighbouring cells weakly misoriented. At substantially larger deformations the discrete twisting by some tens of degrees is due to dynamic recrystallization mechanisms.

According to the described mechanisms of the formation and development of the cell structure at the beginning of deformation, in copper and nickel a mincing of cells and an increase of the misorientation angle between them are supposed. If the minimumn cell size (0.2 µm) and the maximum misorientation of 10° are reached, the cell walls lose their ability to further admit dislocations. Usually, the further increase of the density of dislocations in the cell volume leads to a rapid increase of the internal stress level and to the destruction of the sample. In our experiments the high external pressure prevents the destruction. Thus, the dislocations are accumulated in the cells and due to the extremely high internal stresses the cell walls are reconstructed into large-angle grain boundaries. This dynamic recrystallization is probably an athermal process. The deformation of nickel and copper further advances according to mechanisms known for structural superplasticity. According to [6] the main contribution to the deformation results from plastic rotations.

The size of microcrystallites is determined by the length of the slip path of the dislocations. With decreasing deformation temperature in nickel the equilibrium size of the microcrystallites decreases correspondingly. Alloying results in a remarkable decrease of the dislocation mobility which is accompanied with much smaller grain sizes. According to the SFE values, in nimonic 80A e.g., the microcrystallites are one order of magnitude smaller than those of nickel. Discontinuities, which inevitably appear at grain boundary sliding, are restored due to the grain boundary diffusion and due to the very high level of external pressure.

After such high strains in copper and nickel postdynamic recrystallization occurs at room temperature. In copper the velocity of this process is much higher than that for nickel, which reflects the differences in the micro-hardness. In nickel aged for 3 years at room temperature after deformation, the grain size becomes 3 times larger.

If samples of nickel and copper are annealed at the relatively low temperatures of 270 to 370 K, usual recrystallization starts resulting in annealing twins of about 1 µm in size. In nimonic 80A during recrystallization at 100 K homogeneous precipitation takes place and a dual structure is produced with grain sizes of the phases of 0.1 and 0.5 µm.

SUMMARY

1. High plastic deformations by combined torsion and hydrostatic pressure transform single crystals of f.c.c. metals (Cu, Ni, nimonic 80A) into the polycrystalline state. A weak torsion texture is formed, which depends on the deformation temperature and on the chemical composition, but which is independent of the initial orientation.

2. The transition between the single crystalline and polycrystalline structure includes the development and perfection of the cell structure and is finished by dynamic recrystallization. In nimonic 80A the transition is associated with the formation and distribution of shear bands.

REFERENCES

1 H. Mecking, A. Grinberg, In: Strength of Metals and Alloys, P. Haasen, V. Gerold, and G. Kostorz (Eds.), vol. 1, Toronto 1979, p.289-294

2 P.W. Bridgman, In: Studies of Large Plastic Flow and Fracture, Mc Graw Hill, New York 1952

3 J. Saunders and J. Nutting, Metal Sci. 18 (1984) 571-5

4 J. Nuttal, J. Nutting, Metal Sci. 12 (1978) 430-7

5 V.A. Pavlov, Phys. Met. and Metallogr. (Sov.) 58 (1984) 177-84

6 B.B. Rybin, In: High Plastic Deformation and Fracture in Metals (in Russian), Metallurgija, Moscow 1986

ACOUSTIC INVESTIGATION OF DISLOCATION PINNING BY ORDERED ATMOSPHERES OF OXYGEN ATOMS IN NIOBIUM SPECIMENS DEFORMED IN SITU

V.V. Zoninashvili, I.A. Naskidashvili, V.A. Melik-Shakhnazarov

Institute of Physics, Academy of Sciences of the Georgian SSR,
6 Guramishvili str., Tbilisi, 380077, U.S.S.R.

ABSTRACT

Using internal friction techniques, we have investigated the processes of the formation and decomposition of oxygen atom atmospheres in niobium polarized within the elastic field of dislocations. It is shown that the formation of such atmospheres leads to a decrease in both the relaxation time and the strength for low-temperature relaxation processes connected with the dislocation motion. We have tried to determine the potential relief due to the dislocation interaction with the atmosphere of oxygen atoms on the basis of acoustic measurements.

INTRODUCTION

It is well known that the long-range interaction of dislocations with point defects of tetragonal symmetry (for example the interstitial impurities of O, N or C in b.c.c. metals) leads to the orientational ordering of the defects within the elastic field of the dislocations. Such polarized atmospheres (so-called Snoek atmospheres) form during a period of time sufficient for the impurities to jump only one or two interatomic distances and manifest themselves by an increase of the flow stress on the one hand [1, 2] and by a decrease of the dislocation damping on the other [3, 4].

In the present paper we show the results of acoustic investigations of the processes of the formation and decomposition of oxygen atom atmospheres polarized within the elastic field of dislocations, as well as the influence of these atmospheres upon the low-temperature relaxation processes in niobium connected with the dislocation motion. We also try to determine the potential relief due to the dislocation interaction with the atmospheres of oxygen atoms on the basis of acoustic measurements.

EXPERIMENTAL METHOD

The investigations were carried out over the temperature range of

100 to 300 K using the apparatus for the acoustic measurements of specimens during their elastic loading or during slow plastic deformation described in [5]. Internal friction (IF) was determined by the decrement of the attenuation of free torsional oscillations (at a frequency of about 10^3 Hz) with an accuracy better than 1.5 %. The specimens used in the experiments were of pure zone-melted niobium containing $1.5*10^{-2}$ at% oxygen. To increase the mobile dislocation density, the specimens were predeformed in situ by 5 to 7 % at room temperature. Their yield stress at $T \approx 300$ K after such a predeformation was equal to 280 to 300 MPa. Since the stresses applied afterwards did not exceed 200 MPa, i.e. they were within the quasi-elastic range, dislocation displacements caused by them were of reversible character.

EXPERIMENTAL RESULTS

Fig. 1 shows the dependence of the IF on the elastic stress with subsequent recovery. The measurements were carried out at room temperature. The loading rate was $\dot{\sigma} \approx 0.5$ MPa/s. As the figure shows, after maintaining the specimen at zero stress for 24 hours, the IF grows in the process of loading, reaching a saturation state (curve 1). After the active loading is stopped at stresses of $\approx$110 MPa and the specimen is kept at such a stress, nearly complete recovery of the sound absorption is observed in $\approx 10^4$s (curve 2). Subsequent unloading again leads to an increase of Q^{-1} (curve 3) with its nearly complete recovery in the process of holding the sample at $\sigma = 0$ (curve 4). The recovery kinetics may well be described by the expression

$$(Q^{-1} - Q_o^{-1})/(Q_m^{-1} - Q_o^{-1}) = \exp\left[-(t/\tau)^{1/3}\right]$$

suggested in [4] to describe the process of the Snoek atmosphere formation of oxygen atoms in niobium. The relaxation times τ obtained for the IF recovery at $\sigma = 110$ MPa and at $\sigma = 0$ are nearly the same and equal to $7.0*10^3$ s and $6.7*10^3$ s, respectively. These values are sufficiently close to the relaxation time due to the oxygen atom re-orientation in the matrix, $\tau_0 \approx 1.4*10^4$ s, which makes it possible to connect the recovery process of the sound absorption of dislocations presented in Fig. 1 (curves 2, 4) with the formation of ordered or polarized oxygen atom atmospheres around the dislocations.

Thus, Fig. 1 demonstrates the process of dislocation unlocking from the atmospheres of oxygen atoms polarized in the dislocation field (curves 1, 3) and the new formation of such atmospheres around the new positions of the dislocations at $\sigma = 110$ MPa and $\sigma = 0$ (curves 2 and 4).

Fig. 2 shows the curves of Q^{-1} vs. σ plotted in the process of the

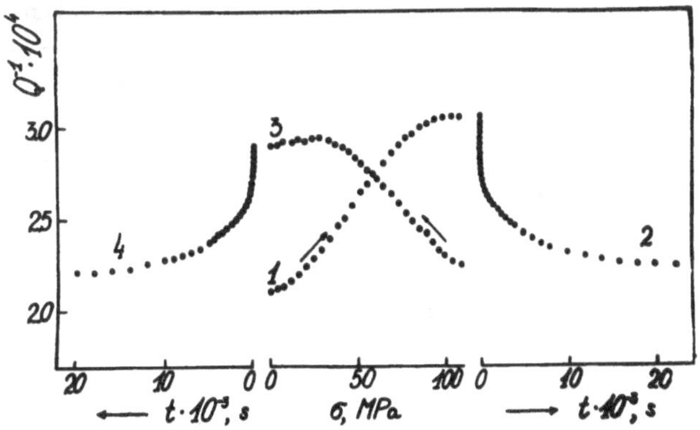

Fig. 1 Dependence of the sound absorption of dislocations on the elastic loading with subsequent recovery at T ≈ 295 K.

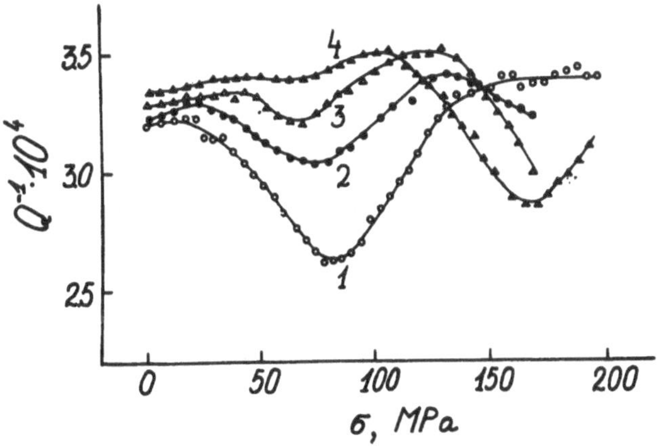

Fig. 2 Dependence of Q^{-1} on σ in the process of decomposition of the atmospheres formed at 80 MPa

decomposition of the Snoek atmospheres formed when the specimen was kept at σ = 80 MPa, with their simultaneous formation at σ = 170 MPa. The experiments were carried out at T ≈ 295 K. To obtain these dependences, the specimen was kept at σ ≈ 80 MPa for 20 hours, then unloaded and the dependence of Q^{-1} on σ was recorded during subsequent loading at a rate of 0.5 MPa/s (curve 1). After reaching the elastic stress σ ≈ 170 MPa, the loading was stopped and the specimen was kept under this load for a certain time t with subsequent unloading and repeated loading. The dependences of Q^{-1} on σ recorded during loading after t = 0 (curve 1), t = $3.6*10^3$ s (curve 2), t = $9*10^3$ s (curve 3), and t = $3.6*10^4$ s (curve 4) demonstrate the presence of a characteristic Q^{-1} minimum for the positions of dislocations defined by the loads of σ ≈ 80 MPa and σ ≈

170 MPa, i.e. the loads corresponding to the Snoek atmosphere formation. It follows from Fig. 2 that the Snoek atmospheres formed when the specimen was kept at $\sigma \approx$ 80 MPa completely decompose during the time of their formation at $\sigma \approx$ 170 MPa.

Fig. 3 presents temperature spectra of Q^{-1} recorded under the same load of $\sigma \approx$ 130 MPa for two different cases - when the Snoek atmospheres have formed around the dislocations, and for their absence. As the figure shows, in both cases there are two relaxation processes in the temperature spectrum, and their temperature positions in the strained niobium at a frequency of about 10^3 Hz are characteristic of the well-known α and β relaxations [6]. The results presented in Fig. 3 show that the formation of a Snoek atmosphere of oxygen atoms (curve 1) around mobile dislocations leads to a decrease in the relaxation strength and to a shift of both the α and β processes towards lower temperatures.

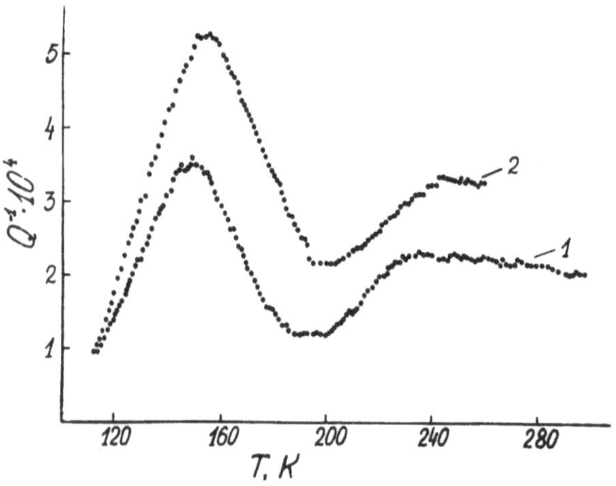

Fig. 3 Temperature spectrum of Q^{-1}

DISCUSSION

The presence of a fixed polarized impurity atom atmosphere around a dislocation is equivalent to the existence of a wide potential well (of the order of 5 to 7 interatomic distances) the profile of which was computed in [1, 7]. This means that in order to shift the dislocation by a distance x with respect to the atmosphere centre, an additional force is required, and the force gradient arising at such a shift, $\partial F_{SN}/\partial x$, decreases with growing x [7] (here F_{SN} is the force per unit length of the dislocation). Accordingly, the existence of an atmosphere increases the effective rigidity of the dislocation line C_{eff} which, according to

[4], in this case equals $C + \partial F_{SN}/\partial x$ (C – line tension of the dislocation independent of x). As Chambers has shown [6], independent of the nature of the relaxation process connected with the dislocation motion in b.c.c. metals, both the relaxation time τ and rate Δ of this process are inversely proportional to C_{eff}. In the presence of a fixed polarized impurity atom atmosphere around the dislocations, the expressions for τ and Δ obtained in [6] become

$$\tau = 1/\mu(C + \partial F_{SN}/\partial x) \quad \text{and} \quad \Delta = \alpha b^2 \Lambda G/(C + \partial F_{SN}/\partial x) \tag{1}$$

where Λ is the mobile dislocation density, G the shear modulus, α the orientation factor and μ the mobility of dislocations.

According to [1], in the absence of the atmospheres, i.e. at $\partial F_{SN}/\partial x = 0$, both τ and Δ of any process should grow, which was observed experimentally (Fig. 3). One should keep in mind that neither the α nor the β band corresponds to a single relaxation process, and therefore the expression (1) being applied to α or β processes yields a certain set of close relaxation time and strength values; moreover, the process has proved to be badly resolved. Within the 70 K range 230 K < T < 300 K, Q^{-1} is almost independent of T. This allows one to connect Q^{-1} (at T $\gtrsim$ 230 K) with $\partial F_{SN}/\partial x$ by a simple expression [4]:

$$Q^{-1} \approx \Delta/2 = \alpha b^2 \Lambda G/(C + \partial F_{SN}/\partial x). \tag{2}$$

On the basis of eq. (2), the results presented in Figs. 1 to 3 may be explained as follows. Holding a specimen under a certain load for a certain time leads to a temporary increase of the number of ordered oxygen atoms around the positions of the dislocations corresponding to the given load. This, in turn [7], increases $\partial F_{SN}/\partial x$, which leads to a decrease in Q^{-1} at a fixed temperature (curves 2 and 4 in Fig. 1). The subsequent load increase or decrease leads to a displacement of the dislocations with respect to the atmosphere centres, i.e. to an increase in x, which causes a gradual decrease in $\partial F_{SN}/\partial x$ down to zero and, according to (2), to an increase of Q^{-1}, reaching a plateau (curves 1, 2 in Fig. 1). The results given in Fig. 2 may be explained in a similar way.

The expression (2) allows us to determine rather simply (at least qualitatively) the potential profile of the dislocation interaction with a fixed atmosphere of polarized impurity atoms on the basis of the dependence of Q^{-1} on σ, for example, such a dependence as presented in Fig. 1 (curve 1). In fact, since Q^{-1} in Fig. 1 (curve 1) reaches its maximum value Q_m^{-1} at $\partial F_{SN}/\partial x = 0$, it follows from (2) that

$$\partial F_{SN}/\partial x = C(Q_{max}^{-1}/Q^{-1} - 1) \tag{3}$$

Using (3) and assuming that C is defined only by the line tension ($C \approx 6Gb^2/L^2$, where L is the average dislocation segment length), we may obtain the dependence of F_{SN} on x for reasonable values of L by evaluating the results in Fig. 1 (curve 1). The dependence of U_{SN} on x is then given by $-\partial U_{SN}/\partial x = F_{SN}$. The obtained dependences $F_{SN}(x)$ and $U_{SN}(x)$ for L values typical of the present case, $L \approx 10^{-5}$m, are presented in Fig. 4.

As the figure shows, the maximum value of F_{SN} equals $1.75*10^{-3}$ N/m, which coincides well with the result of theoretical calculations of the oxygen concentration of 0.015 at % ($1.85*10^{-3}$ N/m) [8]. The dislocation displacement x along which this value of F_{SN} is achieved ($\approx$2 nm) also corresponds with the displacement necessary for the dislocation to be released from the atmosphere ($\approx$1.75 nm) as computed in [7].

Thus, it is shown that IF connected with the dislocation motion with respect to a fixed Snoek atmosphere is defined by the force gradient acting on the dislocation due to the atmosphere. This allows the direct determination of the potential relief of the dislocation interaction with the atmosphere from the experimental dependence of Q^{-1} on σ. The formation of Snoek atmospheres in Nb decreases the relaxation time and strength of both the α and the β processes.

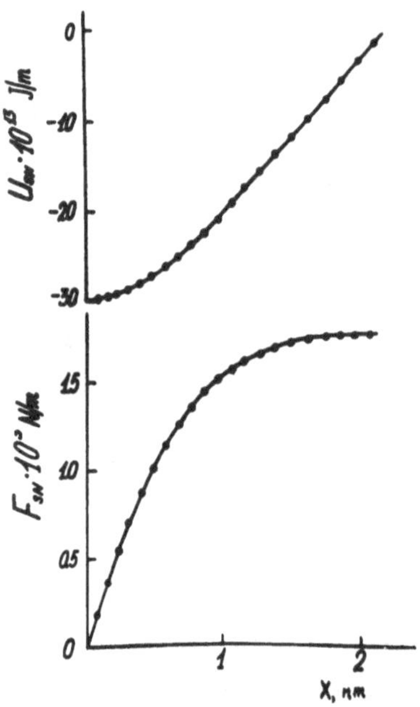

Fig. 4 The dependence of U_{SN} and F_{SN} on x

The authors express their deep gratitude to Z.K. Saralidze for useful discussions, to L.M. Kolesnikova, N.M. Yastrebova, V.A. Kocherov, and A.M. Uvarov for the specimen preparation and help in the tests.

REFERENCES

1 J.T. Evans and R.M. Douthwaite, Acta Met. 21 (1973) 49-54

2 D. Delobelle, D. Varchon and C. Ottana, Phys. Stat. Sol. (a) 83 (1984) 235-43

3 O.P. Quist and S.H. Carpenter, Acta Met. 23 (1975) 321-8

4 V.V. Zoninashvili, I.A. Naskidashvili, V.A. Melik-Shakhnazarov, Fiz. Tverd. Tela 31 (1989) 161-6 (in Russian)
5 V.V. Zoninashvili, U. Messerschmidt, V.A. Melik-Shakhnazarov, I.A. Naskidashvili, and N.M. Yastrebova, Pribor. Tekh. Exp. 27(6) (1984) 175-8 (in Russian)
6 W. Chambers, In: Physical Acoustics, Vol. 3 A, W. Mason (Ed.), Mir, Moscow 1969, p. 578
7 R.E. Reed-Hill, In: Dislocat. Model. Syst, Proc. Conf., Gainsville FA 1980, p. 163-7
8 G. Schoek and A. Seeger, Acta Met. 7 (1959) 469-77

2. MICROSTRUCTURE AND DEFORMATION OF STEELS

PECULIARITIES OF STRAIN HARDENING AND FAILURE OF SINTERED MATERIALS

S.A. FIRSTOV

The Institute of Problems of Materials Science,
Krjijanovskogo str., 3, Kiev 252680, U.S.S.R

ABSTRACT

The influence of structure features of powder materials on the mechanical properties is analysed. It is shown that the strain hardening of porous solids can be described by a universal dependence when certain parameters are normalized by the modulus of elasticity. It is found that the fracture mechanism changes as a function of porosity which results in a non-monotonic dependence of the fracture toughness on the porosity.

INTRODUCTION

The main peculiarities of the structure of sintered materials are determined by the technology of their production and processing. It allows one to vary porosity in a wide range (volume, dimensions, shape, and distribution of pores) and to alter the quality of the contacts between the sintered particles.

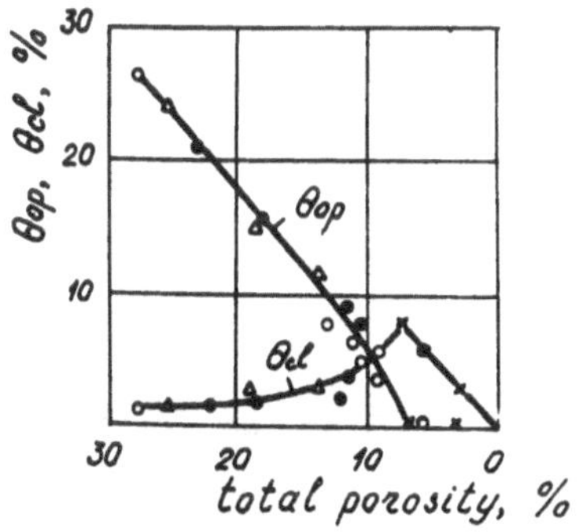

Fig. 1. Change of the fraction of open Θ_{op} and closed porosity Θ_{cl} in dependence on the total porosity for sintered iron.

Porosity is one of the most important parameters of the structure of sintered materials. The so-called closed and open porosity may be distinguished. The first case is characterized by a system of isolated pores (usually spherical or ellipsoidal), in the second case a branched system of communicating pores exists (Fig. 1). The morphology of the pores changes with the pore volume. In a similar way, the morphology and distribution of second phases may change in two-phase materials. The distribution of pores in the sintered materials may be characterized by a parameter like the structure connection, which usually changes abruptly in a narrow interval of the porosity or the content of second phases [1- 6].

GENERAL IDEAS OF THE INFLUENCE OF THE POROSITY ON THE MECHANICAL PROPERTIES OF SINTERED MATERIALS

The respective questions have been considered in detail earlier, mainly in investigations of powder metallurgists [1 - 10]. They studied the ratio between a property of porous bodies and that of the compact material as a function of the relative density. According to [2, 3], $\varkappa \sim \delta^m$, where $\varkappa$ is the relative property, for example the ratio of the strength of a porous body to that of a compact one $\sigma(\theta)/\sigma_0$, or between the moduli $E(\theta)/E_0$, δ is the relative density, and m an exponent between 0 and 10. Such an approach allows a classification of the dependences of special properties on the relative density (porosity) into four groups:

1. Independent of the porosity (m = 0): melting temperature, thermal expansion coefficient.
2. Additive properties (m = 1): heat capacity, magnetic susceptibility.
3. Properties of the conductivity type (m = 1,5-2): electric conductivity, modulus of elasticity.
4. Properties sensitive to structure (m > 2): strength, plasticity and others.

The mechanical properties of sintered materials are drastically reduced when the porosity increases. In particular, the following formulae have been derived in a number of publications [2 - 7]:

$$\sigma_y = \sigma_{y0}\,(1-\theta)^m; \quad (1) \qquad \sigma_p = \sigma_{p0}\, e^{-b\theta}; \quad (2)$$

$$\sigma_y = \sigma_{y0}\,\frac{(1-\theta)^2}{\sqrt{4-3\theta}}; \quad (3) \qquad \sigma_p = K_y d^{-1/2} e^{-b\theta}; \quad (4)$$

$$E = E_0\,(1-\theta)^m; \quad (5)$$

These dependences agree with the experiment as to the porosity values of $\theta \leqslant 30\%$. Some more complicated and more exact dependences of the properties on porosity have been proposed for the range of $\theta \leqslant 60\%$ [1, 4 - 8 and others].

Balshin supposed [2, 3] that the dimensionless coefficient $\alpha = E(\theta)/E_0$ characterizes the relative fraction of load carrying (critical) material. Balshin suggested that

$$\alpha = \frac{E(\theta)}{E_0} \approx \frac{S(\theta)}{S_0}; \quad (6)$$

where $S(\theta)$ and S_0 are the properties of porous and compact bodies, which are proportional to the "critical" fraction (for example, tensile strength).

Krasovsky [9] considers, however, that the relation $E(\theta)/E_0$ characterizes a mean value of the "critical" fraction, but that $\sigma(\theta)/\sigma_0$ is determined by weakest fraction, therefore, $E(\theta)/E_0 \geqslant \sigma(\theta)/\sigma_0$.

There is a somewhat unusual influence of the dimension and shape of the specimens (scale factor) on the properties of sintered porous bodies. As it has been shown in [10], the strength is reduced if the specimen length increases. But it is slightly enhanced with increasing cross-section area.

STRAIN HARDENING AND FAILURE OF THE POROUS SINTERED MATERIALS

The important conclusion was drawn from the studies in [11 - 13] that it is impossible to establish a single theory, for example, for strain-hardening or the influence of different factors on failure on the basis of a single elementary mechanism of deformation or failure. The whole variety of mechanisms of deformation and failure are to some extent reflected in the corresponding maps constructed by Ashby (e.g. [14]).

In agreement with the analysis of strain-hardening the essential conclusion may be drawn that the hardening mechanism changes drastically for large plastic deformations. It has been shown that at the beginning of plastic deformation the hardening may be described by the relationship $\Delta\sigma(\varepsilon) = \alpha G b \sqrt{\rho(\varepsilon)}$ in connection with the creation of dislocations of different arrangement and density, but for large plastic deformations hardening is described by the relationship

$$\Delta\sigma(\varepsilon) = K d_{,}^{x} \qquad (7)$$

where x=1 or 1/2, d - size of disoriented cells.

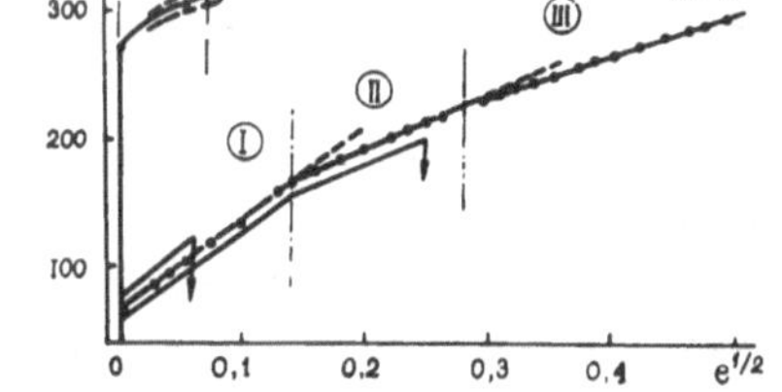

Fig. 2. Typical "true stress - true deformation" curve in coordinates of σ vs. $e^{1/2}$ in the parabolic hardening stage.

According to structural state diagrams [13] hardening occurs at distinct stages. This has been revealed in [15] by plotting the strain hardening curves in coordinates of S and $\sqrt{e}$, where S is the true stress and e the true strain. The different strain-hardening stages in parabolic hardening curves in coordinates of S and $\sqrt{e}$ had been observed long ago by Bell [16], but he did not interpret them in structural terms. Usually parabolic hardening curves look like broken lines (Fig. 2), where each linear segment corresponds to a different stage of strain hardening. Usually, at least three sections are revealed in such curves. If a limited part of the curve is analysed not all the stages may be obtained. Table 1 summarises values of the yield stress and the ultimate strength for sintered iron as a function of porosity according to data [6]. As follows from the table data, the dependence on porosity of yield strength and ultimate strength related to the modulus of elasticity for a given porosity are practically independent of the porosity. This agrees with the above-mentioned hypothesis of Balshin. This circumstance has been considered by us in studies of strain hardening in porous iron [17, 18].

Table 1

θ,%	E, GPa	$\sigma_{0.2}$ MPa	σ_b, MPa	$\frac{\sigma_{0.2}}{E}$ x10^3	$\frac{\sigma_b}{E}$ x10^3
15	151.6	76	172	0.5	1.12
18	132.3	65	144	0.49	1.08
21	113.7	60	121	0.53	1.06
26	88.35	51	93	0.58	1.05
33	65.2	39	62	0.59	0.95

Fig. 3a compares the strain-hardening curves of compact polycrystalline iron and sintered iron of different initial porosity.

Fig. 3 shows that in the coordinates used two linear regions are distinctly revealed for the compact material whereas they are absent for the porous materials. Furthermore, an increase in the porosity results in a reduction of both the flow stress and the strain to failure (critical deformation).

The differences between the individual curves are reduced if σ is related to the modulus of elasticity $E(\theta)$. This is plotted in Fig. 3b for the initial modulus $E_o(\theta)$ of undeformed specimens. The curves coincide except for large strains. A special analysis has shown that the porosity changes during deformation (in the case of tension it increases up to failure by approximately by 4%).

If the strain-hardening curves are plotted as $\Delta\sigma/E(\theta)$ vs. $\sqrt{\varepsilon}$, where $E(\theta)$ is the calculated value of the modulus for the current porosity, it turns out that the strain-hardening of the studied materials may be described by two linear regions like that of the compact material [17] (Fig. 3c).

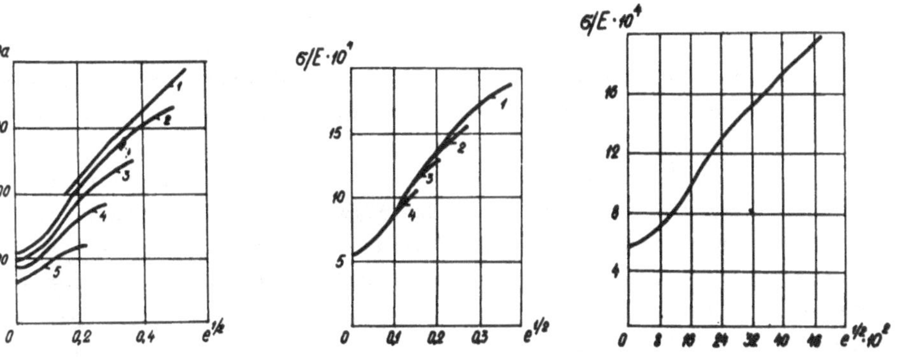

a b c

Fig. 3. a) "Stress-true strain" curves of compact armco-iron (I) and powder iron having a porosity of 6.9%(2); 9.4%(3); 15.7%(4) and 23.7% (5); b) the same as in a) for the redused values $\sigma/E_0(\theta)$, where $E_0(\theta)$ is the modulus of the porous, undeformed material; c) the same as in b) for values $\sigma/E(\theta)$, where $E(\theta)$ is the actual modulus depending on θ and e.

Let us consider some peculiarities of the deformation behaviour of sintered dispersion-hardened materials for example sintered iron containing titanium oxides [19]. As electron microscopical investigations have shown, there exists a remarkably wide distribution of the particle dimensions in the materials studied. The materials were tested over a wide range of temperatures (from -196° C to 700° C) with a strain rate 10^{-4} s^{-1}. At liquid nitrogen temperature the flow stress is high while the material shows a low plasticity. This is a consequence of an abrupt change of the flow stress over the temperature range considered. With the temperature increasing up to room temperature, a significant plasticity occurs. The plastic properties are maintained almost constant up to 400° C, and above 400° C the plasticity strongly increases with the hardening coefficient decreasing. Three types of fracture behaviour have been observed: brittle cleavage at liquid nitrogen temperature, tough dimple fracture initiated by particles in the temperature range below 400° C, and dimple failure with no particles being observed directly in the dimples (at testing temperature above 400° C).

On dimple fracture surfaces only particles have been observed having dimensions of 2000 Å or more and smaller particles which were not the nucleation sites of micro-voids. It may be supposed that this "critical" size of the particles is connected with a change of interaction mechanism between particles and dislocations - small particles are cut by the dislocations or are overcome by a mechanism of "fast climb" while at larger particles the dislocations accumulate which leads to a decohesion between the particles and the matrix. Decohesion has been observed after a critical strain of $\mathcal{E} \approx 10-12\%$, which corresponds to literature data for dispersion-hardened materials [20].

In such a way, we revealed two interesting facts. First: particles are not always the centres of primary micro-void nucleation (which is attributed to particles of smaller dimensions and also to test temperatures above 400° C, at which intensive polygonization processes occur and a disoriented cell structure is formed). Second: dispersion-hardened material deforms first (in the medium temperature range) at retaining coherent material but later on at the deformation-induced porosity. It is obvious that after porosity has been formed, the deformation behaviour of the material follows the relationships considered above for porous materials.

It may also be expected that the characteristics of the hardened state are non-monotonous after critical deformation has been reached.

Fig. 4 shows the change of internal stresses of the second kind $\Delta a/a$ (the broadening of X-ray line) with deformation for dispersion-hardened materials. One can see that the internal stresses decrease after a critical deformation.

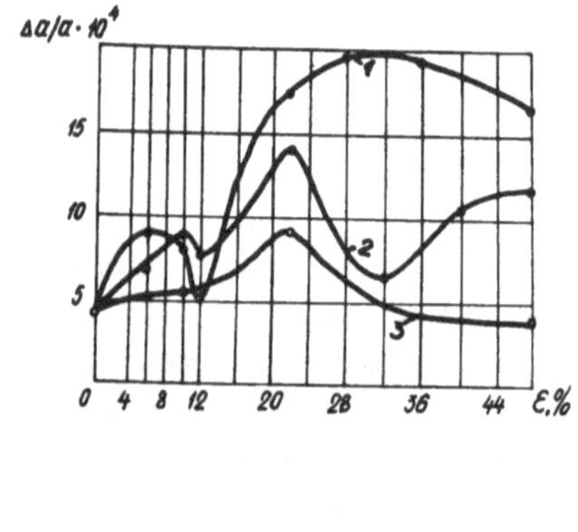

Fig. 4. Change of the parameter $\Delta a/a$ as a function of the deformation extent for Fe - $3\%TiO_2$. The curves correspond to: 1 - rolling; 2 - rolling + aging at 600° C; 3 - rolling + aging at 700° C.

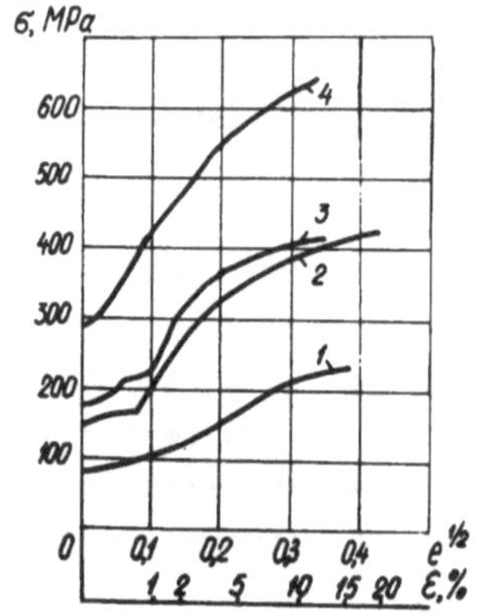

Fig. 5. Strain hardening of armco-iron (1), Fe - $1.5\%TiO_2$ (2), Fe - $3\%TiO_2$ (3), Fe - 13%Cr - $3\%TiO_2$ (4).

The strain hardening of pure iron with an initial porosity $\leqslant$ 2% and of iron with 1.5% TiO_2 and 3% TiO_2 have been studied in [18]. It is demonstrated in Fig. 5 that at the initial stage of deformation the hardening coefficient of dispersion-hardened material is significantly higher in comparison with pure iron, but after the deformation continues up to the decohesion of the particles, the slope of the curves turns out to be almost the same.

Let us finally examine the influence of porosity on some features of the failure of porous materials. As follows from the data in [21 - 23], most interesting results are obtained for the change of the micro-mechanisms of failure and other mechanical properties near the brittle-ductile transition temperature (for iron this temperature range lies between -120...-196°C). According to fractographic data the fracture micro-mechanism changes from cleavage to dimple failure which occurs due to a transition from open to closed porosity. In total correspondence with this fact the fracture toughness depends non-monotonously on the porosity (Fig. 6). A correct theoretical explanation of this phenomenon cannot be given at present. It may be supposed that the right-handed branch of the curve corresponds to an abrupt reduction of the deformable volume in connection with a localization of the deformation in the contact bridges.

According to [24], tough failure can be described by

$$K_{IC} \sim \sqrt{E \sigma_T e_r n^2}, \qquad (8)$$

where E is the elastic modulus, σ_T the yield strength, e_r the maximum deformation and n the strain hardening exponent.

After Hurland [25], for two-phase materials

$$e_r = e_o + \tfrac{1}{2} \ln\left\{1 + \beta\left[\tfrac{2}{3}\frac{1+2f}{f} \exp\left(\ell_T/r - 2\right)\right]^2\right\}, \qquad (9)$$

where f is the volume fraction of inclusions or pores, and for porous bodies $e_o = 0$. Accounting for an abrupt reduction of E, σ_T and e_f, at tough failure K_{Ic} decreases continuously for increasing porosity.

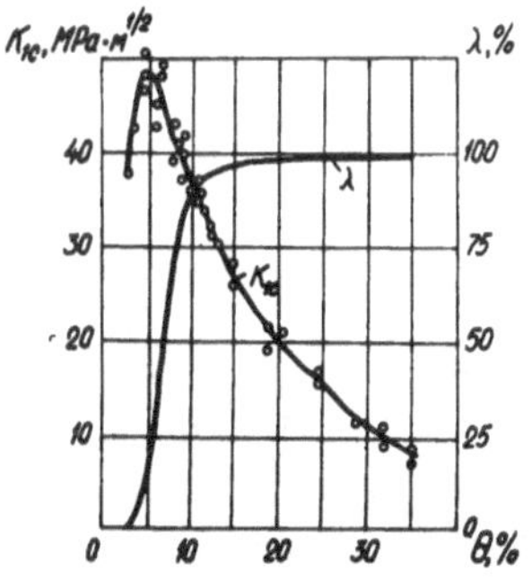

Fig. 6. Influence of the porosity on the fracture toughness K_{1c} and the fraction of dimple surface (λ) of iron, T_{at}=-196°C.

The left-handed branch of the dependence of K_{Ic} on θ shows that for cleavage the fracture toughness increases in comparison with the compact material due to the plastic relaxation enhanced by the porosity. As far as for such a failure mechanism $K_{Ic} \sim \sigma_p^3 / \sigma_T^2$ [26] holds, it may be supposed that the behaviour of $K_{Ic}(\theta)$ results from the more abrupt reduction of the yield stress with porosity in comparison with the breaking stress.

Fig. / presents the dependece of the maximum strain and the impact strength on the porosity for a number of sintered iron materials obtained from different powders [18].

As may be seen from Figs. 7a and b, the curves differ essentially in the small porosity and, as special investigations have shown, the lower values of plasticity and toughness originate from materials having higher contents of interstitial impurities and, correspondingly, particles of the second phase (mainly of oxides).

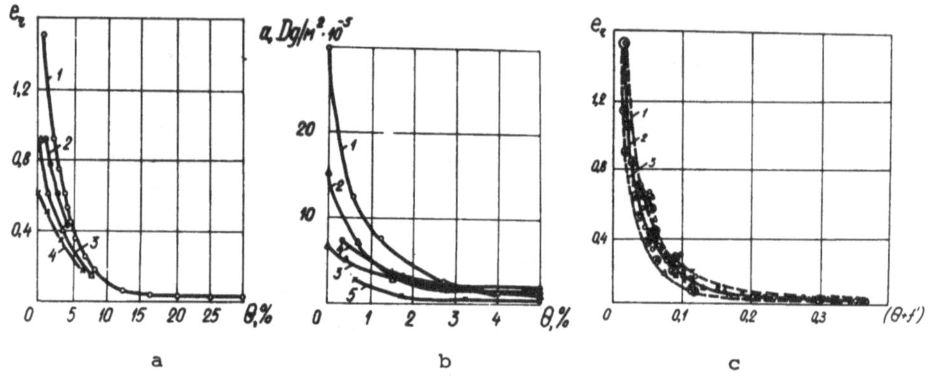

Fig. 7 Influence of porosity on a) the critical deformation and b) the impact strength of sintered iron with different content of ox gen: 1 - 0.11%; 2 - 0.16%; 3 - 0.32%; 4 - 0.42%; 5 - 0.48%; c) critical deformation as a function of the parameter (θ + f) for: 1 - K_2^2/K_1=0.14; 2 - K_2^2/K_1=0.12 and 3 - K_2^2/K_1=0.10.

In Fig. 7c the true deformation to failure is plotted as a function of the parameter (θ + f), where f is the volume fraction of particles of the second phase in the materials investigated. As it follows from results obtained, the curves of e_r vs. (f + θ) coincide for all the studied materials. Thus, the parameter "f + θ" describes more exactly the plasticity of sintered materials.

REFERENCES.

1. I.M. Fedorchenko and R.A. Andrijevsky, Powder metallurgy fundamentals, Naukova Dumka, Kiev 1963, in Russian
2. M.Ju. Balshin, Scientific fundamentals of powder metallurgy and fibre metallurgy, Metallurgy, Moscow 1972, in Russian
3. M.Ju. Balshin and S.S. Kiparisov, Powder metallurgy fundamentals. Metallurgy, Moscow 1978, in Russian
4. V.V. Skorokhod, Powder materials as the base of refractory metals and compounds. Tekhnika, Kiev 1982, in Russian
5. R.A. Andrijevsky.- Powder met., (1982) No.1, p. 37-42, in Russian.

6. Pulvermetallurgie. Sinter - und Verbundwerkstoffe. W. Schatt (Ed.), VEB Deutscher Verlag fur Grundtoffindustrie, Leipzig 1977.
7. Md. Haminddin, Metallurgy internat, Vol. 12 (1986), No.2, p. 73-5.
8. N.A. Fleck and R.A. Smith, Powd. Met. Vol.24 (1984) No.3, 121-25, in Russian
9. A.Ya. Krasovsky, Powder met., (1964) No.4, 167-74, in Russian
10. V.T. Troschenko and V.N. Rudenko, Cermet materials strength and methods of its determination, Tekhnika, Kiev 1965, in Russian
11. V.I. Trefilov, Ju.V. Milman and S.A. Firstov, Physical fundamentals of refractory metals strength, Naukova Dumka, Kiev 1975, in Russian
12. V.V.Rybin, Large plastic deformations and metals failure, Metallurgy, Moscow 1986, in Russian
13. V.A. Lichachev, V.E. Panin and Ye.E. Zasimchuk, Cooperative deformation processes and deformation localization, Naukova Dumka, Kiev 1989, in Russian
14. N.J. Frost and M.F. Ashby, Deformation-mechanism maps: The plasticity and creep of metals and ceramics, Pergamon Press,Oxford 1982.
15. V.I. Trefilov, V.F. Moiseev and E.P. Pechkovsky, Strain hardening and failure of polycrystal metals, Naukova Dumka, Kiev 1989, in Russian
16. J. Bell, Experimental fundamentals of mechanics of deformable bodies, Nauka, Moscow 1984, in Russian
17. A.S. Drachinsky, A.Je. Kuschevvsky and Yu.N. Podrezov, Powder met.(1984), No.10, 78-83, in Russian
18. S.A. Firstov, I.I. Ivanova and A.N. Demidik, Powder met. (1989), No.6, 70-7, in Russian
19. A.N. Demidik, A.S. Drachinsky and I.I. Ivanova, Powder met. (1983) No.7, 69-74, in Russian
20. J.U. Martin, Mechanisms of dispersion hardening of alloys, Metallurgy, Moscow 1982, in Russian
21. A.S. Drachinsky, A.E. Kuschevsky and A.V. Perepelkin, Powder. met. (1982) No.12, 80-4, in Russian
22. S.A. Firstov, Yu.N. Podrezov and A.G. Zherdin, Powder met. (1987) No.10, 90-6, in Russian
23. K. Pelikan, L. Parilak, Yu.N. Podrezov and A.V. Perepelkin, in: Fractografia 83, Vol.7, Celostatna fractograficks konferencia, Vysoke Tatry 1983, p. 107-11
24. C.T. Hahn, R.G. Hoahland and A.R. Rosenfield, Met. Trans. 2 (1971) 537-41
25. J. Hurland and N.M. Parih, Microstructural aspects of two-phase alloys failure, in: Failure, Vol.7, part. I, Mir, Moscow 1979, p. 472-512, in Russian
26. Yu. Ya. Meshkov and G.A. Paharenko, Metal structure and brittleness of steel products. Naukova dumka, Kiev 1985, in Russian.

MICROSTRUCTURE DEVELOPMENT IN THE PLASTIC DEFORMATION OF TWO - PHASE STAINLESS STEEL

P. Klimanek

Department of Materials Science, Academy of Mining,
DDR - 9200 Freiberg, P.O.Box 47

ABSTRACT

Microstructure changes occurring in cold work ($T \leq 293$ K) and hot - deformation ($T \geq 1023$ K) of fine - grained ($d < 10$ µm) two - phase stainless steel X5CrNiTi 26.6 were studied by TEM and other techniques of microstructure analysis. The results show that the deformation behaviour of the material is significantly influenced by the sample pre - history (initial microstructure) and the special deformation conditions. In particular this concerns the effect of a strain - induced martensite transformation of the austenite in cold work and the operating restoration processes of the ferrite and the austenite in hot work.

INTRODUCTION

Plastic deformation of heterogeneous metallic materials is a problem which was frequently studied during the last two decades. In this connection two - phase stainless steels on the base of alloy systems as, for instance, Fe - Ni, Fe - Cr - Ni or Fe - Cr - Ni - Me (Me = Si, Mn, Mo etc., / 1 /) were also intensively investigated. The microstructure of these alloys can be considered in a first approximation as mixture of b.c.c. ferrite (α- or δ- phase) and f.c.c. austenite (γ- phase) with volume fractions of comparable order of magnitude. But though there is much work concerning their mechanical properties (e.g. / 2,3,4 /), particularly the technical deformation behaviour of the steels is not satisfactorily understood until now because of insufficient knowledge of the interrelations between property changes , deformation conditions and microstructure development. Two problems being of special interest in this connection are the role of TRIP phenomena due to a strain - induced martensite transformation of the austenite in cold work and the nature of the deformation and restoration mechanisms occurring in hot work. The present paper gives a brief survey of investigations of such processes which were performed by TEM and other techniques of microstructure analysis with fine - grained ($d < 10$ µm) steel X5CrNiTi 26.6 in both cold and hot work at technically relevant strain rates ($\dot{\varphi} \geq 0.1$ s^{-1}). The deformation behaviour of the sample material, the microstructure of which can be varied within wide limits, should be representative for two - phase stainless steel containing austenite with low stacking fault energy.

EXPERIMENTAL PROCEDURES

The investigated material was a charge of the steel X5CrNiTi 26.6 with the following chemical composition (wt.-%) : 0.04 C , 24.06 Cr, 6.45 Ni, 0.53 Si , 0.91 Mn , 0.40 Ti . Above 1600 K the steel is a pure δ- ferrite, but at lower temperatures decomposition takes place into a thermodynamically metastable austenite of nearly the type X5CrNiTi 18.10 and a ferrite which approximately corresponds to the alloy X5CrNiTi 33.3 / 5 /. Below 1300 K the austenite - volume fraction within reach is 0.4 to 0.5 ; some details of microstructure formation are considered in connection with the results of the deformation experiments.

Plastic deformation of two - phase specimens was carried out at temperatures $T \leqslant 293$ K (cold work, cp. / 6,7 /) and $T \geqslant 1023$ K (hot work, / 8,9 /) by means of either tensile and compression tests or by flat rolling. In all cases strain rates $\dot{\varphi} \geqslant 0.1\ s^{-1}$ corresponding to technical deformation conditions were used. TEM investigations of deformation - induced micro - structure changes could be performed by means of an electron microscope DEM 200 available at the Mining Academy Freiberg and by use of the high - voltage electron microscope JEM 1000 of the I F E Halle / Saale. In addition, X - ray profile analysis, quantitative texture analysis from neutron diffraction pole figures and magnetic measurements were used for integrating microstructure studies (see / 7 /).

RESULTS AND DISCUSSION

Microstructure Formation in Two - phase Steel X5CrNiTi 26.6

The plastic behaviour of a heterogeneous alloy is essentially influenced by the characteristics of the spatial arrangement (i.e. volume fractions, particle geometry, textures, position correlation) of the phases being present / 3,4 / and cannot be discussed therefore without regard to the initial microstructure of the sample material taken under investigation. A classification of such microstructures was given in / 10,11 /. In order to get them in stainless steel of the type X5CrNiTi 26.6 the following procedures can be used / 12,13 /:

(i) slow cooling from the melt (as - grown microstructures),

(ii) heat treatment (HT) with ferritization and either slow cooling to an elevated temperature, austenite precipitation and quenching (HT1) or quenching of the δ- ferrite with subsequent heating to a certain temperature and austenite precipitation (HT 2), and

(iii) thermomechanical treatment (TMT), enclosing again ferritization and then either hot - working at a temperature of the two - phase region of the alloy and annealing for austenite precipitation (TMT 1) or quenching, cold - working of the supersaturated δ - ferrite and heat treatment for austenite precipitation (TMT 2) .

Fine - grained microstructures can be obtained by (ii) and (iii). The technically important TMT 1 leads to microduplex structures ($v_\gamma \approx 0.5$; $d_\gamma \approx 10\ \mu m$, Fig. 1a), HT 2 and especially TMT 2 gives rise to the formation of fine - disperse matrix structures ($v_\gamma = 0.4 \ldots 0.5$, $d_\gamma = 1 \ldots 10\ \mu m$; Fig. 1 b,c,d). A common feature of all microstructures is strong correlation between the orientation distributions of the δ - ferrite and the austenite due to texture inheritance via the Kurdjumov - Sachs relationship. / 7,14 /. Moreover, the character of the phase boundaries is different in TMT and HT, and in all cases the defect content of the crystallites is considerably higher than in corresponding single - phase or coarse - grained heterogeneous structures. Typical for TMT is the formation of numerous extended stacking faults (Fig. 2a,b) and an increased dislocation content of the austenite (The effect is also clearly visible in the results of X - ray profile analysis / 6,7 /; see e.g. Fig. 4), important characteristics of microstructures due to HT below 1300 K are semicoherent δ/γ phase boundaries and a high dislocation density especially of the δ- ferrite (Fig. 2c,d).

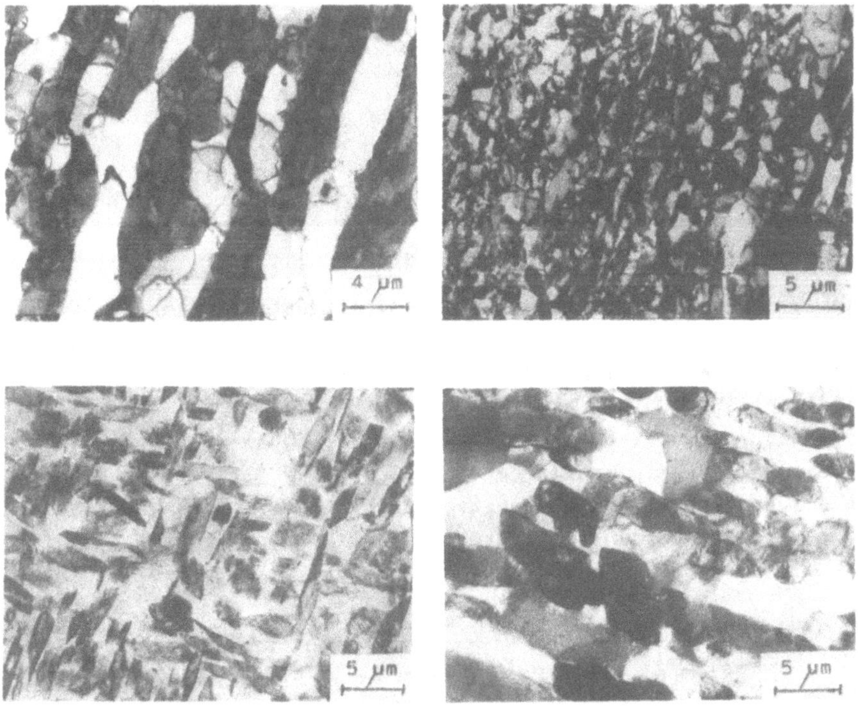

c) H T 2 : 1023 K   d) H T 2 : 1323 K

Fig. 1 : Microstructure formation in X5CrNiTi 26.6

The influence of the microstructure on the mechanical properties of the steel X5CrNiTi 26.6 may be illustrated by a comparison of experimentally measured and calculated by means of a mixing rule / 3,4 / yield stresses of sheets (R_δ , R_γ according to / 15 /). The result (Table 1) can easi-

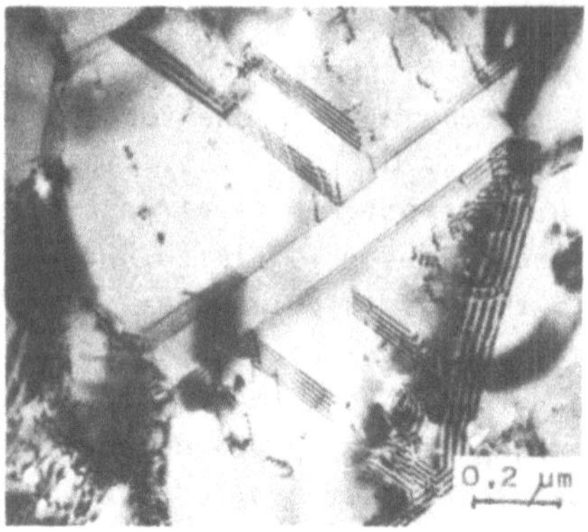

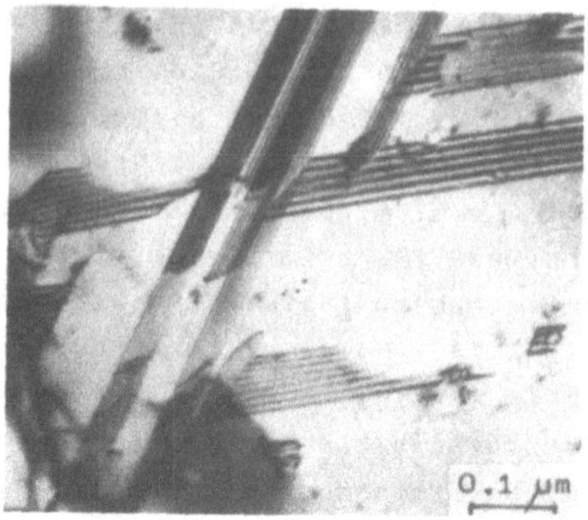

a) TMT 1   b) TMT 2

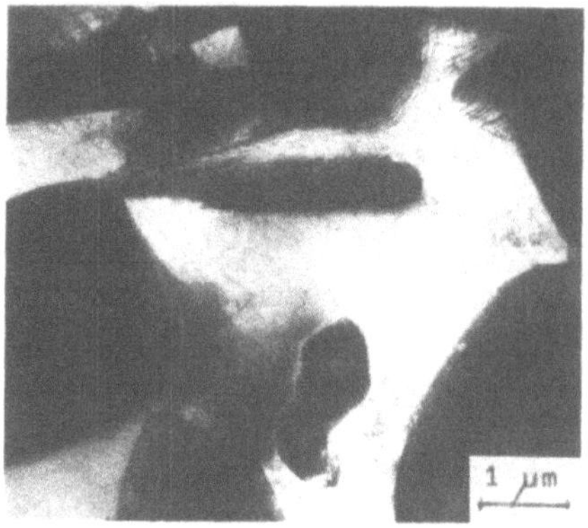

c) HT 2 : 1023 K   d) HT 2 : 1323 K

Fig. 2 : Lattice defects in different microstructures of X5CrNiTi 26.6

ly be explained as follows : If the defect concentrations of the phases are not to high (TMT 1) , the flow-stress ratio $K = R_\delta / R_\gamma$ becomes really large and the initial stage of the deformation is controlled by the austenite. Accordingly one finds $R^{exp}_{0.2p} < R^{calc}_{0.2p}$. After TMT 2 especially the defect content of the austenite is high. This corresponds to a pre-deformation (work hardening) of the austenite which reduces the actual parameter K and favours the validity of a mixing rule for the yield stress.

Table 1

	R_δ (MPa)	R_γ (MPa)	$K = R_\delta / R_\gamma$	$R^{calc}_{0.2p}$ (Mpa)	$R^{exp}_{0.2p}$ (MPa)
TMT 1	66.5	36.5	1.8	50	44
TMT 2	69.5	40.0	1.7	55	55

Microstructure Development in Cold Work

The deformation behaviour of the steel X5CrNiTi 26.6 at temperatures $T \leqslant 293\,K$ may be illustrated by stress - strain curves of rods and also by yield stresses of cold - rolled sheets, which are shown in Fig. 3,4 together with the magnetically measured volume fractions of austenite and deformation - induced martensite being present after different strain. The results indicate that the interrelations between the mechanical properties and the microstructure development are complicated.

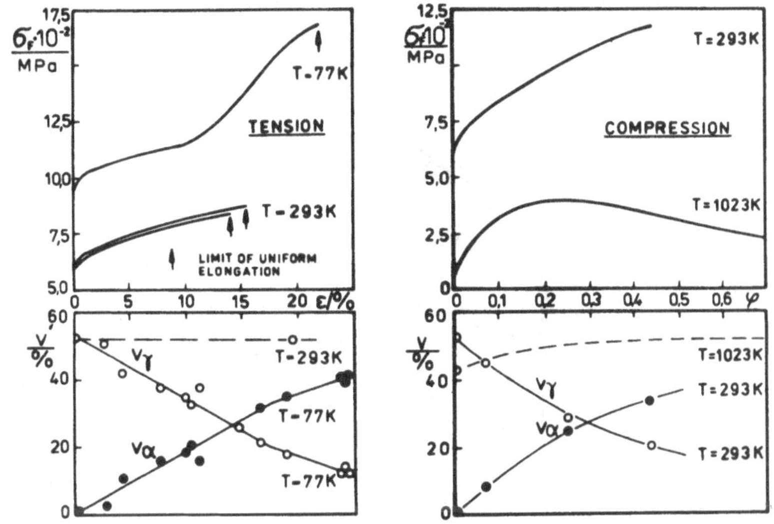

Fig. 3 : Mechanical properties of X5CrNiTi 26.6 at elevated temperatures

TEM investigations of the deformation - induced defect structures of the phases were described in previous work and can be summarized in the following manner / 6,7,14 /:

- Deformation of the δ - ferrite takes place via dislocation slip at all temperatures $T \geqslant 77\,K$ and is characterized by rapid development of a dislocation - cell structure.
- Cold work of the γ - phase, which is metastable at temperatures at and below room temperature and has a low stacking - fault energy of about 20 mJ/m^{-2} / 16 /, is usually based on the co - operation of several deformation mechanisms : planar dislocation slip accompanied by the generation of pile ups and numerous extended stacking faults, deformation twinning, and strain - induced martensite formation. According to both TEM and X - ray analysis the stacking - fault formation is clearly favoured by decreasing grain size. (Fig. 4).

The partial process which essentially influences the mechanical properties in cold work is the strain - induced martensite formation. Its crystallography depends on the stacking - fault energy of the austenite, and the M_d temperature is a function of the alloy content, the deformation mode (stress state), the grain size, and of the texture of the γ- phase. (e.g. / 17,18 /). Under the conditions of the present work the following important features of the phase transition were found :

- In contrast to single - phase materials the transformation starts preferentially in very small austenite grains and, therefore, earlier after TMT 2 than after TMT 1, for instance. Moreover, the beginning of significant martensite formation is accompnied by a drop of the stacking - fault probability measured by X - ray diffraction (Fig. 4), which refers to martensite nucleation at stacking faults.

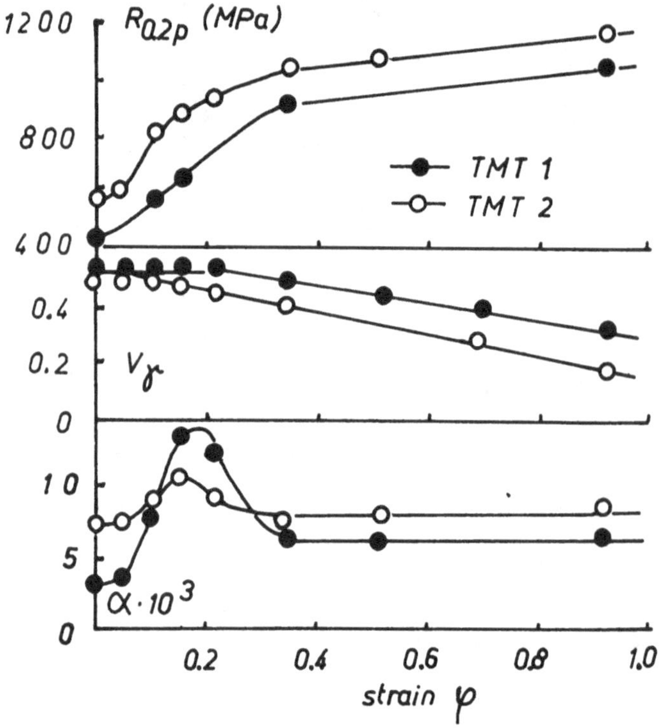

Fig. 4 : Yield stresses $R_{0.2p}$, austenite - volume fractions v_γ and stacking - fault probabilities α of cold - rolled X5CrNiTi 26.6

- In single - phase austenite X5CrNiTi 18.10 a transformation sequence $\gamma \rightarrow \varepsilon \rightarrow \alpha'$ is observed, i.e. the stable b.c.c. α'- phase originates from h.c.p. ε - martensite / 17 /. But though it should be favoured by pressure, the ε - phase could never be detected experimentally even in cold - rolling and in compression tests. That means, the austenite is directly transformed into the α' - phase in the two - phase steel.
- As in single - phase materials / 17 / the stress - induced α' - phase is a lath (or dislocation) martensite and connected with the f.c.c. austenite

via the Kurdjumov - Sachs orientation relationship . / 19 /. However, as it follows from the contrast phenomena in the TEM micrographs of the α' - grains (Fig. 5), the microstructure of the martensite is clearly modified by substructure inheritance, i.e. lattice defects (stacking faults, microtwins) of the austenite are transformed into the martensite lattice (e.g. / 20 /). In addition, texture inheritance takes place / 14,21 /.

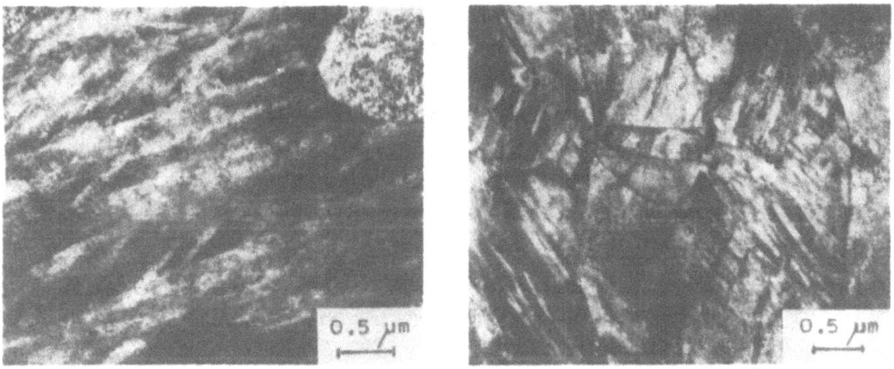

Fig. 5 : Martensite (α' - phase) microstructures in cold - rolled steel X5CrNiTi 26.6

That the γ-α' transition is closely related to the stacking faults of the austenite becomes evident in TEM and may be demonstrated by Fig. 6. Because of the contrast effects due to substructure inheritance in the bright - field image it can easily be seen that, in agreement with observations in single - phase materials / 22,23 /, different martensite variants are formed at intersections of stacking faults. Since such intersections should be frequent especially in small austenite grains (Fig. 2), martensite nucleation at stacking failts explains also the unexpected dependence of the deformation - induced phase transition on the grain size.

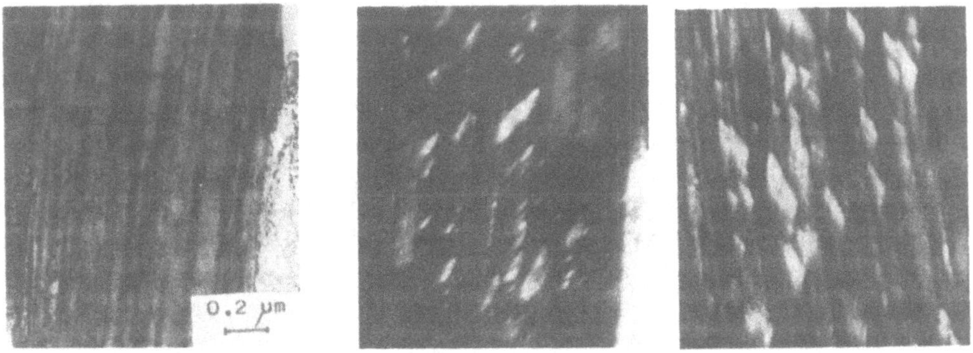

Fig. 6 : Martensite formation at stacking faults (a - bright field, b,c - dark field with (002) and (211))

A transformation sequence $\gamma \rightarrow \varepsilon \rightarrow \alpha$ is typical for martensite formation in austenite with very low stacking - fault energy / 17 /. Suppression of the ε - phase as mentioned above can therefore be a consequence of the fact that the γ - phase of the two - phase steel X5CrNiTi 26.6 and single - phase austenite X5CrNiTi 18.10 have different stacking - fault energies. With regard to / 29,17 /, as a possible reason for such an effect a strong grain - size dependence of the actual stacking - fault energy may be suggested. In order to check this idea, stacking - fault energies of carefully prepared two - phase microstructures with different austenite - grain diameters were determined from splitted dislocation nodes. The work was done in cooperation with the Institute of Metal Physics of the Ukrainian Academy of Sciences at Kiev, USSR. The results are summarized in Table 2. From the data is visible that, indeed, the stacking fault energies of the fine - grained austenites are significantly increased and out of the region where formation of ε - phase is observed.

Table 2

Grain size d (μm)	2.8	3.2	7.5	16.2
SFE (mJ / m^{-2})	41	41	40	32

Microstructure Changes in High - Temperature Deformation

The deformation behaviour of fine - grained two - phase steel of the type X5CrNiTi 26.6 at temperatures T $\geqslant$ 1023 K was intensively studied by Küchhold / 8 /. Flow curves measured in both compression and tensile tests showed clearly a maximum especially at such deformation temperatures where the austenite content of the samples is high. (Fig. 3). Similar results were obtained in hot - torsion of coarse - grained microstructures with different austenite content and interpreted as a consequence of dynamic recrystallization of the f.c.c. γ - phase / 26 /. However, in the present work both analysis of the X - ray diffraction line broadening and quantitative texture analysis from neutron diffraction leads to the conclusion that the flow - stress changes must essentially be caused by dynamic restoration of the b.c.c. δ - ferrite. Fig. 7 illustrates this for the case of compression at 1323 K. Corresponding to the flow curves the physical line broadening of the ferrite reflections goes through a maximum, while that of the austenite diffraction peaks remains practically unchanged after an increase at low deformation steps. Simultaneously the texture of the δ -ferrite is changed from $\langle 111 \rangle$ (deformation texture) to $\langle 100 \rangle$ (annealing texture), but in the austenite only a $\langle 110 \rangle$ deformation texture is found. / 27 /. Analogous results were obtained in hot - rolling of the steel. / 8 /.

The results of the integrating microstructure investigations are clearly confirmed by TEM (Fig. 8). At higher strains the dislocation content of the ferrite is reduced and contrast phenomena due to high - angle boundaries can locally be observed. This indicates that the restoration of the ferrite includes dynamic recrystallization. Simultaneously the character of the δ/γ - phase boundaries is modified (lost of coherence, probably by absorption of lattice dislocations of the ferrite), but the defect structure of the austenite shows no considerable changes.

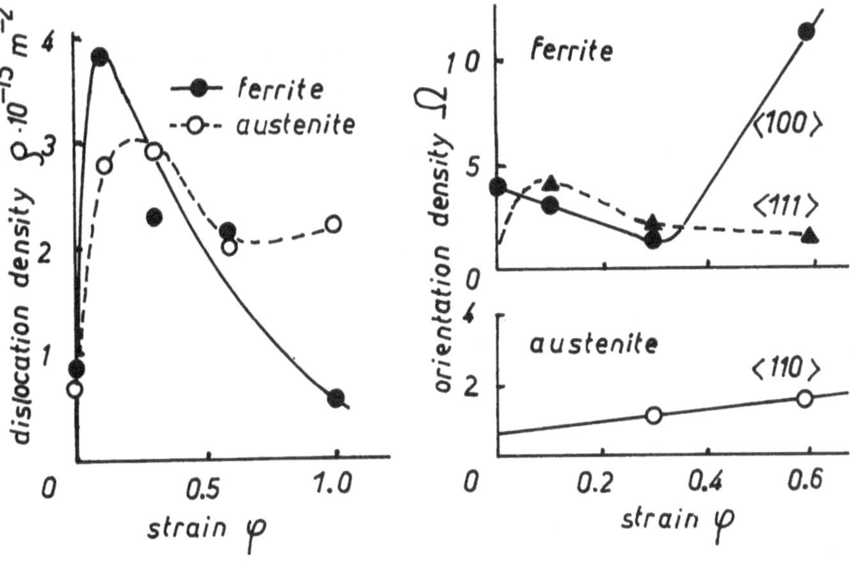

Fig. 7 : Microstructure changes in hot - compression of the steel X5CrNiTi 26.6 at 1323 K

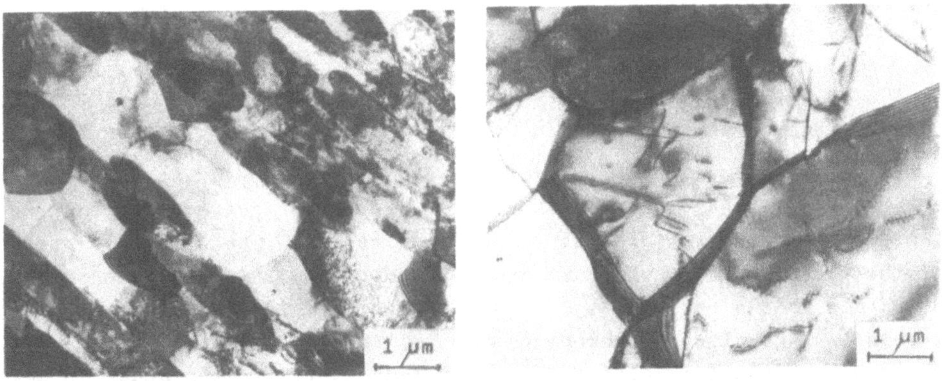

a) T = 1023 K , φ = 1.0 b) T = 1323 K , φ = 1.0

Fig. 8 : Formation of new grains of the δ - ferrite in hot compression

A qualitative explanation of the experimental results seems to be possible in the following manner :

- Above 973 K the austenite becomes the hard phase of the microstructure and the yield-stress ratio $R_{\gamma} / R_{\delta} = K(T)$ quickly increases to values $K > 2$ / 8,9 /. Therefore it can be expected that the γ-phase does not take part considerably in the plastic deformation especially at small strains and the driving forces for dynamic restoration remain small. Additionally such restoration should be retarded since the diffusion rate in the austenite is lower than that in the ferrite by about two orders of magnitude. / 28 /.
- On the other hand, the driving forces for restoration of the b.c.c. phase may be high already at the beginning of the hot-deformation because of the significant increase of the dislocation density during the formation of the initial microstructure (austenite precipitation, see Fig. 1,2) which corresponds to a pre-deformation. Moreover, an increased generation rate of dislocations in the weak phase must be expected even in the initial stage of the deformation because of the necessity of adjusting the different deformation behaviour of the phases. At least it must also be taken into account that in the ferrite cross-slip of screw dislocations is easy.

But whereas the general role of the δ-ferrite in the mechanical softening of the fine-grained samples can well be understood, the mechanism of the restoration is less clear. Taking into account (Fig. 1) that δ/δ-boundaries are very rare at the beginning of the hot-deformation (Austenite precipitation in H T 2 takes place within large ferrite grains.) and, moreover, grain-boundary migration should be very difficult because of the fine-disperse distribution of the phases, as the dominating process an "in situ"-recrystallization due to local dislocation redistribution (formation of δ/δ - boundaries, absorption of lattice dislocations by phase boundaries) is suggested. In order to confirm this interpretation, further work is necessary.

CONCLUSIONS

Technically interesting (i.e. fine-grained) two-phase microstructures of Fe-Cr-Ni steels as shown in Fig. 1,2 are far from equilibrium and important parameters (e.g. stacking-fault energy, yield-stress ration) of their components can significantly be modified in comparison to those of single-phase materials. For this reason the plastic deformation of the steels is a complicated phenomenon which, in general, is strongly influenced by the initial microstructure of the material taken under investigation and cannot easily be related to the behaviour of corresponding single-phase steels. Moreover, since the relationships between mechanical properties and microstructure changes do not be unambiguous, a reliable nrediction of the nature and the contribution of actually operating deformation or restoration mechanisms is difficult. Therefore, a physically realistic explanation of the macroscopical working behaviour of a given sample material requires systematic microstructure analysis. In this connection TEM is inevitable for both the identification of the mechanisms and a well-defined interpretation of integrating measurements of microstructure changes.

REFERENCES

1 I.Ja. Sokol : Dvuchfaznie stali. Metallurgija, Moskva 1974

2 Y. Tomota, K.Kuroki, T.Mori, I.Tamura, Mat.Sci.Eng. 24 (1976) 85

3 H.Fischmeister, B.Karlsson, Z.Metallkde. 68 (1978) 311

4 Y.Tomota, I.Tamura, Trans. ISIJ 22 (1982) 665

5 F.Koch, U.Koch, Dissertation A, Academy of Mining, Freiberg 1977

6 D.Bergner, P.Klimanek, R.Scherke, In: Proc. ICSMA 5 , Aachen 1979, Vol. 2, Pergamon Press, Toronto 1979, p. 861

7 R.Scherke, P.Klimanek, D.Bergner, In: Freiberger Forschungshefte, B 225 (1882) 65

8 G.Küchhold, Dissertation A, Academy of Mining, Freiberg 1990

9 P.Klimanek, G.Küchhold, Submitted to Neue Hütte

10 K.Lücke, Z.Metallkde. 75 (1984) 948

11 E.Hornbogen, Acta Metall. 32 (1984) 615

12 K.Bungardt et al., DEW Techn. Berichte 10 (1970) 85

13 C.Brock et al., Archiv EHW 47 (1976) 513

14 P.Klimanek, R.Scherke, K.Hennig, A.Mücklich, In: Proc. 4-th RISØ Int. Symp. Deformation of Multi-Phase and Particle Containing Materials, RISØ Nat. Lab., Roskilde, Denmark, 1983, p. 131

15 F.B.Pickering, as / 14 /, p. 103

16 C.G.Rhodes, A.W.Thompson, Metall. Trans. 8 A (1977) 1901

17 Ju.A.Petrov: Defekty i bezdiffuzionnoe prevraščenie v stali. Naukova Dumka, Kiev 1978

18 G.Benkißer, Dissertation B, University Rostock 1984

19 K.Wakasa, T.Nakamura, Scripta Metall. 11 (1977) 1067

20 K.-E.Hensger, P.Klimanek, Archiv EHW 55 (1984) 381

21 P.Klimanek, A.Mücklich, K.Hennig, In: Proc. 6-th ICOTOM, Vol.2 , Japan Inst. Metals, Tokyo 1981, p. 901

22 S.S.Hecker et al., Metall. Trans. 13 A (1982) 619

23 L.E.Murr et. al., Metall. Trans. 13 A (1982) 627

24 R.M.Latanision, A.W.Ruff, Metall. Trans. 2 (1971) 505

25 P.Klimanek, Ju.N.Petrov, I.A.Jakubcov, To be published

26 F.E.Al-Jouni, C.M.Sellars, as / 14 /, p. 131

27 P.Klimanek, G.Küchhold, U.Scherch, A.Mücklich, In: Annual Report 1983, ZfK-Publ. 530, ZfK Rossendorf 1984, p. 66

28 M.T.Malik, Dissertation A, Academy of Mining, Freiberg 1984

29 P.Ju.Volosevic, V.N.Gridnev, Ju.N.Petrov, Fizika Metallov i Metalloved. 40 (1975) 554

MICROSTRUCTURE AND MECHANICAL PROPERTIES OF LASER PROCESSED STEELS AND CAST IRONS

A. Luft, B. Brenner, W. Löschau, W. Reitzenstein

Akademie der Wissenschaften der DDR, Zentralinstitut für Festkörperphysik und Werkstofforschung, DDR-8027 Dresden

ABSTRACT

The optimization of laser surface treatment techniques requires an understanding of the relationships between process variables, microstructures and mechanical properties. The present paper reports on the electron microscopic characterization of the microstructures generated by transformation hardening, remelting of cast iron and alloying of steel with TiC. The microstructural observations are correlated with measured mechanical properties of processed surface layers.

INTRODUCTION

It is well established that the properties of the near-surface region of a material can determine its performance in many technological applications. The recent availability of high power lasers as directed energy sources opened up new possibilities for the surface modification of metals and alloys. The advantages of laser treatment include precise control of energy delivery at high power densities, precise control and limitation of process interaction time, accurate control of the location of energy deposition, the ability to operate out-of-vacuum in a variety of gaseous atmospheres, and the adaptability of lasers to automation. These characteristics of the laser have led to various surface treatment technologies such as surface heating for transformation hardening, surface melting for homogenization or generation of rapid solidification structures, and surface alloying and cladding for improvement of corrosion and wear resistance. Reviews on laser processing can be found in /1 - 4/.

These advanced technologies require not only a knowledge of the processing techniques themselves but also a detailed understanding of the factors which control the microstructure and the mechanical behaviour of the modified surface region. As laser processing represents a short time heat treatment with high heating and cooling rates fine grained and multiphase microstructures develop which cannot be identified sufficiently by optical metallography. Hence the present paper will report on the application of electron microscopy in the characterization of laser induced microstructures. Specific surface treatments to be considered include transformation hardening, remelting of cast iron and alloying of steel with TiC. From the microstructural observations conclusions are drawn on the underlying metallurgical processes and the resulting mechanical properties.

LASER TRANSFORMATION HARDENING

Laser transformation hardening is achieved by passing a laser beam over the surface of a hardenable material - usually a carbon steel or a pearlitic cast iron. A limited surface layer is thereby rapidly heated to above the austenitizing temperature in the short time the beam is incident. Once the beam has passed, this surface layer is quenched by the conduction of heat into the cold bulk material. The high cooling rates give hardening by the transformation of the austenite to martensite. The power, size and shape of the beam and the speed at which it scans the workpiece must be carefully adjusted so that the absorbed energy is sufficient to austenitize the steel to the desired depth without melting the surface. Among the most significant features of laser surface hardening are high heating and quenching rates (10^3 ... 10^5 K/s), high temperature gradients (10^5 ... 10^7 K/m), and short times (<1 s) for which the steel is heated above the critical temperature /5, 6/. We investigated the pearlite to austenite transformation under these nonequilibrium conditions prevalent in laser hardening /7 - 9/.

In the near-eutectoid plain carbon steel C70 there has been found a sharp boundary between the laser hardened zone and the pearlite of the matrix, as shown in Fig. 1 by SEM and TEM observations. The laser treatment produced a fully martensitic structure leaving only a little undissolved cementite in the outskirts of the hardened zone. The formerly existing α/γ-interface exhibits distinct protrusions into the interlamellar ferrite /7/. According to a two-dimensional diffusion model /8/ this microscopic shape of the advancing α/γ-interface gives evidence that the diffusion path of the carbon goes from the cementite to the α/γ-interface through the ferrite. Since carbon diffusion in α-iron is fast the pearlite to austenite transformation can occur at a relatively low temperature at which the available time between exceeding the A1-temperature and the arrival of the austenitization front is not sufficient for nucleation and growth processes. Hence these observations suggest that in plain carbon steel with fine lamellar pearlite the austenitization induced by a moving laser beam proceeds by transformation front propagation and not by nucleation and growth of isolated austenite nuclei. The rate controlling process is thereby the carbon diffusion in ferrite /8, 9/.

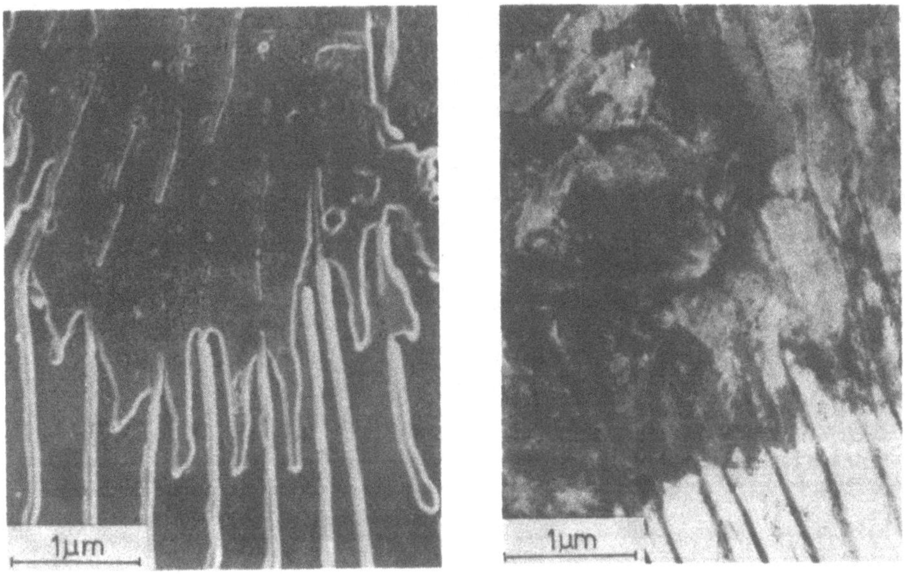

Fig. 1 Boundary region of laser hardened zones in the pearlitic plain carbon steel C70

a) "Frozen-in" α/γ-interface with distinct protrusions into the interlamellar Ferrite (SEM)

b) Transformation of pearlite to austenite and finally to martensite (TEM)

The presence of substitutional alloy elements may alter significantly the hardening response of a pearlitic structure to laser heat treatment because such elements as Si and Cr suppress carbon diffusion through the ferrite and retard the dissolution of the cementite. Fig. 2 shows SEM micrographs from the boundary region of the laser hardened zone in a pearlitic cast iron containing 2.2 % Si and 0.67 % Cr. Austenite formation occurred either by the propagation of a transformation front (Fig. 2a) or by nucleation and growth processes (Fig. 2b). In this material, unlike in plain carbon steel, the α/γ-interface does not exhibit protrusions into the interlamellar ferrite. Instead, the microscopic shape of the interface is of such a type which is expected when the propagation of the transformation front is controlled by the slower carbon diffusion through the freshly formed austenite /7, 8/. Consequently, a higher superheat of the pearlite over the A1-temperature is necessary in order that austenite formation proceeds to the required extent. Under these conditions nucleation and growth processes can occur ahead of the moving transformation front. The other characteristic feature in Fig. 2 is the high amount of undissolved cementite left in the austenitic (now martensitic) zone. This may be attributed to the partitioning of carbide stabilizing elements to the cementite. The dissolution of cementite is then controlled by the very slow diffusion of the alloying elements in austenite. Homogeneous austenite (and martensite) can be achieved only in the near-surface region to a depth to which the thermal cycle is of sufficient strength for alloying element diffusion controlled dissolution of the alloyed carbides /10/.

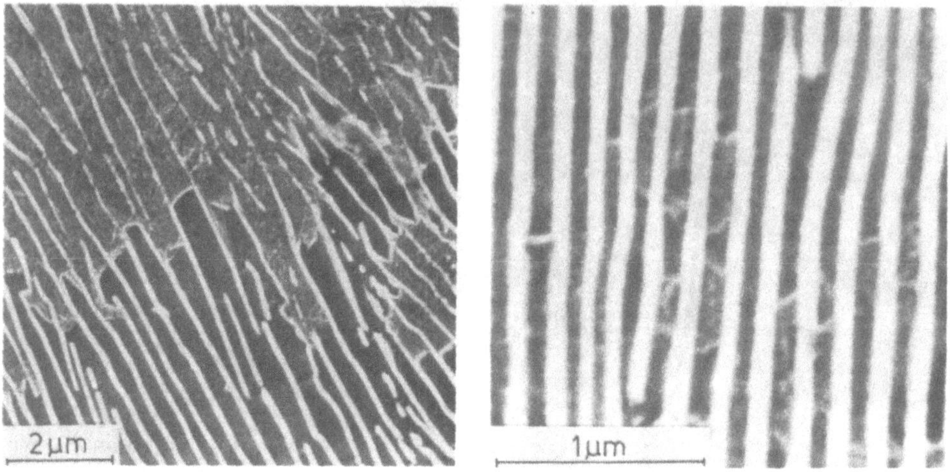

Fig. 2 Boundary region of laser hardened zones in the pearlitic cast iron GGL320Cr2

a) Pearlite to austenite transformation by front propagation. Note the high amount of undissolved cementite (SEM)

b) Nucleation and growth of austenite between cementite lamellae ahead of the transformation front (SEM)

Laser transformation hardening has been successfully used to improve wear resistance /1, 11 - 13/ and to enhance fatigue life /1, 11, 14 - 16/. A recent application to be outlined here is the laser heat treatment of turbine blades for reducing the erosion caused by the attack of impinging water droplets /17 - 20/. Fig. 3 shows the results of erosion tests on the martensitic stainless steel X20Cr13 for various heat treatments /19, 20/. The tempered state which is commonly used in the operating turbine blades is compared with induction hardening and two laser processing regimes. In regime I a pattern of several parallel passes was produced, whereas in regime II the whole eroded surface was hardened by a single pass in order to avoid tempering of previously treated area. It is evident that the erosion rate can be reduced significantly if the laser treatment is carefully optimized such that a high hardness and a uniform hardness profile over the whole eroded surface are achieved. TEM investigations revealed that the base material comprises elongated and globular carbides predominantly of the $M_{23}C_6$ type which are characteristic for the highly tempered martensitic structure (Fig. 4a). On the contrary, the fully hardened state achievable by laser heat treatment is free of these carbides and contains a highly dislocated lath martensite (Fig. 4b). Hence the increase of hardness is mainly determined by the extent of carbide dissolution controlling the carbon content of the matrix and by the dislocation density of the formed martensite.

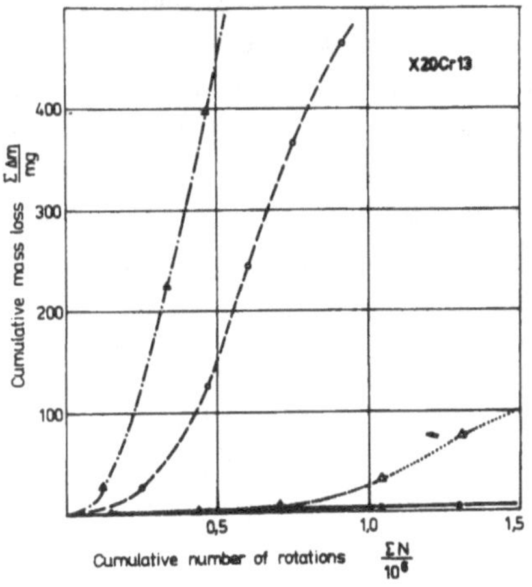

Fig. 3 Results of erosion tests on the steel X20Cr13

- ▲ conventionally harde ned and tempered state
- △ induction hardening
- ○ laser regime I overlapped trace pattern
- ● laser regime II single pass

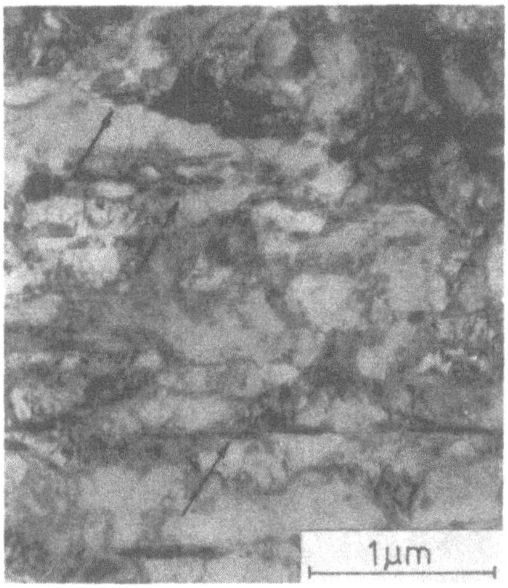

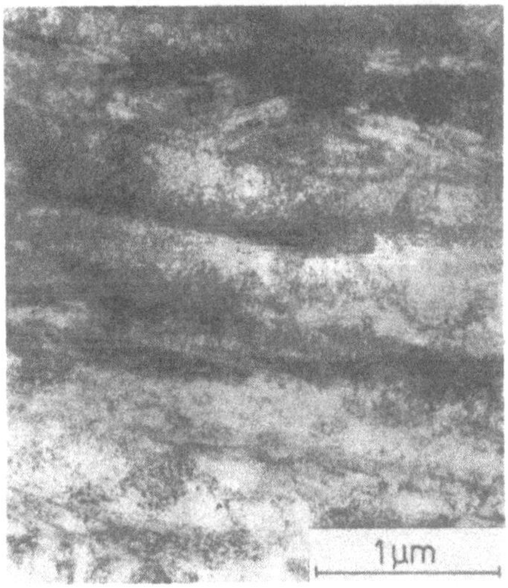

Fig. 4 Microstructures of the stainless steel X20Cr13 (TEM)

a) Base material in the tempered state; some $M_{23}C_6$ carbides are marked by an arrow

b) Laser hardened state; highly dislocated martensite

Laser surface processing at higher power densities and/or longer interaction times leads to the melting of the surface followed by rapid solidification of the molten material. Laser surface melting of grey cast iron has received considerable attention because ledeburitic surface layers of high hardness and improved wear properties can be produced /21 - 27/. The main metallurgical processes during the thermal cycle are the dissolution of graphite in the melt pool and the rapid solidification of the carbon-rich iron melt to a ledeburitic structure with a high content of carbides.

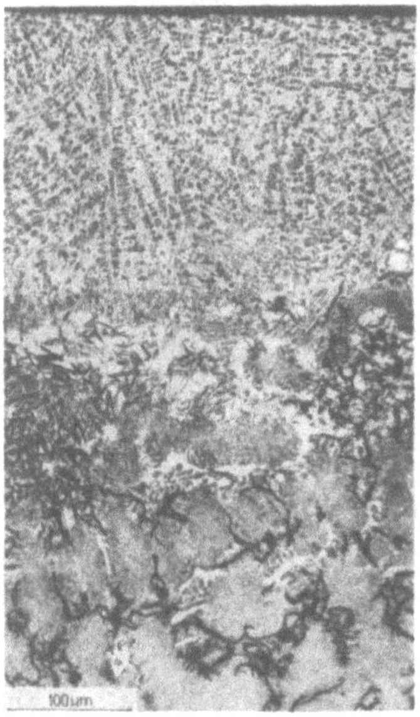

Fig. 5 Cross section of laser melted zone in the pearlitic grey cast iron GGL320Cr2

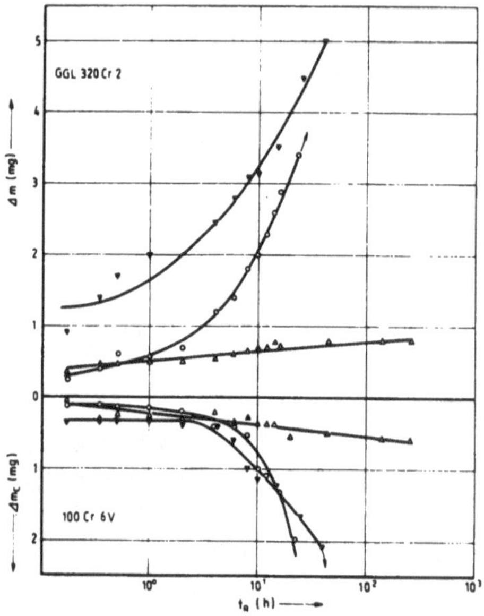

Fig. 6 Mass loss versus friction time for lubricated sliding wear of cast iron GGL320Cr2
○ as-cast state, p_a = 7 MPa
△ laser melted, p_a = 7 MPa
▼ laser melted, p_a = 12 MPa

We investigated the effect of a laser melt treatment on the microstructure and the wear behaviour of the cast iron GGL320Cr2 which contained in the as-cast state a pearlitic matrix and flake graphite /28, 29/. Fig. 5 shows a cross section of a melt zone and part of the underlying heat affected zone. It is evident that the graphite completely dissolved in the melt zone. The microstructure in the melt zone comprises primary austenite dendrites together with an interdendritic eutectic (ledeburite). Crack formation could be avoided by adjustment of the processing parameters to achieve relatively low cooling rates. SEM observations revealed that under these conditions a great part of the primary austenite has been transformed to fine lamellar pearlite and to a less extent to plate martensite (Fig. 7a). The eutectic regions contained densely spaced acicular cementite. Higher quenching rates produced finer dendrites and finer carbides within the eutectic /29/.

Wear tests were conducted under conditions of lubricated sliding friction using a hardened 100Cr6 steel as counterpart /28, 29/. In Fig. 6 the cumulative mass loss is plotted versus cumulative friction time for as-cast and laser melted samples, respectively. Untreated samples failed at a nominal contact pressure of 7 MPa after 25 hours, whereas the laser melt treated surfaces experienced an extremely low mass loss even up to 256 hours after which the test was stopped. Furthermore, it has been shown that the laser melted surfaces can endure a significantly higher pressure (12 MPa) for longer friction times than as-cast samples without suffering severe wear.

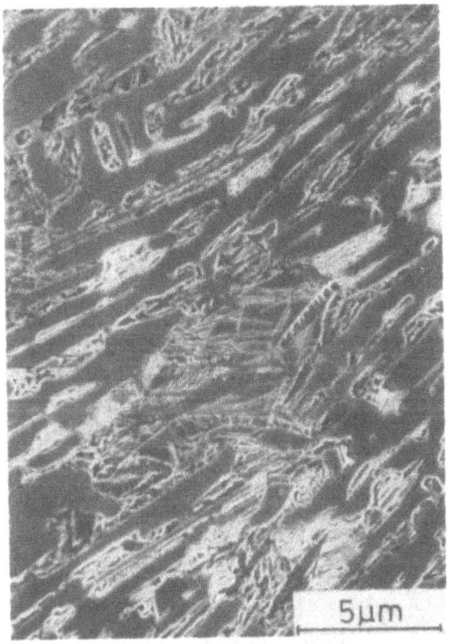

Fig. 7a Ledeburitic microstructure of laser melted cast iron GGL320Cr2 (SEM)

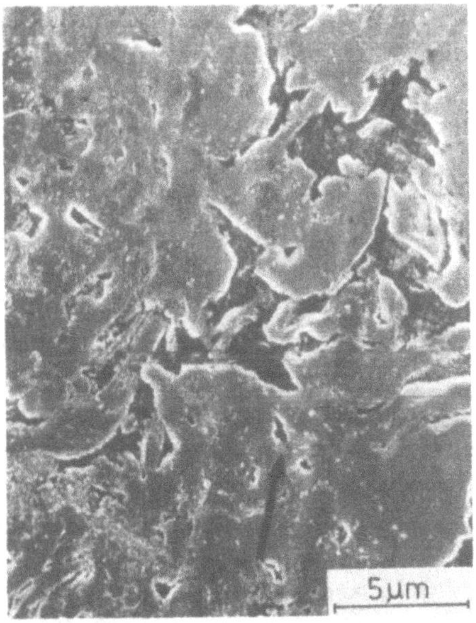

Fig. 7b Microstructure at the wear surface after lubricated sliding wear at p_a = 12 MPa for 50 hours (SEM)

In order to gain information on the wear mechanisms prevalent under the imposed conditions the wear surfaces were examined by SEM. Fig. 7b represents a typical example of the microstructure developed at the wear surface of the laser melted samples. The sliding direction which is recognizable by some shallow grooves is indicated by an arrow. It is evident by comparing Figs. 7a and 7b that the material underneath the surface suffered large plastic deformation without cracking or spalling. In particular, the acicular cementite of the eutectic flattened and broadened plastically such that almost all the surface is covered by a continuous cementite layer. Hence the excellent wear properties obtained by laser surface melting of grey cast iron result from the elimination of the graphite flakes as predominant sites of crack formation and the generation of a finely dispersed eutectic cementitic network of high hardness and sufficient plastic adaptability.

LASER SURFACE ALLOYING OF STEEL WITH HARD CARBIDES

Surface alloying with a laser is similar to surface melting except that another material is added to the melt pool. The alloy ingredient may be either deposited on the substrate before the laser melt treatment or placed immediately into the melt pool by powder blowing. A wide variety of surface alloys can be produced /1, 30, 31/. A processing variant pursues the aim to introduce hard particles into the surface region of metals in order to improve their resistance against abrasive wear /32 - 34/.

We have investigated the possibility to produce hard and wear resistant surfaces on tool steels by laser melting of pre-placed layers containing TiC or WC particles /35, 36/. On purpose to improve the wetting behaviour of the carbide particles the carbide powders were mixed with an Fe-B alloy. Microstructural observations revealed that under the chosen conditions nearly all carbide particles dissolved in the melt pool and new TiC or WC crystals of small size formed on cooling from the melt. The solidification of the melt proceeded by cellular or celldendritic growth of γ-crystals (austenite) associated with the formation of an interdendritic eutectic comprising mainly cementite. Fig. 8 shows SEM micrographs of the solidification structure produced by laser melting of a TiC/Fe-4%B layer into 90MnV8 steel. The main features are (1) a few undissolved TiC particles, (2) many re-grown TiC crystals ($\lesssim 0.5\,\mu m$) and (3) an eutectic cementitic network with mesh sizes of 1 - 2 μm. The austenite within the cells has been mainly transformed to high-carbon martensite. Volume concentrations of re-grown TiC crystals were in the range of 10 - 50 % depending on the thickness of the pre-placed layer and the depth of the molten substrate. The microhardness was higher than 1 200 $HV_{0.05}$ exceeding substantially that achievable by transformation hardening.

Preliminary results on the response of the TiC alloyed surface layers to abrasive wear are shown in Fig. 9 /35/. Samples of 90MnV8 steel laser processed with overlapped passes at various scan speeds were pressed against a rotating disk covered with dry emery paper (SiC, grain size about 70 μm). The mass loss was measured as function of friction time. It can be seen that the laser alloyed samples show a significant reduction in wear compared to that of the conventionally

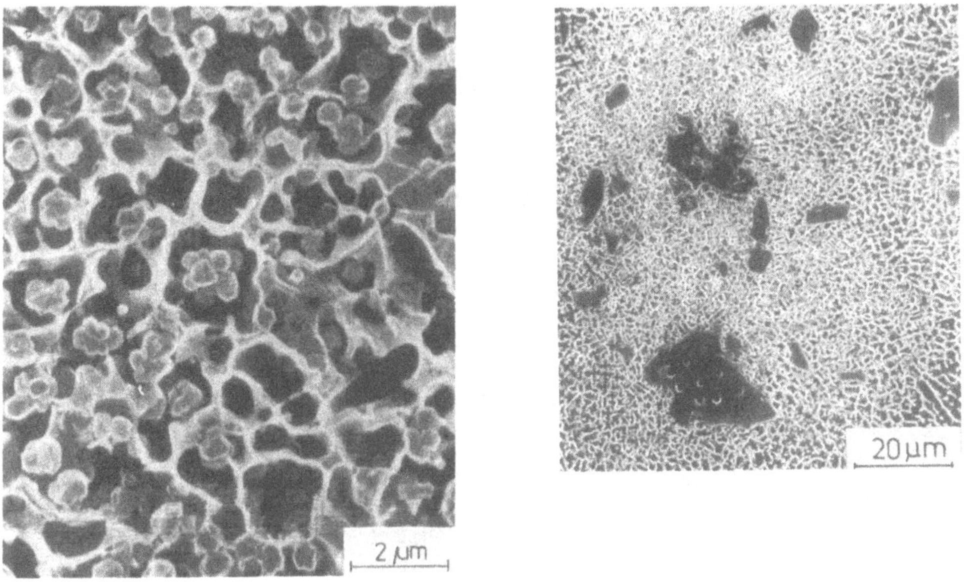

Fig. 8 Solidification microstructure formed by laser melting of a preplaced TiC/Fe-4% B layer in 90MnV8 steel

a) A few undissolved TiC particles are embedded in the solidification structure (SEM)

b) Re-grown TiC crystals and eutectic cementite network (SEM)

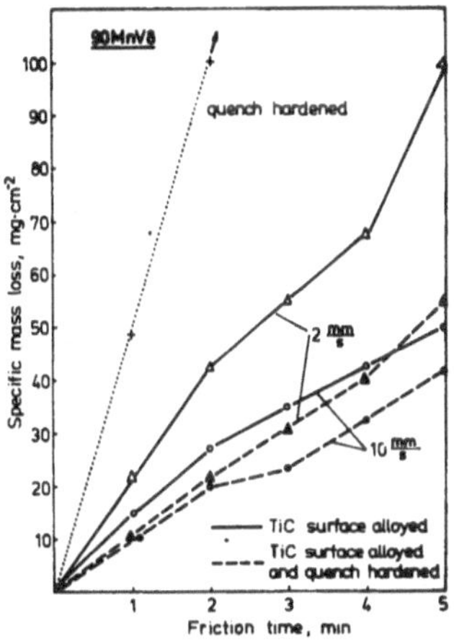

Fig. 9 Abrasive wear test (SiC-emery paper, 70 µm grains) of 90MnV8 steel after conventional heat treatment and after laser surface alloying with TiC. Laser scan speeds are noted.

quench-hardened steel. The wear rate was found to be lower for the samples processed with higher laser scan speed. This may be attributed to the higher volume concentration of TiC and the lower tempering effect on previously processed areas. A further decrease of wear rate could be achieved by a subsequent quench-hardening treatment of the laser processed samples by which the tempered areas were re-hardened.

FINAL REMARKS

Laser processing offers the possibility to taylor the surface properties of tools and machine components to the technological requirements without sacrificing the bulk properties. The application of the various techniques to practical problems is expected to grow as more knowledge is gained from fundamental research studies. One approach to a better understanding of the relationships between processing technology and resulting properties is the detailed characterization of the produced microstructures. It has been shown in this paper that scanning and transmission electron microscopy can help substantially to identify the microstructures, to understand their formation, and to evaluate their effect on the mechanical behaviour. However, it should be taken into account that mechanical properties such as wear and fatigue resistance are influenced not only by microstructure but also by other factors, e.g. internal stresses, environment and loading conditions.

REFERENCES

/1/ D.S. Gnanamuthu, Optical Engineering 19 (1980) 183 - 92

/2/ W. Amende and H. Zechmeister, VDI-Z. 124 (Aug. 1982) 581 - 90

/3/ S.L. Ream, In: Surface Engineering, NATO Adv. Study Inst. Ser. E, Vol. 85 (1984) p. 285 - 99

/4/ W.M. Steen, Metals and Materials 2 (Dec. 1985) 730 - 6

/5/ S. Kou, D.K. Sun and Y.P. Le, Metallurg. Trans. 14A (1983) 643 - 53

/6/ M.F. Ashby and K.E. Easterling, Acta Met. 32 (1984) 1935 - 48

/7/ A. Luft, D. Lepski, B. Brenner and W. Reitzenstein, Wiss. Berichte des ZFW Dresden, Nr. 36 (1988) 111 - 23

/8/ D. Lepski, A. Luft and W. Reitzenstein, In: Proc. 5th Int. Congr. Heat Treatment of Materials, Budapest 1986, Vol. III, p. 1560 - 67

/9/ D. Lepski, A. Luft, W. Reitzenstein and H.-G. Kusch, In: Laser-Technology and Applications, Yearbook, H. Kohler (Ed.), Vulkan Verlag, Essen 1988, p. 278 - 286

/10/ J.R. Bradley and S. Kim, Metallurg. Trans. 19A (1988) 2013 - 25

/11/ M. Kikuchi, H. Hisada, Y. Kuroda and K. Moritsu, In: Proc. 1st Joint US/Japan Int. Laser Processing Conf., Laser Institute of America, Toledo, Ohio 1981, Paper No. 12

/12/ T. Bell, Metallurgia 49 (1982) 103 - 11

/13/ P.A. Molian and M. Baldwin, Wear 118 (1987) 319 - 27

/14/ H.B. Singh, S.M. Copley and M. Brass, Metallurg. Trans. 12A (1981) 138 - 40

/15/ H. Peiyan, L. Chun, Z. Qiguang, L. Guagyia and H. Lee, Fatigue Fracture of Engineering Materials and Structures, Vol. 8 No. 1 (1984) 89 - 99

/16/ B. Winderlich and B. Brenner, Härt. Techn. Mitteil. 44 (1989) 166 - 73

/17/ M. Roth and M. Cantello, In: Proc. 2nd Int. Conf. Lasers in Manufacturing, M.E. Kimmitt (Ed.), Birmingham 1985, p. 119 - 128

/18/ V. Bedogni, M. Cantello, W. Cerri, D. Cruciani, R. Festa, G. Mor, F. Nenci and F.P. Vivoli, In: Proc. of Laser Advanced Materials Processing, Osaka 1987, p. 567 - 572

/19/ B. Brenner, W. Reitzenstein, G. Wiedemann, W. Storch and H. Junge, 4. Lasertechnologietagung Jena 1987

/20/ W. Storch, G. Blum, B. Brenner and G. Wiedemann, Mitteil. Kraftwerksanlagenbau der DDR, 28 (1988) 2, 16 - 17

/21/ A. Blarasin, S. Corcoruto, A. Belmondo and D. Bacci, Wear 86 (1983) 315 - 25

/22/ C.H. Chen, C.J. Altstetter and J.M. Rigsbee, Metallurg. Trans. 15A (1984) 719 - 28

/23/ H.W. Bergmann, Surf. Eng. 1 (1985) 137 - 55

/24/ W.M. Steen, Z. Chen and D.R.F. West, In: The Industrial Laser Annual Handbook 1987, D. Belforte and M. Levitt (Eds.), Penn Well Publ., Tulsa, Oklahoma 1987, p. 80 - 96

/25/ C.H. Chen, C.P. Ju and J.M. Rigsbee, Mat. Sci. Technology 4 (1988) 161 - 6

/26/ C.P. Ju, C.H. Chen and J.M. Rigsbee, Mat. Sci. Technology 4 (1988) 167 - 72

/27/ P.W. Leech, Wear 113 (1986) 233 - 45

/28/ B. Brenner, B. Winderlich, R. Franke and W. Reitzenstein, In: Laser-Technology and Applications, Yearbook, H. Kohler (Ed.), Vulkan-Verlag, Essen 1988, p. 264 - 7

/29/ A. Luft, B. Brenner, R. Franke and W. Reitzenstein, Proceed. Conf. Special Technologies 90, SKODA-Konzern, Plzen, CSSR (April 1990)

/30/ C.W. Draper, J. Metals 34 (1982) 6, 24 - 32

/31/ C.W. Draper and J.M. Poate, Int. Metal Reviews 30 (1985) 85 - 102

/32/ J.D. Ayers, R.J. Schaefer and W.P. Robey, J. Metals 33 (1981) 8, 19 - 23

/33/ R.P. Cooper and J.D. Ayers, Surface Eng. 1 (1985)4, 263 - 72

/34/ H.W. Bergmann and B.L. Mordike, Z. Werkstofftechn. 12 (1981) 142 - 50

/35/ W. Löschau, A. Luft and H. Heinze, to be publ.

/36/ A. Luft and W. Löschau, Veröffentl. 12. Tagung Elektronenmikroskopie, Dresden (1988), Bd. II, 347 - 8

ON THE COMBINATION OF ELEMENTARY STRENGTHENING MECHANISMS IN A MICROALLOYED STEEL

D. Aurich, W. Österle

Bundesanstalt für Metallforschung und -prüfung (BAM)
Unter den Eichen 87, D-1000 Berlin 45

ABSTRACT

In a simple vanadium bearing HSLA steel several microstructures were produced by different heat treatments and microstructural parameters were measured by means of light optical as well as transmission electron microscopy. Using elementary strengthening equations the contributions of different mechanisms to the yield stress were calculated and compared with the yield stresses measured by tensile tests. The results showed that substructure strengthening in addition to a friction stress and solid solution strengthening can explain the measured yield stress in a satisfying way. In some cases a Hall-Petch relationship was observed, which is attributed to a correlation between dislocation density and grain size.

INTRODUCTION

It is generally accepted that, besides grain refinement, substructure strengthening by fine carbonitride particles and dislocations is an essential strengthening mechanism in high-strength low-alloy steels. In the literature the interpretation of experimental results is often based on the assumption that the different strength contributions can be added independently. The assumption is not easily verified, since the grain size of such steels cannot be varied in a wide range without changing the substructure as well.

The aim of this work was to check the validity of strengthening equations if several mechanisms superimpose.

The grain boundary strengthening σ_g is described by the well-known Hall-Petch relationship [1]:

$$\sigma_g = k/\sqrt{D} \qquad (1)$$

k is a constant and D the mean grain diameter.

The contribution of the dislocations σ_d is given by the work hardening relationship

$$\sigma_d = \alpha\ Gb\sqrt{\varrho} \qquad (2)$$

with α being a constant of about unity, G the shear modulus (82 700

MPa), b the Burgers vector of the gliding dislocations (0.248 nm), and ϱ the dislocation density.

The third mechanism to be considered is particle strengthening σ_p, which was originally modelled by Orowan and modified by Ashby [2]. Taking loop interaction into account and adapting the model for tensile loading of bcc polycrystals, the following equation was derived

$$\sigma_p = 2.75 \frac{0.83}{2\pi(1-\nu)} \frac{Gb}{L} \ln \frac{x}{b} . \quad (3)$$

ν is Poisson's ratio (0.3), L the interparticle spacing, and x the distance between the bowed out dislocation segments.

In addition to these mechanisms controlled by the microstructure a friction stress σ_f arising from the iron matrix, and the stress part σ_s of the solid solution strengthening should be considered.

EXPERIMENTAL PROCEDURE

The material used for this investigation was a microalloyed steel with the following chemical composition (mass fractions in %).

C	Si	Mn	P	S	Al	Cu	Cr	Ni	V
0.03	0.22	1.41	0.01	0.01	0.03	0.01	0.03	0.02	0.11

Specimens with the dimensions 15 x 15 x 70 mm^3 were cut from the center of a 18 mm thick hot-rolled sheet. After a solution heat treat-

Table 1: Heat treatments and microstructures

ser.	heat treatment	microstructure	grain size optical microscopy
I_1	1300 °C, 10', wq + 700 °C, 100 h, ac	ferrite with coarse V(CN) paticles and carbide films on grain boundaries	coarse
I_2	1300 °C, 10' + 700 °C, 1 h, wq + 700 °C, 100 h, ac		
I_3	1050 °C, 5' + 700 °C, 1 h, wq + 700 °C, 100 h, ac		fine
II_1	1300 °C, 10' + 700 °C, 1 h, wq	ferrite with fine V(CN) particles and 2 % martensite	coarse
II_2	1050 °C, 5' + 700 °C, 1 h, wq		fine
III_1	1300 °C, 10', wq	martensite	coarse
III_2	1050 °C, 5', wq		fine

wq: water quenching, ac: air cooling

ment at 1300 °C followed by water quenching three different series of microstructures were produced by subsequent heat treatments as outlined in Tab. 1.

The ferrite grain size was measured semi-automatically by lineal analysis. In the martensite the grain size was defined as the mean slip line length as suggested by Naylor [3]. More details about the determination of the martensite grain size will be published in another paper [4].

Dislocation densities were measured on TEM micrographs according to Ham's intersection method [5]. The foil thickness at the respective sites of evaluation was determined by appropriate convergent beam electron diffraction patterns as suggested by Kelly et al. [6].

The particle size and the volume density N_v can be obtained from TEM micrographs provided that the foil thickness is known. With these data the parameters needed in eq. (3) can be calculated in the following way: If the particles are almost spherical, as in the microstructures of series I, a relationship suggested by Törönen [7] can be used to calculate the particle spacing L

$$L = 1.23 \, (N_v d)^{-1/2} - d \, . \qquad (4)$$

d is defined as mean caliper diameter and in eq. (3) x = d is assumed.

Eq. (4) is no longer valid if the shape of the particles deviates considerably from the spherical one. This is the case for microstructures of series II. Then the particles have adopted the morphology of thin, square shaped plates. As shown in a previous paper [8], the following relationship has to be applied

$$L = 1.273 \, (N_v d)^{-1/2} - x \, . \qquad (5)$$

d is now defined as the edge length of the square shaped plates and $x = 0.572\,d$.

Tensile tests were performed on cylindrical specimens of 5 mm in diameter and 50 mm gauge length with a strain rate of approximately $2 \times 10^{-4} s^{-1}$

RESULTS

Fig. 1 shows stress-strain curves referring to the different microstructures. The microstructures of type I typically developed a yield point. The corresponding Lüders band elongation increased with decreasing grain size. Obviously, a certain amount of work hardening is necessary before macroscopic yielding can start. Therefore in the following the lower yield stress was correlated to the microstructure which had developed at the end of the Lüders band elongation. If continuous yielding occurred (Ser. II and III) the stress at 0.2 % plastic strain was taken as yield stress and the corresponding microstructures were evaluated on undeformed specimens.

Tab. 2 summarizes the structure parameters, the related stress values calculated by eqs. (1) to (3), and the stress values measured in tensile tests. The constants $k = 19\ Nmm^{-3/2}$ and $\alpha = 1.1$ were derived from Figs. 3 and 4 shown in the next section.

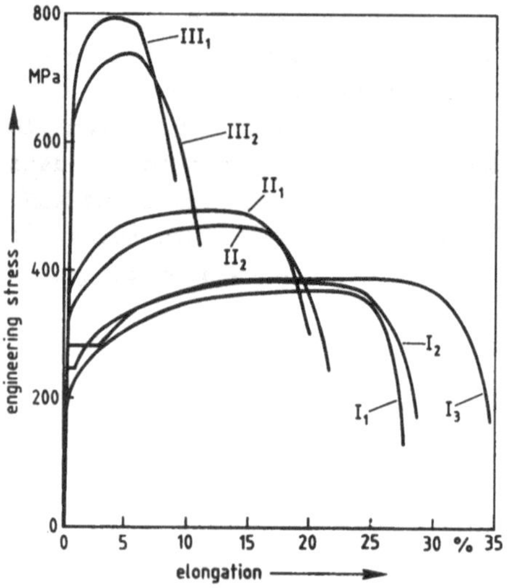

Fig. 1 Stress-strain curves belonging to different microstructures

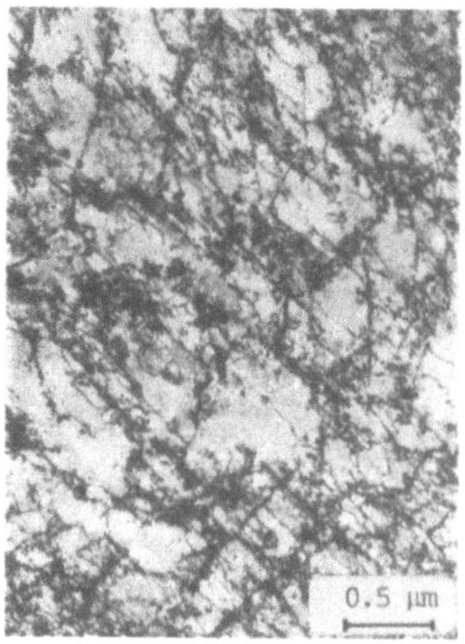

Fig. 2 TEM micrograph of a sample of series I_3 after 2.2% Lüders band elongation

Table 2: Results of quantitative measurements

ser.	grain size D [µm]	dislocation density $\rho \cdot 10^{-7}$ [mm^{-2}]	particle diameter d [nm]	particle distance L [nm]	σ_g equ. (1) [MPa]	σ_d equ. (2) [MPa]	σ_p equ. (3) [MPa]	R_{el}, $R_{p0.2}$ [MPa]
I_1	100	1.6 *	33	644	60	90	81	209
I_2	28	2.9 *	39	885	113	121	61	249
I_3	18	8.4 *	40	432	142	207	125	286
II_1	45	2.0	13	221	90	101	164	352
II_2	16	2.0	13	233	150	130	155	321
III_1	1.5	53.0	-	-	490	519	-	662
III_2	1.7	34.0	-	-	461	416	-	564

* measured at the end of Lüders elongation

DISCUSSION

From Table 2 it becomes obvious that the yield stress cannot be interpreted as the sum of the individual contributions σ_g, σ_d, and σ_p because already this would exceed the measured stress values considerably without taking σ_f and σ_s into account. In Figs. 3 and 4 correlations are proved between the measured stress and structure parameters. Surprisingly, most microstructures as well as some additional ones not described here, fit to a Hall-Petch relation (Fig. 3). The slope is very similar to that reported for pure iron. The intercept can readily be explained by a contribution arising from a friction stress and from the solid solution strengthening by manganese and silicon.

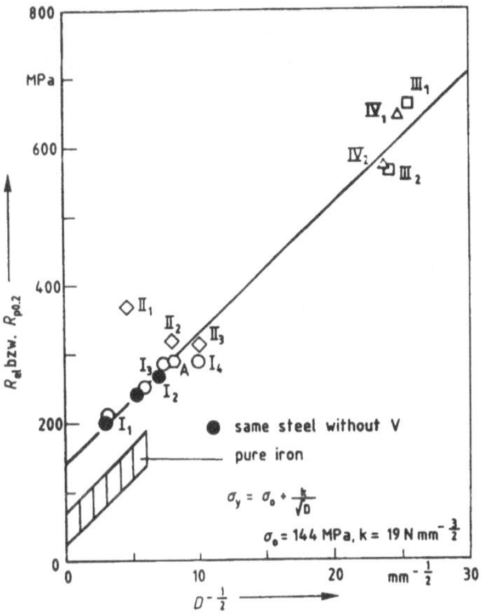

Fig. 3 Yield stress as a function of the grain size D

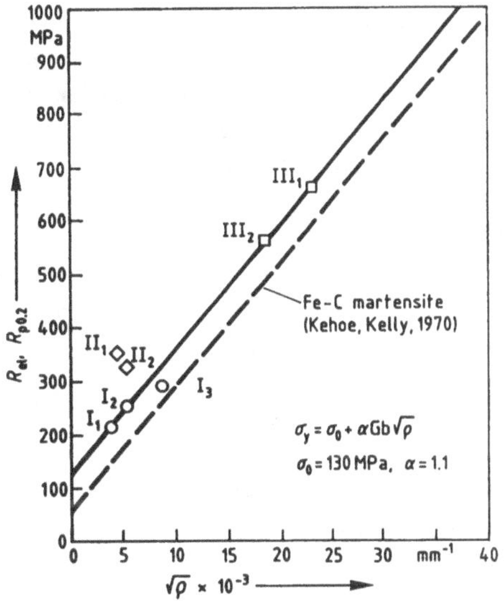

Fig. 4 Yield stress as a function of the dislocation density ϱ

As shown in Fig. 4 the work hardening relationship holds for most of the microstructures, too. The intercept is about the same as in Fig. 3 and consequently it has the same meaning. The different strength values of the martensitic specimens correlate with the corresponding dislocation densities. This result agrees well with the paper of Kehoe and Kelly on Fe-C martensites of different carbon content (broken line) [10]. But also the ferrite samples which were more or less work hardened during the Lüders band elongation fit to the same line. Only the results of the series II_1 and II_2 deviate considerably, indicating that an additional mechanism (particle strengthening) is operating.

The yield stress related to the microstructures of series I and III can be described alternatively by eqs. (1) or (2). Consequently there should exist a correlation between these two mechanisms. For $\rho \sim 1/D$ eq. (1) is converted into eq. (2). This result is consistent with the assumption that certain grain size dependent stress concentrations, produced by dislocations, have to be formed before macroscopic yielding can start. While in the ferrite the critical dislocation arrangement is inhomogeneously formed during the Lüders band elongation, it is already present in the martensite. In principle this assumption is very similar to the pile up theory of the Hall-Petch relationship though classical dislocation pile ups have not been observed.

Since the stress related to the microstructures of series I_1 can entirely be described as the sum of σ_f, σ_s and σ_d (or σ_g, respectively), it has to be explained why a contribution of about 100 MPa from particle strengthening is predicted by eq. (3). A reasonable explanation could be that particle strengthening will diminish if the interparticle spacing becomes equal to the mean free path of moving dislocations. The latter parameter will correspond to the mean diameter of the dislocation cells. Fig. 2 shows that the majority of particles indeed are entangled in cell walls. Measurements of the cell size on micrographs similar to Fig. 2 yielded a value of 0.65 μm irrespective of the grain size.

Since only for microstructures of series II the value of L is considerably smaller than the cell size, particle strengthening is restricted to this series. As pointed out by Hornbogen [9], the two contributions σ_p and σ_d have to be combined as $(\sigma_p^2 + \sigma_d^2)^{1/2}$. The relationship giving the yield stress of microstructures of series II can therefore be written as

$$\sigma_y = \sigma_f + \sigma_s + (\sigma_p^2 + \sigma_d^2)^{1/2} \qquad (6)$$

For $\sigma_f + \sigma_s$ = 137 MPa, the calculated stress values of 330 and 322 MPa, respectively, are in satisfying agreement with the measured ones of 352 and 321 MPa.

Finally, it can be concluded that the yield stress of this microalloyed steel was entirely interpreted by substructure strengthening in addition to a friction stress and solid solution strengthening. The grain size dependence observed of some specimens was due to a correlation between the grain size and the dislocation density.

ACKNOWLEDGEMENTS

The authors would like to express their graditude to the Hoesch Stahl AG for providing the material and to the Deutsche Forschungsgemeinschaft for financial support.

REFERENCES

1 E.O. Hall, Proc. Roy. Soc. B 64 (1951) 747
2 M.F. Ashby, Acta Met. 14 (1966) 679
3 J.P. Naylor, Metall. Trans. 10 A (1979) 861

4 W. Österle, In: Proc. Int. Symp. on The Martensitic Transformation in Science and Technology, Bochum 1989
5 R.K. Ham. Phil. Mag. 6 (1961) 1183
6 P.M. Kelly, A. Jostons, R.G. Blake, and J.G. Napier, Phys. Stat. Sol. A 31 (1975) 771
7 K. Törrönen, Metallography 13 (1980) 329
8 W. Österle, Pract. Met. 21 (1984) 377
9 E. Hornbogen, Z. Metallkde. 68 (1977) 455
10 M. Kehoe and P.M. Kelly, Scripta Met. 4 (1970) 473

MICROSTRUCTURE DEVELOPMENT IN HIGH-SPEED HOT ROLLING OF STEEL RODS

M. Barthel+, P.Klimanek*, K.-E Hensger*

+Institute of Solid State Physics and Electron Microscopy, Academy of Sciences of G.D.R., Weinberg 2, DDR-4050 Halle/Saale
*Section of Materials Technology, Academy of Mining, P.O. Box 47, DDR-9200 Freiberg

ABSTRACT

An analysis of the microstructure changes in high-speed hot-working of ferrite X8CrTi17 by pass-rolling was carried out with the systematic qualitative and quantitative interpretation of TEM-micrographs. Stereological methods of the determination of grain or subgrain sizes and the analysis of the dislocation content enables one to distinguish between the volume fractions of microstructures connected with different restoration processes. The results are in good agreement with the fundamental ideas of the hot-working at high deformation speeds.

INTRODUCTION

High-speed pass rolling of steel rods and wires at high temperatures is a modern hot-working technology with considerable economical importance. An essential characteristic of the process is high adiabatic overheating which often complicates the microstructure formation in the pass-rolled material [1, 2]. Accordingly, in order to be able to understand the mechanical behaviour of a material in pass-rolling, systematic microstructure analysis is necessary. The present paper informs on the results of such an analysis, which was carried out by transmission electron microscopy (TEM) on wires of a ferritic (b.c.c.) stainless steel X8CrTi17 after hot-rolling in a four-step groove sequence of the type round-oval.

EXPERIMENTAL PROCEDURE

The sample material is characterized by the following alloy content: 0.05 wt% C, 15.98 wt% Cr, 0.52 wt% Ti.

The deformation experiments were performed on steel bars having a circular initial cross-section (d=10mm), using a research pass-rolling mill available at the Mining Academy of Freiberg [3]. The deformation rates were in the range $\dot{\varphi} < 2\cdot 10^3\ s^{-1}$. All deformation conditions are summarized in Table 1. The scheme presents the deformation $\Delta\varphi$, the deformation rate $\dot{\varphi}$ and the contact time Δt_r between the wire and the rolls for each rolling pass in dependence on the final rolling speed v_f. Moreover, interpass times Δt_p and the total time $t_{tot} = \Sigma\Delta t_r + \Sigma\Delta t_p$ of the deformation process are given.

Table 1 Deformation conditions

v_f (ms^{-1})	n	1	2	3	4
	$\Delta\varphi$	0.435	0.465	0.420	0.456
5	$\dot{\varphi}$ (s^{-1})	70	80	110	140
	Δt_r (s)	$6.6\cdot10^{-3}$	$6.3\cdot10^{-3}$	$3.9\cdot10^{-3}$	$3.3\cdot10^{-3}$
	Δt_p (s)	0.57	0.47	0.37	
	t_{tot} (s)			1.420	+ 0.021
12	$\dot{\varphi}$ (s^{-1})	160	180	270	350
	Δt_r (s)	$2.7\cdot10^{-3}$	$2.6\cdot10^{-3}$	$1.6\cdot10^{-3}$	$1.3\cdot10^{-3}$
	Δt_p (s)	0.24	0.195	0.155	
	t_{tot} (s)			0.590	+ 0.0082
30	$\dot{\varphi}$ (s^{-1})	400	450	670	870
	Δt_r (s)	$1.1\cdot10^{-3}$	$1.0\cdot10^{-3}$	$6.0\cdot10^{-4}$	$5.0\cdot10^{-4}$
	Δt_p (s)	0.095	0.078	0.062	
	t_{tot} (s)			0.235	+ 0.0032
50	$\dot{\varphi}$ (s^{-1})	670	750	1110	1450
	Δt_r (s)	$7.0\cdot10^{-4}$	$6.0\cdot10^{-4}$	$4.0\cdot10^{-4}$	$3.0\cdot10^{-4}$
	Δt_p (s)	0.057	0.047	0.037	
	t_{tot} (s)			0.141	+ 0.0020

In order to obtain the microstructures corresponding to the deformation states after rolling to the n-th step (n=1,2,3,4) of the groove sequence, the wires were directly introduced into a water bath. That means, that the time for postdynamic or static restoration was relatively small ($\Delta t \leq 0.5$s).

TEM-investigations of the pass-rolled steel X8CrTi17 were performed using a high-voltage electron microscope JEM 1000 at an accelerating voltage of 1000 kV. The imaged microstructures are representative for the core of the pass-rolled wires. In order to characterize them quantitatively, the following quantities were determined by stereological analysis:

(i) pair-correlation functions (PCF) of the dislocation arrangements of individual grains [4, 5],

(ii) the mean grain or subgrain size and the grain or subgrain size distribution, respectively, by means of a modified SALTYKOV method [6,7], and

(iii) the mean volume weighted volume as proposed by GUNDERSEN [8,9] and the recrystallized portion on the basis of the volume weighted particle size distribution.

A discussion of methodical aspects of the analysis will be given [10]; here only some results are presented.

EXPERIMENTAL RESULTS AND DISCUSSION

Qualitative Microstructure Analysis

The steel X8CrTi17 is a material with a high stacking-fault energy, the only observed restoration process of which is dynamic recovery in hot-work at low strain rates ($\dot{\varphi} \leq 20\ s^{-1}$) [7]. Under the conditions of the present work a similar behaviour was observed for strain rates $\dot{\varphi} < 700\ s^{-1}$ (single-pass rolling) or $\dot{\varphi} < 200\ s^{-1}$ (multi-pass rolling), respectively [2]. At higher deformation rates inhomogeneous microstructures were found (Fig. 1a, see also [2,7]) which indicate both recovery and recrystallization.

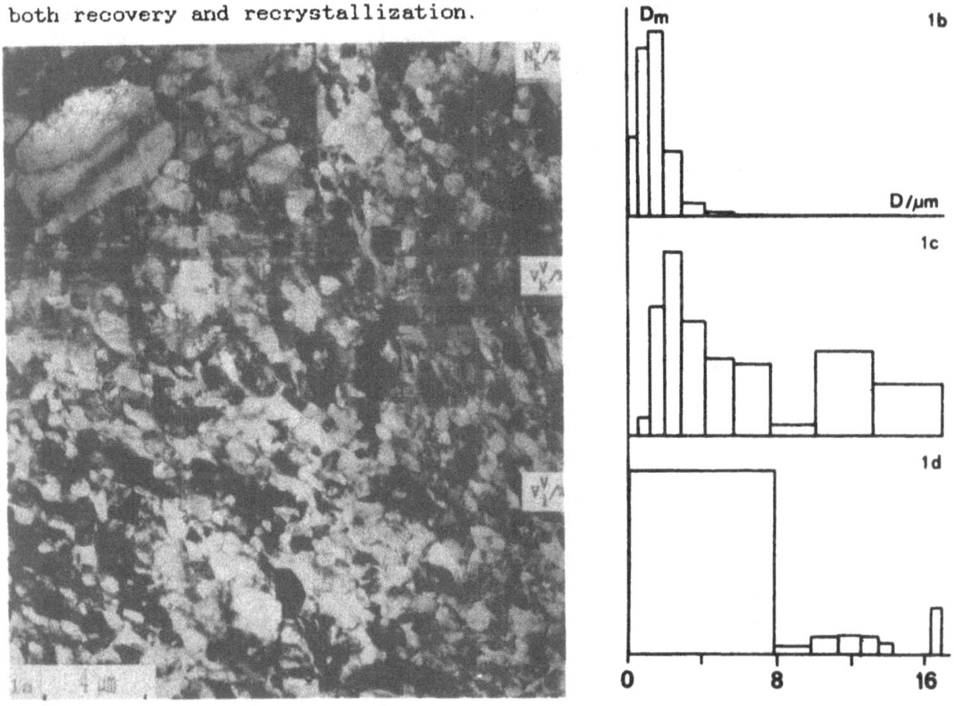

Fig. 1 Stereological grain and subgrain size analysis of ferritic steel X8CrTi17 ($\dot{\varphi}=750 s^{-1}$);
1a) TEM micrograph, 1b) spatial diameter distribution,
1c) volume fractions, 1d) volume weighted volume distribution

The detailed knowlegde of the mean grain and subgrain size, the grain and subgrain size distribution and the volume fractions is necessary to characterize the microstructure development connected with different restoration processes.

Quantitative Microstructure Analysis

For the quantitative characterization one has the option of either obtaining mean values (mean grain size, mean volume) or of determining distributions of these parameters (grain size distribution, volume fractions). If a suitable measuring system is used there is only a slight difference with respect to the work necessary to collect the relevant data. The information provided by mean sizes alone is, however, of relatively low-value and poses considerable difficulties when heterogeneous structures have to be interpreted. In such a case it is thus particularly important to determine size distributions.

In the present paper the microstructure analysis of the mean values was first carried out using Gundersen's method. It does not require an assumption concerning the particle shape. The result is the mean volume-weighted volume $\overline{V^V}$ (Table 2) and the distribution of the volume fractions against the volume V_i^V (Fig. 1d). The grain size region of the investigated microstructure covered by the first class extends because of the usual division of the abszissa for the Gundersen process to approximately 8µm, i. e. the information contained in the TEM micrograph (Fig. 1a) especially about the subgrains on the differentiated structural description could not be used in this interval. The information already gained was therefore supplemented by using the Saltykov method (Fig. 1b and 1c). It should be noted that the region of Fig. 1d under consideration of Saltykov's method again contains two maxima. Peaks in the histograms of the frequency distribution of the spatial grain diameters as a function of the diameter (grain size distribution) N_k^V and of the corresponding volume fractions V_k^V are a hint for the existence of different grain size fractions. For determinating the mean subgrain diameter only values of the first classes have to be used (Tab. 2).

In order to correlate the grain size fractions to the different restoration mechanisms the dislocation content of the individual grains has to be considered. In this connection three microstructure fractions can be distinguished:

- subgrains due to dynamical polygonization with high dislocation density (peak in the grain size distribution N_k^V, Fig. 1b)
- relatively large and irregularly shaped grains with low dislocation density which should be formed by dynamic recrystallization[1], and

---

[1] The approximate portion of the dynamic recrystallized grains can be determined by Gundersen's method. On the basis of the results of the stereological analysis it was assumed that all grains which are not in the first class of the volume weighted volume distribution V_i^V must be a consequence of the dynamic recrystallization.

- small, globular crystallites with very low dislocation content, which are considered to be a consequence of postdynamic (metadynamic) recrystallization (the first peak in the volume fraction V_k^V, Fig. 1c).

The small grains are present especially after medium deformation rates [1,2].

Table 2 Mean volume weighted volume $\overline{V^V}$, corresponding linear size $\sqrt[3]{\overline{V^V}}$, mean subgrain diameter $\bar{d}_{SK}$ and recrystallized portion V_{RX} in dependence on the deformation rate φ

n	1	2	3	4
$\dot{\varphi}(s^{-1})$	670	750	1110	1450
$\overline{V^V}(\mu m^3)$	13.60	385.70	720.50	2865.70
$\sqrt[3]{\overline{V^V}}$ (µm)	2.39	7.28	8.96	14.20
$\bar{d}_{SK}$ (µm)	0.92	1.22	1.38	1.41
V_{RX}(%)		72.40	34.90	88.20

As shown in Tab. 2 the mean volume weighted volume, the mean subgrain diameter and the recrystallized portion increase with increasing deformation rate.

In order to explain the observed microstructure changes at high strain rates the deformation process may formally be interpreted as follows.

Each pass of the groove sequence can be related to a flow curve associated with a ZENER-HOLLOMON parameter $Z_n = \varphi_n \exp(U/kT_n)$ (U - activation energy of hot-working, $T_n = T + \Delta T_n$ - effective deformation temperature; ΔT_n - adiabatic overheating). At a given flow-stress σ_1 and strain φ_1 the further deformation depends on the material behaviour during the interpass time and the change of the parameter Z. Obviously the following situations are possible:

(i) no interpass restoration, dominating strain rate effect (Fig. 2a)
In this case it is $Z_2 > Z_1$ and $\sigma_1 = \sigma_{2.0} < \sigma_2(\varphi_1)$, i.e. by the transition to Z_2 the driving force for dynamic restoration is reduced and the critical value σ_{cr} for dynamic recrystallization cannot be reached. (By interpass restoration the reduction is enlarged yet.)

(ii) interpass restoration, dominating overheating (Fig. 2b)
Now $Z_2 < Z_1$, but though $\sigma_{2,0} < \sigma_2(\varphi) < \sigma_1$ the driving force for dynamic restoration increases in the 2nd pass and can attain the value σ_{cr} for dynamic recrystallization. Moreover, after the 2nd rolling pass postdynamic or metadynamic recrystallization may be possible.

(iii) no interpass restoration, dominating overheating (Fig. 2c)
On such circumstances it is $Z_2 < Z_1$, again. But because of $\sigma_{2,0} = \sigma_1 > \sigma_2(\varphi_1) > \sigma_{cr}$ dynamic recrystallization starts immediately at the beginning of the 2nd rolling step.

Case (ii) is valid for low and medium strain rates and case (iii) is valid for large strain rates. This result is in a good agreement with the behaviour of the rolling forces observed in the present work. The rolling force decreases in the second deformation step [11].

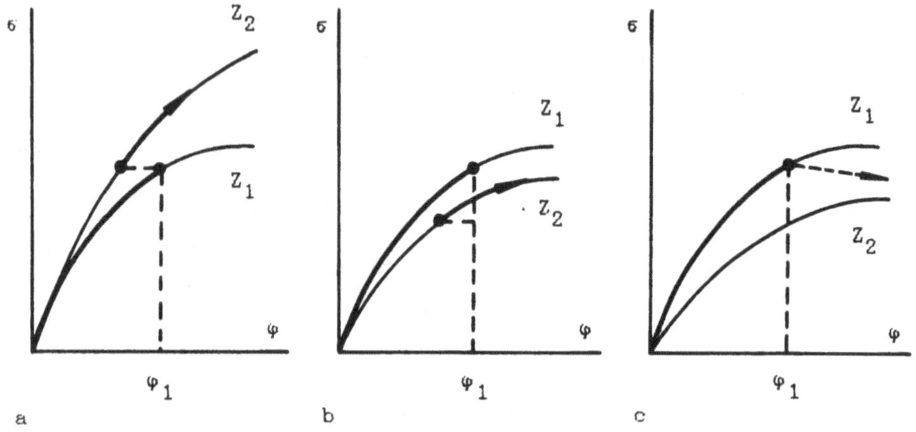

Fig. 2 Dependence of Z, σ and φ on the restoration conditions

REFERENCES

1 A. Schubert, M. Barthel, K.-E. Hensger, P. Klimanek and M. Koch, Proc. 5-th Conference "Metallkundliche Probleme der Werkstoffentwicklung-Festgkeit und Plastizität", Freiberg 1986, 98-102

2 P. Klimanek, K.-E. Hensger, A. Schubert and M. Barthel, Czech. J. Phys. B38 (1988) 373-376

3 A. Hensel, G. Oelstöter and K. Lietzmann, Neue Hütte 27 (1982) 37

4 K.-H. Hanisch, P. Klimanek and D. Stoyan, Cryst. Res. Technol. 20 (1985) 921-930

5 P. Klimanek and D. Stoyan, in the present book

6 M. Barthel and D. Baither, Pract. Met. 26 (1988) 261-271

7 M. Barthel, P. Klimanek and K.-E. Hensger, Czech. J. Phys. B38 (1988) 377-379

8 H.J.G. Gundersen and E.B. Jensen, J. Microsc. (Lond.) 131 (1983) 291-310

9 H.J.G. Gundersen and E.B. Jensen, J. Microsc. (Lond.) 138 (1985) 127-142

10 P. Klimanek and M.Barthel, in prep.

11 A. Schubert, Dissertation A, Bergakademie Freiberg 1985

ANALYTICAL ELECTRON MICROSCOPY IN RESEARCH AND DEVELOPMENT OF MICROALLOYED HIGH-STRENGTH, LOW-ALLOY STEELS

A. Güth, L. Kaun, A. Köthe, J. Richter

Akademie der Wissenschaften der DDR, Zentralinstitut für Festkörperphysik und Werkstofforschung Dresden, DDR-8027 Dresden

ABSTRACT

The capabilities of the analytical electron microscopy in research and development of microalloyed high-strength, low-alloy steels are demonstrated by investigations of size, distribution, morphology, and chemical composition of finely dispersed precipitates. In steels of the types C-Mn-Nb, C-Mn-Nb-V, and C-Mn-Ti-V the dissolution and precipitation of the microalloying elements were studied. The content of microalloying elements in individual complex particles (related to the metallic content) were quantitatively estimated in as-cast condition, after reheating and after thermomechanical treatment.

INTRODUCTION

The analytical electron microscope (AEM) is capable of direct microstructure observations and crystallographic analyses as well as of quantitative chemical analyses of microregions. The high spatial resolution and the simple quantification procedure of the EDX-analysis qualifies it as a powerful tool in steel research and materials science /1-5/.

The advantages of the AEM are widely used in research and development of microalloyed high-strength, low-alloy (HSLA) steels. Superior yield strength, high ductility and toughness and a good weldability are the main features of HSLA steels achieved by microalloying and thermomechanical treatment. The unique combination of properties arises from precipitation of carbides, nitrides and carbonitrides and their effect on recrystallization and/or grain growth in austenite and precipitation hardening in ferrite. The alloying elements added in amounts less than 0.2 wt.% are typically niobium (Nb), vanadium (V), and titanium (Ti) either individually or in combination. The combined addition of microalloying elements leads to the formation of finely dispersed complex carbonitrides. Size, distribution, morphology and chemical composition of these particles depend on the thermal and thermomechanical history (casting, reheating, controlled rolling, cooling, welding ...). This paper gives some examples illustrating the use of AEM for the characterization of small particles in microalloyed HSLA steels.

EXPERIMENTAL PROCEDURE

The analyses of the concast C-Mn-Nb, C-Mn-Nb-V, and C-Mn-Ti-V steels used in this investigation are given in Table 1.

Table 1 Compositions of steels used in present investigation (wt.%)

steel	C	Si	Mn	Nb	V	Ti	Al	N
St 355 Nb	0.12	0.30	1.45	0.05	-	-	0.05	0.015
St 460 NbV	0.10	0.30	1.40	0.04	0.08	-	0.05	0.014
St 355 TiV	0.13	0.36	1.32	-	0.08	0.01	0.03	0.012

The EDX-analysis of small particles (size 2-200 nm) in thin foils is extremely difficult due to the excitation of the surrounding matrix. Therefore, the precipitates must be extracted from the matrix. They were studied by AEM on carbon extraction replicas prepared by standard techniques. The investigations were performed at an operating voltage of 200 kV in a scanning transmission electron microscope JEM-200CX equipped with a KEVEX 7000 energy dispersive X-ray analysis system. The STEM/EDXS-microanalysis of individual particles was possible for a particle size > 10 nm. Smaller precipitates were analyzed as particle agglomerates produced on extraction replicas by a special method.

EDX-spectra were quantitatively interpreted by a standardless thin film analysis using the Ti-K_α, V-K_α, and Nb-L_α or Nb-K_α peaks to estimate the content of the microalloying elements in complex precipitates. The results are given in at.% of the metallic component of the particles.

DISSOLUTION OF CARBONITRIDES

During reheating of microalloyed steels, grain growth is strongly inhibited by fine carbonitrides. For thermomechanical treatment the dissolution behaviour at reheating temperature of these particles is of great interest /6-9/.

The C-Mn-Nb and C-Mn-Nb-V steels were reheated at 1100, 1170 and 1250 °C for 0.5 hr and quenched into water. In Fig. 1 individual undissolved precipitates are shown together with the corresponding EDX-spectra for the Nb-steel. Despite the fact that no Ti was intentionally added to the steel, all particles contain Nb and Ti. When the specimens were held at 1100 °C, the particles are Nb-rich containing less than 10 at.% Ti (Fig. 1b). With increasing temperature the Ti/V ratio increases. A reheating temperature of 1250 °C leads to cuboidal particles with a Ti concentration up to 70 at.% (Fig. 1a). The portion of precipitated Nb decreases from 30 % at 1100 °C to 3-4 % at 1250 °C.

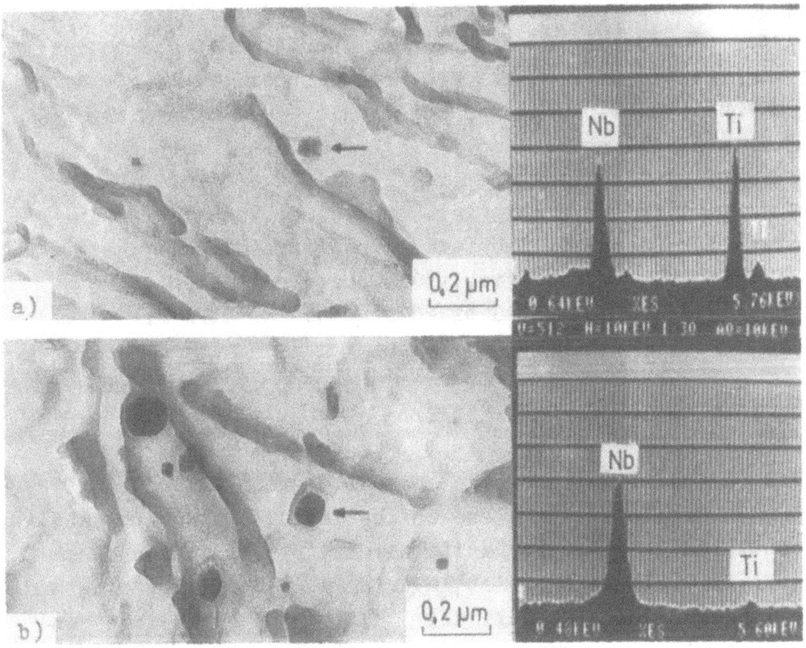

Fig. 1 EDX-microanalysis of individual particles (marked by an arrow) in C-Mn-Nb steel after reheating and quenching
a) reheating temp. = 1250 °C; Ti-rich cuboidal particle
b) reheating temp. = 1100 °C; Nb-rich particle

Similar results were also achieved for the Nb/V steel. The ratio of the microalloying elements in the carbonitrides changes from 93Nb-3Ti-4V at 1100 °C to 30Nb-60Ti-10V at 1250 °C. The Ti-content in the undissolved particles is caused by a small amount of about 20 wt.ppm Ti in the steels introduced by scrap during melting. The effect of trace amounts of Ti on the precipitation and dissolution behaviour has been observed recently also in other HSLA steels /4,10,16/.

PRECIPITATION DURING CONTROLLED ROLLING

Matrix and grain boundary precipitation during controlled rolling of C-Mn-Nb-V steels has been studied by STEM/EDX-analyses /9-14/. The results provide a better understanding of precipitation processes in deformed austenite thus providing a base for optimization of composition and thermomechanical processing of modern HSLA steels. Within interrupted rolling experiments, the steel plates were quenched into water after reheating (1150 °C/ 0.5 hr), pre-rolling, and after each pass of the finish rolling procedure /9,14/.

Large carbonitrides which remained undissolved during reheating had the following contents of microalloying elements: (55-65)Nb-(30-40)Ti-(5-10)V. Thus, undissolved particles could easily determined by their Ti content. After pre-rolling (two passes at 1060-1040 °C and quenching) no strain-induced precipitation could be observed. Nb-rich carbonitrides (50-70Nb)(30-50V) were formed at prior-austenite grain boundaries after pre-rolling, air-cooling to 880 °C and subsequent quenching. During finish rolling (rolling temp. 880-760 °C, five passes) the V content in the precipitates increases with decreasing rolling temperature, and matrix precipitation occurs. Finally, after five passes and air-cooling, the finely dispersed particles (size <10 nm) have an average composition of 20 at.% Nb- 80 at.% V.

PRECIPITATION IN C-Mn-Ti-V STEEL

With regard to the refining of the austenite grain size during recristallization controlled rolling and to the improvement of weldability by means of thermal stable TiN particles, the precipitation behaviour in C-Mn-Ti-V steels is of great importance /9,14,17-19/. The (Ti,V)(C,N) particles were systematically studied in the as-cast condition, after reheating, and subsequent recrystallization controlled rolling with respect to their size, morphology, and Ti/V ratio. In continuously cast steel dendritic and ellipsoidal particles are observed (Fig. 2a).

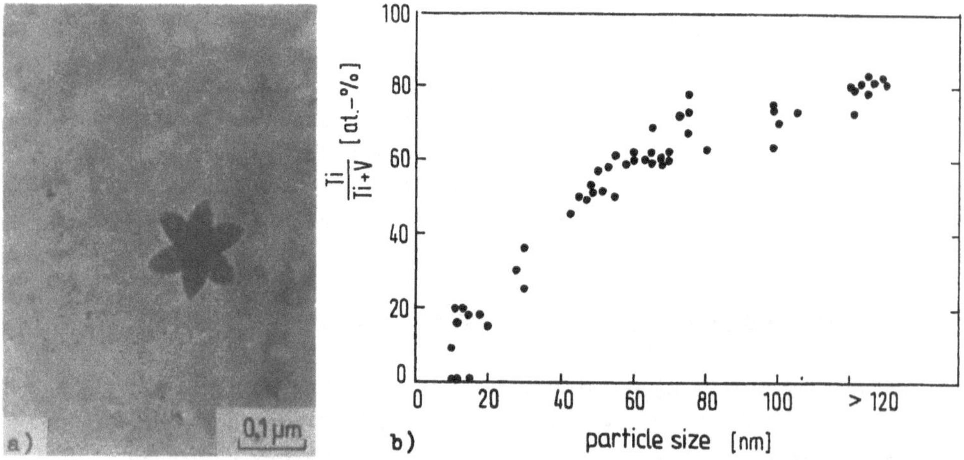

Fig. 2 (Ti,V)(C,N) particles in C-Mn-Ti-V steel (as-cast condition)
a) extraction replica, dendritic and ellipsoidal particles
b) Ti content in particles plotted vs particle size

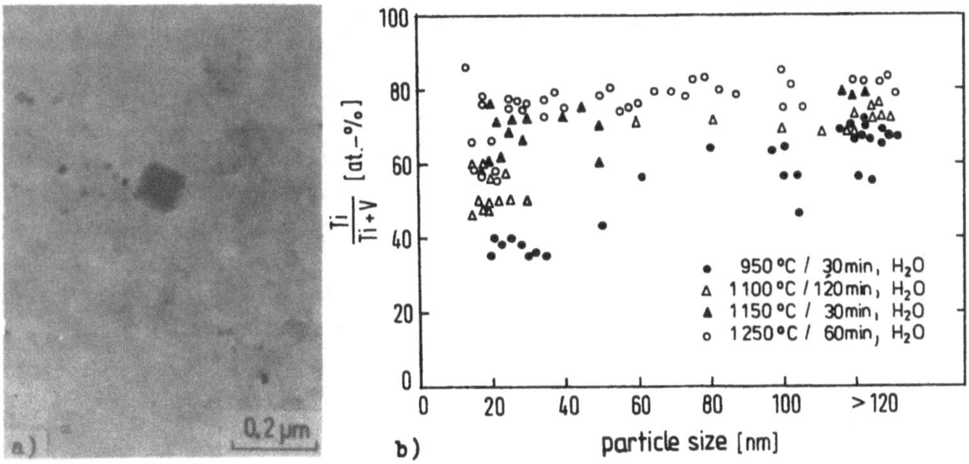

Fig. 3 (Ti,V)(C,N) particles in C-Mn-Ti-V steel after reheating
a) extraction replica, temp. = 1250 °C, cuboidal particles
b) Ti content in particles plotted vs particle size

Fig. 2b shows the composition of mixed $(Ti_yV_{1-y})(C_xN_{1-x})$-particles. Reheating to 950-1250 °C causes a change in morphology and composition. The precipitates get a cuboidal shape, and the Ti/V ratio increases whereas the particle density decreases (Fig. 3). After recrystallization controlled rolling (finish rolling temp. 900 °C) the particle compsition is influenced by the reheating temperature.

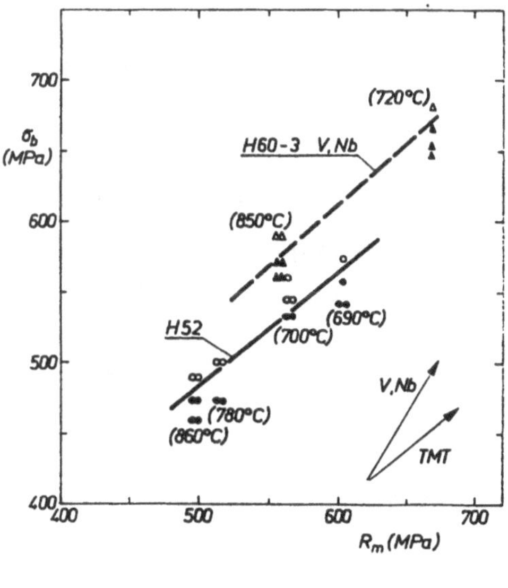

Fig. 4 Fatigue limit and tensile strength of HSLA steels after controlled rolling

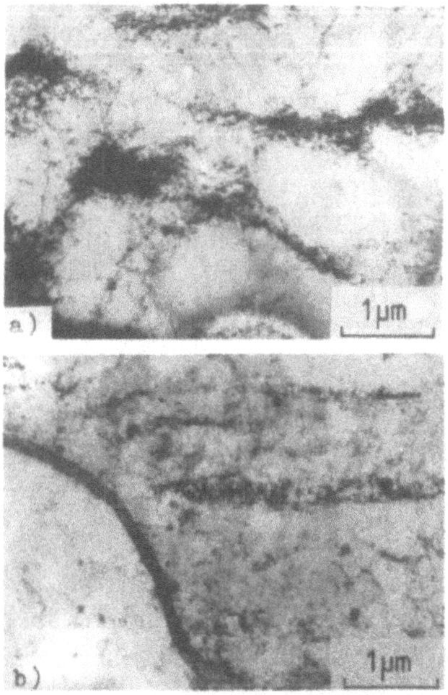

Fig. 5 Dislocation structure after fatigue tests to failure: a) C-Mn steel, b) C-Mn-Nb-V steel

MICROALLOYING AND FATIGUE PROPERTIES

Tensile strength and endurance limit of microalloyed HSLA steels can be enhanced simultaneously by lowering the finish rolling temperature (Fig. 4). This effect is attributed to some stabilization of the dislocation structure by finely dispersed precipitates /20/. In the C-Mn-Nb-V steel the carbonitride particles considerably delay the formation of dipolar dislocation walls or fatigue cells compared with the dislocation arrangement in an unalloyed C-Mn steel (Fig. 5). It is assumed, that the existence of finely dispersed particles is very effective in preventing the rearrangement of dislocations during cyclic deformation as it was also observed in a Cu-1.5%Fe alloy containing precipitates /21/.

REFERENCES

1 Z. Horita, T. Sano and M. Nemoto, ISIJ International 29 (1989)179-90
2 J.R. Michael, K.A. Taylor and J.G. Speer, J. Met. 40 (1988) 8-12
3 M.H. Loretto, physica status solidi (a) 107 (1988) 673-80
4 P. Schwaab, Pract. Met. 23 (1986) 15-25
5 H. Fischmeister, I. Liem and S. Karagöz, Pract. Met. 25 (1988)568-77
6 R. Kleemaier, M. Michelitsch, A. Kneissl and F. Jeglitsch, Pract. Met. 26 (1989) 248-60
7 J.G. Speer and S.S. Hansen, Metall. Trans. A 20A (1989) 25-38
8 A. Güth, L. Kaun, A. Köthe, D. Müller and J. Richter, Scripta Metallurgica 21 (1987) 163-8
9 A. Güth, J. Kunze, L. Kaun and J. Kunath, 3rd Conf. "Thermal Treatment", Karl-Marx-Stadt 1988, TU Karl-Marx-Stadt, Vol. 13/1988, p.77
10 J.R. Michael, Mat. Res. Soc. Symp. Proc. Vol. 62, p. 263-72
11 J.G. Speer, J.R. Michael and S.S. Hansen, Metall. Trans. A 18A (1987) 211-22
12 M.J. Crooks, A.J. Garratt-Reed, J.B. Vander Sande and W.S. Owen, Metall. Trans. A 12A (1981) 1999-2013
13 Xu Wenchong and Sun Fuyu, Acta metallurgica sinica 19 (1983)A483-89
14 L. Kaun, A. Güth, J. Richter and A. Köthe, 12. Tagung "Elektronenmikroskopie", Dresden 1988, Vol. 2, p.335-6
15. L. Kaun, A. Güth, J. Richter, D. Müller and A. Köthe, 16. Metalltagung in der DDR "Thermomechanical Treatment of Steel", Dresden 1987, ZFW-Report Nr. 34, p.18-21
16 J.D. L'Ecuyer, G.L. L'Espérance, M.G. Akben and B. Bacroix, Acta metall. 35 (1987) 1149-58
17 B. Loberg, A. Nordgren, J. Strid and K.E. Easterling, Metall. Trans. A 15A (1984) 33-41
18 J. Strid and K.E. Easterling, Acta metall. 33 (1985) 2057-74
19 Cui Wenxuan and Fang Weipeng, Thermec '88, Tokio 1988, p.122-29
20 M. Schaper, 16. Metalltagung in der DDR "Thermomechanical Treatment of Steel", Dresden 1987, ZFW-Report Nr. 35, Vol. 2, p.395-416
21 M. Kato, A. Nishimura, T. Matsuyama and H. Horie, Materials Science and Engineering A111 (1989) 17-25

IN SITU OBSERVATIONS OF PLASTIC DEFORMATION OF Fe-Si SINGLE CRYSTALS IN THE SEM

J. Prohászka[+x] and Kristyák G. Maróti[x]

[+]Technical University Budapest, Műegyetem rkp 3, Hungary
[x]Department of Metals Technology of the Hungarian Academy of Sciences, Budapest, Műegyetem rkp 3, Hungary

ABSTRACT

Fe-Si single crystals were stretched in a SEM and the slip lines were observed during tension. The active slip planes have been determined at the beginning of the deformation. The primary slip system was [111](112) and the secondary one was [111](123). The observations showed that the fracture surface contains both ductile and glided chisel parts.

INTRODUCTION

The plastic deformation of b.c.c. metal crystals is more complicated than that of f.c.c. or hexagonal ones [1-3]. However, they are very important for the industry because many metals and one of the phases of steels have a b.c.c. structure. Inspite of the large number of experiments carried out on this structure, the results are controversial and there are different opinions about the slip systems of metals having a b.c.c. structure. Most of the authors believe that the slip planes are the most densely packed {101}, {112} or {123} planes. Moreover, according to Pugh, slip can take place on each crystal plane containing one of the <111> slip directions and having the maximum critical resolved shear stress.

In the last decades some results have been published showing unambiguously that under certain conditions Nb single crystals deform in easy-glide. This fact shows that in certain cases slip may take place on defined slip planes in b.c.c. crystals, too. But there are further conditions under which the slip planes are not well-defined (wavy slip) so that their determination is very difficult or quite impossible. However, it is generally agreed that the slip directions are of <111> type.

In the following we would like to present some observations made during the deformation of Fe-Si single crystals in the SEM.

EXPERIMENTAL

The single crystal specimens were cut by spark machining from recrystallized dynamo sheet material having a grain size of some centimetres. The thickness of the sheet was 0.3 mm. After cutting, the samples were annealed in H_2 at 1273 K.

The Fe-Si alloy contained 2.89 per cent silicon and 0.03 per cent carbon. The concentrations of the other impurities remained below those determined by the standards. Only very few orientations of the specimens could be selected since the sheet orientation was very close to the ideal Goss texture. In addition, the choice of the orientation was limited by the shape of the crystallite constituting the base of the samples.

The orientation of the sample is indicated in Fig. 1 in which $\underline{F}$ shows the tensile axis at the beginning of deformation while $\underline{n}$ and $\underline{s}$ are the normals of the front and the side faces. The loading direction is very close to the 001 pole and therefore the resolved shear stresses are almost equal for some of the slip systems. For example, the maximum differences of the angles between $\underline{F}$ and the four <111> slip directions are smaller than 1.5°. There exist more than ten slip systems with Schmid factors greater than 0.45.

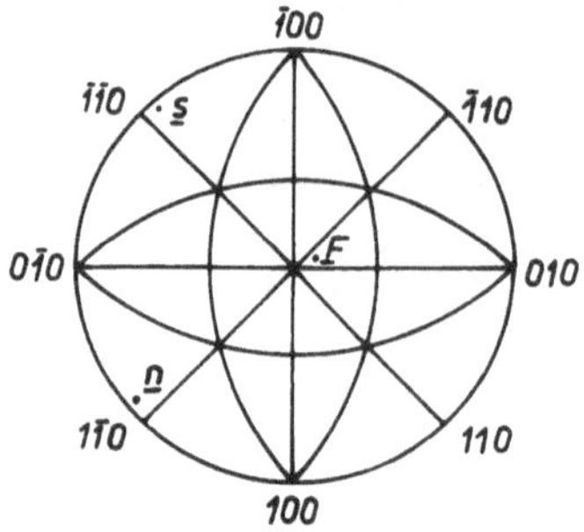

Fig. 1 Orientation relationships

The single crystal samples were stretched in an equipment attached to a Cambridge SEM. The strain rate was 0.00156 min^{-1}. During the tension test the surfaces of the samples were observed and recorded photographically and the force-elongation curves were plotted.

RESULTS

During the tension test we never observed an easy-glide region on the stress-elongation diagram. On samples of the above-mentioned orientation this is improbable. First of all we concentrated upon the slip lines appearing on the front face of the samples in order to get information about the active slip systems and their occasional role in the rupture events.

<u>The systems of slip lines.</u> The deformation was inhomogeneous along the samples. Consequently, the slip lines appeared in different parts of the samples at different times. There were no areas with only one slip line system. Fig. 2 shows an area in which only two parallel slip line systems are imaged. This picture was taken at the beginning of stretching.

The presented slip lines belong to two slip systems, the primary system $[\bar{1}11](1\bar{1}2)$ and the secondary one $[\bar{1}11](1\bar{2}\bar{3})$. The secondary system displaced some of the primary lines. Two such places are marked by arrows. The originally straight lines became curved and wavy after a very small deformation and looked like pencil glide lines. Such a picture is shown in Fig. 3. By further deformation the lines became more curved and their number increased.

Fig. 2 Slip lines of primary and secondary slip systems

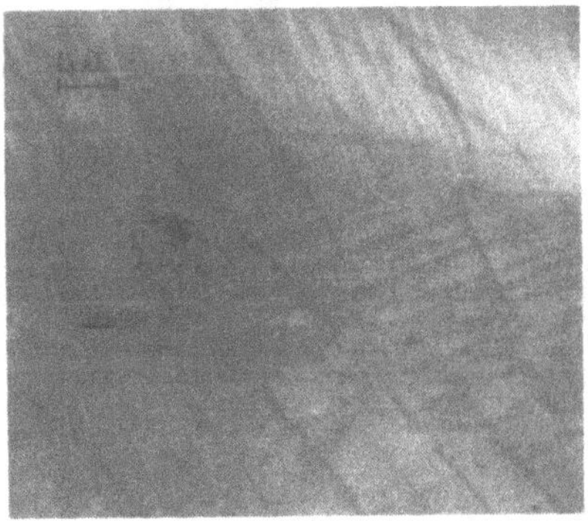

Fig. 3 Wavy slip lines on the front side

<u>Deformation.</u> The curvature of the lines was caused by the very high level of micro-inhomogeneity of the deformation. But in some places it was possible to measure the deformation inside a small area. This made it possible to see the micro-faults on the surfaces.

Fig. 4 shows some etch pits on the side face of a sample. They are imaged in Fig. 5 after some deformation. The elongations of the pits (from left to right) are about 60 %, 40 % and 30 %. These very great differences in the local deformation are characteristic features of this material under these conditions.

<u>Fracture.</u> After a certain time cracks are formed in the places of necking. Afterwards they grow. Fig. 6 shows a starting crack, the tip of which is very sharp. It is interesting that in the vicinity of the crack

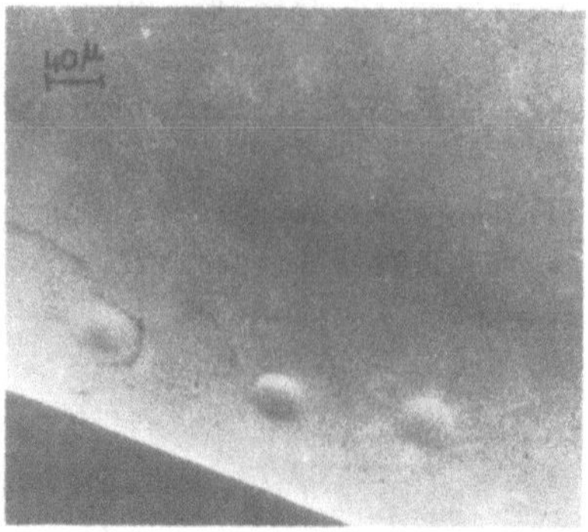

Fig. 4 Etch pits on the lateral side

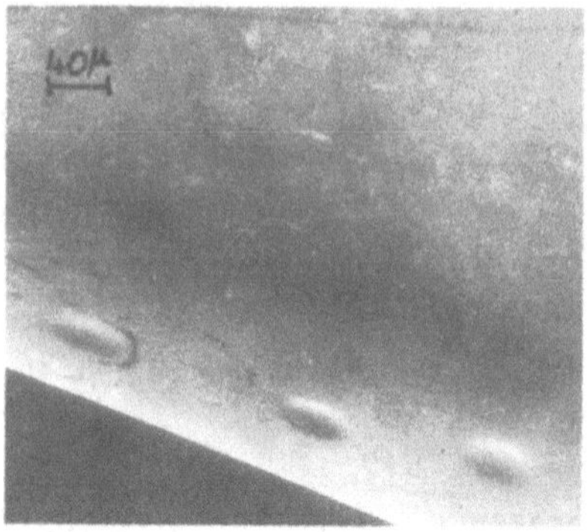

Fig. 5 The same as Fig. 4 after deformation

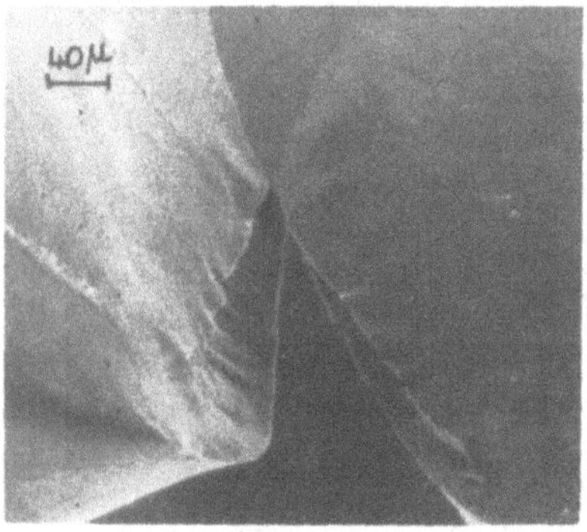

Fig. 6 Appearance of a crack with a sharp tip

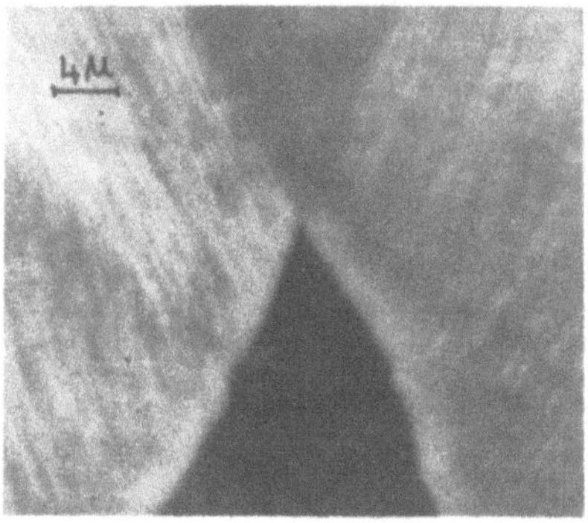

Fig. 7 The same as Fig. 6 at higher magnification

tip three regions are formed with different orientations of the slip lines. They are separated by two well-defined borders (Fig. 7).

Fig. 8 illustrates again the vicinity of the crack and the two bordering lines separating the three areas. The deformation in the two parts being symmetrically arranged with respect to the crack is higher than in that in front of the crack. We can deduce the difference in the deformation between the regions from the two small surface faults which are marked by arrows. As the crack propagates, the border lines also change their positions and one of the faults enters the zone of higher deformation (Fig. 9). During this period the faults being in the symmetrical parts undergo an elongation of 44 % and in the other one only of 28 %.

Near the crack the slip lines become very complicated as demonstrated in the last figure. This determines the structure of the fracture surface. Fig. 10 illustrates part of a fracture surface, which

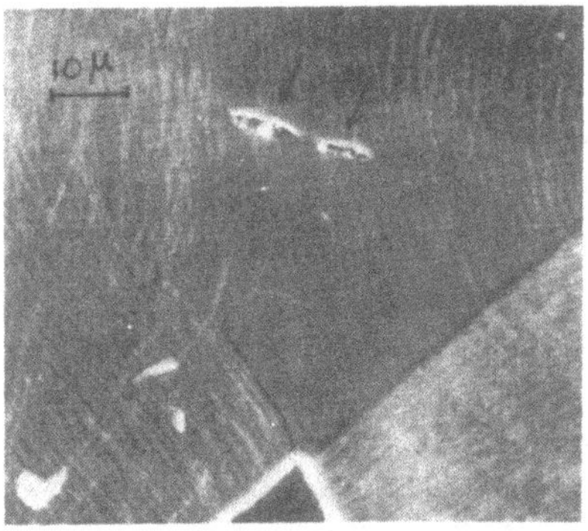

Fig. 8 Zones in the vicinity of a crack

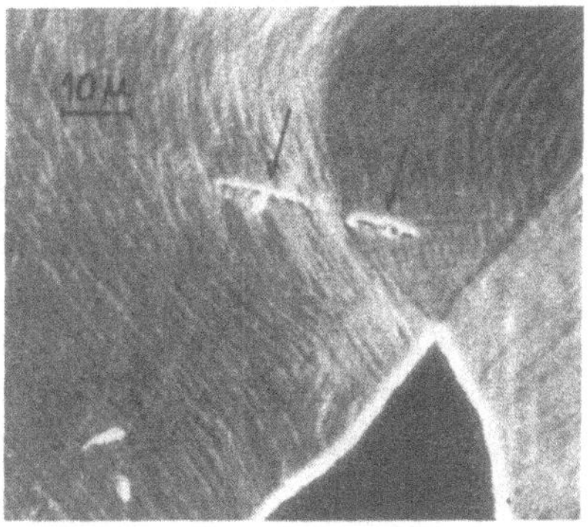

Fig. 9 The same as Fig. 8 after deformation

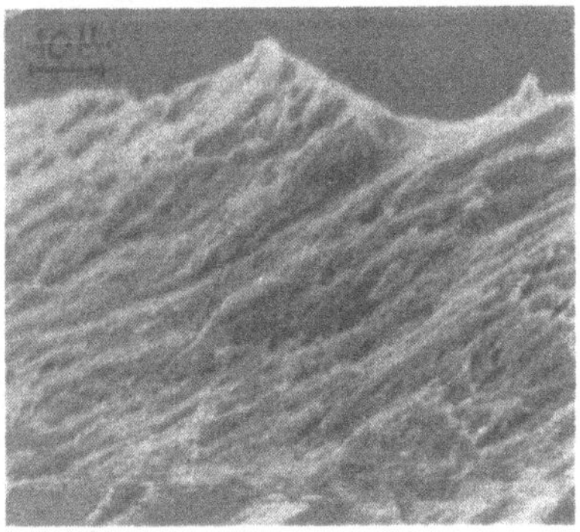

Fig. 10 Ductile fracture surface

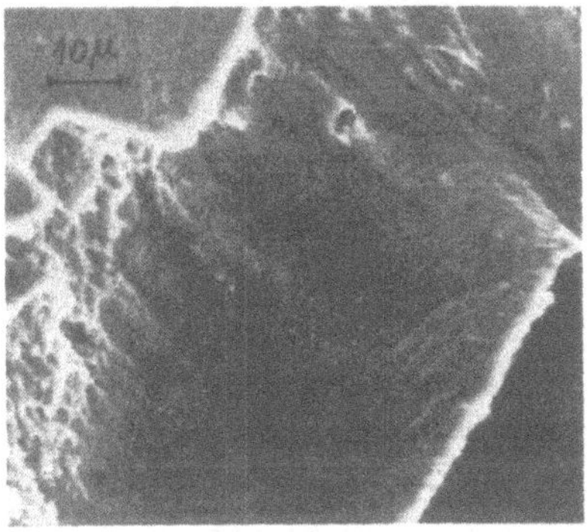

Fig. 11 Knife edge glided surface

corresponds to the ductile fracture of a polycrystalline metal. However, there are other characteristic parts of the rupture surface which to some extent are similar to a cleavage surface. Fig. 11 shows a surface with fine plane parts similar to a cleavage face. This is, however, not a cleavage surface but a $(1\bar{1}2)$ slip plane along which the sample was gliding into two parts.

DISCUSSION

The appearance of straight slip lines at the beginning of deformation unanimously shows that well-defined slip planes play an important role in b.c.c. metal crystals, too.

In our opinion, the reason for the large number of slip systems is that the resolved shear stresses in the different slip system are nearly equal. Only very small changes in the orientation after a small deformation may change the order of the shear stresses in the different slip systems. This is due to the fact that in f.c.c. crystals the maximum value of the orientation factor is the same as in b.c.c. crystals but the minimum values are very different. This value is only 0.28 for f.c.c. crystals but in b.c.c. metals it is not smaller than 0.42. This means that in these crystals the deformation on a single slip system is very improbable and usually takes place as multiple slip.

For example in the present case the orientation factors for the primary and the secondary system are 0.475 and 0.466, respectively. This is similar for a number of slip systems because slip planes belong to a given slip direction and orientation of the tensile axis which include small angles with each other. In our opinion, this is the main reason of the numerous active slip systems in b.c.c. metals.

We should also point out another fact which affects the numerous slip systems. The single crystals may contain inhomogeneities, e.g. an inhomogeneous distribution of impurity atoms. This may result in a changing critical resolved shear stress with another slip system probably becoming active. Small-angle grain boundaries have similar effects. The orientations of the grains differ by 1-2 degrees which also affects the number of active slip systems and leads to multiple slip already at the beginning of deformation.

We may also discuss the fracture. The smooth surface which is imaged in Fig. 11 cannot be a {100} cleavage surface, because in this case it would be perpendicular to the tensile axis. As mentioned earlier, we think that this surface is a $(1\bar{1}2)$ plane. It could be formed because the thickness of the sample was only 0.3 mm. The rupture surface, however, shows the same features as the ductile fracture in a polycrystalline material, except for the knife edge plane.

REFERENCES

1 B. Sestak and A. Seeger, Z. Metallkunde 69 (1978) 194

2 V. Novak, B. Sestak and S. Kadeckova, In: Strength of Metals and Alloys, Vol. 1, P. Haasen, V. Gerold and G. Kostorz (Eds.), Pergamon Press, Oxford 1979, p. 133

3 M. Bucki, V. Novak, Ye. M. Stavitsky, G.E. Burkhanov, and V.M. Kirillova, ibid., p. 145

EFFECT OF MICROALLOYING ON THE STRUCTURE OF RAPIDLY QUENCHED Fe-Ni ALLOYS

I.A. Rusakova, A.V. Suyazov and M.P. Usikov

Central Research Institute of Ferrous Metallurgy, Moscow, U.S.S.R.

Microcrystalline materials rapidly quenched from the melt have intensively been studied recently. These materials show some striking peculiarities compared with conventional products of metallurgy. In particular, they possess an ultrafine grain structure that can be extremely useful in attaining high levels of mechanical and, in some cases, other physical properties.

The purpose of the present study was to examine the structure of Fe-Ni alloys containing 33 wt per cent of Ni with additions of about 0.1 per cent of Ti or V, without and with carbon (approximately 0.1 per cent). The technique used was thin foil transmission electron microscopy. The ribbon specimems, about 50 um thick, were obtained by spinning, the cooling rate being about 10^5 K/s.

After rapid cooling, all alloys had an ultrafine grain structure (Fig. 1). The grain size ranged from 1 to 3 μm and the predominant grain

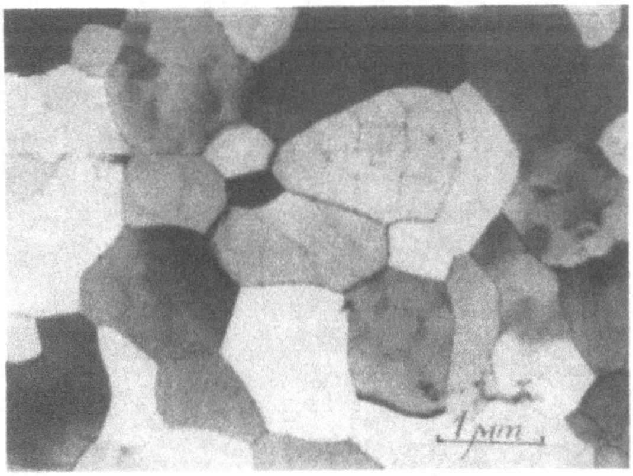

Fig. 1 Alloy Fe-33Ni-0.1Ti (wt.%) quenched from the melt

orientation was such that <001> was almost normal to the ribbon plane, i.e. an axial <001> texture was formed as the result of rapid quenching. The dislocation density in the ribbon specimens was rather low; a cellular structure was observed caused by the segregation of alloying elements and impurities. It should be noted that the segregation cell boundaries sometimes crossed the grain boundaries.

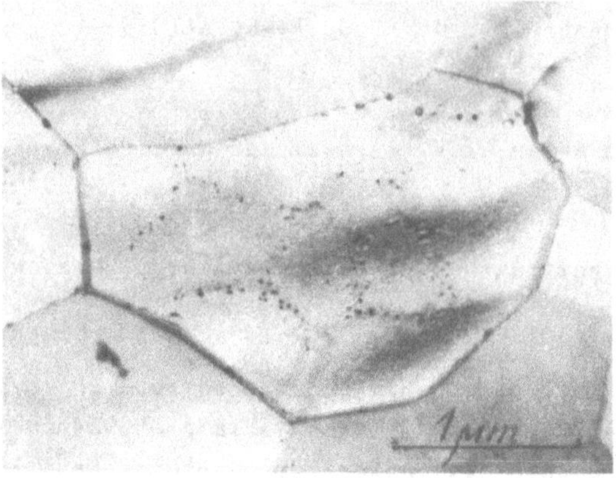

Fig. 2 Alloy Fe-33Ni quenched from the melt

The cell boundaries in the Fe-Ni base alloy and in the carbon-free microalloyed specimens were decorated with prismatic vacancy dislocation loops and voids (Fig. 2). Similar defects appeared at the grain boundaries. Heating the specimens resulted in the annealing of these defects: in Fe-33Ni and Fe-33Ni+Ti alloys these defects disappeared at approximately 900 °C (Fig. 3), but in Fe-33Ni+V alloys they were more stable and were still observed at 900 °C.

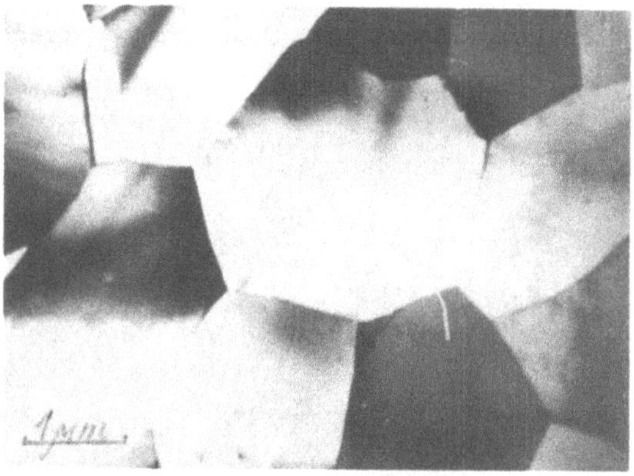

Fig. 3 Alloy Fe-33Ni-0.1Ti quenched from the melt and annealed at 900 °C for 30 min

The loops and voids were not observed in rapidly quenched ribbons of Fe-33Ni+{Ti or V} alloys containing carbon (Fig. 4). When these alloys were annealed at 600 °C or higher temperatures some very fine precipitates had formed at the grain boundaries and at segregation cell boundaries. The second-phase particles were identified as carbonitrides Ti(C,N) and V(C,N) (Fig. 5). The size of such particles in rapidly quenched alloys was several times smaller than that in microalloyed

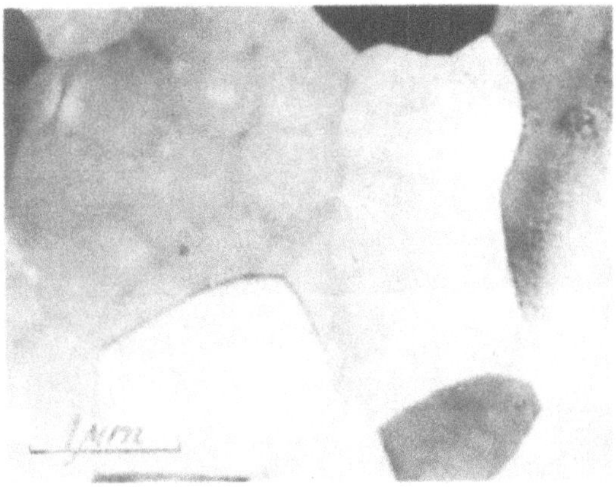

Fig. 4 Alloy Fe-33Ni-0.1Ti-0.1C quenched from the melt

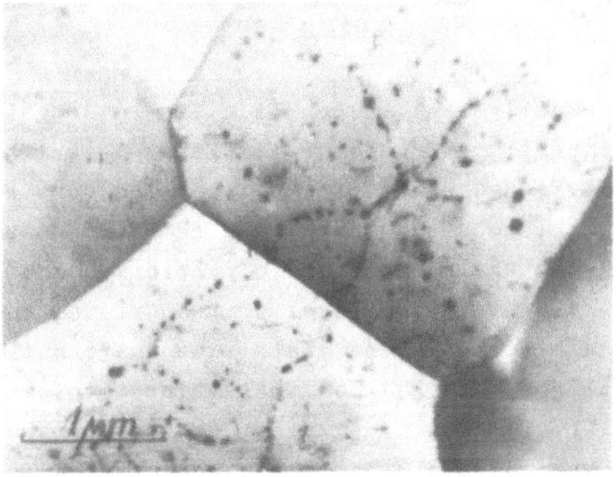

Fig. 5 Alloy Fe-33Ni-0.1Ti-0.1C quenched from the melt and annealed at 1100 °C for 30 min

metals produced by conventional methods: 6nm in rapidly quenched specimens compared with 20nm in conventional specimens of similar composition. The grain structure of rapidly quenched alloys features a high thermal stability: the grains do practically not grow during annealing at temperatures as high as 1100 °C. In Fe-33Ni-0.1C specimens polygonal dislocation tangles appeared at cell boundaries (Fig. 6).

The results of our study show that an excessive amount of loops and voids is formed from vacancies in the rapidly quenched alloys. In alloys containing carbon, the carbon atoms interact with vacancies to form complexes which retard the vacancy coalescence which leads to dislocation loops and voids. Quenching defects of this type become the preferential location for the nucleation of carbonitride particles during annealing, and the large number of nucleation centres predetermines the fine dispersion of these particles.

The precipitates of impurities and carbonitride particles in rapidly quenched specimens are rather stable, and therefore their tendency to

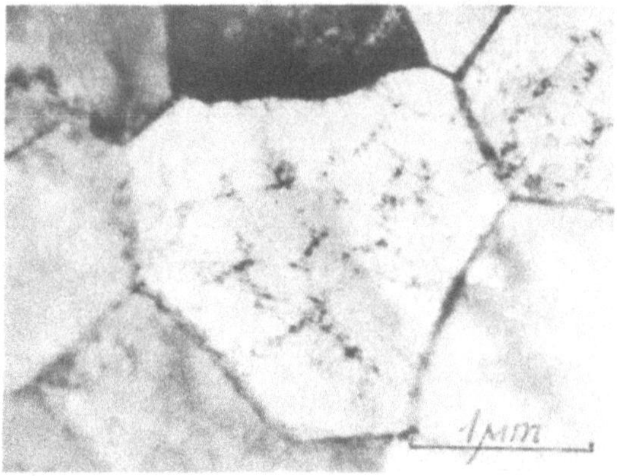

Fig. 6 Alloy Fe-33Ni quenched from the melt and annealed at 600 °C for 30 min

coalesce is small. It is exactly that factor (i.e. the dispersion stability at high annealing temperatures) which essentially determines the critical feature of rapidly quenched specimens, viz. the very slow grain growth during annealing. It is obvious that there are also other reasons for the high stability of the grain structure of rapidly quenched alloys, such as: (1) the relatively low level of accumulated energy compared with the strained state, (2) the grain size uniformity maintained up to high temperatures, and (3) the small ribbon thickness and prevailing axial texture (i.e. the maximum possible grain size is that of the ribbon thickness, whereas a similar grain orientation and size suppress the growth stimulus).

ON THE ORIGIN OF HARDENING IN AGED FeCr ALLOYS: RELATION BETWEEN UNMIXING AND MECHANICAL PROPERTIES

A. Triki, F. Bley, Y. Brechet, and F. Louchet

Laboratoire de Thermodynamique et Physico-Chimie Metallurgiques, E.N.S.E.E.G, BP. 75, 38402 Saint Martin d'Heres, France

ABSTRACT

FeCr alloys containing 20%, 35% and 50 at% Cr have been aged up to 300 h at 500 °C. A decomposition into two Fe rich and Cr rich coherent phases is followed by low-angle neutron scattering. It is associated with a high increase of the conventional yield stress $\sigma_{0.2}$ and a strong modification of the dislocation microstructures. The dislocations become straighter and straighter as the Cr concentration and annealing time are increased. Twinning can appear for high Cr concentrations and long ageing times. This behaviour is discussed in terms of classical precipitation hardening (misfit, modulus, step formation) in the case of the 20% Cr alloy, and of disordering (i.e. re-mixing of phases) which is shown to be predominant in the case of the spinodally decomposed 35% and 50% Cr alloys.

INTRODUCTION

The phase diagram of the FeCr system exhibits a miscibility gap at low temperatures. Therefore, FeCr alloys with a Cr concentration between 15 and 80 at% decompose into a Fe rich and a Cr rich phase. Depending on the volume fraction of phases the topology of phases is equivalent either to small precipitates dispersed in the matrix, or to a percolated structure, in which both phases coexist. This makes this system particularly attractive for studying the effect of topology on structural hardening.

EXPERIMENTAL PROCEDURE

The three investigated alloys contain 20, 35 and 50 at% Cr, respectively. After homogenization and quenching, they have been heat-treated at 500 °C for times between 0 and 500 hours. The evolution of the decomposition has been followed by small-angle neutron scattering at the ILL, in order to determine both the decomposition wave length and the amplitude of concentration fluctuations through the integrated intensity. Compressive tests at 1.3×10^{-4} s^{-1} and at room temperature have been performed in an Instron machine in order to determine the conventional yield stress $\sigma_{0.2}$. Deformed specimens have been observed by TEM at 200 kV. In situ deformation tests have been conducted in a shearing straining stage operating in the JEOL 200CX microscope.

EXPERIMENTAL RESULTS

Mechanical Properties

Fig. 1 shows the increase of the conventional yield stress $\sigma_{0.2}$ as a function of the ageing time t for the three alloys, compared with $\sigma_{0.2}$ results obtained on a Fe 48% Cr by Marcinkowski [1]. The increase occurs for ageing times of some tens of hours, which is also the time required to observe a well-developed peak in the small-angle neutron scattering results, which will be published elsewhere.

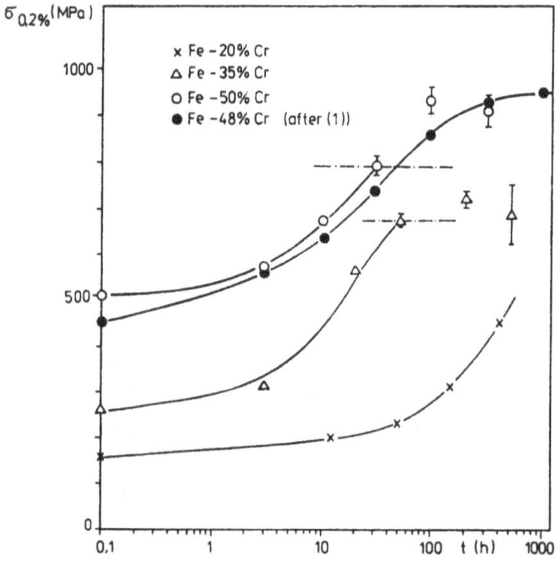

Fig. 1 Conventional yield stress $\sigma_{0.2}$ vs. ageing time for Fe 20%, 35% and 50% Cr, compared to Fe-48% Cr (after Marcikowski [1])

Dislocation Microstructures

The evolution of the observed microstructures is schematically shown in Fig. 2. When the Cr concentration and (or) ageing time are increased, the cell structures observed in the as-quenched 20% Cr alloy (Fig. 3a) transform into a typical landscape of a low-temperature BCC metal with straight screw dislocations (Fig. 3b), which can gather into bands of localized deformation (Fig. 3c). In extreme cases of high Cr concentrations or (and) long ageing times, twinning can occur (Fig. 3d).

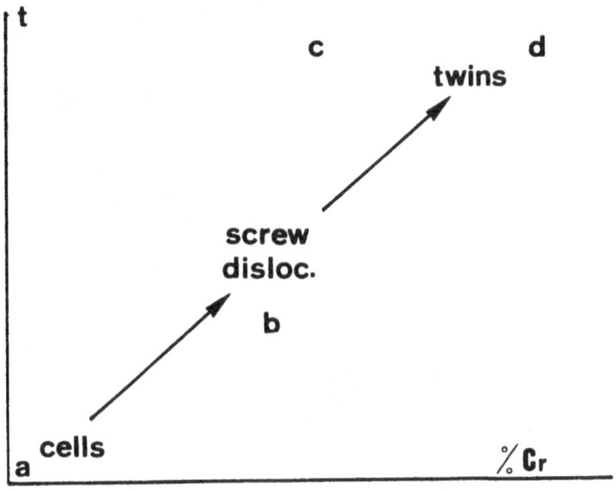

Fig. 2 Schematic evolution of microstructures vs. Cr concentration and ageing time t. The regions labelled a, b, c, d correspond to the four microstructures shown in Fig. 3a, b, c, d

In situ Experiments

The tendency mentioned above that dislocations become straighter and straighter as the Cr content and (or) the ageing time are increased is confirmed by in situ experiments. Fig. 4 shows an as-quenched Fe-50% Cr alloy where straight screw dislocations move just like in a BCC metal at low temperatures. The wavy motion of screws and the wavy shape of slip lines reveal a permanent cross-slip, which often results in the generation of open and closed loops.

The same tendency is found in the 35% Cr alloy aged for 50 h (Fig. 5), whereas the same alloy aged for 500 h exhibits high strain localization and extensive twinning (Fig. 6).

DISCUSSION AND CONCLUSION

Hardening

We shall discuss separately the hardening due to isolated precipitates obtained by a nucleation-growth mechanism (20% Cr alloy) and that due to the percolated structure obtained by spinodal decomposition (35% Cr and 50% Cr alloys). In effect, in the first case, we are close to the miscibility curve, and the hardening contribution of the chemical energy

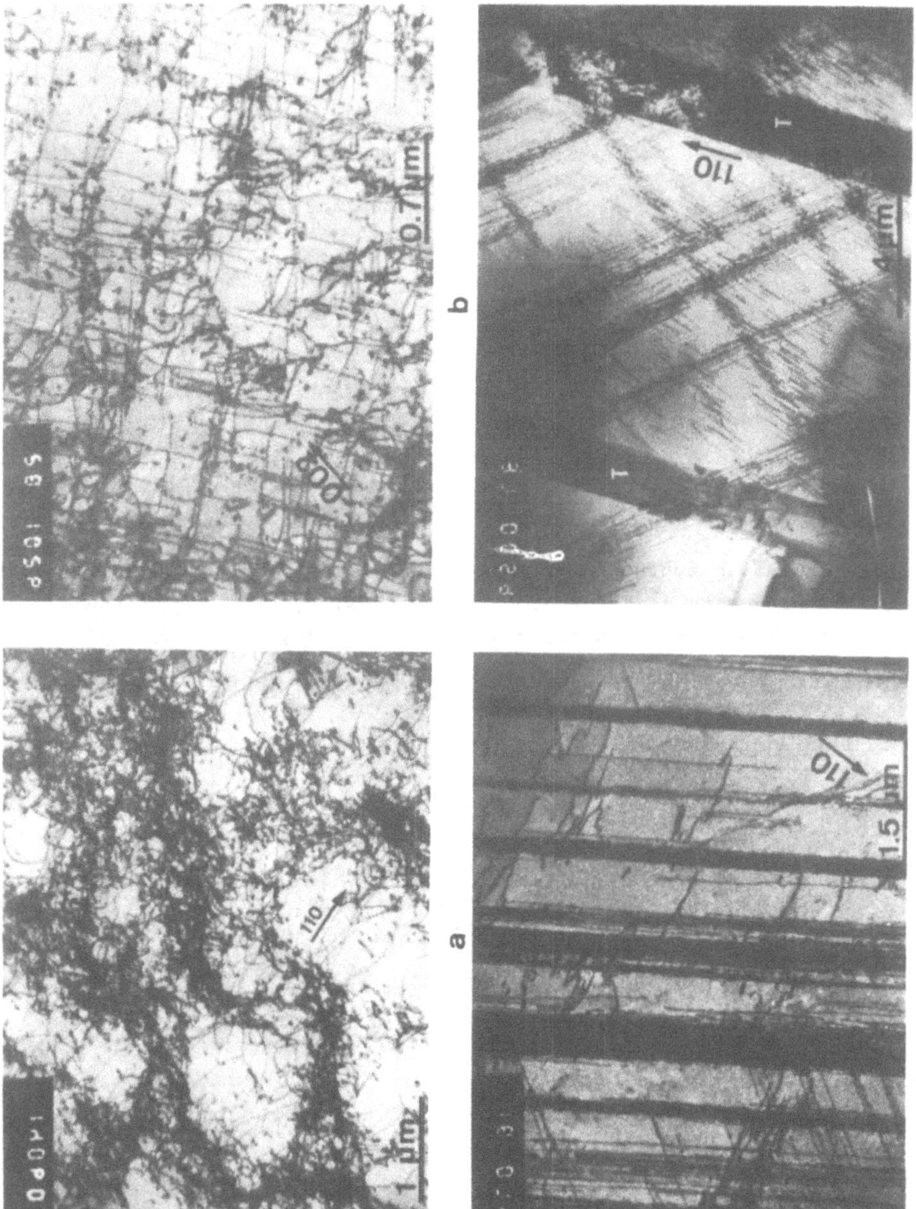

Fig. 3 Microstructure of strained specimens:
a) Fe20% Cr as-quenched. b) Fe35% Cr aged 12 h 500°C
c) Fe35% Cr aged 500 h 500°C. d) Fe50% Cr aged 500 h 500°C

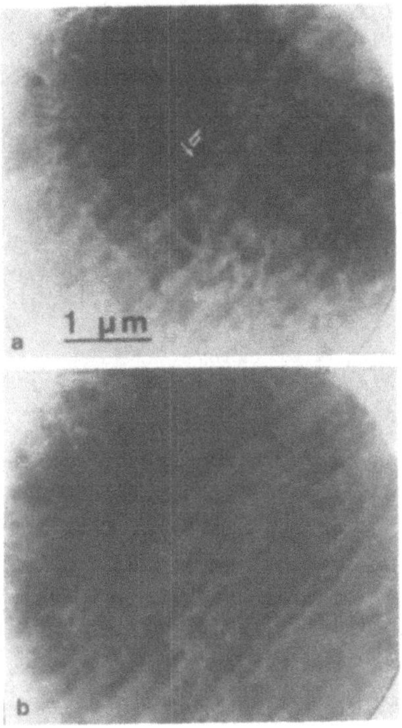

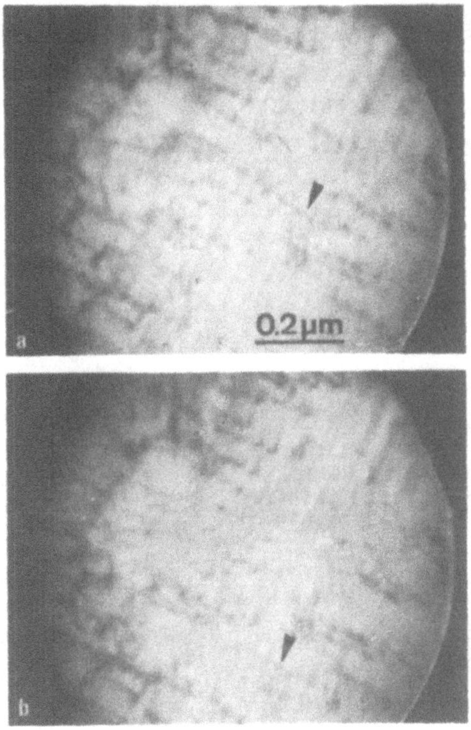

Fig. 4 In situ straining of as-quenched Fe50%Cr
a) movement of long screw dislocations
b) screw dislocations and wavy slip traces (st)

Fig. 5 In situ straining of Fe35%Cr aged for 50 h at 500 °C. Jump of a screw dislocation (arrow) between a) (t = 0) and b) (t = 300 ms)

for unmixing is small compared to the contribution of classical effects as modulus, coherency and surface energy effects, which have been extensively investigated in the literature as "precipitation hardening" [2]. In opposition to that, the case of a continuous precipitation percolating a continuous matrix has to be treated in quite a different way, for two main reasons:

(i) It is unreasonable to assume that the precipitated phase is made of spherical particles.

(ii) In the phase diagram, since we are now deep below the miscibility curve, the energy difference ΔG between the solid solution and the unmixed structure is much larger than it should be close to the miscibility curve.

In the following, we shall define the hardening $\Delta\sigma$ as the difference between the conventional yield stresses of the aged and of the as-quenched specimens.

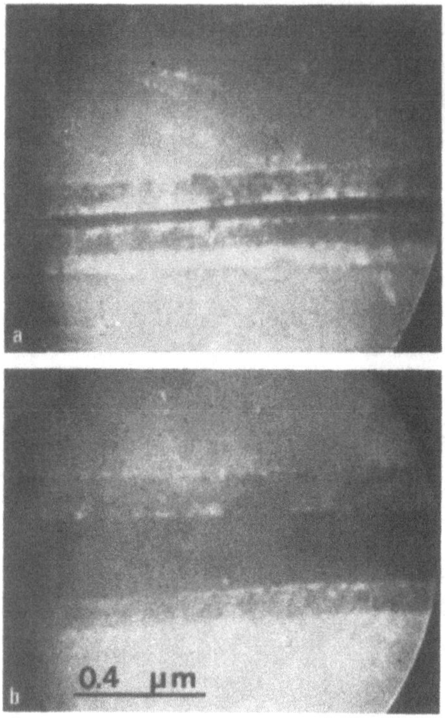

Fig. 6 Twin growth (Δt = 10 min between a) and b))

<u>20% Cr alloy.</u> The different possible contributions to the hardening have been computed from [2, 3] and the numerical results are summarized in the following table, where a Taylor factor of 2.5 has been taken. The computed hardening is in reasonable agreement with the experimental one.

<u>35% and 50% Cr alloys.</u>
Since the pioneering work of J.W. Cahn [4], most papers have tried to account for hardening in spinodally decomposed structures by internal stresses arising from lattice distortions [5, 6], or by the interfacial energy due to step formation [7]. As mentioned above, however, the energy of unmixing is no longer negligible. When numerous dislocations shear the unmixed structure, they reverse the system into the initial solid solution.

Ageing time (h)	12	50	150	405
Misfit (MPa) [2]	10	12	17	24
Modulus (MPa) [3]			50	70
Step formation (MPa) [2]	11	8.5	7.5	7
$\Delta\sigma_{theor}$ (MPa) (Taylor factor N 2.5)	52.5	51.2	186	252
$\Delta\sigma_{exp}$ (MPa)	40	60	145	290

This disordering requires an amount of energy which has to be provided by the work of the applied stress. The strain necessary to disorder the system can completely be estimated from Fig. 7 to be about 1. Equating the work of the applied stress (Peach Koehler force) to the increase in energy ΔG of a volume of matter $2(\lambda/2)^3$, one gets:

$\Delta\sigma\, b(\lambda/2)$	$2(\lambda/2)$	$\lambda/(2b)$	$\simeq$	2.5	$\Delta G\ 2(\lambda/2)^3$
Peach Koehler force	dislocation path	number of dislocations		average Taylor factor	

which gives:

$$\Delta\sigma \cong 2.5\,\Delta G \qquad (1)$$

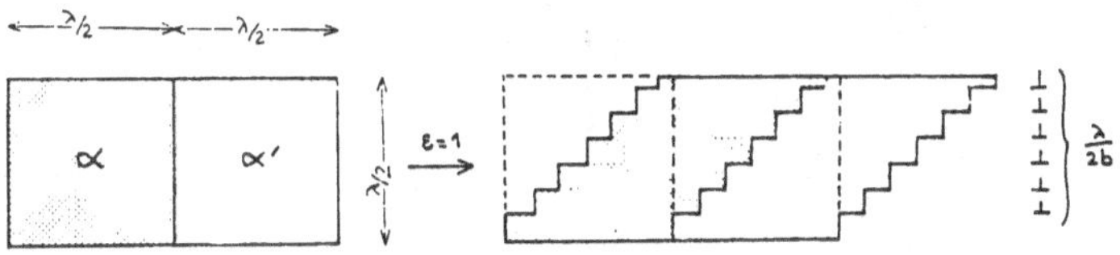

Fig. 7 Schematic shear (ε = 1) of two neighbouring domains of α and α' phases by $\gamma/2b$ dislocations travelling a distance γ

After a Taylor expansion of the free energy G around the nominal concentration c_o, the hardening can be written:

$$\Delta\sigma \cong -2.5 \quad G''(c_o)(\Delta c)^2 \qquad (2)$$

where Δc is the concentration amplitude of the unmixed structure. Since $(\Delta c)^2$ is proportional to the integrated intensity of the small-angle scattering experiments I_{int}, one expects:

$$\Delta\sigma \cong (-2.5/K)\; G''(c_o)\; I_{int} \qquad (3)$$

K can be computed from the scattering factors of Fe and Cr, and $G''(c_o)$ can be estimated from a model of regular solutions corrected by a magnetic contribution:

$$G(c) \cong c(1-c)\; L^{bcc}_{Cr,Fe} + kT\,[c\ln c+(1-c)\ln(1-c)] + kT\,\ln(\beta+1)f(T/T_c) \qquad (4)$$

where f, β and L are defined in [8] and where T is the room temperature at which the straining tests are performed. The theoretical hardening is plotted vs. the integrated intensity in Fig. 8 for the 35% and the 50% alloys and compared to the experiment. It can be noticed that, considering that there is no fitting parameter and that this model is actually rather rough, the agreement between theory and experiment is satisfactory. However, the extrapolation of the experimental curves intersects the axis at a negative value. This could be explained by the fact that in the model the reference state used to calculate the relation between $\Delta\sigma$ and $(\Delta c)^2$ is a completely disordered state, whereas the reference state used to determine the experimental value is the as-quenched state, which is short-range ordered, or even perhaps slightly unmixed. The negative value of $\Delta\sigma$ at $(\Delta c)^2 = 0$ could then be attributed to the contribution of the actual initial state to the hardening. At high integrated inten-

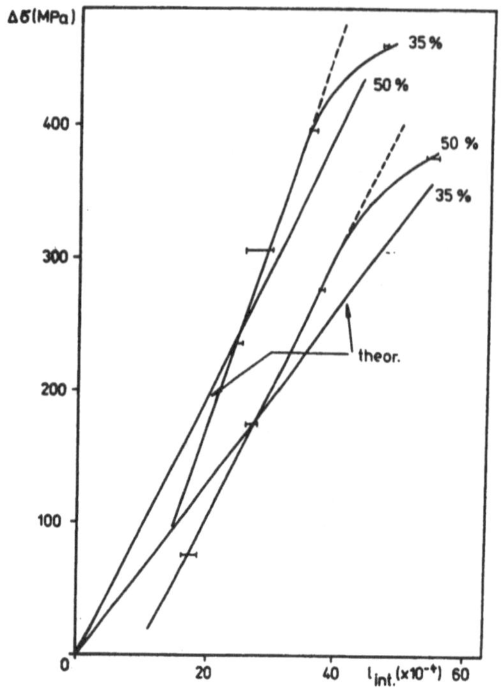

Fig. 8 Experimental and theoretical hardening $\Delta\sigma$ vs. the integrated intensity for Fe35%Cr and Fe50%Cr

sities, the experimental curves depart from the straight lines, which can be attributed to a failure in the Taylor development as unmixing proceeds.

Stiffness of Screw Dislocations and Strain Localization

The strain localization observed as unmixing proceeds can easily be explained on the basis of the preceding mechanism: as precipitates are sheared by dislocations, they become easier and easier to shear. This is essentially the same mechanism as for classical structural-hardened materials as Ni base superalloys or Al base alloys. It is also the same explanation as for channel formation in irradiated metals, where loops are removed by mobile dislocations.

The explanation of the increased stiffness of screw dislocations as unmixing proceeds is perhaps less straightforward, since the more heterogeneous the structure is, the straighter are the screws. In pure BCC metals, a similar increased stiffness is observed as the temperature is decreased. The reason is that, in order to keep the strain rate constant, the stress must be increased to allow dislocations to overcome the Peierls potential. The kinks travelling along screws pile up on forest obstacles, giving rise to cusps the radius of curvature of which is of the order of Gb/σ , which decreases as the temperature decreases. The screws appear then to be longer [9]. In the present case, the origin of the increase of stress is quite different, but the increased stiffness of screws could be explained in a similar way. Moreover, the heterogeneities of concentration have a wavelength of about 50b, much too small to have any observable effect on the apparent stiffness of screws.

REFERENCES

1 M.J. Marcinkowski, R.M. Fischer and A. Szirmae, Trans. Am. Inst. Min. Engs. 30 (1964) 676
2 L.M. Brown and R.K. Ham, In: Strengthening methods in crystals, A. Kelly and R.B. Nicholson (Eds.), 1971, p. 9
3 E. Nembach, Phys. Stat. Sol. (a) 87 (1983) 571
4 J.W. Cahn, Acta Metall. 11 (1963) 1275
5 M. Kato, T. Mori and L.H. Schwartz, Acta Metall. 28 (1980) 285
6 R. Wagner, Czech. J. Phys. B31 (1981)
7 Y. Hanai, T. Miyazaki and H. Mori, J. Mater. Sci. 14 (1979) 599
8 J.O. Anderson and B. Sundman, Calphad 11, 1 (1987) 83
9 F. Louchet, L.P. Kubin and D. Vesely, Phil. Mag. A39, 4 (1979) 433

MICROSTRUCTURE CHANGES IN HIGH-SPEED COMPRESSION TESTS OF METALLIC MATERIALS

U. Trinks, K.-E.Hensger, P. Klimanek, U. Martin, K. Cyrener

Department of Materials Science, Mining Academy of Freiberg, Freiberg 9200, G.D.R.

ABSTRACT

Transmission electron microscopy (TEM) was used to study the microstructure changes occurring in the hot-deformation of high-alloy stainless steel X8CrTi17 (ferrite) and X8CrNiTi18.10 (austenite).

Plastic deformation was carried out by means of a pneumatic-mechanical plastometer, which allows one to perform conventional and so-called groove compression tests. The deformation rates are in the range of $10^3 s^{-1} < \dot{\varphi} < 5*10^3 s^{-1}$. Thus, the present study may contribute to the simulation of pass rolling.

The investigations indicate that at higher deformation rates modified substructures occur due to adiabatic overheating, the generation of excess vacancies and extremely low deformation times.

INTRODUCTION

In future, high-strain rate deformation becomes an increasingly important process, especially for the production of steel rods and wires. Therefore, it is necessary to investigate the mechanical behaviour (flow stresses) and the microstructure development under various deformation conditions, but it is very difficult and expensive to do this in wire and rod mills.

EXPERIMENTAL PROCEDURES

The chemical analyses of the steels used are shown in Tab. 1.

Table 1 Chemical composition of the materials

	%C	%Si	%Mn	%P	%S	%Cr	%Cu	%Ni	%Ti
X8CrTi17	0.03	0.48	0.52	/	0.0092	18.10	0.13	0.19	0.74
X8CrNiTi18.10	0.086	0.61	1.42	/	0.02	17.83	0.32	9.81	0.54

The initial grain sizes of the ferrite and the austenite were 80µm and 5µm, respectively.

The materials were plastically deformed in high-speed compression tests by means of a plastometer which allows both the registration of the flow curves and microstructure investigations of differently hot-worked specimens. For this purpose the machine is equipped with a sample ejector, which allows quenching after a minimum holding time of about half a second or, of course, after other defined holding times. The design of the plastometer is shown in Fig.1.

The plastometer allows one to carry out compression tests in two modes:

- conventional compression tests on cylindrical specimens and
- groove compression tests, which simulate the 1st step of pass rolling of rods or wires, e.g. in a round-oval sequence [1,2].

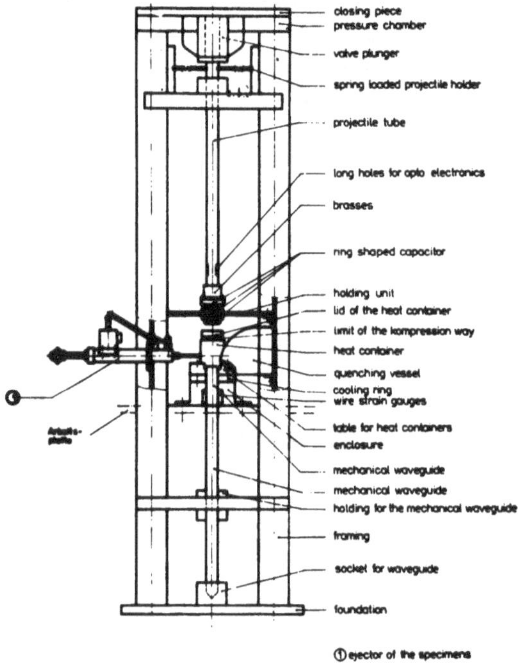

Fig.1 Outline of the high-speed compression test plastometer

Fig. 2 presents the schemes of both procedures together with the experimentally measured flow curves.

The groove compression test was introduced by KLIMANEK, SCHUBERT and HENSGER [1,2] and carried out at low strain rates ($\dot{\varphi}<20s^{-1}$) to simulate the first step of pass rolling of steel rods. It was demonstrated by quantitative texture analysis (QTA) and microstructure investigations that this compression test is suitable for simulatinq the one-step pass rolling [1,3,4].

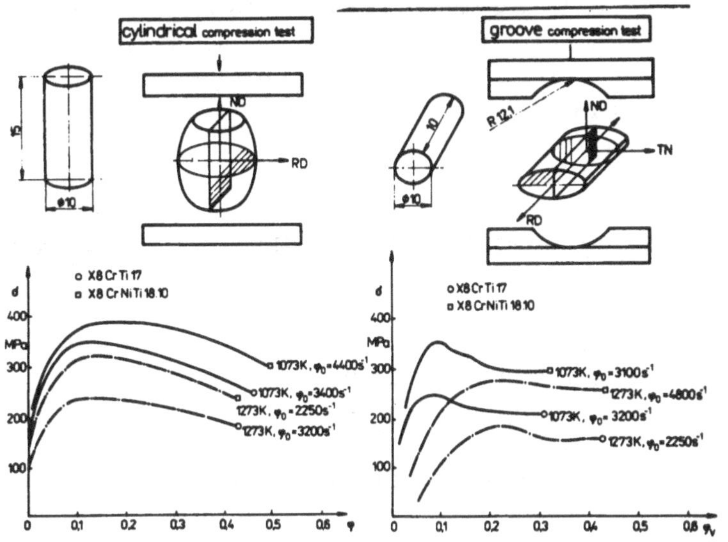

Fig.2 The principle of the compression tests and stress strain curves of X8CrTi17 and X8CrNiTi18.10 at elevated strain rates and 1073K

In order to calculate stress-strain curves of the groove compression test, the method of visioplasticity [5] was used to investigate the respective local flow and deformation state. As usual, the flow curves were formally corrected for adiabatic overheating, which was directly measured for conditions of $\dot{\varphi}=4000s^{-1}$, T=1073K. For a deformation of $\varphi=0.7$ the temperature increase was ΔT = 80 K for the austenitic steel and ΔT = 65 K for the ferritic material.

The above mentioned strain rates are valid for the initial stage of the compression tests and rapidly drop with increasing deformation. For this reason, it is also difficult to use corrected stress strain curves for predicting the structure.

The results of the microstructure development of the high speed compression tests were compared with those of lower strain rates. The deformation conditions are illustrated in Tab. 2.

Table 2 Deformation conditions

low strain rates [1] (tensile or compression testing machine and pass rolling)	high strain rates (high speed compression test plastometer)
$0.007s^{-1} < \varphi < 670s^{-1}$	$2500s^{-1} < \dot{\varphi} < 4000s^{-1}$
1073K < T < 1473K	1073K < T < 1273K

$0 < \varphi < 0.8$

The TEM investigations were performed in the microscopes JEOL 200 CX (Central Institute of Solid State Physics and Materials Science Dresden) and JEM 200 or PHILIPS CM 30 (Mining Academy of Freiberg). The investigated foils were taken from the centre of the groove compressed specimens parallel to the transverse direction.

RESULTS OF THE MICROSTRUCTURE INVESTIGATIONS AT HIGH STRAIN RATES

Microstructure Development of X8CrTi17 (b.c.c.)

The initial microstructure of the steel X8CrTi17 is characterized by a low dislocation density.
After deformation at 1073K and $\dot{\varphi}=4000s^{-1}$ a polygonized structure is identified with sharp subboundaries (Fig.3). The subgrain sizes are 0.5 to 3µm. The total dislocation density determined by X-ray Powder Diffraction Analysis (XPDA) increases up to $5*10^{11}cm^{-2}$ for a deformation of $\varphi=0.5$. However, after higher deformations it decreases to a level of about $5*10^{10}-1*10^{11}cm^{-2}$. The dislocation density inside the subgrains was determined by a computer-aided line-intersection method [6] to amount to $1.5*10^{10}cm^{-2}$ for $\varphi=0.4$. This indicates that most dislocations are inside the subgrain boundaries. In comparison with the slow deformation rates [4], it is an important feature that the cell structure does not show any globular subgrains.

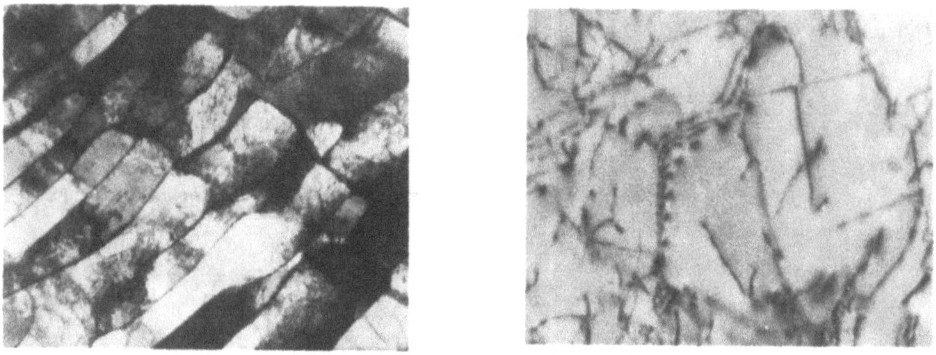

(a) (b)

Fig.3 Microstructure development of X8CrTi17 after deformation at 1073K, $\dot{\varphi}=4000s^{-1}$, $\varphi=0.4$ / (a) subgrains, (17600:1), (b) arrangement of dislocations within subgrains, (88000:1)

QTA unambiguously showed that in agreement with investigations of pass-rolled samples [1,3] a deformation texture of the type {111}<uvw> develops up to φ=0.5. Higher deformations (φ =0.8) cause the orientation of a cube texture {001}<uvw>.

SCHUBERT [1] has found that this component is connected with recrystallization, but, according to Fig.3 no recrystallization has been observed in the present study. This indicates that the {001} <uvw> texture component may be connected with the generation of a strongly oriented substructure.

The microstructure development in groove compression tests can be compared with results obtained after pass-rolling of steel in a round-oval sequence, if only the first deformation step is considered. At deformation rates of $70s^{-1}$ to $670s^{-1}$, (T=1073K) only recovered structures have been found at this stage [1,4,7]. The microstructure shows globular subgrains the sizes of which increase with increasing the strain rate.

The different microstructures can be explained by the fact that the deformation time at low $\dot{\varphi}$ is sufficient for a more complete polygonization of the ferrite.

According to the differences in the microstructure development the question should be answered, which mechanism leads to the sharp substructure in deformation times smaller than 1ms? Since the deformation time is negligible in comparison with the minimum holding time (before quenching) it is assumed that the microstructure mainly forms after the deformation. The possible mechanisms of dislocation motion or rearrangement may be affected by dislocation climb or cross-slip, supported by overheating and a high concentration of excess vacancies. Such an idea is suggested by the investigations of SCHUBERT [1] who found activation energies of dynamic recovery corresponding to the migration energy of vacancies even for strain rates of $\dot{\varphi}=0.1s^{-1}$.

Only at higher deformation temperatures (1273K) and higher strains recrystallization can be expected, in agreement with the investigations at low strain rates [1,4,7].

Microstructure Development of X8CrNiTi18.10 (f.c.c.)

In the austenitic material a very dense dislocation cell structure with thick cell walls has developed during deformation at $\dot{\varphi}=4000s^{-1}$ and 1073K (Fig.4).

The total dislocation density determined by XPDA increased from $1*10^{11}cm^{-2}$ (φ=0.1) to $5*10^{11}cm^{-2}$ at φ=0.5. In some regions it was possible to resolve single dislocations inside the cells and a dislocation density of $7.6*10^{10}cm^{-2}$ was determined by the line-intersection method [6]. There exist isolated areas of low dislocation density within the microstructure, which may have been formed by dynamic or metadynamic recovery, but in spite of the existing driving forces due to the high dislocation density recrystallization has not been observed. This is in agreement with results of LEVIT [8] who investigated the same steel under similar conditions. This confirms the assumption that the available time was not sufficient for diffusion processes.

In the austenite approximately equal numbers of screw and edge dislocations have been observed. The contribution of cross- slip is low (in the ferrite mainly screw dislocations are expected and therefore cross slip should be the dominant process), the slow dislocation motion due to climb processes determines the development of the microstructure. Due to the extremly short time sharp subgrain walls have not formed.

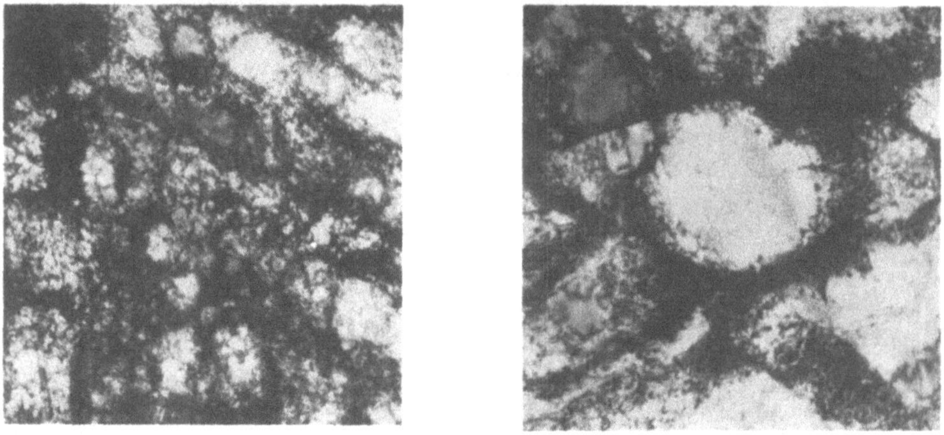

(a) (b)

Fig.4 Microstructure development of X8CrNiTi18.10 after deformation at 1073K, $\dot{\varphi}$=4000s^{-1}, φ=0.4, (a) 30 000 : 1 (b) 66 000 : 1

Similar to the ferrite, after high deformations the cell structure is sharpened. The effect becomes visible if $\dot{\varphi}$ decreases and/or the overheating increases. After smaller deformation the austenite texture mainly contains the components {011}<111>.....{113}<110> which were also found by SCHUBERT [1,3] but for φ>0.5 the GOSS-orientation {011}<001> was observed. At low $\dot{\varphi}$ SCHUBERT [1,3] then found the cube texture. The GOSS-orientation frequently occurs in austenite microstructures with shear bands [9] if the deformation is restricted due to high friction [10]. For this reason it may be assumed that the development of both substructure and texture may be related to the development of shear bands. Some deformation bands were observed in light microscope images.

At low $\dot{\varphi}$ and T=1073K the austenite is restored only due to recovery. The microstructure shows well-developed subgrains [4].

SUMMARY

A pneumatic-mechanical high speed plastometer was used to perform compression tests on both a ferritic and an austenitic steel, which are suitable for simulating the first deformation step in pass-rolling. The results can be summarized as follows:

- The increasing strain rate leads to a decreasing influence of dynamic processes on the formation of the microstructure but significantly enhances the driving forces for meta-dynamic restoration processes. Due to the differences in the stacking fault energy and the diffusion parameters of the materials this trend is more pronounced in the austenite than in the ferrite.
- In the ferrite a well-developed substructure with sharp subgrain boundaries is observed, which is accompanied with the formation of a $\{001\}\langle uvw\rangle$ texture.
- The cell structure of the austenite is far less well-formed. The dislocation densities are higher. From the texture measurement it can be concluded that the austenite texture is closely connected with the formation of shear bands at higher deformations.

REFERENCES

1. A. Schubert, Dissertation A, Academy of Mining, Freiberg (G.D.R.), (1985)
2. A. Schubert, P. Klimanek and K.-E. Hensger, Neue Hütte, Leipzig, 32 (1987) 423-425
3. A. Schubert, P. Klimanek and K.-E. Hensger, Textures and Microstructures, Vol.8,9 (1988) 207-222
4. P. Klimanek, K.-E. Hensger, A. Schubert and M. Barthel, Czech. J. Phys. B38 (1988) 373
5. H. Weber, J. Leopold and G. Schmidt, "Visioplastizität", Wissenschaftliche Schriftenreihe der TU Karl-Marx-Stadt 13 (1985)
6. K.-H. Hanisch, P. Klimanek and D. Stoyan, Cryst. Res. Technol. 20 (1985)
7. A. Schubert, M. Barthel, K.-H. Hensger, P. Klimanek and M. Koch, Proc. V. Conference "Metallkundliche Probleme der Werkstoffentwicklung", Academy of Mining Freiberg (1986) 98-102
8. V.J. Levit, M.A. Smirnov and N.P. Bachteeva, Fiz. met.i metallovedenie 45 (1978) 1244
9. Recryst. of metallic materials, Frank Haessner (Ed.), Dr. Riederer Verlag G.m.b.H., Stuttgart (1978)
10. Y. Shimizu, Y. Ito and Y. Iida, Met. Transactions 17A (1986) 1323

3. DEFORMATION OF IONIC CRYSTALS AND SEMICONDUCTORS

INTERSTITIAL PLASTICITY IN INHOMOGENEOUS STRESS FIELDS

V.N. Rozhanski

Institute of Crystallography, Academy of Sciences of the USSR, Leninsky pr. 59, 117333 Moscow, USSR

ABSTRACT

In the field of high hydrostatic pressure gradients the point defects and their complexes change their position and configuration. Particularly the interstitial movement can result in the plastic deformation of crystals. Some cases are considered in which the interstitial plasticity prevails over the plasticity caused by dislocations. In crystals consisting of different sorts of atoms, the interstitial plasticity may be accompanied with the occurrence of local non-stoichiometry and by chemical decomposition. The results obtained by computer simulation, high resolution transmission electron microscopy and scanning cathodoluminescence microscopy are discussed.

The movement and the generation of defects resulting from mechanical forces cause irreversible changes in the crystal shape. The homogeneous shear stress field gives rise to the displacement and multiplication of dislocations. And this elementary mechanism of crystal plasticity seems to be most widely observed. However, there are some cases of plastic deformation occurring due to the movement of point defects. First of all, the Nabarro-Herring mechanism [1, 2] for polycrystals is well-known which consists in the change of the grain shape due to the flow of point defects directed from one grain boundary to another one (Fig. 1).

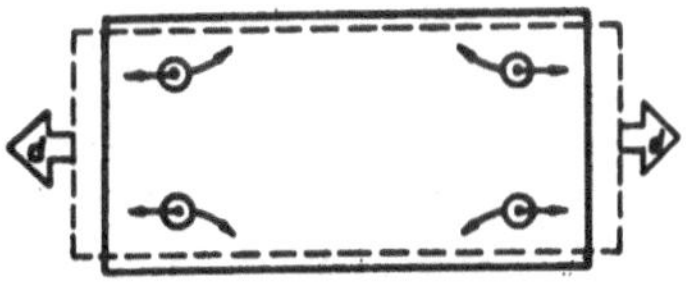

Fig. 1 The directed diffusion motion of vacancies and interstitials under external loading

Another mechanism is the diffusion-dislocation plasticity when deformation occurs due to the redistribution of point defects between dislocations [3]. The evolution of a dislocation loop ensemble in a Ge-crystal under tensile loading illustrates this mechanism [4] (Fig. 2). The force causing the directed point defect motion arises in the inhomogeneous stress field under hydrostatic pressure gradients.

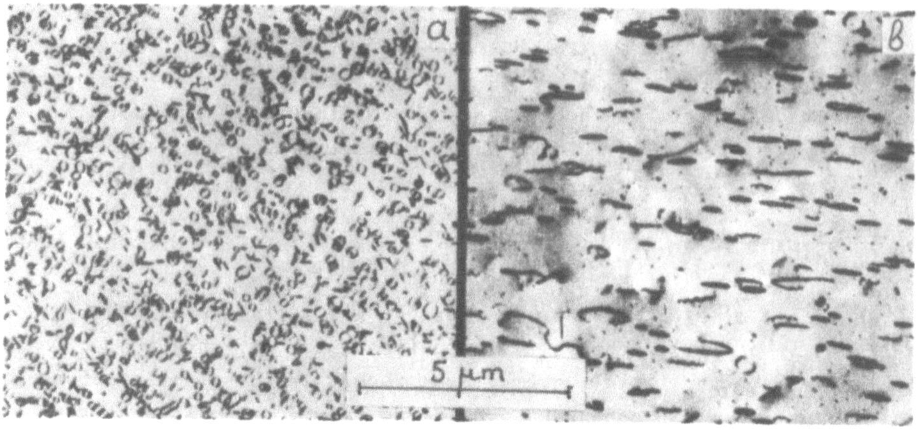

Fig. 2 The diffusion-dislocation evolution of an interstitial dislocation loop ensemble in Ge under tensile stress
a - before deformation
b - after deformation during 20 min in vertical direction
P = 3.6 kg/mm^2, T = 560°C

This force is the reason of the absence of equal probabilities of point defect jumps in opposite directions and results in the directed motion and therefore in the plastic deformation. If these gradients are high enough this does not only lead to the directed flow of point defects but also to the generation of a large number of Frenkel pairs.

Computer studies of defect motion enable one to analyze the behaviour of single point defects.

In contrast to the dislocations the motion of which can be observed directly in the electron microscope, the point defect motion is mostly studied by indirect methods as a result of a directed mass displacement by point defects. However, the motion of individual point defects can be observed by computer simulation.

For the computer simulation of the interstitial atom migration [5] a model crystal of α-Fe with BCC structure was formed. It consisted of 559 atoms and had a cubic shape with an edge length of 12 l_o, with l_o = 1.43 Å being half the lattice constant. The interaction between the atoms was described by the Erginsoy-Vineyard-Englert twin potential [6]. The calculations were carried out by the usual scheme of molecular dynamics at the temperature of 0 K under flexible boundary conditions. Two kinds of the inhomogeneous stress fields of deformation were used: (1) stress fields caused by the action of concentrated loading and (2) non-uniform fields with a hydrostatic stress gradient along the Z axis [7, 8].

Qualitatively, both stress fields reduce the potential relief of the migration of point defects and thus facilitate their mi-ration (Fig. 3).

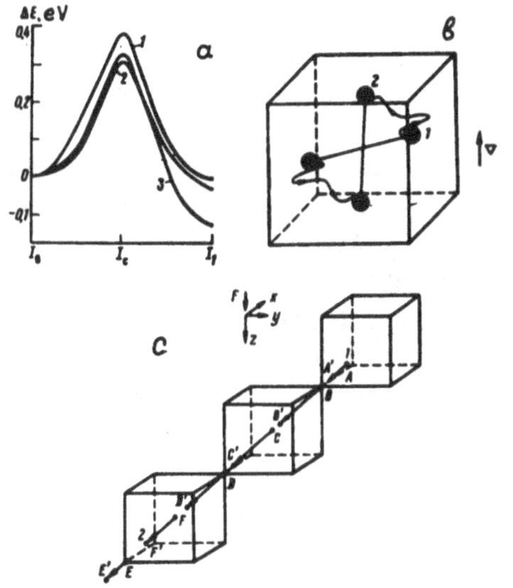

Fig. 3 a - Decrease of the potential barrier of the interstitial migration in normal stress gradient: 1 - original crystal, 2 - in the stress field of concentrated load, 3 - in the field of a uniform stress gradient

b - Orientation change of the interstitial dumbbell in the direction of the gradient

c - Crowdion migration when the strain is about 3 %

The change of two discrete configurations of an interstitial atom was investigated: the dumbbell and the crowdion. When the gradient was large enough the dumbbell axis was rotated in the direction of the compression gradient. The crowdion transformed to the dumbbell when the deformations were small and was migrating along the axis of the crowdion configuration when the deformations were large. It may be stressed that this occurs along the crowdion axis but not along the compression stress gradient which forms a certain angle with the axis. These examples show a possible realization of plastic deformation due to interstitial migration in an inhomogeneous stress field.

Originally, such a possibility was experimentally illustrated by the plastic deformation of a corundum crystal under concentrated load [9].

In the general case when a crystal surface is indented the hole forms by both interstitial displacement and dislocation motion. For investigating the plasticity mechanism caused by interstitial displacement it is worthwhile experimenting with crystals in which dislocations do not move. For example, silicon crystals may be indented at temperatures near room temperature. In this case the hole emerging under the indentation is obviously of plastic origin. Its shape perfectly follows that of the diamond indenter. But dislocations are not revealed in the region near the hole.

High resolution electron microscopy was used to investigate point defects and their complexes near this hole in a Si crystal [10]. The experiment was performed in the following way. After chemical polishing a (110) Si disk, 3 mm in diameter and 400 µm thick, was indented in a net-like way by a diamond indenter PMT-3 under a load of 0.015 N at temperatures of 20, 100 and 200 °C. After a few seconds of chemical polishing on both sides, the polishing was continued on the side opposite to the indentation surface by jet polishing down to a thickness suitable for the transmission microscope investigation. The point defects were studied in a JEM-100C electron microscope, equipped with a top-entry goniometer and a high resolution pole piece, on the thinnest parts of the crystal (5-10 nm thick) by many beam dark-field imaging using 3 diffracted beams [11].

In the many beam dark-field images of crystals indented at 200 °C, defects of a size of about 1 nm were observed in an area of about 1 µm radius around the place of indentation. They are imaged as dark spots (Fig. 4). It may be supposed that they are of vacancy type because of their dark contrast in the many beam dark-field images. The fact that these defects are similar to those in gold emerging under 2 MeV electron irradiation [12] allows one to identify them in a first approximation as very small stacking fault tetrahedra. In the region far from the indentation mark such defects have not been observed. They can appear due to the creation of Frenkel pairs in a zone with a radius of 1 µm from the identation mark. The vacancies and the interstitials migrate to the crystal surface and into the crystal bulk, respectively. The interstitials can take part in the formation of the hillocks and the vacancies participate in the formation of the hole below the indenter. The coagulation of the vacancies, which are less mobile than the interstitials, may result in the formation of the tetrahedral vacancy complexes observed (Fig. 5). The silicon-oxygen complexes available in the crystal may be both the centres of the nucleation of Frenkel pairs and those of the coagulation of excess vacancies.

In crystals with mobile dislocations the interstitial plasticity can take place side by side with the usual dislocation plasticity, thus complicating the hole formation mechanism. Moreover, the nucleation and motion of both dislocations and interstitials determine the plasticity process under deformation conditions of concentrated load leading to a non-uniform stress state. The processes of the dislocation-point defect interaction are also activated. In most experiments only the dislocation behaviour is studied and a number of methods are developed for their direct observation. It is, however, more complicated to observe the point defects. And we get something of a funny story, when somebody is

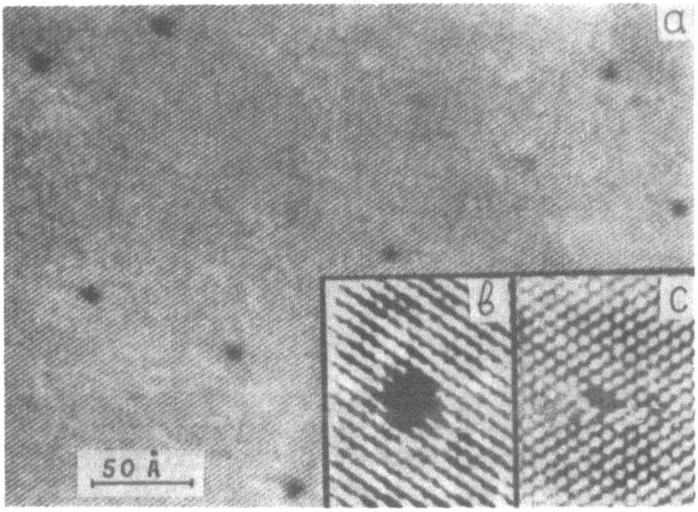

Fig. 4 a, b – Microdefects observed in Si near the indentation mark
c – Microdefects in an undeformed crystal. Dark-field three beam electron microscope image

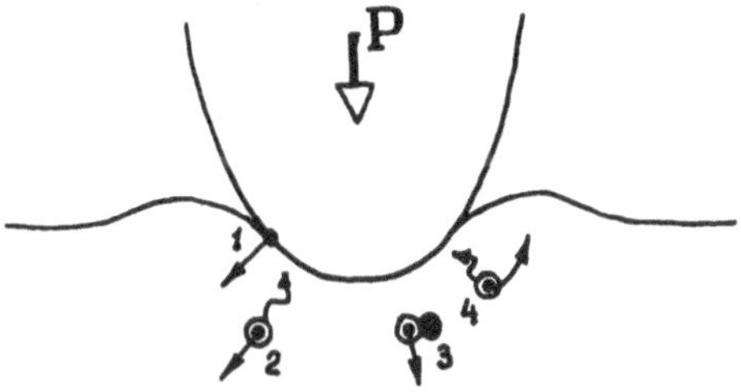

Fig. 5 Localization scheme of points of emergence of interstitials near the indenter.

1 – interstitial creation near the surface contact between the indenter and the crystal
2 – Frenkel pair creation in the bulk near the indentation mark
3 – interstitial creation near Si-O complexes
4 – interstitials moving to the surface, realizing the contribution of interstitials to the hillock formation near the hole

looking for something that he has lost in the bright light under the lamp and not where it actually happened. Due to the lack of direct methods of observing single point defects it is necessary to investigate the processes connected with them more carefully.

The plastic deformation processes become more complicated if the crystal consists of more than one type of atoms.

In the field of hydrostatic stress gradients different types of interstitials are influenced by different forces, they will move at different velocities. This may result in their spatial separation and, in ionic crystals, in the occurrence of spatially distributed charges and currents which neutralize these charges. And this is nothing but a substance decomposition under the influence of the mechanical load. There are numerous indirect evidences of such a decomposition.

It seems to me that the change in the luminescence properties of a number of semiconductor and ionic crystals near the indentation mark is connected with the mechanical decomposition. In CdS, GaP and ZnS the cathodoluminescence rosette disappears near the indentation mark. In MgO, LaO, SrO, LiF, NaF, KCl, Al_2O_3, and $I_3Al_5O_{12}$, a brighter cathodoluminescence rosette appears. The cathodoluminescence spectrum in the rosette [13-20] differs from that in the undeformed crystals. The cathodoluminescence rosette near the place where a concentrated load is acting can well be observed in the scanning microscope in the cathodoluminescence mode. The change of electron energy allows one to determine the distribution of cathodoluminescence centres in the specimen depth [21]. By indentation during the observation the motion of these centres can be observed during loading and immediately afterwards [22]. The nature of the luminescence centres can be identified by an analysis of the cathodoluminescence spectra.

In MgO, the cathodoluminescence spectra near the indentation mark were compared with those caused by direct implantation of Mg^{2+} and O^{2-} ions [23] (Fig. 6).

This comparison shows that Mg atoms are preferentially displaced from the place of identation (Fig. 7).

As I have already pointed out, in spite of their importance, these phenomena connected with interstitial plasticity have insufficiently been investigated. This is due to the lack of direct methods of observing single point defects but there are no general limitations preventing the development of those techniques. The most promising perspective is the combined application of different methods of electron microscopy. The present report is aimed at attracting the attention and efforts of research workers to these important problems.

We have considered the results of the local action of external forces forming hydrostatic stress gradients in the bulk near the places of action. In real crystals containing structure defects inhomogeneous stress fields are always present also under conditions of most accurate loading and even if there are no external forces. Thus, under real conditions local inhomogeneities of the state of stress always occur so that plastic deformation can take place by a directed movement of point defects.

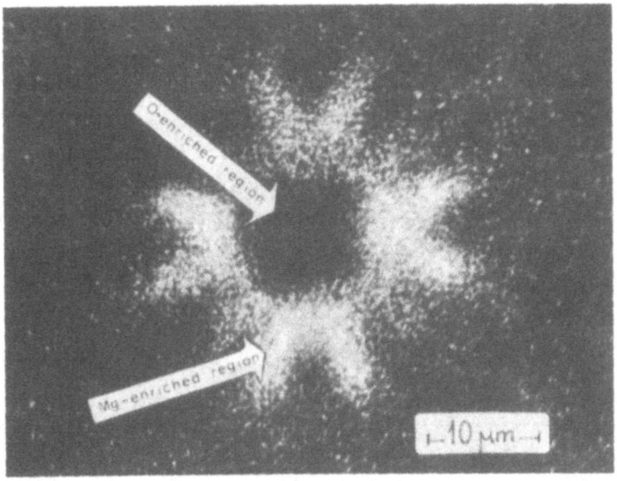

Fig. 6 Comparison of the changes of cathodoluminescence spectra of MgO crystals caused by the action of a concentrated load and by ion implantation of Mg^{2+} and O^{2-}

a - original spectrum

b - spectrum in the rosette near the indentation mark

c - spectrum of the crystal implanted with oxygen ions

d - the same after magnesium ion implantation

Dose: 3...6 * 10^{14} $ions/cm^2$

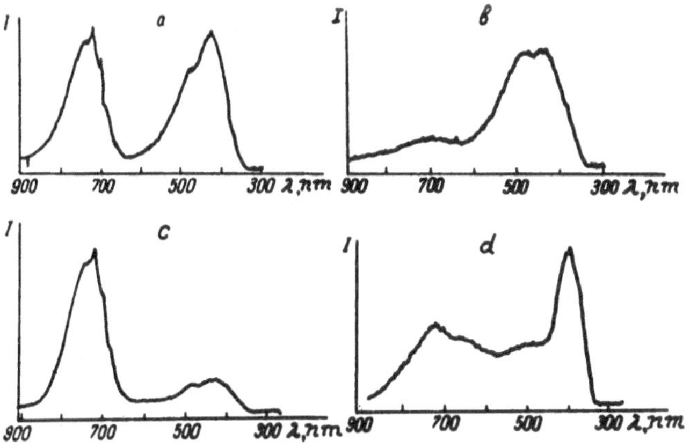

Fig. 7 Cathodoluminescence rosette in an MgO crystal ({100} face) around the indentation mark. Arrows point to crystal regions enriched with oxygen and magnesium near the indentation mark

REFERENCES

1 F.R. Nabarro, In: Conf. on Strength of Solids, Phys. Soc., London 1948, p. 75

2 C. Herring, J. Appl. Phys. 21 (1950) 437

3 A.M. Kosevich, Z.K. Saralidze and V.V. Slezov, Zhurn. Exper. i Teor. Fiz. 50 (1966) 958

4 V.N. Rozhanski, N.D. Zakharov, V.V. Slezov, and G.A. Osipov, Fizika Tverdogo Tela 17 (1975) 381-8

5 V.V. Kirsanov, S.B. Kislitsin, E.M. Kislitsina, and V.N. Roshanski, Kristallografiya 33 (1988) 167-71

6 G. Erginsoy, G.H. Vineyard and A. Englert, Phys. Rev. 133 (1964) A595

7 L.D. Landau and E.M. Lifshits, In: Teoriya Uprugosti, Nauka, Moskva 1965, p. 202

8 D.R. Olander, J. Nucl. Mat. 92 (1980) 163

9 V.N. Rozhanski, M.P. Nazarova, L.L. Svetlov, and M.V. Kalashnikova, Phys. Stat. Sol. 41 (1970) 579-90

10 V.N. Rozhanski, B.M. Pushkash, P. Werner, and N.D. Zakharov, In: Struktura, Mehanicheskie Svoistva i Razrushenie Realnykh Kristallov, IPM AN USSR, Kiev 1988, p. 23-8

11 N.D. Zakharov, M. Pasemann and V.N. Rozhanski, Phys. Stat. Sol. (a) 71 (1982) 275-81

12 J. Takai, N. Ajika and H. Hashimoto, In: Point Defects and Defect Interaction in Metals, Tokyo University Press, Tokyo 1982, p. 702-4

13 O. Bruemmer and J. Schreiber, Kristall und Technik 7 (1972) 683

14 O. Bruemmer and J. Schreiber, Mikrochim. Acta Suppl. 6 (1975) 331

15 V.N. Rozhanski and M.A. Velednitskaya, Kristallografiya 19 (1974) 1111-2

16 M.A. Velednitskaya, V.N. Rozhanski, L.F. Komolova, G.V. Saparin, J. Schreiber, and O. Bruemmer, Phys. Stat. Sol. (a) 32 (1975) 123-32

17 J. Chen, M.M. Abraham, T.J. Turner, and C.M. Nelson, Phil. Mag. 32 (1975) 99

18 M.S. Akchurin and V.G. Galstjan, Izv. Akad. Nauk SSSR, Ser. Fiz. 44 (1980) 2134-7

19 M.S. Akchurin and V.R. Regel, In: Defekty v Segnetoelektrikakh, Nauka, Leningrad 1981, p. 88-96

20 M.S. Akchurin, E.N. Vasev, E.J. Mikhina, and V.R. Regel, Fizika Tverdogo Tela 30 (1988) 760-4

21 M.S. Akchurin, V.G. Galstjan, J.I. Krjuchkov, E.J. Mikhina, and V.R. Regel, Kristallografiya 33 (1988) 523-5

22 M.S. Akchurin, V.G. Galstjan, V.R. Regel, and V.N. Rozhanski, Poverkhnost 3 (1983) 119-23

23 V.N. Rozhanski, M.A. Velednitskaya, J. Schreiber, and V.V. Krasnopevtsev, Fizika Tverdogo Tela 19 (1977) 1980-2

DISLOCATION INTERSECTION AND THE FORMATION OF JUNCTIONS IN IONIC CRYSTALS OF THE NaCl STRUCTURE

F. Appel

Institute of Solid State Physics and Electron Microscopy
Academy of Sciences of the G.D.R.
Weinberg 2, Halle/Saale, 4050, G.D.R.

ABSTRACT

The formation of dislocation junctions in ionic crystals of the NaCl structure is investigated by considering the dependence of the line energy on the character of the intersecting dislocations. By using different electron microscopy techniques the process is experimentally studied on NaCl and MgO single crystals. Accordingly, dislocations intersecting on oblique [110]{$\bar{1}$10} systems frequently react and form obstacle structures with strong long-range stress fields.

INTRODUCTION

Gliding dislocations have various interactions with forest dislocations, which may remarkably contribute to the flow stress. One of the most important hardening mechanisms of forest dislocations arises from the formation of junctions by dislocation reactions. Since the junctions are mostly immobile, they represent strong obstacles to the gliding dislocations. The obstacle strength has been estimated by several authors [1 to 4]. The main result is that the overcoming of junctions results in an athermal stress part, which is proportional to the square root of the forest dislocation density. Because of the complexity of the problem and the approximations used further details can hardly be deduced from the theoretical models. Direct experimental data are rare, on the other hand. To get some information on the hardening mechanism of dislocation reactions the problem was investigated by electron microscopic slip line studies and in situ deformation experiments in the HVEM. The investigations were performed on NaCl and MgO single crystals. The advantages of the materials are their simple slip geometries and the fact that the dislocations are not dissociated. Thus the situation is rather well-defined.

DISLOCATION REACTIONS IN THE NaCl STRUCTURE

The possible formation of junctions depends on a balance of the energies and interaction energies of the dislocation segments involved in the dislocation reaction. The simplest criterion for predicting the occurrence of dislocation reactions given by Frank [5] assumes a constant dislocation line energy so that only the squares of the Burgers vectors of the reacting dislocations have to be considered. Because of the simplifications used the criterion can only be applied for rough estimates; it does not give any information as to e.g. dislocations with perpendicular Burgers vectors. Since the line energy considerably

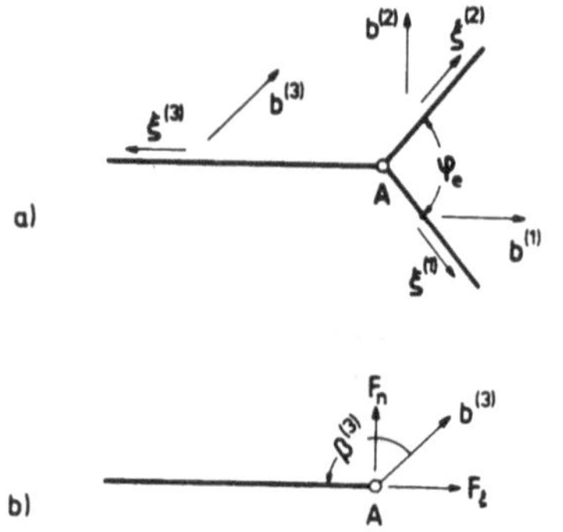

Fig. 1: a) Configuration of the dislocation segments joined by one of the two triple nodes formed by a dislocation reaction; b) the forces $\mathbf{F}_\xi$ and $\mathbf{F}_n$ acting on one of the dislocation segments

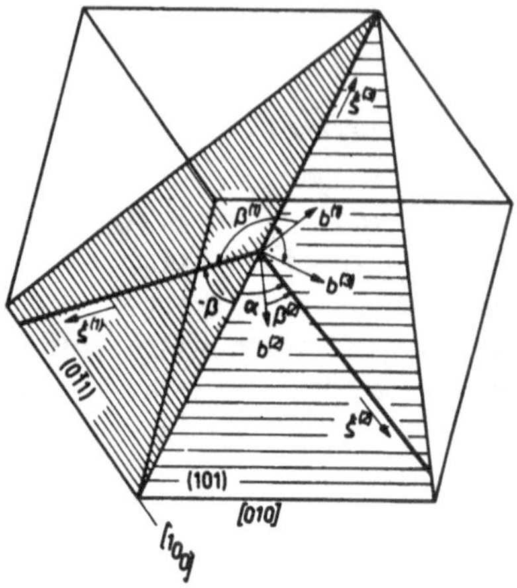

Fig. 2: The coordinates used for investigating the reactions between dislocations on oblique slip systems

varies with the dislocation character, this dependence should be taken into account for a more elaborated treatment. Fig. 1a shows the configuration of the dislocation segments joined by one of the two triple nodes formed by a dislocation reaction. Vectors $\xi^{(i)}$ and $\mathbf{b}^{(i)}$ are the line vectors and Burgers vectors of the individual segments, respectively, with $\xi^{(3)}$ and $\mathbf{b}^{(3)}$ representing the junction. The equilibrium position of each segment is determined by the forces $\mathbf{F}_\xi$ and $\mathbf{F}_n$ (Fig. 1b) [6, 7]. $\mathbf{F}_\xi^{(i)} = W_s^{(i)}\xi^{(i)}$ prevents the segment from contracting whereas $\mathbf{F}_n^{(i)} = (dW_s^{(i)}/d\beta^{(i)})\mathbf{n}^{(i)}$ prevents it from turning into a low-energy configuration. $\beta^{(i)}$ describes the dislocation character and $\mathbf{n}$ is a unit vector being normal to the dislocation line. $W_s^{(i)}$ is the dislocation line energy with [7]

$$W_s^{(i)} = \frac{\mu}{4\pi}\left[(\mathbf{b}^{(i)}\xi^{(i)})^2 + \frac{(\mathbf{b}^{(i)} \times \xi^{(i)})^2}{1-\nu}\right] \ln \frac{R}{r_o} \qquad (1)$$

μ is the shear modulus and ν is Poisson's ratio. R and r_o are the inner and outer cut-off radii, respectively. The total force acting on the segment (i) is

$$F_t^{(i)} = \xi^{(i)} W_s^{(i)} + n^{(i)} dW_s^{(i)}/d\beta^{(i)} \quad (2)$$

In the equilibrium position the total force $F_t = \Sigma F_t^{(i)}$ acting on the dislocation node is zero [6, 7]. After equating the logarithmic terms of the individual segments one obtains the reduced form

$$f_t = \sum (b^{(i)})^2 \, [1 - \nu \xi^{(i)} \cos^2\beta^{(i)} \xi^{(i)} + 2\nu \cos\beta^{(i)} \sin\beta^{(i)} \frac{\xi^{(i)} \times (\xi^{(i)} \times b^{(i)})}{|\xi^{(i)} \times b^{(i)}|}] = 0 \, . \quad (3)$$

The junction continues to form as long as the force component

$$\Delta w = (f_t \xi^{(i)}) \quad (4)$$

acts in the direction of $-\xi^{(3)}$. To predict dislocation reactions this quantity may therefore be evaluated for dislocation configurations to be expected.

Table 1: Dislocation reactions in the NaCl structure for (a) orthogonal and (b) oblique $\langle 110 \rangle \{\bar{1}10\}$ slip systems

	$b^{(1)}$	$b^{(2)}$	$b^{(3)} = b^{(1)} + b^{(2)}$	$\xi^{(3)}$
(a)	$(a/2)[011]$	$(a/2)[0\bar{1}1]$	$a[001]$	$[\bar{1}00]$
(b)	$(a/2)[011]$	$(a/2)[10\bar{1}]$	$(a/2)[110]$	$(\sqrt{3}/3)[\bar{1}11]$

In the NaCl structure slip occurs on $\langle 110 \rangle \{\bar{1}10\}$ systems. Possible reactions between dislocations on orthogonal and oblique systems are summarized in Tab. 1 and were evaluated by using eq. 4. The coordinates necessary for the oblique slip systems are shown in Fig. 2. The angles α and β describe the orientation of the intersecting dislocations with respect to $\xi^{(3)}$. The result is shown in Fig. 3. According to Tab. 1 the reduced value $2\Delta w/a^2$ is plotted for different α and β values. a is the lattice constant. For dislocation orientations belonging to negative Δw, junctions should be formed. The corresponding ranges of α and β are obtained by solving $\Delta w = 0$. Fig. 3 summarizes the results for NaCl and MgO. The diagram includes the results of similar calculations for NaCl made by Püschl [8], who additionally considered the elastic anisotropy of the material. According to Fig. 4, the orientation range of intersecting dislocations remarkably differs from that expected from Frank's criterion [5], which predicts the range between the respective edge orientations. The evaluation performed for orthogonal slip systems showed that reactions should not occur

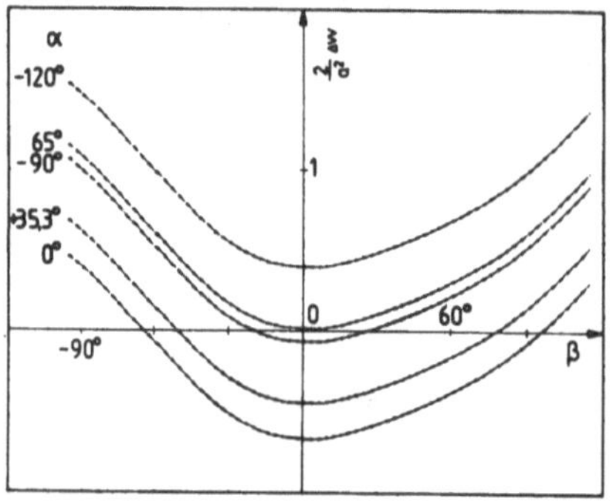

Fig.3: Energy changes for the reaction of dislocations intersecting on oblique <11Ø>{$\bar{1}$1Ø} slip systems

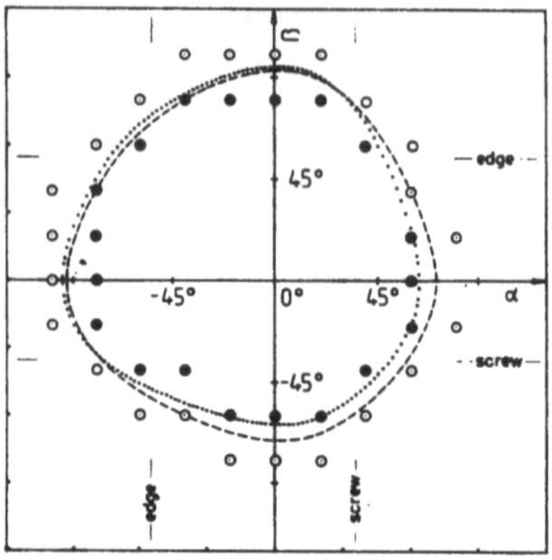

Fig. 4: Ranges of dislocation orientations α and β (included in the lobes) for which reactions are expected: .... NaCl, ___ MgO. Results of calculations on NaCl [8] considering the elastic anisotropy: ● reaction, o no reaction

EXPERIMENTAL RESULTS AND DISCUSSION

Dislocation intersection was studied on MgO single crystals by in situ deformation experiments in the HVEM [9]. Fig. 5 shows the intersection of oblique <11Ø>{$\bar{1}$1Ø} slip systems. The vertically oriented screw dislocations belonging to b_1 are intersected by gliding half loops the leading part of which is of edge character. At the sites marked by arrows the dislocations react. Because of the [ØØ6] reflex used for imaging, the junctions are visible only by residual contrast.

High resolution slip line investigations by means of the heavy-metal decoration technique [1Ø, 11] were combined with latent hardening experiments to investigate the intersection of oblique screw dislocations [12]. The slip geometry is shown in Fig. 6. A primary deformation along [ØØ1] introduced forest dislocations on the [1Ø$\bar{1}$](1Ø1) system. Afterwards, new compression samples with a perpendicular deformation axis were prepared by cleavage. In that way monatomic cleavage steps s_c arise from the forest dislocations. Neighbouring forest dislocations of different signs, e.g. (2) and (3) in Fig. 6, terminate the cleavage steps. A secondary deformation activated the [Ø11](Ø$\bar{1}$1) system with the dislocations trailing the slip steps s_g. To identify the different stages of the experiment, the cleavage steps were decorated with gold (large black particles) whereas the slip steps were decorated with palladium (small gray particles). According to Fig. 4, a gliding dislocation designated by (1) in Fig. 6 should react with only one of the forest dislocations (2) or (3) [12]. This difference should be indicated in the structure of the trailed slip step thus enabling one to detect the junction formation. Fig. 7 demonstrates an example probably suggesting the situation sketched in Fig. 6. The gliding dislocations 1', 1'', and 1''' can pass the forest dislocation (2). With the forest dislocation (3), however, dislocation (1') probably reacts finally being

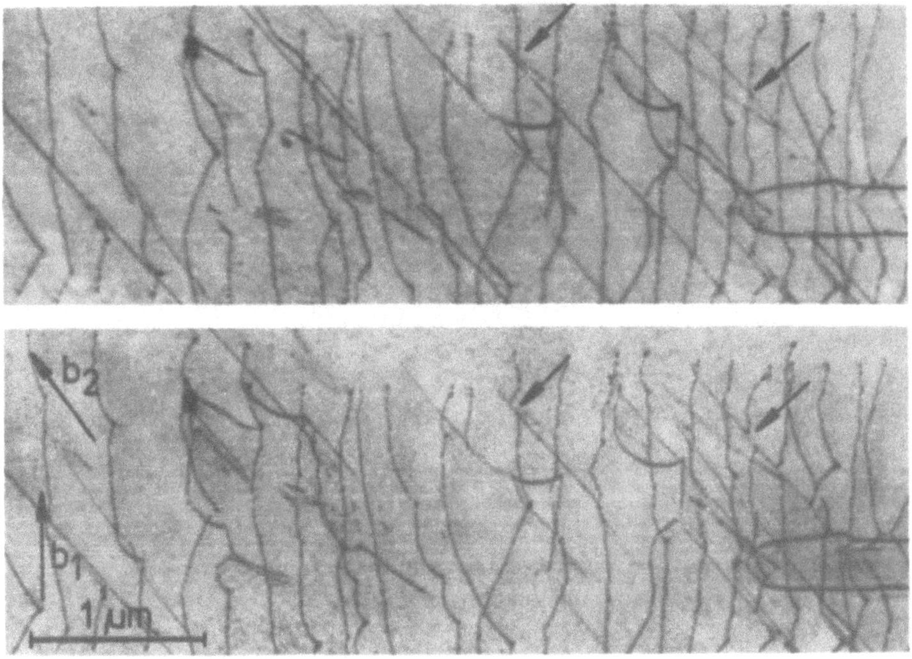

Fig. 5: Junction formation observed during the in situ deformation of an MgO single crystal in the HVEM

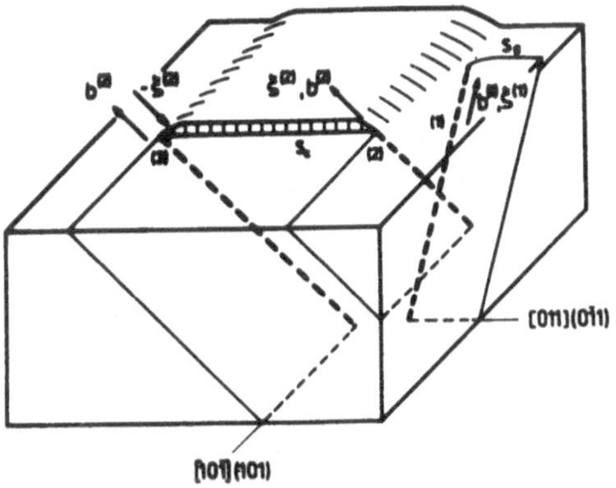

Fig. 6: The slip geometry used in latent hardening experiments and slip line studies to investigate dislocation intersection in NaCl

immobilized. Dislocations 1'' and 1''' cross glide, here. The reactant dislocation is of pure edge type. Since the related slip plane of type {1$\bar{1}$2} is not a usual slip plane in the NaCl structure the junctions are expected to be immobile and to contribute to the formation of very complex dislocation structures. Fig. 8 shows numerous cross slip and immobilization processes in the neighbourhood of two forest disloca-

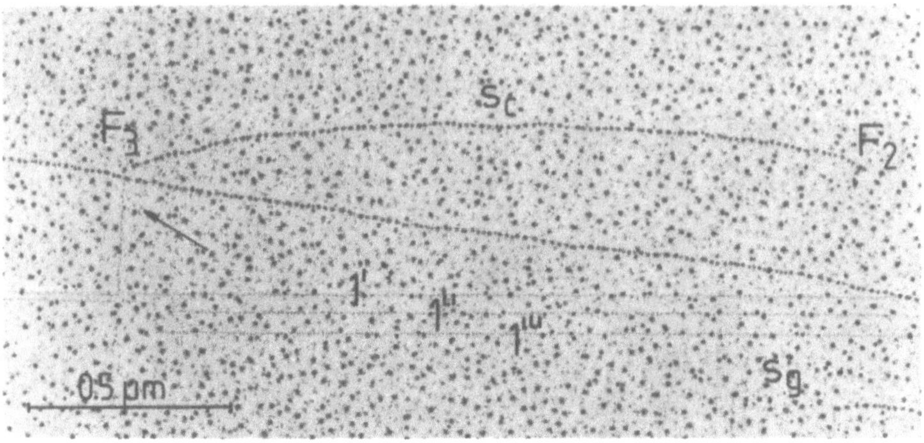

Fig. 7: Junction formation (--→) between dislocations on oblique <11Ø> {$\bar{1}$1Ø} slip systems in NaCl. s_c gold-decorated cleavage step connecting the forest dislocations F_1, F_2 of different signs. s_g slip steps of the dislocations 1', 1'', and 1''' gliding during secondary deformation

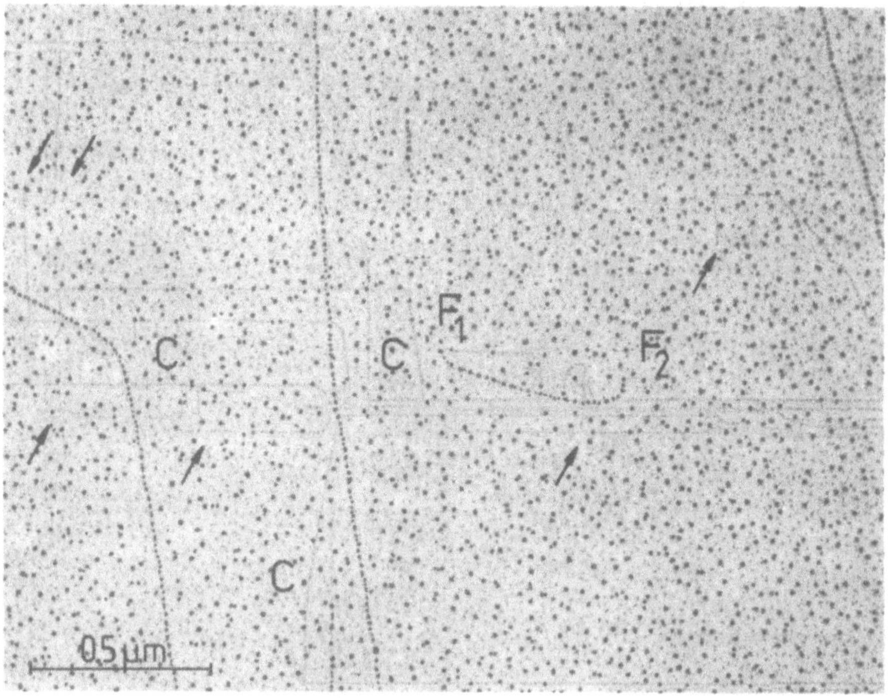

Fig. 8: Cross slip (C) and immobilization (--→) of gliding dislocations by the forest dislocations F_1 and F_2

tions, which are probably involved in dislocation reactions.

Slip structures like that are often observed near forest dislocations. Therefore, intersecting dislocations, often seem to react. Extended cross slip and immobilization of gliding dislocations show that the structures formed by junctions are very strong obstacles having long range stress fields. Their overcoming by gliding dislocations probably results in an athermal stress part. This conclusion agrees with the experimental finding [12] that the activation volume of the thermally activated dislocation motion is not changed by the presence of forest dislocations.

CONCLUSION

Dislocations intersecting on oblique [11Ø]{$\bar{1}$1Ø} slip systems frequently react. The reactant dislocations are immobile and contribute to the formation of dislocation structures with long-range stress fields. The overcoming of the obstacle structures yields an athermal stress part.

REFERENCES

1 G. Saada, Metaux et Corrosion, July-Sept.196Ø

2 G. Schöck and R. Frydman, phys. stat. sol. (b) 53 (1972)

3 R. Frydman and H. Strunk, Scripta met. 7 (1973) 471

4 W. Püschl, R. Frydman, and G. Schöck, phys. stat. sol.(a) 74 (1982) 211

5 F.C. Frank, Physica 15 (1949) 131

6 Y.T. Chou and J.D. Eshelby, J. Mech. Phys. Solids 1Ø (1962) 27

7 J.P. Hirth and J. Lothe, Theory of Dislocations, McGraw-Hill, New York 1968

8 W. Püschl, private communication

9 F. Appel, H. Bethge, and U. Messerschmidt, phys. stat. sol. (a) 42 (1977) 61

1Ø G.A. Basset, Phil. Mag. 3 (1958) 1Ø42

11 H. Bethge, phys. stat. sol. 2 (1958) 775

12 F. Appel, to be published

A T.E.M. IN SITU STUDY OF DISLOCATION GLIDE IN SEMI-INSULATING GaAs

D. Caillard*, N. Clément*, A. Couret*, Y. Androussi**, A. Lefebvre**, G. Vanderschaeve+

*Laboratoire d'Optique Electronique du CNRS, BP 4347, 31055 Toulouse-Cedex, France
**Laboratoire de Structure et Propriétés de l'Etat Solide, USTL, 59655 Villeneuve d'Ascq, France
+Laboratoire de Physique des Solides, INSA, 31077 Toulouse-Cedex, France

ABSTRACT

The velocities of dislocations have been measured in semi-insulating undoped GaAs by in situ straining at 350°C in an electron microscope. Experiments are performed either in a conventional electron microscope operated at 200 kV or in a high voltage electron microscope operated at 1 MV. For a constant local stress, the velocity of screw dislocations is proportional to their length, with no indication of a maximum velocity, up to dislocation length of $\simeq 3\ \mu m$. The kink formation and migration energies, F_k and W_m, respectively, are estimated : $0.65\ eV < F_k < 0.9\ eV$; $W_m < 0.5\ eV$.

INTRODUCTION

Dislocation motion in semiconductors under stresses lower than the Peierls stress is governed by the nucleation and migration of kinks. According to the Hirth and Lothe theory [1], two regimes of dislocation velocity are expected when dislocations are submitted to a lattice friction, depending on whether the length of the moving segment is much smaller, or much larger than the mean free path of kinks along the dislocations. In the first case the dislocation velocity is proportional to the dislocation length, whereas in the second case it is length-independent. Indeed, in elemental semiconductors Si and Ge [2, 3] a transition between the length-dependent regime (short dislocation segments) and a "saturation" regime (long dislocation segments) was observed by in situ straining experiments in an electron microscope. From these experiments, the nucleation energy and the migration energy of kinks, F_k and W_m, respectively, were estimated :

- Si (800 K) : $F_k \simeq 0.5$ eV , $W_m \simeq 1.2$ eV
- Ge (700 K) : $F_k \simeq 0.5$ eV , $W_m \simeq 0.8\text{-}0.9$ eV

In these crystals the kink migration energy is quite large, as compared to the kink formation energy, and this is confirmed using different experimental techniques [4-8]. This could probably be ascribed to a strong tendency to reconstruction for dislocation cores and kinks in elemental semiconductors.

In III-V compounds, the reconstruction appears to be unlikely, since the bonding would take place between two identical atoms. Therefore it is expected that the kink migration energy is lower in these compounds than in Si and Ge.

As pointed out above, kink formation and migration energies can be obtained from the study of dislocation motion under applied stresses. The mobility of dislocations in

III-V compound semiconductors has been extensively studied using different techniques : double etch [9-12], X ray topography [13]. In most experiments, velocities are averaged over travel distances of a few 10 μm, the initial diameter of loops being larger than 50 μm, and the dislocation behaviour in the early stages of source working (short dislocation segments) cannot be analyzed. That is why we have performed in situ straining experiments in an electron microscope on semi-insulating undoped GaAs samples and the results of this work are reported in the present paper.

EXPERIMENTAL TECHNIQUES

GaAs samples, oriented for single slip, (compression axis $[\bar{3}\bar{1}2]$) were predeformed by creep at either 200°C or 300°C up to $\Delta l/l_o$ = 1.6 %. Slices parallel to $(\bar{1}54)$ faces were deformed in situ at 350°C either in a JEOL 200 CX electron microscope operated at 200 kV, or in a high voltage electron microscope (HVEM) operated at 1 MV. An important advantage of using HVEM for in situ work is the increased specimen penetration, resulting in a decrease of the influence of the free surfaces : the behaviour of long dislocation segments (L > a few μm) can be analyzed. However irradiation damages may affect the propagation of dislocations, as it has been observed in [14]. To minimize their influence, dislocation velocities were measured in the early stages of observation.

Another effect of the electron beam was reported to occur in compound semiconductors, that is the enhancement of dislocation glide by electron irradiation [15]. Then the possible influence of the electron irradiation on the velocities of dislocations (cathodoplastic effect) was checked. The ratio between the velocity under irradiation and the velocity in the dark was found close to unity.

One of the main advantages of in situ deformation experiments, as compared with the above mentioned techniques, is that local shear stresses acting on the moving dislocations can be evaluated. In the present work, stresses were calculated :

(i) using Gottschalk's calculations [16], which relate the radii of curvature of dislocations to the local stress in covalent crystals.

(ii) by measuring the critical screw dipole separation above which non screw parts can escape ; this method applies when screw dislocations are much less mobile than non screw ones [17]. Both methods gave similar results.

Finally, dislocation velocities were measured on straight segments which did not intersect the sample surface. (This point will be discussed below in more detail).

DISLOCATION VELOCITIES

Sources were found to work either in the primary slip system 1/2 $[1\bar{1}0](111)$ or in the secondary slip system $1/2[0\bar{1}1](111)$. In the latter case, the screw direction is approximately parallel to the sample surface ; this particular geometry makes it possible to compare the relative velocities of 60°α [As(g)] and 60°β [Ga(g)] dislocations. Two examples of sources emitting dislocations on the secondary slip system are shown on fig. 1 (accelerating voltage 200 kV) and fig. 2 (accelerating voltage 1 MV).

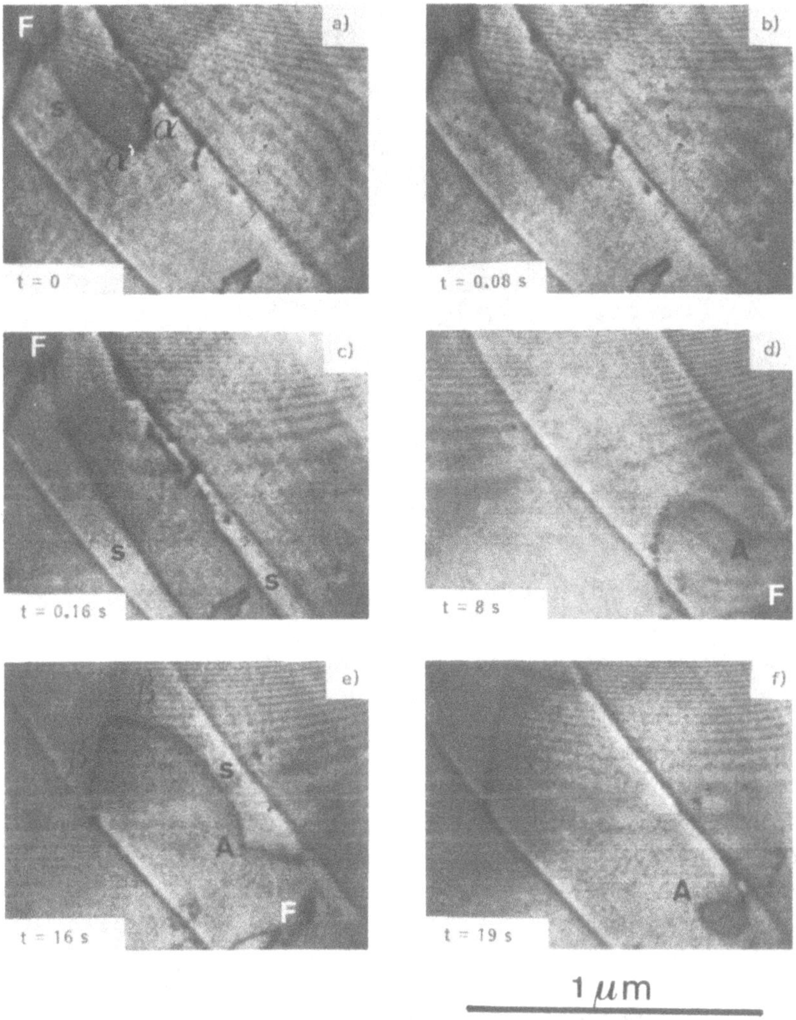

Fig. 1 : Dislocation source in GaAs ; secondary slip system (accelerating voltage 200 kV) ; T = 350°C, $\tau = 50 \pm 15$ MPa. A : anchoring point, F : fixed point. Note the large difference between the velocities of α and β dislocations.

In both examples, the local stress τ, evaluated as described above, is found equal to 50 ± 15 MPa. Dislocations are moving smoothly parallel to the <110> rows, indicating that they experience a strong lattice friction. At T = 350°C and τ = 50 MPa, there is no significant difference between the velocities of 60°β and screw dislocations, which are about 500 times lower than the velocity of 60°α dislocations [18]. Ninomiya [12] suggested that the velocity of 60°β and screw dislocations could be controlled by the lowest mobility of the 30°β partial dislocations. This is consistent with the results of a recent investigation on the mobilities of partial dislocations at room temperature under high stress [19, 20].

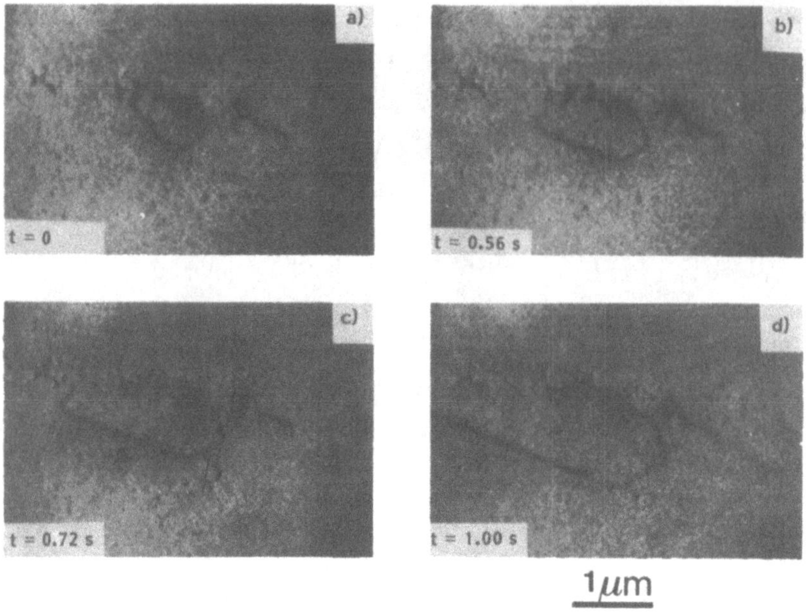

Fig. 2 : Dislocation source in GaAs ; secondary slip system (accelerating voltage 1 MV) ; T = 350°C, τ = 50 ± 15 MPa.

It is noteworthy that the velocities of 60° β and screw dislocations are proportional to their length for a given stress. For example, the results obtained for the source shown on fig. 2 (accelerating voltage 1 MV) are reported on fig. 3. There is no indication of a maximum velocity of screw dislocations, up to L ≃ 3 μm. A previous investigation by the present authors [18] has shown that the velocity of 60° β dislocations is also length-dependent up to L ≃ 0.5 μm (these experiments were performed at 200 kV, and segments of larger length could not be observed).

Note that surface effects may considerably affect the velocity of dislocations in semiconductor materials. In GaAs, the velocity of screw segments lying close to -or emerging at- the sample surface was found to be

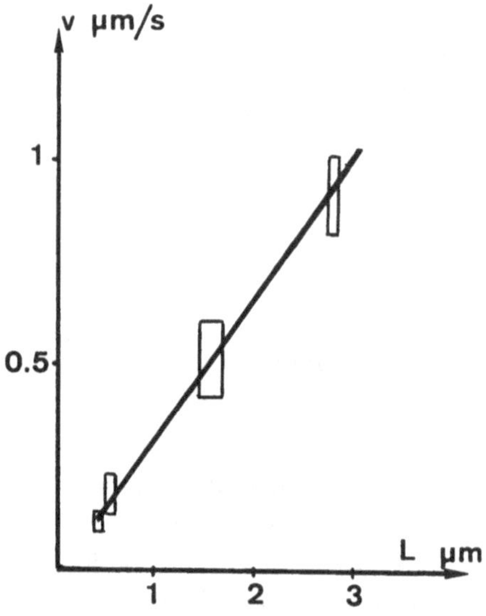

Fig. 3 : Velocity of screw dislocations versus their length in GaAs T = 350°C ; τ = 50 ± 15 MPa.

constant and equal to 250 nm/s in the length range 1-2 μm [18], The results of the HVEM experiments, as reported on fig. 3, definitely show that this was an artefact due to surface effects. A drastic change in velocities has also been observed in silicon when a dislocation emerges at a free surface [21]. Therefore, care should be taken when dislocation velocity measurements are made on segments which are close to a sample surface.

ACTIVATION ENERGIES FOR DISLOCATIONS GLIDE

When the dislocation propagation is controlled by the nucleation and migration of double kinks along the dislocation, two different velocity regimes have to be considered [1].

(i) For a short dislocation segment of length L, the kinks can move along the dislocation line without being annihilated. The dislocation velocity v is proportional to L ; the activation energy for dislocation movement is $F_{dk}(\tau) + W_m$.

(ii) For long dislocation segments, kinks are annihilated with kinks of opposite sign before reaching the ends of the dislocation. v is proportional to the mean free path of kinks, i.e. it is length-independent ; the activation energy for dislocation movement is $1/2\ F_{dk}(\tau) + W_m$. $F_{dk}(\tau)$ is the energy for double kink nucleation, related to the kink formation energy F_k by : $F_{dk}(\tau) = 2F_k - (\mu \tau b^3 h^3/2\pi)^{1/2}$. (μ is the shear modulus, b is the Burgers vector and h the distance between two Peierls valleys).

The activation energy for the length-dependent regime, $F_{dk}(\tau) + W_m$, is estimated from the slope of the v(L) curve. Limiting values of the activation energy for the length independent regime, $1/2\ F_{dk}(\tau) + W_m$, are obtained by considering that the maximum observed velocity is lower than the "saturation" velocity. One has, for screw dislocations (T = 623 K, τ = 50 MPa).

$$F_{dk}(\tau) + W_m = 1{,}6\ \text{eV} \ ;\ 0.55\ \text{eV} < 1/2\ F_{dk}(\tau) + W_m < 1.05\ \text{eV}\ ;$$

$$W_m < 0.5\ \text{eV} \ ;\ 0.65\ \text{eV} < F_k < 0.9\ \text{eV}$$

The kink migration energy W_m is rather low, as compared with the reported values for elemental semiconductors, that are about two times larger [2 - 4]. The relative importance of W_m in the activation energy for dislocation glide is lower in GaAs than in Si and Ge. This might be an indication that dislocations and kinks are not reconstructed in this compound.

Only a few data on the motion of screw dislocations in GaAs under high stress can be found in the literature. Osvenskii and Kholodnyi [9] have investigated the dependence of the velocity of individual dislocations in undoped GaAs upon temperature and stress, by means of the double etch method. The measurements were made on half loops, the initial diameter of which was higher than 200 μm, i.e. when the dislocation velocity is length-independent. The activation energy for the motion of screw dislocations (τ = 50 MPa) is 0.98 ± 0.05 eV. Our own result : $0.55\ \text{eV} < 1/2\ F_{dk}(\tau) + W_m < 1.05\ \text{eV}$ is consistent with this estimation.

REFERENCES

1. J.P. Hirth and J. Lothe, Theory of Dislocations, McGraw-Hill, New York, 1968.
2. F. Louchet, Inst. Phys. Conf. Ser., 60 (1981) 35-38.
3. F. Louchet, D. Cochet-Muchy, Y. Bréchet and J. Pélissier, Phil. Mag. A, 57 (1988) 327-335.
4. P.B. Hirsch, A. Ourmazd and P. Pirouz, Inst. Phys. Conf. Ser., 60 (1981) 29-34.
5. H. Gottschalk, H. Alexander and V. Dietz, Inst. Phys. Conf. Ser., 87 (1987) 339-347.
6. B. Ya Farber, Yu. L. Iunin and V.I. Nikitenko, Phys. Stat. Sol. (a), 97 (1986) 469-478.
7. H.J. Möller and U. Jendrich, in : Deformation of ceramic materials, R.E. Tressler and R.C. Bradt (Eds.), Plenum Press, New York 1984, p. 25-35.
8. P. Gadaud, Thesis, University of Poitiers, 1987.
9. V.B. Osvenskii and L.P. Kholodnyi, Sov. Phys.-Solid State, 14 (1973), 2822-2825.
10. S.A. Erofeeva and Yu. A. Osip'yan, Sov. Phys.-Solid State, 15 (1973) 538-540.
11. S.K. Choi, M. Mihara and T. Ninomiya, Japan. J. Appl. Phys., 16 (1977) 737-745.
12. T. Ninomiya, J. Physique, 40 (1979) C6, 143-145.
13. J. Di Persio and R. Kesteloot, J. Physique, 44 (1983), C4, 469-474.
14. D. Caillard, N. Clément, A. Couret, Y. Androussi, A. Lefebvre and G. Vanderschaeve, 6th Int. Conf. Microsc. S. Cond. Mat., Oxford (1989), in press.
15. K. Maeda and S. Takeuchi, J. Physique, 44 (1983) C4 375-385.
16. H. Gottschalk, J. Physique, 44 (1983) C4 69-75.
17. M. Fnaiech, F. Reynaud, A. Couret and D. Caillard, Phil. Mag. A 55 (1987), 405-423.
18. D. Caillard, N. Clément, A. Couret, Y. Androussi, A. Lefebvre and G. Vanderschaeve, Inst. Phys. Conf. Ser. 87 (1987) 361-366.
19. A. Lefebvre, Y. Androussi and G. Vanderschaeve, Phil. Mag. Letters, 56 (1987) 135-141.
20. Y. Androussi, G. Vanderschaeve and A. Lefebvre, Phil. Mag. A, 59 (1989) 1189-1204.
21. F. Louchet, Phil. Mag. A, 43 (1981) 1289-1297.

IN SITU HVEM OBSERVATIONS OF THE MICRODEFORMATION OF SILICON AND GERMANIUM BICRYSTALS

A. Jacques*, A. George*, H.M. Michaud**, X. Baillin**, J. Pelissier**

* Laboratoire de Physique du Solide, CNRS : URA n°155, Institut National Polytechnique de Lorraine, Nancy, France

** CEA, IRDI, DMECN, Département de Métallurgie, Centre d'Etudes Nucléaires de Grenoble, Grenoble, France

ABSTRACT

In situ straining experiments in the high voltage electron microscope have been designed in order to study dislocation transmission by coïncidence grain boundaries in Si and Ge. Transmitted dislocations appear at pile-up tips, either as perfect 1/2⟨110⟩ dislocations or as Shockley partials trailing extended stacking faults. In coincidence boundaries, the tip of a pile-up can be cleared of residual dislocations by the emission of glissile DSC dislocations and/or the activation of secondary slip.

THE PROBLEM : DISLOCATION TRANSMISSION ACROSS GRAIN BOUNDARIES

Although the title looks more general, the only problem which is addressed in this paper is that of dislocation transmission across grain boundaries (GBs). The starting situation is very commonly encountered in crystalline materials. A dislocation driven by a force meets a GB, which introduces a discontinuity in the geometry of slip. A possibility is that the GB acts as a sink. Since most GBs have ordered structures, the dislocation can dissociate into interfacial dislocations (GBDs) whose Burgers vectors belong to the so-called DSC lattice. Dispersion of such GBDs usually requires climb motion and may be difficult at low temperature. Then, pile-ups form at the GB plane, creating large stress concentrations, so that yielding can locally take place through the interface : new dislocations with a different Burgers vector are created in the opposite grain.

Let us describe in more detail a possible scheme for dislocation transmission across symmetric tilt coincidence boundaries in elemental semiconductors (1-5).

The incoming dislocation, with Burgers vector b^{I}, is dissociated in its slip plane P^{I} in two Shockley partials bordering an intrinsic stacking fault. The leading partial, b_{l}^{I}, enters the GB first. Because GBs are symmetric, there is always a possible slip plane, P^{II}, which meets the GB along the same line as P^{I}. In P^{II}, there are three possible Shockley partials which have the right sign to be pushed away from the GB under the total force (applied plus pile-up) which is locally exerted. On the other hand, there is usually one unit DSC Burgers vector, b^{g}, which makes GBDs glissile in the GB plane. If climb motion is not possible, four reactions are considered :

one of the type $b_l^{I} \rightarrow b^{g} + b_{r1}$ (1)

and three like $b_l^{I} \rightarrow b_l^{II} + b'_{r1}$ (2)

where b_r represent residual GBDs. The respective probabilities of these reactions can be estimated to a first approximation by the Σb^2 criterion. Reaction (1) can be spontaneous while, in most cases, reactions (2) require extra energy. It is of course possible that b'_{r1} further decomposes in a b^{g} plus a smaller residue.

As the next step, the trailing dislocation b_t^{I} enters the GB, reacts with the residue to form a b_{r2} dislocation, which in turn will decompose according to one of the reactions 1 or 2, and so on with next incoming partial dislocations of the pile-up.

Two remarks can be made at this point.

(i) A Shockley partial transmitted in crystal II is still linked to the GB plane by a stacking fault. It is possible that the next transmitted partial has the proper Burgers vector to end up the fault so as to form a full dislocation with $b^{II} = 1/2\langle 110\rangle_{II}$, glissile, and dissociated, in P^{II}. But it is also possible that the second transmitted partial cannot close up the fault and has to trail a new one.

(ii) When several similar dislocations are transmitted from the same pile-up tip, the Burgers vector of the residual GBD tends to increase, which would soon stop the transmission process. The GB can, however, be cleared at some steps because the corresponding residues are glissile either in the GB plane or in slip planes that are available in either grain.

The crossing of high angle GBs by lattice dislocations is likely to be a difficult process, which can take place only when very favourable circumstances are realized, so that in most usual situations, GB crossing is probably bypassed by other dislocation mechanisms. While many observations by conventional TEM can be misleading (6), in situ straining experiments have recently revealed to be the most suitable way to gain clear evidence about dislocation transmission across GBs. We give here a classification of the results obtained in oriented bicrystals of Si and Ge, which show that the theoretical scheme given above can account for the main observed features.

EXPERIMENTAL

Three bicrystals were used. $\Sigma = 9$, tilt axis : [011], rotation angle $\theta = 38.94°$, GB plane : $(1\bar{2}2)_I/(12\bar{2})_{II}$; $\Sigma = 51$ [011], $\theta = 16.10°$, $(1\bar{5}5)_I/(\bar{1}\bar{5}5)_{II}$ and $\Sigma = 25$ [001], $\theta = 16.26°$, $(710)_I/(7\bar{1}0)_{II}$.

In situ straining experiments were performed in the 1 MeV electron microscope of the CENG operated at 400 kV, using the high temperature deformation stage described in (7). Thanks to the two tilt axes available, convenient diffracting conditions could be obtained for each grain simultaneously. Different stress axes were chosen, depending on the transmission reaction which was looked for in any case. The amount of pre-strain was fixed to indroduce few dislocations at the GB plane, confined in one an the same crystal. Thinning of the area close to the GB was done by ion milling. See (3,5) for details. In situ straining experiments were carried out at rather low a temperature (0.5-0.6 T_m, T_m : melting temperature) in order to allow the velocity of dislocations to be measured in real time, without the need of a frame-by-frame analysis and to make sure that dislocation climb is negligible. Dislocation velocities provide a direct estimate of the actual shear stress. Typically, the shear stress which sustained the pile-ups was always found to be about 10^{-3} μ (μ:shear modulus) i.e., in the range of 50-100 MPa. The stress is further reinforced at the tip of pile-ups.

When possible, in situ experiments were interrupted prior to failure and samples were cooled down with the load applied, so that the Burgers vectors and details of the slip geometry could later be determined by conventional TEM.

RESULTS

Transmission of perfect dislocations by the $\Sigma = 51$ GB

The simplest case of dislocation transmission which we have observed is depicted on Fig.1. Activated dislocations were mainly $b = 1/2[\bar{1}\bar{1}0]_I$ (= CD^I in Thompson's notation), gliding in $(1\bar{1}1)_I$. Transmission of perfect dislocations took place in the $(1\bar{1}1)_{II}$ plane. The first dislocation was emitted from the tip of a six-dislocation pile-up and, in the next 90 s., three other dislocations were transmitted, while four new dislocations had joined the pile-up, out of which four dislocations cross-slipped in $(\bar{1}11)_I$ planes. Transmitted dislocations moved fast out of the field of view but were clearly revealed by their slip traces, which vanished out only after some seconds. The observed reaction is very probably :

$$CD^I \rightarrow 1/34\ [\bar{4}3\bar{3}]_I + CD^{II} \qquad (3)$$

which, according to the scheme proposed above, has the lowest energy barriers, at least till the number of transmitted dislocations does not exceed four. Observations suggest that the trailing partial is transmitted just after the leading one, since no abnormal dissociation width is observed in crystal II.

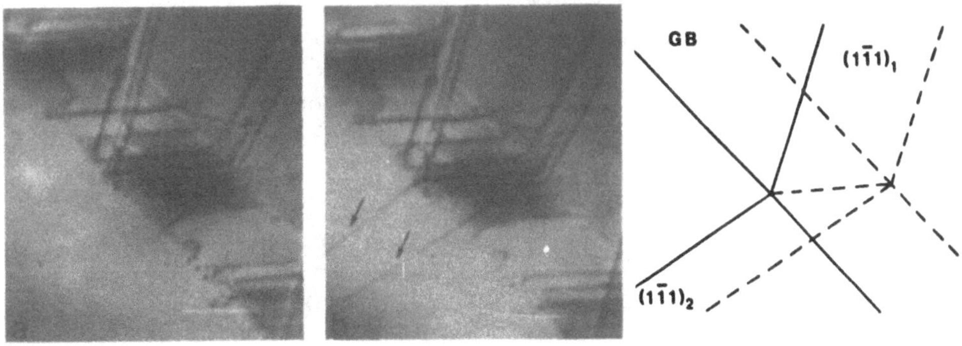

Fig.1 Transmission of perfect dislocations across a $\Sigma = 51$ in Ge (743 K, 85 MPa, stress axis : $[532]_I/[523]_{II}$, $\bar{g}$:022. Marker : 1 μm)

Transmission of partial dislocations by the $\Sigma = 9$ GB

An alternative observation where all transmitted dislocations appear to be Shockley partials is illustrated on Fig.2. Up to nine separate partials were emitted in about 2 mins, while twenty dislocations have joined the pile-up. Incoming dislocations had $b^I = 1/2[110]_I = DC^I$, gliding in $(\bar{1}11)_I$, and most transmitted partials were D_δ^{II} dislocations, gliding in $(\bar{1}11)_{II}$, so that they cannot recombine to form perfect dislocations. It has been checked that GB dislocations with $b^g = 1/18[41\bar{1}]_I$ were also created during in situ straining and that the D_δ^{II} partial was the most highly stressed at the pile-up tip. On the other hand, a cyclic chain of reactions can be determined with the lowest energy steps, which converts two DC^I dislocations into three D_δ^{II} Shockleys and one b^g GBD. Such a cycle leaves the GB free of any residual GBD at the pile-up tip since the $\Sigma = 9$ coïncidence reltionship exactly gives :

$$2DC^I = b^g + 3\ D_\delta^{II} \qquad (4)$$

Such a reaction could lead to micro-twinning, provided that stacking of transmitted dislocation slip planes is the right one, which has not been determined.

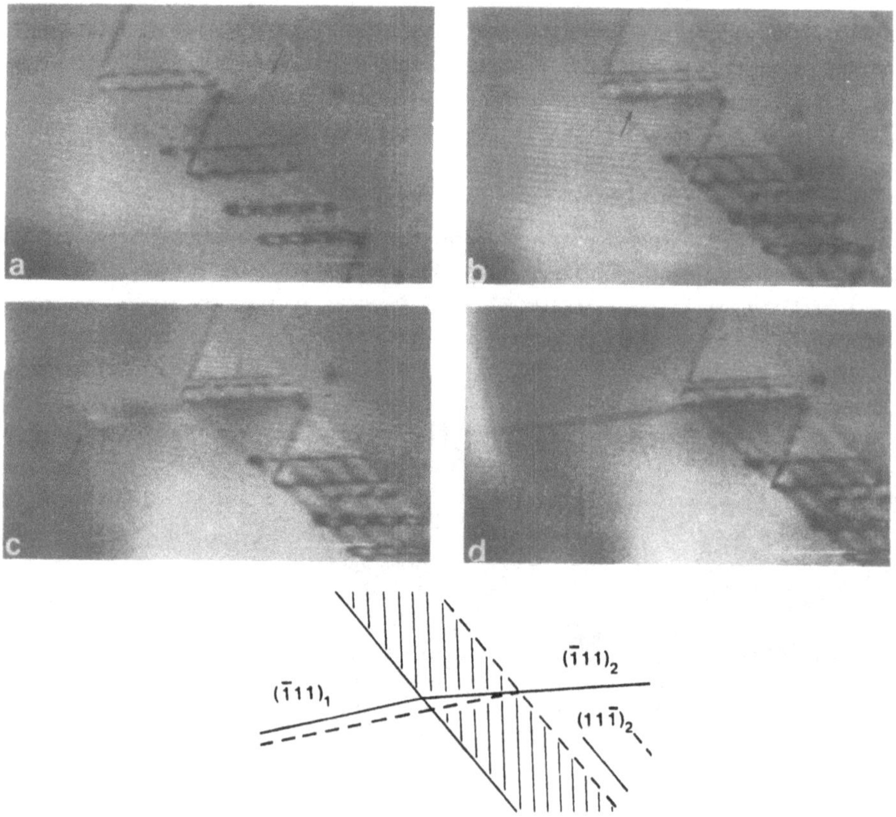

Fig.2 Transmission of Shockley partials across a $\Sigma = 9$ GB in Ge (650 K, 150 MPa, stress axis : $[243]_I/[234]_{II}$, $\bar{g}$: 022. Marker : 1 µm)

Transmission of partial and perfect dislocations with abnormal dissociations by the $\Sigma = 9$ GB.

An intermediary case was observed in a $\Sigma = 9$ bicrystal of Si, stressed along $[1\bar{2}2]_I/[\bar{1}\bar{2}2]_{II}$ at 840 K, 100 MPa. A detailed description of this intricate sequence is given elsewhere (5). Fig.3 gives salient features. Dislocations of the primary slip system $(11\bar{1})_I[101]_I$ pile-up against the GB plane where they lie parallel to the tilt axis. On the selected frames one can see the transmission of a γA^{II} Shockley dislocation in $(11\bar{1})_{II}$. This partial trailed an extended intrinsic fault till a second partial with $b = D\gamma^{II}$ was emitted from the same line. This partial was seen to close the fault, catch up the leading γA^{II} and form with it a perfect DA^{II} dislocation.

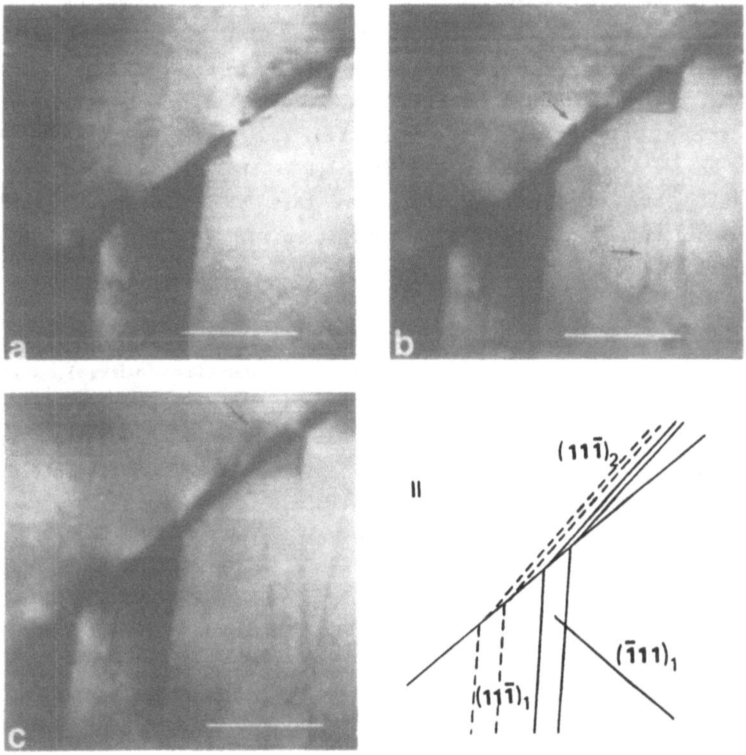

Fig.3 Transmission of partial and perfect dislocations across a $\Sigma = 9$ GB in Si (840 K, 100 MPa, stress axis : $[1\bar{2}2]_I/[\bar{1}\bar{2}2]_{II}$, $\bar{g}$: $2\bar{2}0_I/02\bar{2}_{II}$. Marker : 1 μm)

An additional interesting feature was the observation of CA^I dislocations ($b = 1/2[0\bar{1}1]_I$) gliding in (111) planes of crystal I. Such loops expanded from the pile-up tip immediately after a few transmission events. A possibility is that one pile-up dislocation, DA^{I}, plus the next leading partial, γA^{I}, were converted into one DA^{II} dislocation ($\gamma A^{II} + D\gamma^{II}$), one b^{g} dislocation, glissile in the GB plane, and one perfect CA^{I} dislocation, which can be re-emitted in crystal I in the $(100)_I$ cube plane and then in $(111)_I$, since the $\Sigma = 9$ coincidence ensures that :

$$DA^{I} + \gamma A^{I} = DA^{II} + b^{g} + CA^{I}$$

The new finding here is that secondary slip systems may be activated in order to clear the GB plane of residual dislocations.

Transmission of dislocations across the $\Sigma = 25$ GB

Our purpose with that bicrystal, stressed along $[430]_I/[340]_{II}$, was to check the possibility of transmission of $(\bar{1}11)_{II}[101]_{II}$ dislocations from $(\bar{1}11)_I[101]_I$ pile-ups. Such a reaction is favoured by different factors : the difference between the incoming and transmitted Burgers vectors is very small and glissile in the GB plane. However, contrarily to the previous cases, the corresponding slip planes do not match at the GB plane, their lines of intersection with the GB are at an angle of 8.21°.

Instead of the expected reaction, a complicated configuration was observed, involving several slip systems in the grain where the first pile-up forms. Transmission was indeed observed, but in planes which do match at the GB.

We take this as a strong indication that slip plane matching at the GB is a necessary condition for dislocation transmission, at least at such moderate temperatures.

The Σb^2 criterion neglects the stacking fault energies. In order to explain why transmitted dislocations look sometimes perfect sometimes partial, a slightly more elaborate treatment was done, taking into account the string action of the stacking faults on partials and also the fact that dislocation interactions depend on the character of dislocation lines. This calculation shows that, for the first dislocation in the $\Sigma = 51$ bicrystal, the force wich is required to transmit the leading partial is much larger than the forces needed either to incorporate the trailing partial in the GB or to transmit the trailing partial in grain II. In the $\Sigma = 9$ case, on the contrary, the first step is not the most difficult. It is then believed that extended faults should form in the $\Sigma = 9$ cases and not in the $\Sigma = 51$ case.

1 D.A. Smith, J. de Physique 43 (1982) C6-225.

2 A.H. King and F.R. Chen, Mater. Sci. Engng 66 (1984) 227.

3 X. Baillin, J. Pelissier, J.J. Bacmann, A. Jacques and A. George, Phil. Mag. A55 (1987) 143.

4 A. Jacques, A. George, X. Baillin and J.J. Bacmann, Phil. Mag. A55 (1987) 165.

5 X. Baillin, J. Pelissier, A. Jacques and A. George, Phil. Mag.A (to be published).

6 M. Martinez-Hernandez, H.O.K. Kirchner, A. Korner, A. George and J.P. Michel, Phil. Mag. A 56 (1987) 641.

7 J. Pelissier, J.J. Lopez and P. Debrenne, Electron. Microsc. 4 (1980) 30.

DISLOCATION STRUCTURE IN LiF CRYSTALS DEFORMED AT LOW TEMPERATURES

H.-J. Kaufmann*, S.V. Lubenets**, T.V. Samoilova***, B. I. Smirnov***

*Institute of Semiconductor Physics, Academy of Siences of the DDR, Low Temperatures Lab., Berlin 1040, DDR
**Physico-Technical Institute of Low Temperature Physics, Academy of Siences of the Ukrain SSR, Kharkov 310164, USSR
***A.F.Ioffe Physico-Technical Institute, Academy of Siences of the USSR, Leningrad 194021, USSR

ABSTRACT

Selective etching and electron-microscopic replica method were used to study the dislocation structure of LiF crystals deformed in the range 4.2-293 K. In case of the single slip the densities of both the screw and edge components of dislocations increased linearly with strain. The screw dislocation multiplication coefficient was proportional to the yield stress (which increased considerably as a result of reduction of the deformation temperature) and the probability of cross slip of these dislocations was proportional to the square of the yield stress. The distribution of the edge dislocation dipole heights roughly agrees with the Wiedersich model.

INTRODUCTION

Transmission electron microscopy has found a wide application nowadays in studies of dislocation structure in crystals /1/. However, this method is optimal not in all cases. In particular, it is not widely used for alkali halide crystals which are convenient model materials in the field of physics of strength and plasticity. First of all, this is due to the fact, that the irradiation damages during their TEM investigation are formed, and as a result the initial dislocation structure is completely distorted. This effect can be partly avoided by studying these crystals under an HVEM at very low temperature. However, such experiments may be single ones (e.g., /2,3/). For this reason the dislocation structure of alkali halide crystals are commonly studied by selective etching and then imaging the etch pit replica pattern in a conventional electron microscope.

In the present work we use this method to study the dislocation structure of LiF crystals deformed at the temperatures in the range 4.2-293 K. The paper is concerned with the determination of (i) screw and edge dislocation components density, (ii) the pair distribution of dislocations in edge dipoles in respect of the internal distance in pairs and (iii) screw dislocation cross-slip and multiplication parameters.

EXPERIMENTAL

A large Kyropoulos-grown pure single crystal was used. The specimens 3 x 4 x 10 mm^3 in size were cleaved after thermal treatment which involved annealing at 1020 K for 48 h and then cooling at a rate of 5 K/h. As a result of such treatment the dislocation density was lowered to 10^4 cm^{-2}. The specimens were deformed in conventional compression /4/ or by single slip /5/ at constant rate of 10^{-4} s^{-1}.

The dislocation structure in deformed crystals was revealed by selective etching at RT. Screw and edge dislocation components were revealed in different side faces of the specimen. The density of the etch pits N was deduced from electron micrographs obtained for replicas deposited on etched surfaces. A possible surface effect on the dislocation density was avoided by cleaving a sample along (100) und (010) planes after the deformation but before etching /6, 7/. In the case of edge dislocation components the dislocation dipoles were of particular interest.

RESULTS AND DISCUSSION

The stress-strain diagrams of the crystals deformed by single slip at all temperatures consist of two regions: a yield plateau where the shear stress $\tau = \tau_y$ = Const (τ_y is the yield stress) and a linear hardening region. The value of the yield stress increased by an order of magnitude when the temperature decreased from 293 to 4.2 K.

Throughout the investigated range of temperatures the plastic flow of the crystals at the yield stress involves formation and growth of slip bands. The density of the etch pits in the edge slip bands, N_{ey}, is considerably higher than in the screw bands, N_{sy}. The relationship between N_y and τ_y is approximately linear /8/.

At the hardening region the plastic deformation develops homogeneously and the dislocation density increases gradually with the strain. The relationship between the screw dislocation density $\rho_s = \sqrt{2}\, N_s$ and the strain ε may be expressed as $\rho_s = \beta_s \cdot \varepsilon$, where β_s - the coefficient which characterizes screw dislocation multiplication. The dependence $N_e(\varepsilon)$ is the same, $N_e = \beta_e \cdot \varepsilon$, but values of β_s and β_e increase by about an order of magnitude when the deformation temperature changes from 293 to 4.2 K.

The dependence of the screw dislocation multiplication coefficient $\delta_s = 2\, b\, d\rho_s/d\varepsilon = 2\, b\, \beta_s$ (b is absolute value of the Burgers vector) on τ_y is plotted in Fig. 1 for various deformation temperatures. It is seen

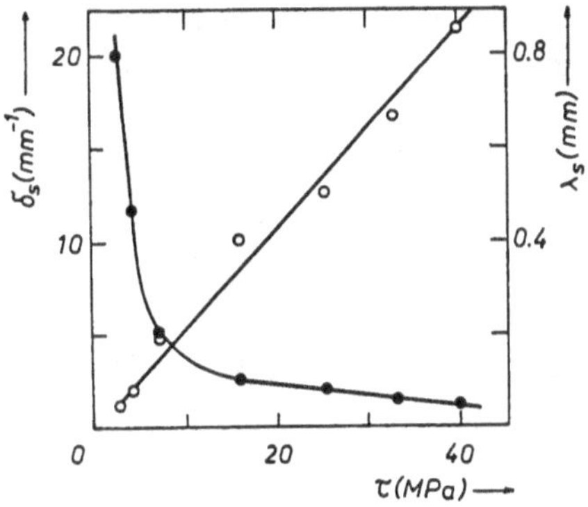

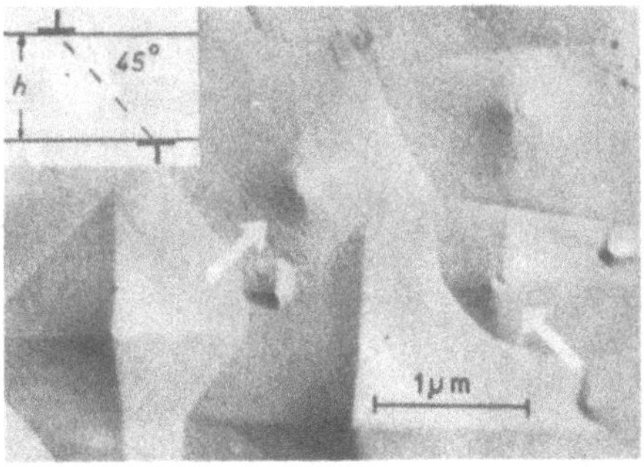

Fig. 1 Dependences of δ_s (1) and of λ_s (2) on the yield stress at temperatures in the range 4.2 - 293 K.

Fig. 2 Schematic drawing and etch pit pattern of the edge dipoles. Slip planes are horizontal (along the dia gonals of etch pits).

that the experimental points fit approximately a straight line passing through the origin of the coordinate system, i.e.,

$$\delta_s = B\tau_y, \tag{1}$$

where $B = 5.3 \times 10^{-7}$ m/N. It should be noted that the similar dependence $\delta_s(\tau_y)$ has been observed earlier /9/ in the case when τ_y changes due to a variation of deformation rate. The dependence of the average slip distance of the screw dislocation $\lambda_s = 1/\delta_s$ on τ_y is also shown in Fig. 1. At the single slip the dislocation multiplication coefficient remains constant when the flow stress increases. In the case of multiple slip the value of δ_s increases with strain.

The correlation between the dislocation multiplication coefficient and the yield stress, observed in a wide range of temperatures and strain rates, clearly indicates that the dislocation multiplication mechanism remains constant. In the case of alkali halide crystals this is the double cross slip mechanism. This mechanism explains the difference between the densities N_s and N_e. The difference is due to the fact that the edge components contain not only dislocation but also large numbers of dislocation dipoles formed during the motion of the screw dislocations undergoing cross slip /10/.

In order to describe dislocation multiplication by double cross slip we refer to Wiedersich's formula /11/ for the number Q of ejections of height $y > h$ into the cross slip plane

$$Q\ (y > h) = q \cdot \exp(-ph/b), \tag{2}$$

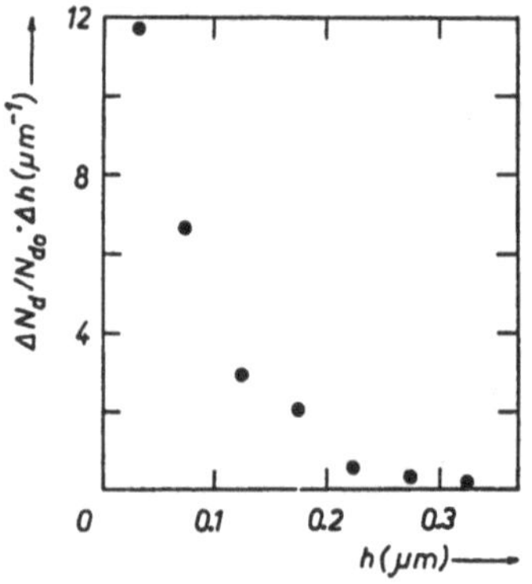

Fig. 3 Distribution of the edge dipole heights h in LiF crystals deformed at 4.2 K; N_{do} = 778.

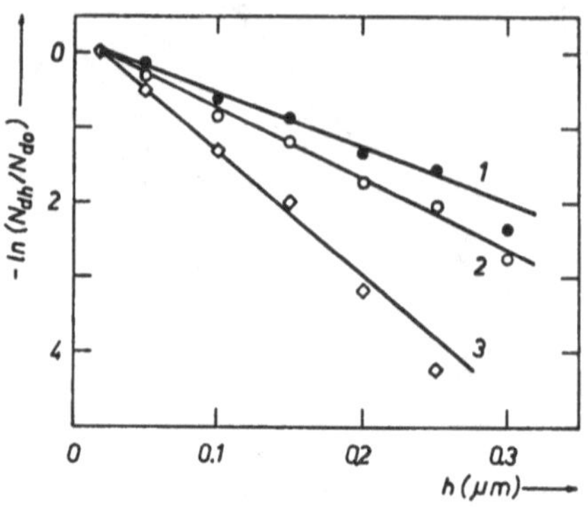

Fig. 4 Dependences of the number of dipoles with height larger than h on h in LiF crystals deformed at 293 (1), 77 (2) and 4.2 K (3); N_{do} = 286 (1), 388 (2) and 778 (3).

q being the number of cross slip events on a srew dislocation per unit area swept by the dislocation, p the probability of transition of the dislocation from the cross slip plane into one of the primary slip planes. It has been shown earlier /9/ that experimental data on the change in N_s and N_e during single slip enable one to calculate q. In the same approximation /9/ we get

$$q \approx \delta_s^2 N_e/2 \sqrt{2} N_s , \tag{3}$$

Substituting Eq. (1) in Eq. (3), we find that

$$q \approx C\tau_y^2 \tag{4}$$

where $C = 9 \times 10^{-13} m^2/N^2$.

The parameter p in Eq. (2) can be determined from the experimental data on the distribution of the cross slip distances of individual dislocations in gold decoration patterns /12, 13/ and also from the distribution of the distances between two dislocations in edge dipoles /9/. The latter method was used in the present work.

The edge dislocation dipole is shown schematically in Fig 2, a. The dipole represents a pair of edge dislocations with opposite signs lying along a line enclosing an angles of 45° or 135° with the slip planes. The distance between these planes is h. Typical etch pattern of the edge dipoles is shown in Fig. 2b.

In order to determine the distribution of the dipole heights h, the number ΔN_d of dipoles within a length class of width Δh was plotted as $\Delta N_d/N_{do} \cdot \Delta h$ versus h where N_{do} is the total number of dipoles observed. Fig. 3 shows a typical distribution of the distances h in LiF crystals deformed at 4.2 K. It is seen that the distribution is caracterized by a rapid decrease of the $\Delta N/\Delta h$ at the beginning, leading to more slowly decreasing with increasing h.

Fig. 4 represents the dependences of the number N_{dh} of dipoles with height larger than h on h in the coordinate system $\ln(N_{dh}/N_{do})$ - h in LiF crystals deformed at 4.2, 77 and 293 K. It is seen that in all cases experimental points coincide with straight lines within the accuracy of experiment. In other words, the relation $N_{dh} \sim \exp(-kh)$ is valid. Since $Q(y>h) \sim N_{dh}$, the obtained results may be considered to agree with Eq. (2). Therefore, the value $p = b \cdot k$ may be determined from the slope of straight lines in Fig. 4. It is seen that the value of p increases with decreasing of deformation temperature and equals to 2.5, 3.2 and 5.6 x 10^{-3} at 293, 77 and 4.2 K, respectively. The results obtained are in qualitative agreement with the theory /14/.

CONCLUSIONS

The present study allows the following conclusions.

(i) The rate of dislocation multiplication strongly increases with decreasing deformation temperature, the multiplication coefficient being approximately proportional to the yield stress of crystals.

(ii) The number of cross slip events on a screw dislocation per unit area swept by dislocation is proportional to the square of the yield stress.

(iii) The distribution of the edge dipole heights roughly agrees with the Wiedersich model. The probability of transition of the dislocation from cross slip plane into one of the primary slip planes increases with decreasing of deformation temperature.

REFERENCES

1. Yamada Conference IX "Dislocation in Solids", H. Suzuki, T. Ninomiya, K. Sumino, s. Takeuchi (Eds.), University of Tokyo Press 1985, 673 p.
2. J. Hesse and L. W. Hobbs, phys.status solidi (a) 4 (1972) 599-604.
3. H. Strunk, Mater. Sci. Eng. 27 (1977) 225-38.
4. H.-J. Kaufmann, S. V. Lubenets and E. I. Ostopchuk, phys.status solidi (a) 66 (1981) 229-233.
5. B. I. Smirnov, Fiz. Tverd. Tela 10 (1968) 2689-96.
6. T. V. Samoilova and B. I. Smirnov, Fiz. Tverd. Tela 20 (1978) 2645-8.
7. O. V. Klyavin, Yu. M. Chernov, N. N. Pravdina and I. I. Rykova, Fiz.

Tverd. Tela 20 (1978) 3100-6.
8. H.-J. Kaufmann, S. V. Lubenets, T. V. Samoilova and B. I. Smirnov, Fiz. Tverd. Tela 26 (1984) 702-7.
9. B. I. Smirnov, Dislocation Structure and Hardening of Crystals, Nauka, Leningrad 1981, 235 p., in Russian.
10. J. J. Gilman and W. G. Johnston, Solid State Phys. 13 (1962) 14-222.
11. H. Wiedersich, J. Appl. Phys. 33 (1962) 854-8.
12. F. Appel, U. Messerschmidt and B. I. Smirnov, Soviet-German Symp. on Solid State Phys. and Mater. Sci., Tbilissi 1976, Moscow 1976, p. 51-5, in Russian.
13. F. Appel, H. Grube, U. Messerschmidt and B. I. Smirnov, Cryst. Lattice Defects 7 (1977) 65-70.
14. A. N. Orlov, V. N. Perevesentsev and B. I. Smirnov, phys. status solidi (a) 32 (1975) 35-42.

PERFECT AND PARTIAL DISLOCATIONS IN MICRODEFORMED GaAs

H. S. Leipner*, J. Schreiber*, M. Surowiec**

*Sektion Physik der Martin-Luther-Universität Halle-Wittenberg, DDR-4010 Halle, GDR

**Instytut Fizyki i Chemii Metali, Uniwersytet Śląsky, Bankowa 12, PL-40007 Katowice, Poland

ABSTRACT

Using electron microscopy the dislocation movement in the strain field of indents and microscratches for (001) and {111} oriented n-GaAs was studied. The dynamics, interaction and splitting of screw-B(g)-screw and screw-A(g)-screw dislocations in dependence on temperature and indentation load determine the observed dislocation distribution.

INTRODUCTION

Microindentation is a commonly used method to characterize the low temperature deformation behaviour of brittle semiconductors /1-5/. A(g) and B(g) dislocations (A(g) corresponds to β, B(g) to α dislocation /6/. g stands for the supposed glide set.) on separate slip systems were identified by selective etching /1/, x-ray topography /7/ and transmission electron microscopy (TEM) /2,5/ in the plastic zone of indents on (001) oriented compound semiconductors. Based on etch pit methods the activation of different slip systems in connection with stress field evaluations was analyzed /3,4/.

In this contribution new experimental material is given about the generation and propagation of dislocation loops in the strain field of indents and microscratches for (001) and ±(111) oriented n-GaAs.

EXPERIMENTAL PROCEDURE

GaAs single crystals grown by the liquid encapsulated Czochralski technique (Si doped, carrier concentration about 3.10^{18} cm^{-3}) were prepared in (001) and ±(111) slices by standard cutting and polishing procedures. The distinction between [110] and [$1\bar{1}0$] directions in (001) oriented samples essential for the determination of the dislocation type /1/ was done using the anisotropic shape of etch pits /8/. This method is also applicable for the orientation determination in the electron microscope because of the occurence of characteristically etched "line defects" with an elongation in [1$\bar{1}$0] direction /9/. In the case of {111} orientation A and B faces were determined by etching with a solution of bromine in methanol. The A surface remains mirrorlike whereas on the B surface etch pits were revealed.

Indentation and scratching were carried out in air using a standard Vickers indenter. The deformation was performed i) at room temperature and followed by a heat treatment at different temperatures T for 1 h in an inert gas atmosphere or ii) at elevated temperatures T. The dislocation distribution was studied using scanning electron microscopy with detection of the integral cathodoluminescence signal (SEM-CL). The preparation of the indents for TEM followed the procedure given elsewhere /2/. The investigations were done with a high voltage electron microscope at an acceleration voltage of 1 MV. The presented TEM micrographs are bright field images.

RESULTS AND DISCUSSION

The characteristic feature of indentation of GaAs is the generation of dislocation loops. The expansion of perfect and partial dislocations in dependence of temperature was investigated for (001) GaAs. It was observed that the defect distribution is qualitatively the same for indents (i) done at room temperature with a following heat treatment at temperature T compared to indents (ii) introduced at high temperature T. For indentations at room temperature with no additional heat treatment no single dislocations can be recognized using SEM-CL. The dislocations are strongly located near the indentation site with a maximum extension of 1 μm (indentation load 0.01 N). For a load of 0.03 N the elongation is up to 5 μm (fig. 1). In fig. 1 perfect screw-B(g)-screw loops with a Burgers vector (BV) $\bar{b}$ = 1/2[110] and partial dislocations with BV $\bar{b}_p$ = 1/6[211] or the corresponding stacking faults are visible. In other cases partial dislocations with 1/6[12$\bar{1}$] are found. The BV are identified with the standard extinction rule $\bar{g}.\bar{b} = 0$, $\bar{g}$ - diffraction vector. The polarity of the leading segments of perfect loops in approximately 60° orientation was determined after the glide prism model /1,2,10/. According to this model A(g) dislocations are generated on ($\bar{1}\bar{1}\bar{1}$) and (11$\bar{1}$) planes gliding in [1$\bar{1}$0] direction and B(g) on ($\bar{1}$1$\bar{1}$) and (1$\bar{1}\bar{1}$) in [110]. The slip planes are confirmed by stereo imaging. Similar loops as in fig. 1 are found in [1$\bar{1}$0] direction whereas the nonscrew dislocations are of the A(g) type. The BV of screw-A(g)-screw perfect loops is 1/2[1$\bar{1}$0], for partials was found 1/6[2$\bar{1}\bar{1}$] and 1/6[1$\bar{2}\bar{1}$]. Stacking faults or partials are in this case in a lower distance to the indentation centre compared to [110].

For temperatures up to 800 K an extension of the dislocation rosette

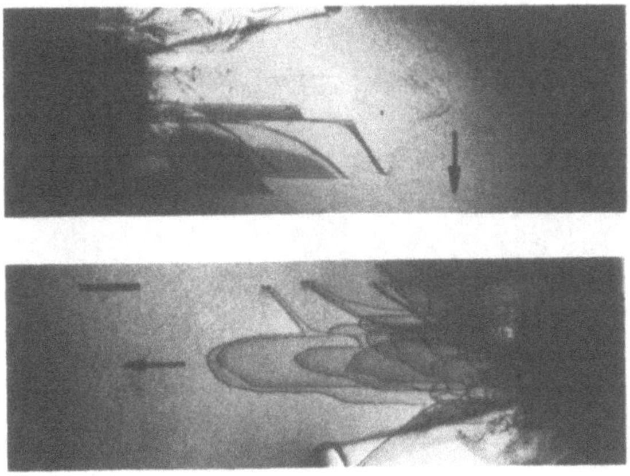

Fig. 1 Dislocations and stacking faults elongated in [110] from the indentation site (dark area on the left). Indentation with 0.03 N at room temperature. Diffraction vector $2\bar{2}0$ (a) and 220 (b). Marker 1 μm. The fringe contrast in the upper part is due to a crack.

up to 50 μm could be observed by SEM-CL imaging. According to the different dislocation mobility the rosette arm in [110] with B(g) is much more elongated than the [$1\bar{1}0$] arm with A(g) dislocations. From measuring the length of different rosette arms it is possible to estimate the mobility ratio for A(g) and B(g) 60° dislocations /11/. Furthermore, dislocations introduced by microdeformation exhibit a radiation enhanced glide observed by SEM and TEM at accelerating voltages lower than 100 kV. The glide mobility is considerable even for liquid nitrogen temperature. The enhancement of dislocation glide could be attributed to the effect of recombination enhanced defect reactions (REDR) /12/.

In the temperature range between 300 and 800 K multiple stacking faults are found preferentially in the [110] direction. In many cases they are confined to the margin of the glide prism (fig. 2). The comparison of the dislocation structure of fig. 1 with fig. 2 shows that in the latter not full perfect loops are visible but shorter cusps and only a few long screw dislocations. This is due to the fact that certain screw parts of loops during further movement reach the surface or annihilate with screws of the second activated slip system inside the glide prism. Fig. 3a gives schematically the model for expansion of screw-B(g)-screw perfect loops in the [110] glide prism. The movement of dislocations inside the [$1\bar{1}0$] prism is similar. The expansion and interaction of "hexagonal" shaped loops leads to the formation of prismatic dislocations. In the original HU model /10/ the generation of prismatic loops is supposed *ab initio*. For temperatures above 800 K only prismatic loops are found in the rosette arms. In no slip system stacking faults are identified. The dynamics of the dislocation loops was studied by *in-situ* heating experiments in the HVEM /11/. These investigations confirm the model of fig. 3.

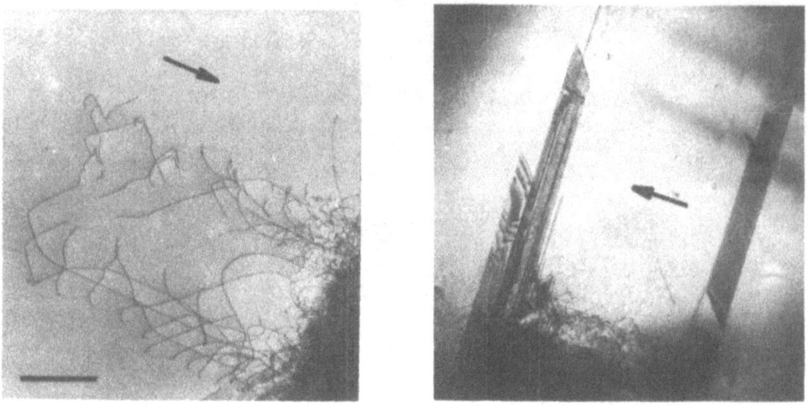

Fig. 2 a) Perfect dislocations in the $[1\bar{1}0]$ glide prism with $\bar{b} = 1/2[1\bar{1}0]$ on $(\bar{1}\bar{1}\bar{1})$ resp. $(11\bar{1})$ slip planes. $\bar{g} = \bar{2}20$. Indentation with 0.15 N. b) Multiple stacking faults elongated in [110], $\bar{g} = 2\bar{2}0$. Indentation load 0.30 N. (001) GaAs, heat treatment 623 K, 1 h. Marker 1 μm.

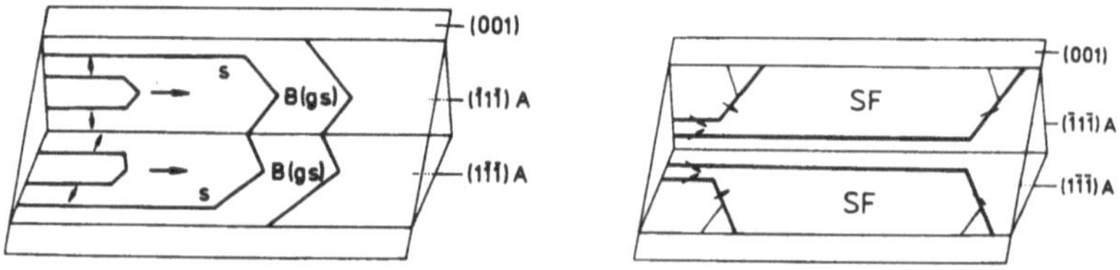

Fig. 3 a) Expansion and reaction of perfect loops inside the [110] prism. The arrows show the glide direction, s - screw, B(gs) $\equiv$ B(g) dislocation. b) Dissociation of loop segments. The arrows indicate the BV of partial dislocations.

Fig. 3b gives a model for the occurence of stacking faults resp. partial dislocations based on the dissociation of dislocations in the sphalerite structure: 0° ---> 30° A(g)/30° B(g), 60° A(g) ---> 30° A(g)/90° A(g), 60° B(g) ---> 30° B(g)/90° B(g).
In regions with very high values of the resolved shear stress the dissociation width is extraordinarily high. The movement of separated partial dislocations is a characteristic feature of the plastic deformation of semiconductors in the low temperature/high stress range /13/. The above mentioned "perfect" loops are dislocations in approximately equilibrium dissociation. In fig. 1 dislocation segments with different dissociation widths suffering different shear stress are directly connected. It should be mentioned that generally in the slip plane **ABC** in most cases only leading partial dislocations with BV **Aδ** and orientation **AC** are visible. The trailing dislocations **δB** are in the highly dislocated area nearby the indent. Using *in-situ* heating experiments the development of dislocations with almost equilibrium dissociation from separated partials for

$T \geq 823$ K was shown /14/. This is in correspondence with the above mentioned fact that for temperatures higher than 800 K no stacking faults are observed.

The leading segments of the widely separated loops are in most cases in the 30° orientation (AC). It is concluded that 90° partials have a higher mobility than 30°. Taking into account the lower extension of dislocations in $[1\bar{1}0]$ we get for the mobilities μ:

$$\mu_{90B} \gg \mu_{30B} \gtrsim \mu_{90A} \gg \mu_{30A}.$$

It has to be noted that the mobility of separated partials is lower than the mobility of "perfect" dislocations (fig. 1).

The generation of hexagonal dislocation loops was also observed for slip systems with inclined BV (fig. 4) activated immediately at the indent. Similar dislocations were found in the deformed area of microscratches in (001) GaAs /11/.

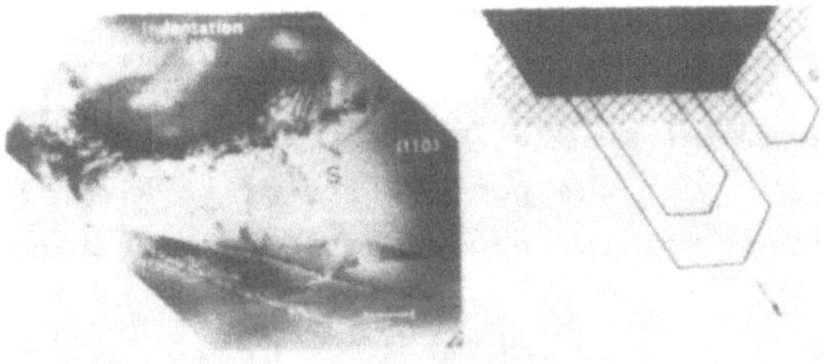

Fig. 4 Cross section micrograph of an indented area in (001) GaAs showing dislocation loops of the $(11\bar{1})[0\bar{1}\bar{1}]$ slip system

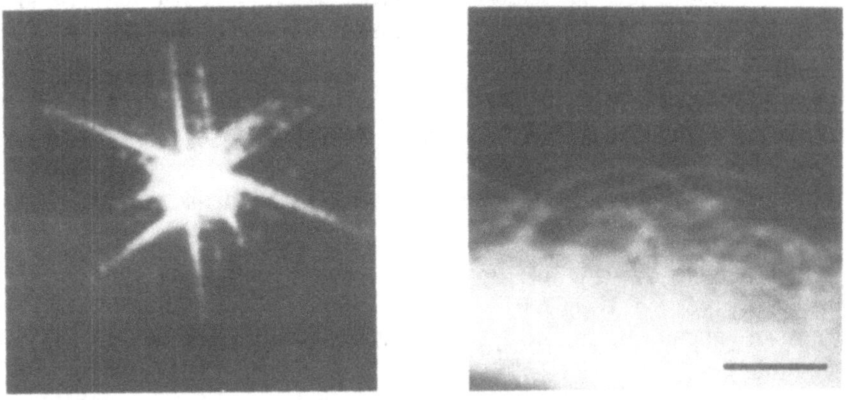

Fig. 5 a) Dislocation rosette of an indent, load 0.20 N. b) Irregular shaped loops at a microscratch in $[\bar{2}11]$ direction. Scratching with a load of about 0.01 N. $(\bar{1}\bar{1}\bar{1})$ GaAs, heat treatment 623 K, 1 h. SEM-CL, inverse contrast, 15 kV. Marker 10 μm.

For indentations and microscratches on $\pm(111)$ GaAs dislocation glide occur on {111} planes forming a tetrahedron with apex external to the specimen an on planes of a tetrahedron with apex internal /3/. According to this model a threefold symmetry of the dislocation rosette around an indent is observed (fig. 5a). It is supposed that for a ±(111) orientation mainly screw-B(g)-screw dislocations are generated /15/. These loops lie in slip planes inclined to the surface as well as in a parallel plane. For loads $\leq$ 0.01 N the loops parallel to the surface are dominating (fig. 5b).

In many cases the surface parallel loops exhibit an anisotropic shape in such a way that one screw segment is missing. This is confirmed by TEM. The explanation for this behaviour follows the model of splitted hexagonal loops. The initial loops having its screw segments split in cross slip planes looses the more mobile segment with 30° B(g) leading by slip towards the surface. An alternative mechanism could be the annihilation of screw segments by interaction with dislocation loops on inclined slip planes.

ACKNOWLEDGEMENTS We are grateful to the Institut für Festkörperphysik und Elektronenmikroskopie Halle, for the possibility of work on the HVEM. Thanks are due to Dr. H. Bartsch for the cross section preparation.

REFERENCES

/ 1/ P.D. WARREN, P. PIROUZ and S.G. ROBERTS: Phil. Mag. A **50** (1984) L23
/ 2/ H.-R. HÖCHE and J. SCHREIBER: phys. stat. sol. (a) **86** (1984) 229
/ 3/ P.B. HIRSCH, P. PIROUZ, S. G. ROBERTS and P.D. WARREN: Phil. Mag. B **52** (1985) 759
/ 4/ S.G. ROBERTS, P.D. WARREN and P.B. HIRSCH: J. Mat. Res. **1** (1986) 162
/ 5/ A. LEFEBVRE, Y. ANDROUSSI and G. VANDERSCHAEVE: phys. stat. sol. (a) **99** (1987) 405
/ 6/ H. ALEXANDER, P. HAASEN, R. LABUSCH and W. SCHRÖTER: J. Physique **40** (1979) C4-foreword
/ 7/ M.S. SUROWIEC and B.K. TANNER: J. Appl. Cryst. **20** (1987) 499
/ 8/ S. KOMIYA and T. KOTANI: J. Electrochem. Soc. **125** (1978) 2019
/ 9/ H.-R. HÖCHE, H. LEIPNER and G. STADERMANN: phys. stat. sol. (a) **98** (1986) 503
/10/ S.M. HU: J. Appl. Phys. **46** (1974) 1470
/11/ H.S. LEIPNER: Dissertation, Halle: Universität 1988
/12/ J. SCHREIBER, H.S. LEIPNER: Sov. J. Quant. Electron. **15** (1988) 2304
/13/ R. MEINGAST and H. ALEXANDER: phys. stat. sol. (a) **17** (1973) 229
/14/ H.S. LEIPNER, H.-R. HÖCHE and J. SCHREIBER: Izv. Ak. Nauk SSSR Ser. Fiz. **51** (1987) 1498
/15/ H.S. LEIPNER and M.R. SUROWIEC: Mat. Sci. Forum (1989) in press

OBSERVATION OF DISLOCATION GLIDE IN SPHALERITE ZnS BY TEM "IN SITU" DEFORMATION

C. Levade*, J.J. Couderc*, D. Caillard**, A. Couret**

* Laboratoire de Physique des Solides, associé au CNRS
INSA, 31077 Toulouse-Cedex, France

** CEMES, Laboratoire d'Optique Electronique du CNRS
BP 4347, 31055 Toulouse-Cedex, France

ABSTRACT

Plastic properties of sphalerite ZnS single crystals are studied by TEM "in situ" deformation. The dislocation velocity is measured near sources at T = 300 - 530 K and the motion of 30° and 90° partials is studied. The velocity of dislocations is proportional to their length and no significant difference can be detected between screw and 60° dislocation rate. The electron beam is found to have a great enhancement effect on dislocation velocity.

INTRODUCTION

Previous work [1] has shown that dislocation movement under stress in sphalerite ZnS single crystals is thermally activated ; moreover, near room temperature, dislocations seem to experience a friction force of the Peierls type and move by nucleation and propagation of kinks like the other II-VI compounds. Results obtained by "in situ" straining TEM experiments [2] on several II-VI compounds (wurtzite type : CdS, CdSe, ZnO, ZnS and sphalerite type : CdTe) can be summarized as follows : i) dislocation glide is generally steady and continuous ; ii) the screw dislocations have the lowest mobility ; iii) cross slips frequently occur in wurtzite crystals and sometimes in CdTe and dislocations exhibit a wavy glide. Furthermore, it has been shown by SEM [3] that the electron beam strongly influences the dislocation mobility.

This paper reports the first quantitative results on dislocation glide in ZnS sphalerite observed by TEM "in situ" deformation.

EXPERIMENTAL TECHNIQUE

Natural sphalerite single crystals were 1 % pre-strained at 573 K by [011] uniaxial compression. Slices parallel to (011) were cut using a wire saw and then chemically thinned. Experiments were performed in the temperature range 300 - 530 K, at 200 kV using a tensile holder [4] and video recording. The tensile axis was either

[100] or [2$\bar{1}$1].

The local value of the shear stress was evaluated from the dislocation radius of curvature R using the classical formula :

$$\tau = \frac{\mu b}{4\pi(1-\nu)R} \log \frac{R}{b}$$

$$\mu = 3 \cdot 10^{10} \text{ Pa} ; \quad \nu = 0.32 ; \ b = 3.82 \cdot 10^{10} \text{ m.}$$

The beam intensity was measured using the current intensity of the photosensitive cells located near the observation screen of the microscope.

EXPERIMENTAL RESULTS

The motion of many perfect dislocations the line of which was parallel to <110> was observed. The dislocation lines were almost straight at T < 373 K and bent for higher temperatures. In all the studied range (300 -530 K), the dislocations glide in a viscous and continuous manner.

We focus here on the study of partial dislocations, the analysis of two sources (300 and 453 K) and the effect of the electron beam.

Partial Dislocations

The movement of widely dissociated dislocations in the (111) glide plane was observed at T = 453 K (Fig. 1). The leading partial ($b_2 = \frac{1}{6}$ [$\bar{1}$2$\bar{1}$]) has an edge character for u//[10$\bar{1}$] and is a 30° partial for u//[$\bar{1}$10] while the trailing one with $b_1 = \frac{1}{6}$[11$\bar{2}$] is a 30° partial for u//[10$\bar{1}$] and 90° partial for u//[$\bar{1}$10].

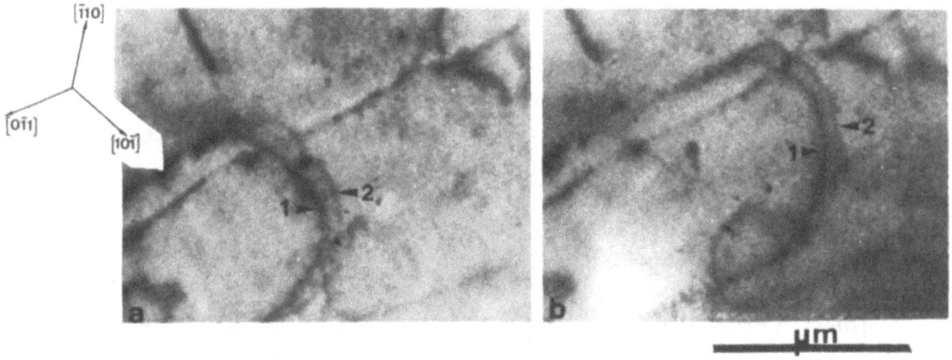

Fig. 1 Motion of a widely dissociated dislocation in plane (111) at T = 453 K.
1 : $b_1 = \frac{1}{6}$ [11$\bar{2}$] 30° partial for u//[10$\bar{1}$] and 90° partial for u//[$\bar{1}$10].
2 : $b_2 = \frac{1}{6}$ [$\bar{1}$2$\bar{1}$] 90° partial for u//[10$\bar{1}$] and 30° partial for u//[$\bar{1}$10]
a = 0 s ; b = 1.72 s.

The width of the fault ribbon ($d = 80 - 120$ nm) is greater than the equilibrium distance ($d_o = 25$ nm) since the force F_2 acting on dislocation 2 is higher than the force F_1 on dislocation 1. Following Wessel and Alexander [5], an analysis of the dissociation width d can be made by defining for each partial i a friction force R_i which balances, during steady state motion, the force F_i, the stacking fault tension γ and the repulsive interaction of the two partials :

$$\frac{d}{d_o} = \left[1 + (f - \delta) \frac{\tau b}{2 \gamma} \right]^{-1} \quad (1)$$

with $f = \frac{F_1 - F_2}{F_1 + F_2}$, $\delta = \frac{R_1 - R_2}{R_1 + R_2}$

Applying this analysis to the observed configuration shows that the friction forces acting on the two partials are of the same order of magnitude ($R_2/R_1 \simeq 0.9 - 0.7$).

Dislocation Sources and Velocity Data

Fig. 2 shows a single ended source recorded at 453 K. It belongs to the secondary slip system $\frac{1}{2}$ $[01\bar{1}](111)$ (theoretical Schmid factor $\simeq$ 0.27). The screw segment is almost parallel to the slip traces.

Measurements of dislocation velocities were performed for different segments of this source. Care was taken to consider only segments which do not emerge at the foil surfaces. The local shear stress was 40 ± 12 MPa near the source. Fig. 3 (velocity data versus length) reveals that :

i) At constant stress, the velocity of a dislocation segment is proportional to its length. The straight lines do not cut the length axis at the origin but at L_o. L_o is probably related to the gradual transition between screw and 60° segments which results in kink accumulation at each end of a segment. The velocity is then proportional to the length ($L - L_o$) which is free from kinks. In the studied length range no evidence of a saturation in the dislocation velocity is observed.

ii) There is no significant difference between the screw segment rate and both types of 60° dislocations.

The same conclusions can be drawn from the analysis of a source working at 300 K with $\tau \simeq 30 \pm 9$ MPa.

Cathodoplastic Effect

The dislocation velocity is strongly enhanced by electron irradiation (cathodoplastic effect). Using a periodic beam deflecting apparatus, it is shown that, at a constant stress and T = 453 K, the dislocation velocity under electron irradiation ($I \simeq 1000$ A/m^2) is $\simeq$ 2500 times greater than in the dark.

From the data of Fig. 3, showing the enhancement of the dislocation velocity with increasing electron beam intensity, the Fig. 4 is drawn revealing that, for a given L (0.15 μm), the velocity is proportional to the beam intensity I.

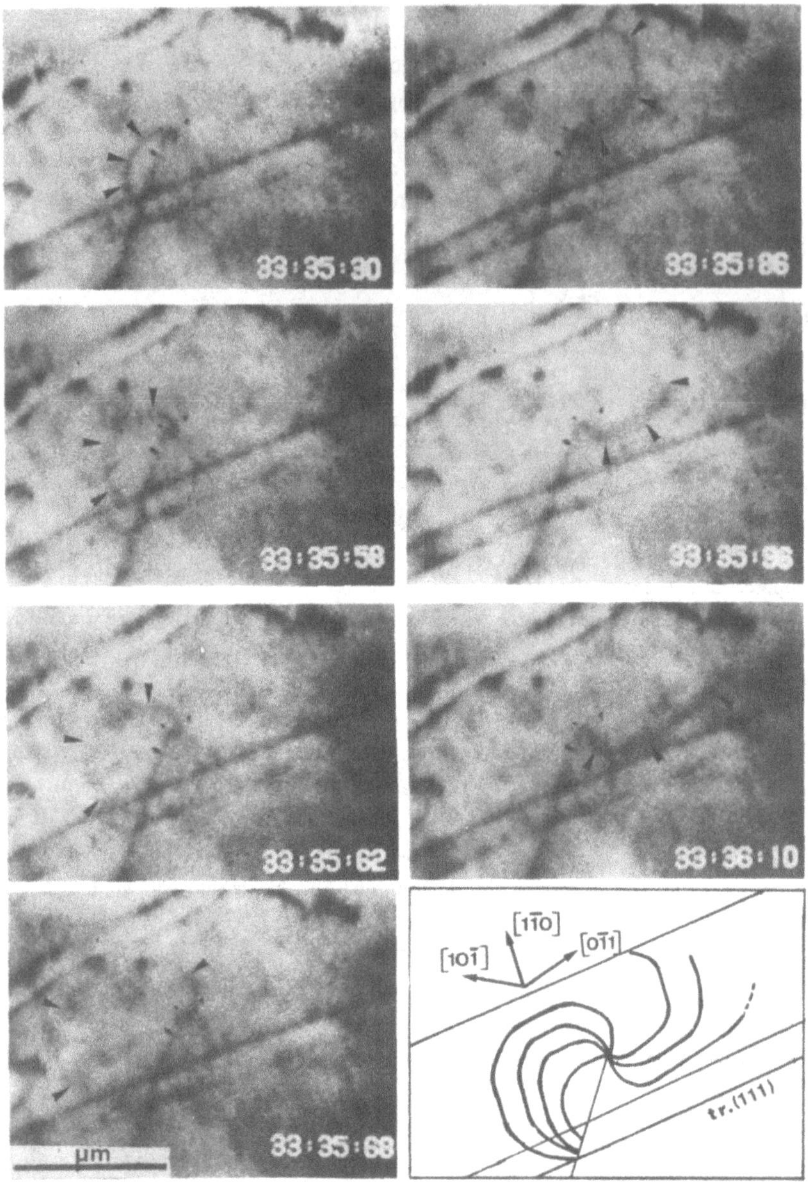

Fig. 2 Dislocation source gliding in plane (111). T = 453 K. Successive positions of the dislocation are arrowed. **b** = $\frac{1}{2}$ [$01\bar{1}$].

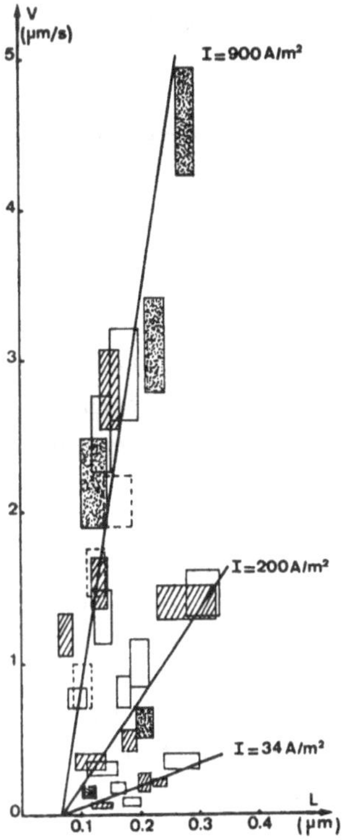

Fig. 3 Dislocation velocity vs length for dislocation source of Fig. 2 :
screws (left and right)
60° right 60° left.

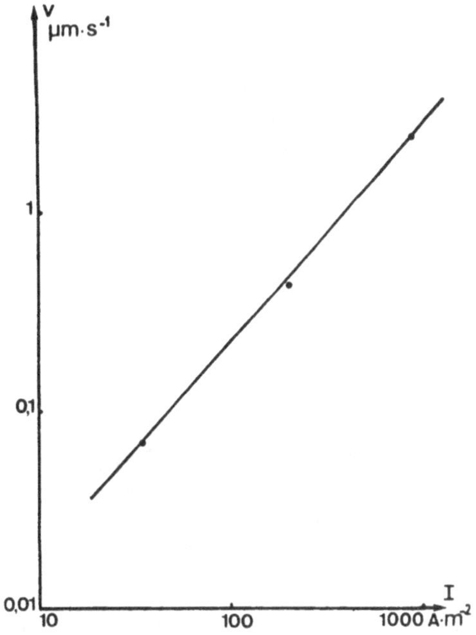

Fig. 4 Dislocation velocity vs electron beam intensity (L = 0.15 μm).

DISCUSSION

1. Several experimental results confirm that the dislocation glide in ZnS crystals is governed by the nucleation and propagation of double kinks : i) the viscous movement of all dislocations is observed even though the dislocation lines are curved during glide ; ii) the cathodoplastic effect enhances dislocation velocity up to 530 K ; iii) evidence of the "length effect" implies that the Hirth and Lothe's model [8] giving the velocity of a dislocation in a Peierls potential is fulfilled.

But a slight difference is noted with III-V compounds in which the dislocation lines remain more linear as the temperature increases [6, 7]. Moreover, non-crystallographic glide reported for CdTe [2] was not observed in ZnS.

From macroscopic deformation experiments, we suggested the occurence of a new controlling mechanism of dislocation glide from 400-450 K, temperatures for which the shear stress versus temperature curve shows a flat level [1]. The present work does not support this assumption so that the Peierls mechanism still controls glide up to 530 K.

2. The velocities of the screw and the two types of 60° dislocations are shown to be the same in ZnS. Consistent with these results, TEM observations of specimens after macroscopic deformation show an equal density of screw and 60° dislocations whatever the temperature [1] ; this indicates that the cause of such a behaviour is intrinsic rather than connected to the electron beam effect.

The difference observed between the velocities of screw and 60° dislocations are generally explained refering to partial dislocation mobilities and particularly to the lower mobility of 30 β partials. In ZnS the friction forces acting on 30° and 90° partials either in the leading or trailing position are found to be more or less equal. This is a very different behaviour than III-V compounds [9] and elemental semiconductors [5].

3. Our results have pointed out a great influence of the electron beam intensity on the dislocation velocity. They do not agree with Suzuki et al [2] who concluded that II-VI compounds do not present enhancement above room temperature ; for ZnS, a strong cathodoplastic effect is observed up to 530 K. The dislocation velocity is proportional to the electron beam intensity as established by Maeda and Takeuchi [3] with very different experimental conditions, but more experiments are needed to say if their model is a good description for the cathodoplastic effect in ZnS.

REFERENCES

1. C. Levade, J.J. Couderc, A. Kara, Phys. Stat. Sol. (a), 112 (1989) 89
2. K. Suzuki, M. Ichimara, S. Takeuchi, K. Nakagawa, K. Maeda, Phil. Mag. A, 49 (1984) 451
3. K. Maeda, S. Takeuchi, J. Phys. C4, 44 (1983) 375
4. L. Kubin, P. Veyssière, Proc. 10th Int. Cong. on Electron Microscopy, Hamburg, 2, (1982) 531
5. K. Wessel, H. Alexander, Phil. Mag. A, 35 (1977) 1523
6. M. Fnaiech, F. Reynaud, A. Couret, D. Caillard, Phil. Mag. A, 55 (1987) 405
7. D. Caillard, N. Clément, A. Couret, Y. Androussi, A. Lefebvre, G. Vanderschaeve, Microscopy of Semiconducting Materials VI, Oxford (1989)
8. J.P. Hirth, J. Lothe, Theory of Dislocations, McGraw-Hill Publ. Co, N.Y. (1968)
9. A. Lefebvre, Y. Androussi, G. Vanderschaeve, Phil. Mag. Letters, 56 (1987) 135.

IN-SITU OBSERVATION OF DISLOCATION MOTION IN COMPOUND SEMICONDUCTORS BY CATHODOLUMINESCENCE MICROSCOPY

N. Maeda*, K. Maeda** and S. Takeuchi*

*Institute for Solid State Physics, The University of Tokyo, Roppongi, Minato-ku, Tokyo 106, Japan

**Department of Applied Physics, Faculty of Engineering, The Universtiy of Tokyo, Hongo, Bunkyo-ku, Tokyo 113, Japan

ABSTRACT

Dislocation glide behavior in compound semiconductors is investigated in-situ by the SEM-CL method, where dislocations are observed as dark spots. In III-V compounds of GaAs and GaP, individual dislocations are observed to glide steadily with almost constant velocity at a constant applied stress at elevated temperatures, and the velocity is remarkably enhanced by the electron beam irradiation. In II-VI compounds of CdS and CdTe, no moving state of individual dislocations is recognized, and instead only an increase of the dislocation density in slip bands is observed. The front of screw bands with high-density dislocations grows steadily at room temperature, while edge-bands are formed almost instantaneously, suggesting that the motion of screws controls the deformation.

INTRODUCTION

Extensive microscopic investigations have so far been performed to understand the deformation mechanisms of plastic deformation of crystals. Most of them are, however, *ex post facto* studies. In order to prove the true mechanism of deformation of a particular crystal under a particular condition, direct evidence by in-situ microscopic observation of the deformation process is required. Main techniques of it are the in-situ TEM method and the in-situ X-ray topography method. These methods have some limitations: in the former method one can use only a tiny specimen with extremely small thickness whose deformation behavior may be different from the bulk specimen, and in the latter method the spatial resolving power is so low that one can use only the crystals with extremely low dislocation density, such as Si. In constrast to these methods, the present method of in-situ cathodoluminescence (CL) microscopy by use of a scanning electron microscope (SEM) does not suffer from the above disadvantages; one can use a bulk specimen, and the spatial resolution is of the order of 1 μm which is an order better than the X-ray topography. Furthermore, since the magnification of observation can be changed over a wide range, one can observe microscopic to macroscopic structures continuously. Of course, the present method can apply only to semiconducting crystals which can be imaged by cathodoluminescence mode of scanning electron microscopy. In this paper, we report some applications of the in-situ SEM-CL method to III-V and II-VI semiconducting crystals.

METHOD

Some details of the in-situ observation of dislocation motion by the SEM-CL method have been described in a previous paper [1]. The principle of the method is as follows. Defects in semiconducting crystals, in many cases, act as non-radiative recombination centers of injected carriers and hence these defects can be imaged by a scanning electron microscope (SEM) in electron beam induced current (EBIC) or cathodoluminescence (CL) mode. It has been confirmed that dark spots or lines observed in SEM-EBIC or SEM-CL images correspond to individual dislocations for various materials (see references in [1]). Thus by combining a deformation apparatus and a scanning electron microscope, one can observe, in-situ, the motion of dislocations under stress by tracing the change in dark spots pattern in SEM-EBIC or SEM-CL images. In the EBIC mode observation, a p-n or a Schottky junction has to be fabricated on the specimen surface, which may affect the deformation process, and hence we have used only the CL method.

Figure 1 shows a sectional view of the specially designed deformation stage on which constant bending stresses can be applied to a specimen at elevated temperatures. The stage consists of the four-point bending jig in a furnace, luminescent light collecting system and the load application system. The specimen, typically of a size $13x3x0.4mm^3$, can be heated up to 700 °C by a platinum heater in a molybdenum furnace. The temperature of the specimen is measured by a thin thermocouple lightly touched on the specimen surface. A constant bending load is applied to the specimen by pushing up the lower bending edges by a vertical rod connected to an electromagnetic driving system, which can produce a pulsed force up to 20 N for any duration, from a fraction of a second to several days. Changes of the cathodoluminescence image during deformation can continuously be recorded with a video-tape recorder.

Crystals used are III-V compounds, Si doped n-GaAs and Te doped n-GaP, and II-VI compounds, n-CdS and n-CdTe. GaAs, GaP and CdS single crystals were purchased from companies and CdTe was grown by the Bridgman method in the authors' laboratory. The

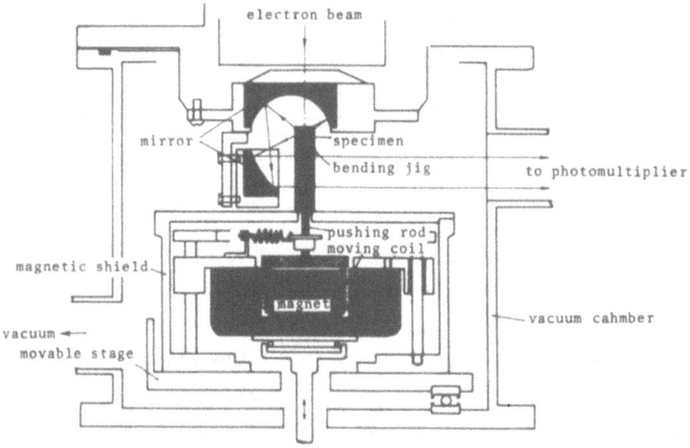

Fig. 1 Sectional view of bending deformation stage in scanning electron microscope for in-situ CL observation

orientation of the III-V compounds was such that the crystal surface is (111), the bending axis [41$\bar{5}$] and the lengthwise direction [2$\bar{3}$1]. For II-VI compounds, two types of orientations were used. For CdTe, one orientation is the same as the above and the other orientation is such that the surface orientation is (1$\bar{5}$2), the bending axis [112] and the lengthwise direction [20$\bar{1}$]. For CdS, one orientation was such that the surface orientation is ($\bar{1}$010) and the lengthwise direction [1$\bar{2}$11] and the other is such that the bending axis is [10$\bar{1}$0] and the lengthwise direction [1$\bar{2}$11]. The dislocations on the primary slip plane that emerge on the specimen surface of the III-V crystals with the above orientation have almost edge character, and thus the orientation will be called the edge-band orientation. The former type orientation of II-VI crystals is also the edge-band one, while the latter type orientation of II-VI crystals is the screw-band one. We have distinguished α-dislocations and β-dislocations from the polarity of the crystals determined by an etching method.

Plate-shaped bending specimens of a size (0.3-0.5mm) x 3mm x (∿10mm), having the above-mentioned orientations, were cut out mechanically and polished chemically. For III-V crystals, to introduce fresh dislocations before testing a scratch was scribed in the lengthwise direction on the surface with a diamond stylus.

RESULTS AND DISCUSSION

III-V Crystals

Some results of the in-situ SEM-CL observations for GaAs have been reported previously [2,3]. Results for GaP are mostly common with those of GaAs. Common features are as follows. (1) Dislocations are emitted preferentially from one side of the scratch, indicating that α-dislocation motion is much more active than β-dislocations. (2) Dislocations move steadily over a long distance almost with a constant velocity under a fixed condition. Figure 2 shows CL micrographs exhibiting a constant and steady motion of a group of dislocations in GaAs. (3) The mobility of dislocations is enhanced by the electron beam irradiation. This phenomenon was first demonstrated by Maeda and Takeuchi for n-GaAs in a quantitative fashion [2] and has been confirmed for various semiconducting crystals [4,5]. (4) The dislocation that has travelled some distance from a source under a constant load usually moves back toward the source after complete unloading, obviously due to the line tension of the dislocation.

The cause of the electron-beam irradiation enhanced dislocation motion has been

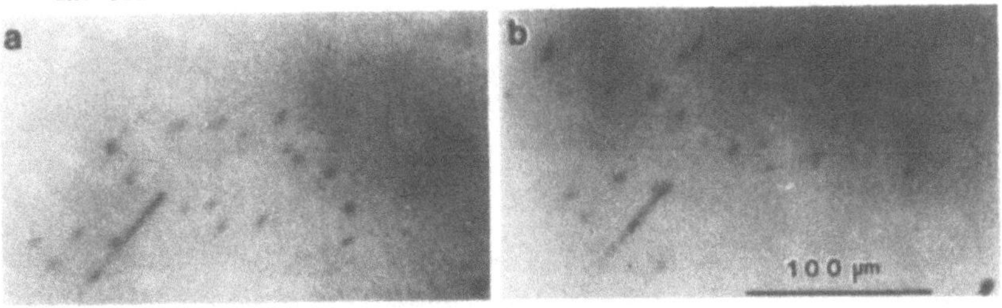

Fig. 2 Successive SEM-CL micrographs which show a constant velocity gliding of a group of dislocations in n-GaAs under τ=40MN/m^2 at 435K. (b) was taken 2min after (a) Note that almost the same dislocation patterns in (a) and (b).

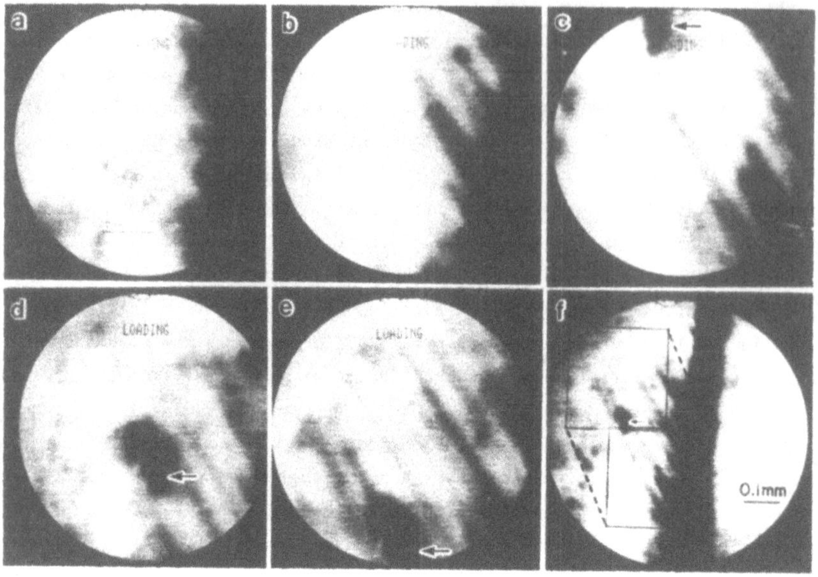

Fig. 3 A series of CL micrographs from a video tape showing continuous motion of dislocations in n-GaP under τ=20MN/m² at 513K. Time interval between frames is 4 sec. (f) is a lower magnification micrograph in which squares indicate initial and final scanning areas.

interpreted by the so-called phonon-kick mechanism where localized phonons excited upon non-radiative recombination of electron-hole pairs at dislocation sites assists the dislocation motion. As a result of this irradiation effect, the dislocation glide becomes more active in the scanning area than outside of the area. Figure 3 gives a series of CL micrographs taken from a video tape which show a continuous motion of dislocations in GaP. The time interval between the micrographs is 4 sec. From (a) to (e) of Fig. 3, the field of view is shifted in order to trace the dislocation motion. (f) shows a lower magnification micrograph after(e), where the initial and final scanning areas are indicated by squares. It is seen that the dislocations have been emitted and moved only in the area where electron beam has been scanned, clearly indicating the enhanced dislocation glide. We should note in Fig. 3 that each dislocation is imaged not as a spotty contrast but as long tailed contrast. This is due to the fact that the enhanced motion of dislocations in the depth direction is limited to the depth of the excitation by electron beam or a few μm; so that, the long dislocation line beneath the surface is also imaged.

II-VI Crystals

The dislocation behavior in II-VI crystals at and above the room temperature is different from that in III-V crystals, e.g. no continuously gliding dislocation was recognized. The behavior is also different between the edge bands and the screw bands. Some results have already been reported previously [1,6].

In the edge band formation in CdTe and CdS (basal slip), the whole multiplied dislocations become visible only after they spread over a long distance, and the change observed with elapse of time is increase in the number of slip bands. Neither individua dislocation motion nor the continuous growth of slip bands was detected. Figure 4 presents an example of the formation of edge bands in CdTe. As in III-V crystals, we see different activities between α- and β-dislocations. Initially slip bands are produced locally around an indentation but then rather homogeneously in the specimen.

In the screw band formation, on the other hand, we could trace the growth process of individual slip bands, whose speed is rather slow. While edge bands are straight, the screw bands are wavy and thick. This fact indicates that cross slip of screw dislocations is frequent. Figure 5 gives a series of CL micrographs showing continuous growth of screw bands of the basal slip in CdS. We see a steady growth of the slip band front. The lengthening appears to take place by successive multiplication of dislocations at the tip of the bands by the double cross slip mechanism.

Above features of slip band formation are consistent with the direct in-situ TEM observations [7], where edge dislocations move much faster than screw dislocations, and the multiplication rate of dislocations from Frank-Read type source is controlled by the velocity of screw dislocation segment. Cross slip was frequently observed in TEM observation and jogs produced by double cross slips retarded the motion of screw dislocations to immobilize them eventually after gliding a certain distance, and hence the mean glide distance of screws was much smaller than edge dislocations.

Since the first discovery by Osip'yan and Savchenko for CdS [8], it has commonly been observed that illumination of light affects,\ usually increases, the flow stress of II-VI crystals: the effect known as the photoplastic effect. Later, it was also shown that irradiation of electron beam produces a similar effect, which is named the cathodoplastic effect [9]. This effect is, however, not significant when the beam intensity is not strong, probably because the affected region is limited to the thin surface layer which is thin enough compared with the depth of the slip band formation.

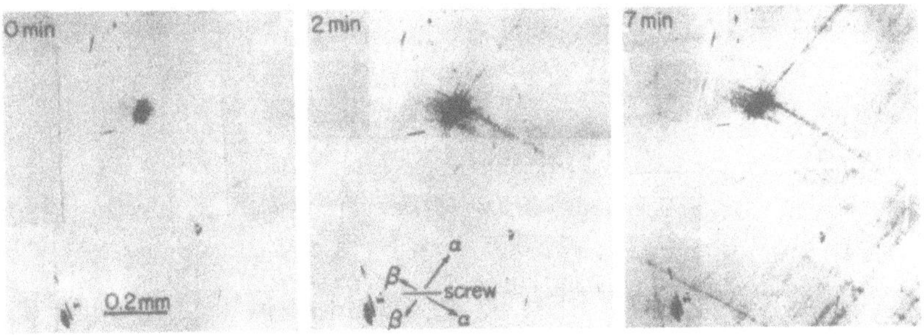

Fig. 4 Seccessive SEM-CL micrographs for edge slip bands formation in CdTe under $\sigma = 17 MN/m^2$ at room temperature

ACKNOWLEDGMENTS

The authors are indebted to Dr. K. Nakagawa, now at Hitachi Central Laboratory, for the experiments on CdTe single crystals. Thanks are also due to T. Hashimoto for his assistance in the experiments.

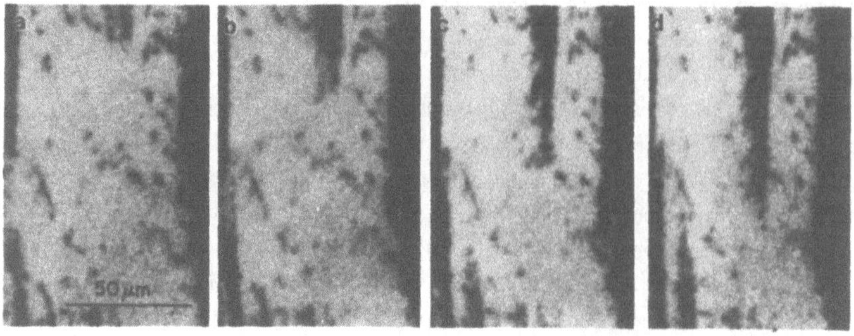

Fig. 5 Successive CL micrographs showing growth of screw slip bands in CdS under $\sigma = 16 MN/m^2$ at 345K.

REFERENCES

1 K. Maeda, K. Nakagawa, S. Takeuchi and K. Sakamoto, J. Mater. Sci. 16 (1981) 927-34

2 K. Maeda and S. Takeuchi, Jpn. J. Appl. Phys. 20 (1981) L165-68

3 S. Takeuchi and K. Maeda, In: Dislocations 1984, P. Veyssiere, L Kubin and J. Castaing (Eds.), C.N.R.S., Paris 1984, p. 331-51

4 K. Maeda and S. Takeuchi, J. de Phys. 44 (1983) C4-375-85

5 N. Maeda and S. Takeuchi, to be published in Proc. Int. Conf. on the Science and Technology of Defect Control in Semiconductors, K. Sumino (Ed.), Elsevier, Amsterdam 1989

6 K. Nakagawa, K. Maeda and S. Takeuchi, J. Phys. Soc. Jpn. 49 (1980) 1909-15

7 K. Suzuki, M. Ichihara, S. Takeuchi, K. Nakagawa, K. Maeda and H. Iwanaga, Philos. Mag. A 49 (1984) 451-61

8 Yu.A. Osip'yan and I.B. Savchenko, JETP Lett. 7 (1968) 100-3

9 K. Maeda and K. Sakamoto, J. Phys. Soc. Jpn. 42 (1977) 1914-17

ON THE POSSIBILITY OF DISLOCATION SUBSTRUCTURE EVALUATION BY TEM IN SEMICONDUCTOR SINGLE CRYSTALS DEFORMED IN THE YIELD REGION

J.P. Michel, A. Oueldennaoua, R. Allem, M. Omri, M. Azzaz and A. George

Laboratoire de Physique du Solide, CNRS : URA n°155, Institut National Polytechnique de Lorraine, Nancy, France

ABSTRACT

The mobile dislocation density and local stresses can be determined from TEM observations in pre-strained Si single crystals deformed in the yield region at low strain rates and moderate temperatures, provided that the applied stress is larger than ≈ 10 MPa. A fair consistency is found with Orowan's formula, most of dislocations appear to be mobile and the applied stress is nearly totally effective at the lower yield point. In GaSb, screw dislocations are rate controlling at low temperature and $\tau_y \geq$ 10 MPa.

INTRODUCTION

The aim of this note is to show that quantitative estimations of quantities such as the mobile dislocation density, ρ_m, or the so-called effective stress, τ_{eff}, which enter any microscopic description of plastic deformation, can be extracted from conventional TEM micrographs in well defined circumstances, specific classes of materials and favourable ranges of temperature and stress.

According to Orowan's classical relation, the shear strain rate $\dot{\gamma}$ is equal to $\rho_m bv$, where b is the Burgers vector modulus and v, the dislocation velocity. The macroscopic $\dot{\gamma}$ results from averaging over local instantaneous quantities. In a homogeneous situation, v the average velocity is that which corresponds to the effective stress at the testing temperature, $v(\tau_{eff},T)$. On conventional micrographs, v can only be derived from measurements of the local stress, provided that the dislocation velocity is already known as a function of τ and T in the range under investigation. Such a stringent requirement is fulfilled in semiconductors, where owing to the large Peierls potential arising from the covalent type of bonding, the glide resistance for a single dislocation is uniform, resulting in smooth motion at constant τ,T (1,2). What can be done in practice is illustrated with the nearly ideal case of pure Si single crystals deformed by compression, in single slip orientation, at constant strain rate, in the yield region. The more difficult case of III-V compounds is discussed, with GaSb as the main example.

EXPERIMENTAL

Details can be found in (3-6). All samples had received a standard pre-straining of a few percent at higher temperature - lower stress, prior to the deformation. To avoid any extrapolation which would introduce large uncertainties, it is essential that the ranges of yield stresses, τ_y, and testing temperatures coincide with these of accurate velocity measurements. This can be obtained by a suitable choice of $\dot{\gamma}$. Typically in Si, $820\ K \leq T \leq 1070\ K$, $5\ MPa \leq \tau_y \leq 50\ MPa$ at $\dot{\gamma} \approx 2 \times 10^{-5} s^{-1}$.

THIN FOIL OBSERVATIONS AND THE DEFORMED STATE

All measurements rely on the assumption that the dislocation structure observed in thin foils is representative of that which prevailed in the bulk crystal when the deformation was interrupted. Thanks to the large temperature dependence of v in Si, it can be taken for granted that no detectable motion can take place at room temperature, either during foil preparation or in the electron beam. The situation is less clear in III-V compounds where dislocation rearrangements can occur in thin foils, especially after high-stress or inhomogeneous deformation (7), because of a lower melting temperature and a marked enhancement of v by the electron beam (8). No motion was detected, however, in the GaSb crystals deformed for this study. The dislocation structure can be frozen-in by cooling with the final stress applied. Figure 1 shows the amount of relaxation which could occur in case of unloading prior to cooling. The two micrographs were taken after the same deformation. Dislocation densities are quite the same but in the unloaded sample (a), preferred crystallographic orientations were lost, while they were preserved in sample (b), cooled with the load applied. As a second point, it is necessary that the dislocation distribution is highly homogeneous in order to make sure that the selection of the observed areas is not biased dur foil

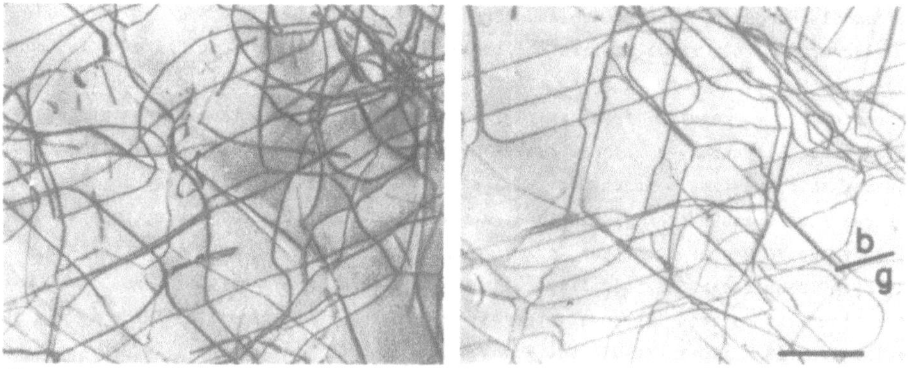

a b

Fig.1 Dislocation configuration in Si after deformation at 823 K, $\dot{\gamma} = 2.5 \times 10^{-5} s^{-1}$, lower yield point. (a) Unloaded before cooling, (b) cooled with the load applied. Foil parallel to the primary slip plane, $g = 20\bar{2}$. Marker : 1µm.

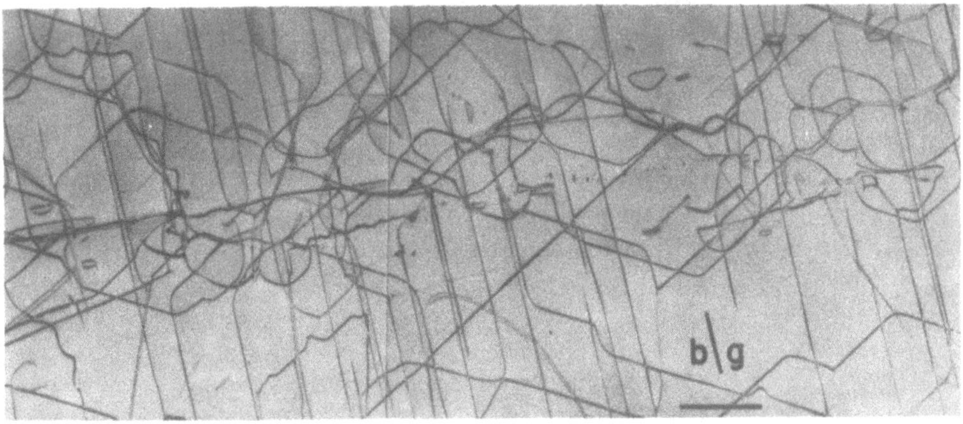

Fig.2 Dislocation configuration in Si after deformation at 823 K, $\dot{\gamma} = 2 \times 10^{-5} s^{-1}$, $\tau = 54$ MPa. Deformation stopped between the upper and lower yield points. Foil parallel to the primary slip plane, g = 20$\bar{2}$. Marker : 1 μm.

preparation. We used to cut several foils at different positions along the gauge length of each sample and were able to check that a homogeneous situation prevailed, not only at the lower yield point but also, at least at low T and $\dot{\gamma}$, all along the yield region of the stress-strain curve. Most probably, pre-straining was very important in that respect.

MEASUREMENTS OF THE MOBILE DISLOCATION DENSITY

It is easy to recognize mobile dislocations and immobile ones in configurations of the kind shown in Fig.2. The majority of dislocations lie along ⟨110⟩ orientations in the slip plane, forming loops made up of screw and 60° dislocations linked by short bends. Dipoles, with a marked edge character although without precise crystallographic orientations, are also visible. Dipolar dislocations have been formed by mutual trapping of passing dislocations and do not contribute to the strain. ⟨110⟩ oriented segments are a clear signature of dislocations whose motion is rate-controlled by the Peierls-Nabarro mechanism and must therefore be taken as mobile. When τ increases, the equilibrium radius of curvature of dislocations, R, decreases and preferred ⟨110⟩ orientations appear when the average distance between dislocations, $\bar{l} \simeq 1/\sqrt{\rho}$, is larger than R. In Si, in the yield region, ρ is in the range of 10^8 cm.cm^{-3} and ⟨110⟩ orientations are visible at $\tau \geq 10$ MPa, which gives the lower bound of the range in which ρ_m can safely be determined. About the same figure would be obtained in GaSb, where configurations are however different. As in Si, edge dipoles are the dominant features of the high T, low τ

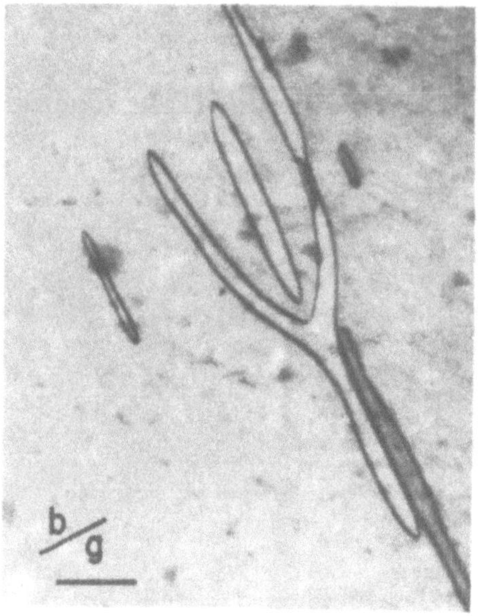

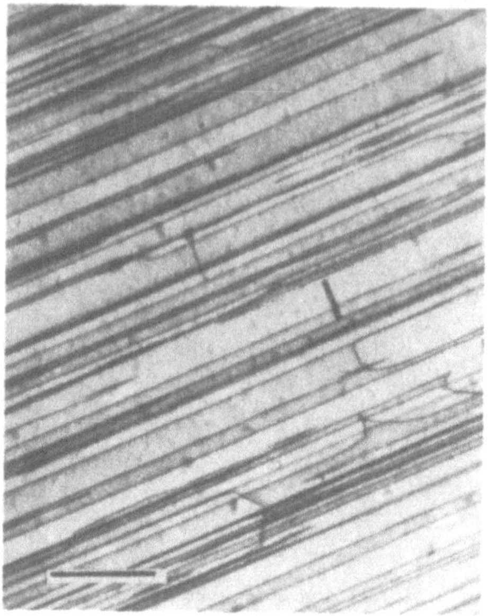

Fig.3 Dislocation configuration in GaSb after deformation at 823 K, $\dot{\gamma}$=2.5x $10^{-5}s^{-1}$, τ = 2.2 MPa, γ = 7 x 10^{-2}. Foil parallel to the primary slip plane, g = $\bar{2}02$. Marker : 0.5 µm.

Fig.4 Dislocation configuration in GaSb after deformation at 473 K, $\dot{\gamma}$ = 5 x $10^{-6}s^{-1}$, τ = 39 MPa. Foil parallel to the primary slip plane. g = $\bar{2}02$. Marker : 1 µm.

substructure (Fig.3) but, while at larger stresses, screws and the two types of 60°s are present in comparable densities in Si, in agreement with their nearly equal velocities, in compounds, screws are predominant due to the much larger velocity of the so-called α 60° dislocations (Fig.4). In configurations of the kind of Fig.4, ρ_m is difficult to determine : close parallel lines cannot always be resolved and it is not clear whether they are dislocations of the same sign or dipoles.

DETERMINATION OF THE LOCAL SHEAR STRESS

As discussed in detail by Gottschalk (9), shear stresses actually exerted on dislocations are related to the radius of curvature at bends by the formula τ = T/Rb, where T is the orientation-dependent line tension. It is unfortunately impossible to derive from such measurements a mapping of the actual stress fluctuations in a given slip plane. Results are best displayed by histograms, which are characterized by an average stress $\bar{\tau}$ and a standard deviation $\Delta\tau$.

A simple check of the relevance of such measurements is that the average value of the local stress τ_{loc} should be equal to the applied stress τ. This was observed to be true within 20 % at the lower yield point. The situation is markedly different before the upper yield point, where $\bar{\tau}$ is always significantly smaller than the applied stress. We believe that a systematic underestimation of $\bar{\tau}$ might have resulted from the inhomogeneous dislocation distribution observed at the micro-scale in those earlier stages (Fig.5) : lower dislocation density areas correspond to those of higher stress and the average is falsified by the larger number of measurements made in higher density areas.

The derivation of the effective stress and of the average velocity is indirect. We have calculated the average velocity of a dislocation gliding in a modulated stress field, and τ_{eff} was defined as this constant stress which would have ensured the same velocity. For this, we assumed that the experimentally determined (10) description of v:

$$v(cm.s^{-1}) = 2.9 \times 10^{5} \ (\tau/\tau_0)^{m} \exp(-2.2eV/kT) \quad (1)$$

with m = ≈ 1.2, is valid everywhere with $\tau = \tau_{loc}$. An essential ingredient of the model is the amplitude of the modulation, which was tentatively related to the standard deviation $\Delta\tau$ of the local stress histograms (4,5). In Si, at $\tau \geq 10$ MPa, the standard deviation is fairly small ($\simeq 0.25\ \bar{\tau}$), which leads to the conclusion that the applied stress is nearly totally effective ($\tau_{eff} \geq 0.75\ \tau$).

In GaSb, the situation is desperately simple : local stresses cannot be measured since the predominance of one type of dislocations suppresses dislocations bends !

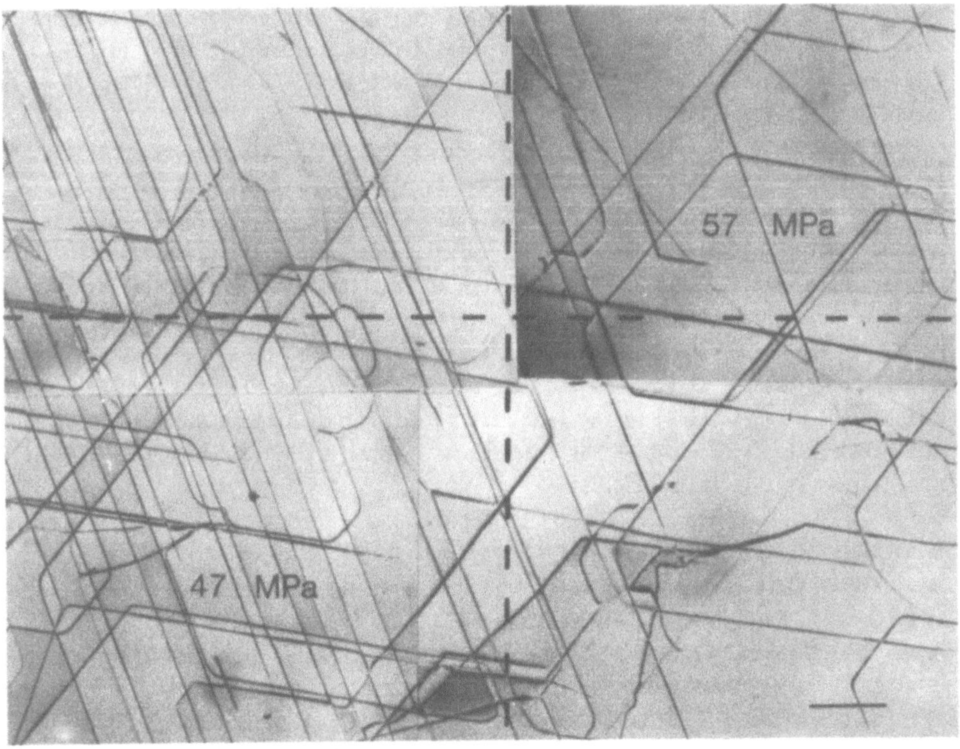

Fig.5 Inhomogeneous dislocation distribution in Si after deformation at the upper yield point (72 MPa) at 823 K, $\dot{\gamma} = 2 \times 10^{-5} s^{-1}$. The average stress derived from measurements of the radius of curvature at bends is larger in low density areas than in high density areas. Marker 1 µm.

CONSISTENCY OF DISLOCATION DENSITY AND STRESS MEASUREMENTS

In Si, once $\bar{v}$ has been determined, it is possible to calculate the mobile dislocation density which would ensure the measured plastic strain rate and to compare it with the measured ρ_m. The relative error on the calculated ρ_m can be estimated as :

$$\Delta\rho_m/\rho_m = \Delta\dot{\gamma}/\dot{\gamma} + \Delta\bar{v}/\bar{v}$$

$\Delta\dot{\gamma}/\dot{\gamma}$ should be smaller than 10 %, with the possible exception of the very earlier stages where the plastic strain rate is much lower than the imposed total $\dot{\gamma}$. $\Delta\bar{v}/\bar{v}$ can be divided in two additive terms. The first one results from the approximation when using eqn.(1) and from the error on the temperature. This term should not exceed 20 % in the investigated stress range. The second term, $m\,\Delta\tau_{eff}/\tau_{eff}$, reflects the approximation in estimating the effective stress and is difficult to evaluate because of our poor knowledge of the amplitude of stress fluctuations. The relative error on the measured ρ_m is also the sum of two errors. One is due to the error on the foil thickness, which does not exceed 10 %. The second, more difficult to estimate, arises from the lack of distinction between mobile and non-mobile dislocations.

Calculated and observed ρ_m usually differ by less than 40 % which, in view of the above considerations, is satisfactory but would seem to leave a large possible fluctuation for τ_{eff} and/or ρ_m. This is not the case since τ_{eff} cannot be increased above the applied stress and ρ_m above the total dislocation density, known within 10 %. Therefore we take the general result that at $\tau \geq 10$ MPa, $\tau_{eff} \simeq \tau$ and $\rho_m \simeq \rho_{total}$ up to the lower yield point, as significant.

In GaSb, only the velocities of 60° dislocations have been measured (11), but it can be estimated that the velocity of screws is close to that of β dislocations (edge type with the opposite sign as α dislocations) as in other undoped III-V compounds. It is clear that β dislocations do not play a significant part in the plastic strain since they are rare and slow. It can also be checked that the density of α dislocations is much too low for that type to contribute more than $\simeq$ 10 % to the strain and strain rate, at $\tau \geq 10$ MPa. Last, one can check that screw dislocations have the right density to ensure the imposed $\dot{\gamma}$ if $v_s \simeq v_\beta$ and $\tau_{eff} \simeq \tau$. As pointed out in (6), the predominant part of screws is explained by the fact that faster α segments of dislocation loops soon reach the sample surface, while the long screws they have trailed behind them have still a large free path in their glide plane.

1 H. Alexander and P. Haasen, Solid St. Phys. 22 (1968) 27.
2 J. Rabier and A. George, Rev. Phys. Appl. 22 (1987) 1327.
3 M. Omri, C. Tête, J.P. Michel and A. George, Phil. Mag. A 55 (1987) 601.
4 A. Oueldennaoua, R. Allem, A. George and J.P. Michel, Phil. Mag. A 57 (1988) 51.
5 R. Allem, J.P. Michel and A. George, Phil. Mag. A 59 (1989) 273.
6 M. Omri, J.P. Michel and A. George, Phil. Mag. A, to be published.
7 K.H. Kuesters, B.C. de Cooman and C.B. Carter, Phil. Mag. A 53 (1986) 141.
8 K. Maeda, M. Sato, A. Kubo and S. Takeuchi, J. Appl. Phys. 54 (1983) 161.
9 H. Gottschalk, J. Phys. Paris, 44 (1983) C4-9, C4-475.
10 A. George and G. Champier, Phys. Stat. Sol.(a) 53 (1979) 529.
11 S.K. Choi, M. Mihara and T. Ninomiya, Japan. J. Appl. Phys., 17 (1978) 329.

MECHANICAL AND STRUCTURAL CHARACTERISTICS OF EUROPIUM-PRECIPITATES IN MONOCRYSTALLINE NaCl

M. Suszyńska, D. Nowak-Woźny

W. Trzebiatowski Institute of Low Temperature and Structure Research, PAS, 50-950 Wrocław, Poland

ABSTRACT

Hardening of $NaCl:Eu^{2+}$ crystals was investigated as function of the dispersion state of the dopant characterized by rentgenographic and optical methods. The contributions to the hardening of free I.V. dipoles, small aggregates and precipitates have been separated by using samples subjected to different thermal treatment.

INTRODUCTION

The sensitivity of the yield stress of alkali halides to the concentration of divalent cations is well established [1]. However, considerations of the hardening effects concern primarily the impurity-vacancy (I.V.) dipoles and small aggregates.

In order to determine the role played by various dispersion forms of the impurity, the yield stress of $NaCl:Eu^{2+}$ crystals was measured as function of the dopant concentration (c_{++}) and the annealing conditions. This crystal system is especially interesting because the obstacles responsible for hardening can be distinguished by means of different methods. EPR and ITC (ionic thermo-current) are most relevant for isolated dipoles whereas X-ray diffraction as well as optical spectroscopy - for precipitates.

It should be mentioned that hardening effects related with the presence of Eu^{2+} ions in NaCl crystals were yet investigated [2-5], and this paper completes our data reported previously [4,5].

SOME CHARACTERISTICS OF THE SYSTEM

The crystals were grown in our Laboratory according to a modified Bridgman method. Table 1 collects the crystallographic data obtained for as received NaCl samples nominally doped with 0.1 % of Eu^{2+}. The stable $EuCl_2$ phase with the orthorhombic ($PbCl_2$-type) structure was detected for crystals with still higher dopant concentration [7].

Taking into account the microstructural data, it was possible to classify the optical spectra of as received crystals into four groups characteristic of various dispersion forms of europium, cp. Table 2.

The appearance of the above mentioned phases (cp. Table 2) could be related either with the increasing dopant concentration in as received crystals or with an appropriate thermal processing of a solution treated sample with sufficiently large amount of the dopant.

Table 1 X-ray diffraction data, acc. to [6]

Characteristics:	Phase: (a)	(b)
stoichiometry	Na_2EuCl_4	$EuCl_2$
structure	CsCl - type(*)	CaF_2 - type
lattice parameter [Å]	7.47	6.969
size of particles [Å]	$\phi < 200$	$\phi = (1.6 - 2.0) \times 10^3$
distribution	at grain boundaries	random
stability	(a) is thermally less stable than (b) and disappears under the action of X-rays	

(*) acc. to [4] planar distribution of quadripoles could be also considered.

Table 2 Classification of the optical absorption spectra, acc. to [4]

Description:	Type of spectrum: (A)	(B)	(C)	(D)
concentration(*) [ppm]	< 51.7	54,1 - 117.5	135.8 - 204.9	> 211.5
position of e_g and t_{2g} bands [nm]	241.1 and 335.7	265.4 and 331.9	259.7 and 331.5	259.6 and 332
$k = \alpha(e_g)/\alpha(t_{2g})$	1.00	1.20 - 1.40	1.50 - 1.70	1.80 - 2.00
europium-related phase	dipoles and aggregates	Na_2EuCl_4	$EuCl_2$ (CaF_2 - type)	$EuCl_2$ ($PbCl_2$ - type)

(*) calculated acc. to c_{++} [ppm] = $4.7 \times \alpha(e_g)$ [cm^{-1}], where $\alpha(e_g)$ is the absorption coefficient of the high energy band.

The increasing dopant concentration and/or the increasing annealing temperature probably yield the europium-related phases in the following sequence:

$(\text{I.V. dipoles + small aggregates})_1 \rightarrow (Na_2EuCl_4)_2 \rightarrow (EuCl_2 - CaF_2 \text{ type})_3 \rightarrow$
$\rightarrow (EuCl_2 - PbCl_2 \text{ type})_4$

It seemed interesting to combine such data with the changes in the strength of crystals subjected to approproate annealing.

EXPERIMENTAL DETAILS

Stress-strain curves were taken at room temperature (RT) by using the INSTRON machine working at the strain rate of 3×10^{-4} [s^{-1}] ; the yield stress value (σ_o) was determined at 0.1 % of plastic strain.

Mechanical testing was perfomed for as received (AR) and solution treated (ST) samples with the dopant concentration ranging between zero and 470 ppm. The solution treatment consisted of 30 min. annealing at 873 K and cooling to RT on a cold metal block.

The hardening induced by precipitates was studied for slightly (c_{++} = 39 ppm) and heavily (c_{++} = 141 ppm) doped crystals of which the ST-samples were annealed at 323 K ($T_{an.1}$), 383 K ($T_{an.2}$) and 493 K ($T_{an.3}$).

RESULTS AND DISCUSSION

Effect of Concentration

Figure 1 presents the $\sigma_0(c_{++})$ dependence of AR- and ST-samples.

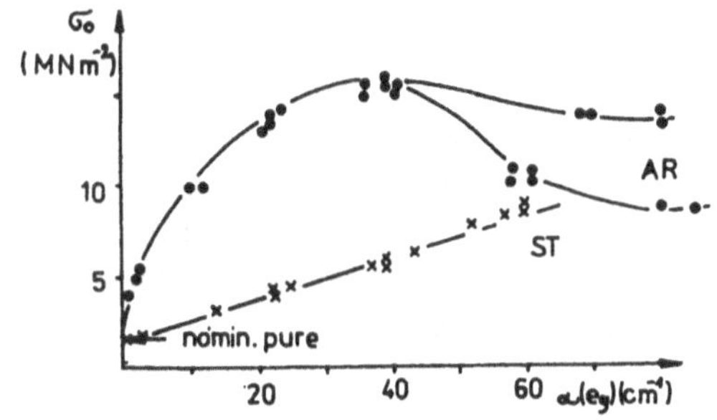

Fig. 1 Concentration dependence of the yield stress of solution treated and as received NaCl:Eu^{2+} crystals.

The most interesting results are as follows:

1. For slightly doped ST samples the $\sigma_0(c_{++})$ - relation is linear suggesting the Snoek--reorientation of I.V. dipoles in the stress field of moving dislocations [8]. The formation of simplest aggregates does not significantly alter the value of σ_0. For these concentrations quench-softening occurs.
2. Nonlinearity of the $\sigma_0(c_{++})$ -plot, observed for heavily doped ST samples, may be related with some ordering of impurities within the matrix, and the corresponding strengthening mechanism could be of Suzuki type [9], for this concentration range quench-hardening is typical.
3. The AR-curves can be understand in terms of precipitation-strengthening and softening mechanisms originally proposed for Al-Cu alloys [10]; one has to consider either the Kelly-Nicholson cutting or the Orowan looping process. The observed two Orowan plots suggest that a complete discussion of the precipitation softening needs informations about the particle size and/or distribution.

Effect of Annealing

Figure 2 shows the isothermal (AN) plots of $\Delta\sigma_0 = \sigma_0(AN) - \sigma_0(ST)$ for slightly (a) and heavily (b) doped crystals.

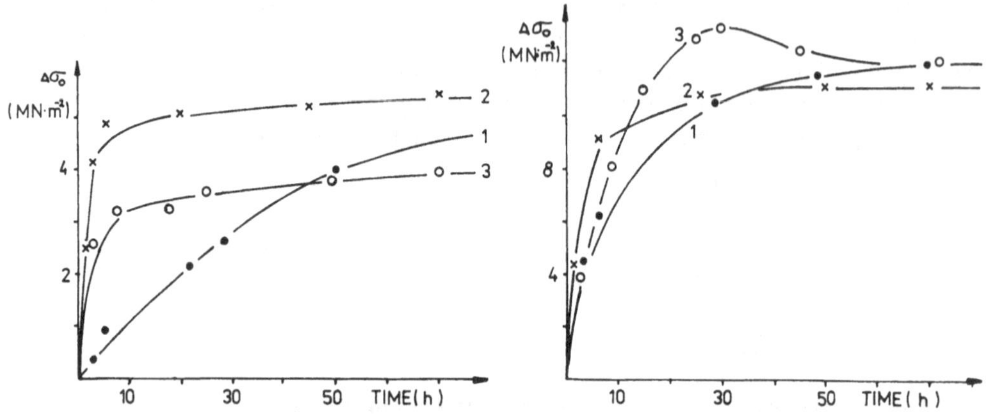

Fig. 2. Time evolution of the yield stress of slightly (a) and heavily (b) doped crystals annealed at 323 (1), 383 (2) and 493 (3) K.

The following findings are typical of the thermal processing performed:

1. The behaviour of both crystals is similar at the lower two temperatures, and the unique difference relates with the hardening rate which is larger for larger c_{++} and higher $T_{an.}$. Small values of $\Delta\sigma_0$ (for $T_{an.\,1.}$) confirm the supposition that the Na_2EuCl_4 particles are formed on the basis of small aggregates (step 1) which do not induce a notable hardening effect.
2. σ_0 - plateau appears when the Na_2EuCl_4 particles grow to such dimensions (step 2) which make possible the transformation into $EuCl_2$ particles with the CaF_2 structure (step 3). This phase is present also after a long annealing period at $T_{an.\,2}$ suggesting that nucleation of the stable $EuCl_2$ phase (step 4) is independent of the thermal history of samples.
3. At $T_{an.\,3}$, typical of the formation of stable $EuCl_2$ phase, the behaviour of both crystals differs qualitatively, cp. the relative softening of heavily doped samples which resembles the $\sigma_0(c_{++})$ - dependence of AR-crystals. In slightly doped samples both $EuCl_2$ phases probably coexist in various proportions.

σ_0(R) - Relationship

According to the growth kinetics [11,12] the dimensions of yet existing precipitation nuclei (e.g. their radius) should change proportionally to t_{an}^{p}. In light of this analysis and the available informations on the isothermal changes of the yield strength of crystals ($\Delta\sigma_0 \sim t_{an.}^{m}$), it is proposed that $\Delta\sigma_0$ is proportional to R^{n} with $n = m/p$. Taking into account that for randomly distributed spheroidal particles, growing under the conditions of bulk diffusion, R is proportional to $t_{an.}^{0.5}$, the empirical relationship between $\Delta\sigma_0$ and the particle radius should be $\Delta\sigma_0 \sim (R^{m/0.5})$ i.e. $\sim R^{2m}$. Table 3 collects the values of n = 2m obtained for ST-samples containing a small and large amount of europium and annealed at three various temperatures. It can be seen that the value of n depends on the kind of particles formed, and for a given type of precipitates it is practically independent of $T_{an.}$

Table 3 Some kinetics -data:

C_{2+} [ppm]	$T_{an.}$ - range	n	Interaction	OA-Spectrum	Phase
141	} high	0.1 ÷ 0.2	looping	} D	$EuCl_2$ of
		0.83	cutting		$(PbCl_2)$-type
	middle	1.3		} C	$EuCl_2$ of
	low	0.96			(CaF_2)-type
39	high	0.83		C÷ D	$(PbCl_2)$
	middle	1.5		C	(CaF_2)
	low	1.5 ÷ 1.9		B	Na_2EuCl_4

Figure 3 presents the $\Delta\sigma_0(R)$ -relationship estimated for heavily doped samples.

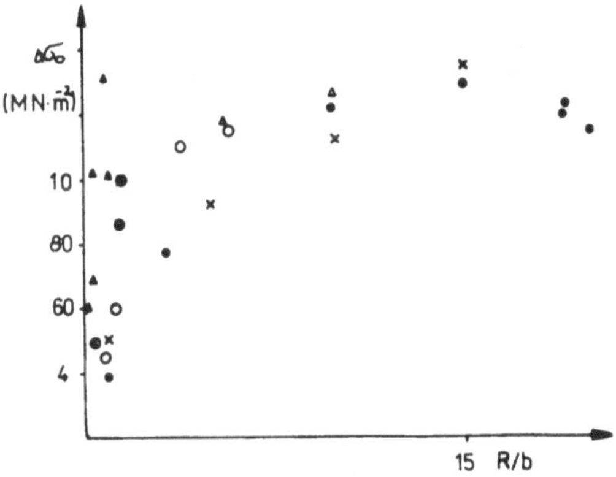

Fig. 3 $\Delta\sigma_0(R)$ -data estimated for heavily doped crystals annealed at various $T_{an.}$[K]: (o) 323, (×) 343, (⊗) 363, (▲) 403, (△) 463, (●) 473.

Two different behaviours have been detected, and the increasing and decreasing parts correspond to the cutting- and looping-interactions, respectively. The critical value of R at which the cutting becomes exchanged for the looping can be estimated on assuming that the volume fraction f of the precipitated phase remains constant during the experiments performed. For spherical particles V. Gerold [13] derived the following equation:

$R/b = G(f/2)^{0.5}/\Delta\sigma_0$,

where G denotes the shear modulus of the matrix and b - the Burgers vector of dislocations. Taking $G = 1.28 \times 10^5$ [kG cm^{-2}] and $\Delta\sigma_0 = 135$ [kG cm^{-2}] $R_{crit.}$ equals to 13.7 ·b. This value can be also estimated from the empirical relation between R and the misfit parameter $\varepsilon = (r_{2+} - r_+)/r_{2+}$, i.e.:

$R = b/(3\cdot|\varepsilon|)$,

where r_i denotes the ionic radius of the impurity and matrix cations, respectively. For $\varepsilon = 0.179$ this relation yields for $R_{crit.}$ a value of $5.6 \cdot b$. Taking into account the approximations made, the agreement of both estimations is surprisingly good. On the other hand, these values are distinctly smaller than those estimated rentgenographically being in agreement with the transparency and turbidity of the annealed and as received samples, respectively.

SUMMARY

The $\sigma_0(c_{++})$ -dependence and the effects of annealing, observed for NaCl crystals doped with europium in a wide concentration range, have substantiated the general conclusion drawn by W. Metag [1] that divalent impurities harden alkali halides intensely. Moreover, it has become clear that differences in the precipitation sequence influence the attainable crystal strength. This is related with the fact that the mechanical response of heavily doped crystals is sensitive to the degree of coherency between the matrix and the precipitated particles.

REFERENCES

1 W. Metag, Z. Physik 78 (1932) 363-366
2 J.J. Martin, V.J. Harness, E.M. Orozco, A.A. Mendoza and J.O. Rubio, Phil. Mag. A 46 (1982) 629-639
3 E.M. Orozco and F. Agullo-Lopez, Acta Met. 34 (1986) 1701-1709
4 M. Suszyńska, A. Gubański and H. Opyrchał, Cryst. Res. Technol. 22 (1987) 1229-1234
5 D. Nowak-Woźny and M. Suszyńska, Bull. Acad. Polon. Sci., sci. chim., to be published
6 J. Stepień-Damm and K. Łukaszewicz, Acta Cryst. A 36 (1980) 54-57
7 J.M. Garcia, J.A. Hernandes, E.H. Carillo and J.O. Rubio, Phys. Rev. B 21 (1980) 5012-5022.
8 G. Schoek and A. Seeger, Acta Met. 7 (1959) 469-477
9 H. Suzuki, In: Dislocations and Mechanical Properties of Crystals, J.C. Fisher, W.G. Johnston, R. Thomson and T. Vreeland, Jr. (Eds.), J. Wiley, New York 1959, p. 361-390
10 W. Hayden, W.G. Maffatt and J. Wulff, In: Structure and Properties of Materials, Vol. III, J. Wiley, New York 1965, p. 173-179
11 C. Zener, J. Appl. Phys. 20 (1949) 950-953
12 F.S. Ham, J. Phys. Chem. Solids 6 (1958) 316-335
13 V. Gerold, Trans. Indian Inst. of Metals 27 (1974) 317-323

GENERATION OF DISLOCATION HALF LOOPS AS THE INITIAL STAGE OF PLASTIC DEFORMATION OF HETEROEPITAXIAL $In_{1-x}Ga_xAs/InP$ LAYERS

G.Wagner+, G.Rohde++, V.Gottschalch++, P.Paufler+

+ Department of Crystallography
++ Department of Inorganic Chemistry, The University,Leipzig,GDR

ABSTRACT

Two types of corrugations have been observed on (001) layer surfaces if the misfit tensile strain exceeded 0.22% and 0.9%,respectively: Broad grooves and fine slip bands parallel to [$\bar{1}$10] directions and cracks parallel to [110] directions. The former are due to slip of partial and/or perfect half loops. TEM work revealed that dislocation half loops were generated at the free surface and moved preferably parallel to (111) and ($\bar{1}\bar{1}$1) planes towards the substrate.

INTRODUCTION

The $In_{1-x}Ga_xAs/InP$ epitaxial layer/substrate system is a promising material for high quantum efficiency, high speed, and low dark-current photodetectors for long wavelength optical communication nets. Dislocations may lead to a localized breakdown of electronic functions,and surface corrugations due to dislocation slip normally disturb subsequent technological processes.

Several authors have dealt with strain relaxation by formation of misfit dislocations. For a summary of the literature cf.[1] . Part of them consider dislocation processes in the layer/substrate interface region. Recently, a model of dislocation generation at the surface of the layer has been put forward by [2] . The present contribution is aimed at providing experimental results for this system which support that idea.

EXPERIMENTAL

By a variation of x in the ternary solid solution $In_{1-x}Ga_xAs$ the lattice parameter of its cubic unit cell (space group F$\bar{4}$3m, zincblende structure) may either be larger ($x < 0.47$) or smaller ($x > 0.47$) than the lattice parameter of the InP unit cell with the same structure. So, the epilayer can be subjected to either compressive or tensile strain.

Results obtained with compressed layers have been reported by [1] . We shall restrict ourselves to layers under tensile strain.

Single heterostructures have been grown by atmospheric pressure metal organic chemical vapour deposition (MOCVD). For details of the preparation cf. [3] .

The layer surface was investigated by optical interference contrast microscopy as well as by a TEM two step replica technique. Lattice defects have been analysed in electron microscopes TESLA BS-500 (90 kV) and JEOL (1000 kV, IFE Halle).

In order to differentiate between [110]/[1$\bar{1}$0] directions a vapour etch method was applied [4] .

Misfit strain was measured with the aid of X-ray double crystal diffractometry (cf. [1]).

RESULTS

Surface Morphology

Two main types of corrugations on the layer surfaces have been observed: The first type is preferably parallel to the [$\bar{1}$10] direction and due to slip processes. The second is parallel to the [110] direction and represents cleavage cracks. We distinguish with the former type between broad deep grooves and fine slip bands. Fig.1 gives an example, where the rather exceptional case of two short grooves parallel to [110] is included.

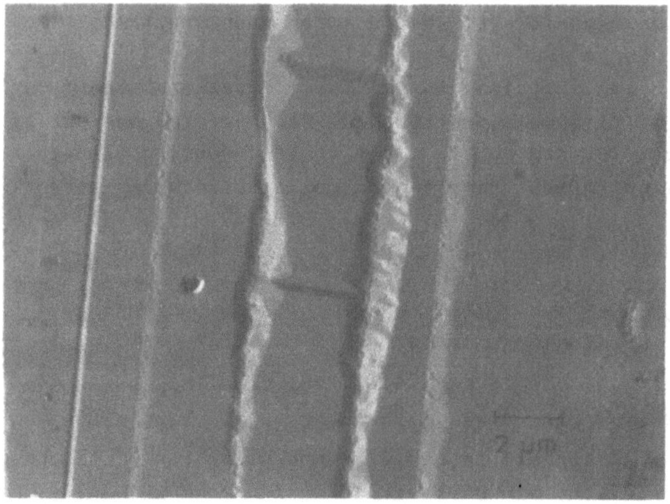

Fig.1 Grooves and slip bands predominantly parallel to [$\bar{1}$10]

The average spacing $\bar{\lambda}$ of the traces and their depths $\bar{h}$ are summarized in Table 1.

Table 1 Geometrical characteristics of slip traces

Specimen No.	Misfit Strain	$\bar{h}$/ nm Grooves	$\bar{h}$/ nm Slip Bands	$\bar{\lambda}$/ μm Grooves	$\bar{\lambda}$/ μm Slip Bands
405	1.8×10^{-3}	-	67	-	4.8
429	2.8×10^{-3}	60	30	-	11.8
474	4.9×10^{-3}	437	61	3.6	1.1
280	9.2×10^{-3}	268	102	10.8	2.1
495	$(0.8..1.0) \times 10^{-2}$	347	74	7.6	2.8
371	1.8×10^{-2}	575	105	3.4	2.4

Dislocation Analysis

TEM work revealed that those grooves were always connected with stacking fault packages and that slip bands were due to dislocations the extension of which could not be resolved.

Moreover, slip occurs predominantly on (111) and ($\bar{1}\bar{1}1$) planes inclined to the (001) surface. Leading partials of stacking faults were always found to be of pure edge type. Dislocations were bent towards the substrate, i.e., they must have been nucleated at the surface of the epitaxial layer.

Knowing that dislocations move in the glide set [5], then In/Ga partials have to move down the (111) or ($\bar{1}\bar{1}1$) planes towards the interface under tensile strain conditions whereas As partials should move up. For ($\bar{1}11$) and ($1\bar{1}1$) planes the opposite behaviour would be expected. Since edge partials have been observed in (111) and ($\bar{1}\bar{1}1$) planes the conclusion must be drawn that In/Ga partials exhibit a lower nucleation energy than As partials.

Critical Thickness/Strain Relation

Epitaxial growth takes place pseudomorphously until the growing layer reaches a critical thickness at which the homogeneous elastic strain and the energy of misfit dislocations become equal. A calculation of the critical thickness/critical strain relation according to the model of half loop nucleation at the layer surface[2] was carried out by [1] and [3] and was found in good agreement with experiments. It is also supported by the present results.

CONCLUSION

Although the nucleation process at the very beginning could not be resolved it can be taken for sure that plastic deformation of the epitaxial layer under tensile strain starts at its free surface. To keep the interface invariant two slip planes have to be activated. Their interplay leads to the formation of groove-like corrugations.

REFERENCES

1 G.Wagner, V.Gottschalch, H.Rhan and P.Paufler, phys.stat.sol.(a) 112(1989)519-531; 113(1989)71-81

2 P.M.J.Marée, J.C.Barbour, J.F.van der Veen, K.L.Kavanagh, C.W.T.Bulle-Lieuwma and M.P.A.Viegers, J.Appl.Phys. 62(1987)4413-4420

3 G.Wagner, P.Paufler and G.Rohde, Z.f.Kristallographie (1989) (accepted for publication)

4 G.Wagner, A.Dreilich and E.Butter, J.Mater.Sci. 23(1988)2761-2767

5 H.Alexander, J.de Physique 40(1959) Suppl. au No.6,C6-1...C6-6

4. MICROSTRUCTURE AND MECHANICAL BEHAVIOUR OF CERAMICS

TEM INVESTIGATIONS OF WHISKER TOUGHENED CERAMICS

by

G. H. Campbell, A.G. Evans and M. Rühle*

Materials Department, College of Engineering, University of California, Santa Barbara, California 93106
* Now at: Max-Planck-Institut für Metallforschung, Institut für Werkstoffwissenschaft, Seestr. 77, 7000 Stuttgart 1, F.R.G.

ABSTRACT

Whisker toughened ceramics have been studied by TEM, with the objective of identifying the mechanism that provides the major contribution to toughness. It is concluded that, for composites with randomly oriented whiskers, bending failure of the whiskers obviates pull-out, whereupon the major toughening mechanisms are the fracture energy consumed in creating the debonded interface and the stored strain energy in the whiskers, at failure, which is dissipated as acoustic waves. The toughening potential is thus limited. High toughness requires extensive pull-out and hence, aligned whiskers with low fracture energy interfaces.

1. INTRODUCTION

The toughening of ceramics by whiskers typically includes contributions from debonding, crack bridging, pull-out and crack deflection.[1-3] It is important to ascertain the relative roles of these contributions, because each mechanism predicts very different dependencies of toughness upon microstructure. Crack deflection is ostensibly governed only by whisker shape and volume fraction[4]: albeit that the relative elastic moduli and thermal expansion coefficients may have implicit effects on the deflection path. Conversely, the other contributions depend sensitively upon the mechanical properties of the interface, the whisker strength/toughness, the whisker radius and the volume fraction, as elaborated below.

Toughening by bridging is induced by debonding along whisker/matrix interfaces[1] (Fig. 1). The debonding allows the whiskers to remain intact within a small bridging zone behind the crack.[1] The magnitude of the toughening involves considerations of the debond extent and the mode of reinforcement failure, as well as of residual stress effects.[1,5] Various attempts have been made to model the toughening and to compare the predictions with experiment. Becher et al.[6] established many of the salient features. However, neither residual stress effects nor the contribution to toughness from the energy of the debonded surfaces were included and, furthermore, explicit determination of the debond extent was not used for comparison between experiment and theory. Campbell et al.[7] examined these and related issues by further experimental investigations and by presenting a fully inclusive model in a form that provided direct physical insight. In the present article emphasize is put on crack/microstructure interactions which are investigated in two whisker-toughened systems: the Al_2O_3/SiC system which has interfaces subject to residual <u>compression</u> and Si_3N_4/SiC with interfaces in residual <u>tension</u>. Microstructure studies are performed by different TEM techniques.

2. MATERIALS

Four materials have been used for the present investigation: two whisker toughened materials and two reference matrix materials. One whisker toughened material is a commercial product, consisting of Al_2O_3 toughened with 20 vol. % SiC whiskers, prepared by hot pressing*. The other toughened material was a SiC toughened Si_3N_4** containing 20 vol. % SiC whiskers, 4 wt-% Y_2O_3 as a sintering aid. It was extruded to align the whiskers and densified by reaction bonding followed by HIPing. Microstructural investigations were performed, not only for the composites, but also for the matrices and those observations were used as reference. The Al_2O_3 matrix material was a hot pressed system containing 0.25 vol.% MgO, heat treated to create the corresponding grain size. The reference Si_3N_4** contained 4 wt % Y_2O_3.

The mode I toughnesses, G_c, of the four materials, evaluated using an indentation method[11], are summarized in Table I. A comparison with models requires evaluation of the increase in toughness caused by the whiskers, ΔG_c, which for Al_2O_3/SiC is 40–80 Jm^{-2} and for Si_3N_4/SiC is 25–55 Jm^{-2} (Table I).

3. CHARACTERIZATION

The microstructures of the different materials were characterized by methods which depend on the level of resolution required for the identification of structures and morphologies. Usually, electron optical instruments are needed. For example, scanning electron microscopy was used for the determination of grain sizes and of different phases in the composite[7] down to dimensions of ~ 0.1 µm, whereas TEM has allowed the identification of phases and defects to the atomic level[8,9]. Analytical TEM has allowed the characterization of chemical composition with high spatial resolution of ~ 4 nm[10,11].

3.1 MICROSTRUCTURE OF WHISKERS

Characterization of two different as-received SiC whiskers*** indicated a broad distribution of length (1µm ... 25µm) and diameter (0.1µm to 1.0µm). Quite frequently, clusters of whiskers and irregular, serrated-edged whiskers were found. The average length and diameter were 17.5µm and 0.5µm, respectively, for ARCO whiskers and 11.7µm and 0.42µm for Tateho whiskers.

The whiskers had a high density of planar faults lying on the close-packed plane (basal plane) perpendicular to the long axis.[12] The faults resulted in a complex arrangement in thin lamellae of different α and β polytypes normal to the whisker axis. Furthermore, the core region of the ARCO whiskers contained a high density of small spherical impurity clusters and/or cavities. Half of the Tateho whisker are hollow. The outer skin of both types of whiskers was covered with a 2nm SiO_2 rich layer as revealed by defocus imaging (Fig. 2)[7,8] and X-ray photoelectron spectroscopy.[13]

3.2 MICROSTRUCTURE OF THE COMPOSITES

Thin foils of each of the materials suitable for TEM have been prepared by mechanical polishing and dimpling followed by ion beam thinning. From conventional TEM micrographs of the SiC *whisker toughened Al_2O_3 material*, a matrix grain size of ~ 1µm was determined. Dark field studies[14] have indicated the presence of a thin (< 5nm thickness) amorphous layer at the interface between the whiskers and the matrix, consistent with previous studies.[1,13] The thickness of the amorphous layer at the whisker/matrix interface depends on the purity of the Al_2O_3. However, even in Al_2O_3–SiC composites processed using ultraclean materials,[15] a thin

* Greenleaf, Inc.
** Norton Company
*** ARCO and Tateho whiskers

(1nm-thick) SiO_2 layer can be detected by TEM imaging. Specifically, by using a range of defocus conditions[14] (Fig. 3), the expected contrast reversal for different signs of defocusing can be observed. Furthermore, it is apparent that the contrast at the interface is not caused by preferential etching. Notably, the interface between the whiskers and the Al_2O_3 matrix in Fig. 3 is perpendicular to the edge of the foil and no preferential etching is visible close to the amorphous region near the edge, because the interface contrast disappears gradually in the amorphous part of the specimen.‡ Occasionally, amorphous pockets also exist at junctions between matrix grain boundaries and the interface. However, within the resolution of dark field and high resolution procedures (~ 1nm), the matrix grain boundaries appear to be devoid of amorphous material.

Occasionally, a whisker is completely embedded in *one* Al_2O_3 grain. In such cases, strain contours become visible under dynamical TEM imaging[7] (Fig. 4). A semi-quantitative evaluation of the strain perpendicular to the whisker diameter is possible by calculating the scattering of modified Bloch waves[16] in the thick foil approximation and evaluating the number and distance of contrast oscillations for different excitation errors. The observations confirm that a homogeneous compressive strain ε occurs in the whisker, with $2 \cdot 10^{-3} < \varepsilon < 6 \cdot 10^{-3}$: a result consistent with the known thermal expansion characteristics of Al_2O_3 and SiC and a cooling range $\Delta T \approx 1000°C$[17].

Studies conducted on the *Si_3N_4 reference material* have revealed several relevant microstructural features. Dark field and high resolution microscopy have indicated that an intercrystalline phase, presumably amorphous, is present as a thin continuous film of equilibrium thickness[7]. Grain pockets are mostly crystalline, except for the thin, intercrystalline outer layer[17]. Analytical microscopy and diffraction studies are consistent with α-$Y_2Si_2O_7$ being the predominant crystalline grain boundary phase. The crystalline pockets are typically 50–100nm in diameter. The *whisker toughened Si_3N_4* had several different characteristics. The whiskers are typically 100–500nm in diameter. About half of the whiskers are hollow, in which case they become filled by the sintering-aid phase. This phase, which is also present at grain pockets, remains entirely glassy upon cooling to room temperature. Analytical microscopy indicates a relatively high SiO_2 content in this phase. Presumably, the excess SiO_2 that prevents crystallization of this phase is introduced by the oxide layer that preexists on the SiC whiskers. High resolution microscopy shows that this intercrystalline phase is present as a thin film at *all* Si_3N_4 grain boundaries and, most importantly, at the interface between the Si_3N_4 grains and the SiC whiskers (Fig. 5).

Finally, in order to clarify ambiguity, it is noted that it has been occasionally reported in the literature that *no* glassy phase can be observed at interfaces between SiC whiskers and the surrounding matrix.[18] A careful inspection of available micrographs[18] suggests that the interface plane in such studies was inclined to the electron beam, whereupon the glassy interphase would not be detected. Specifically, it is evident that an amorphous interface layer can only be imaged if the foil thickness t is such that $t \lesssim \sqrt{aD/2}$, with a being the thickness of the amorphous phase and D the diameter of the whiskers and if the interface is parallel to the electron beam.

3.3 TEM STUDIES OF DEBONDING AND BRIDGING

Observations of debonding and of bridging zones have been made in the TEM. The procedure developed for studying these characteristics consists of indenting the material, mechanically dimpling in the region of the indentation crack tips and then ion thinning to a thickness in excess of the whisker diameter. The procedure ensures debonding and bridging representative of plane strain crack propagation in bulk material and avoids the anomalous crack extension and debonding that can occur upon thinning to less than the whisker

‡ A hole would be visible at the site of the interface for the case of preferential thinning.

diameter. In order to use these foils to investigate near tip phenomena, it is imperative that the specimens be subject to tilting through a large angular range, because the debond and matrix crack opening displacements are small and only detectable when the electron beam is essentially aligned with the crack plane. Some typical views of the bridging zone with schematic drawings used to highlight the salient features.

For Al_2O_3/SiC, the matrix cracks are mostly transgranular and conchoidal, while cracks in the whiskers are planar and always parallel to the basal plane of SiC. These distinctive fracture paths facilitate interpretation of the micrographs. The series of micrographs and schematical drawings summarized in Fig.6 illustrate the salient features. Locations A and B refer to the crack/whisker interactions, which include whiskers having axes either normal or inclined to the crack plane. In both cases, the whiskers are debonded and are fractured. Near the crack tip (location C), a bridging zone with intact whiskers is apparent, with the crack tip located at the arrow. Tilting studies have revealed the extent of debonding at the whisker/matrix interface as illustrated in the accompanying schematic Fig. 6). Such observations are typical of whiskers that exist within a bridging zone that usually extends to about 4 to 6 whisker diameters. *The length of the debonds is variable and ranges between about 2R and 6R*, R being the whisker radius.

For Si_3N_4/SiC, the same general features noted for Al_2O_3 are again evident[7]. The lengths of debonds, which can be measured directly from micrographs, range from R to 3R. Usually, the second whisker behind the crack tip is fractured, indicating that the bridging zone is very small. It is also noted that, because elongated Si_3N_4 grains are present, debonding along the grain boundaries with the equiaxed matrix grains can be observed[7]. Such debonding is exactly analogous to that occurring at whisker interfaces. Consequently, the debonds of the whisker interface are more difficult to detect than in the Al_2O_3 composite. An interesting sequence of events can be deduced from Fig. 7, in which the main crack is clearly visible (large white gap). This crack propagated from the upper right to the lower left. It is surmised that, when the crack first interacted with the whisker, the interface debonded (to a length of about 3 whisker radii). The whisker then fractured within the debond length and pulled out by about 20nm (the width of the gap). While being pulled out, the whisker would be subject to bending, resulting in enhanced stresses, as manifest in strain fringes at the corner of the whisker (arrow in Fig. 7). Consequently, the whisker fractured a second time in the plane of the matrix crack.

It is also of importance to examine whisker failure. In general, whisker fracture can either initiate at the end of the debond by kinking into the whisker,[5] or the debond cracks could remain at the interface such that the whisker fails by the propagation of a pre-existing flaw within the debonded length. The former mode has been rigorously verified in several cases[7] (Figs. 6 and 7) by extensive tilting to confirm the absence of a debonded region beyond the location of the whisker fracture. The latter is much more difficult to unambiguously identify and cannot be substantiated at this juncture.

Within the spatial resolution of the Scanning Electron Microscopy (SEM), *debonds* are only occasionally evident, usually at intact inclined whiskers remote from the crack front[1,7]. Crack surface interlocking can also be identified, caused by debonding around the whisker ends. In some cases, whisker cracks are apparent[1,7], again at inclined whiskers, with the crack often forming near the end of the debond. Measurements of *bridging* zones associated with intact whiskers are not possible by SEM, because of resolution limitations.

Inspection of fracture surfaces by SEM provides complimentary information regarding *pull-out* as well as debonding. Studies on both composites reveal that inclined whiskers fracture on a plane normal to the whisker axis, such that the crack-surface is within one whisker radius of the matrix crack plane. *Pull-out is thus negligible*.

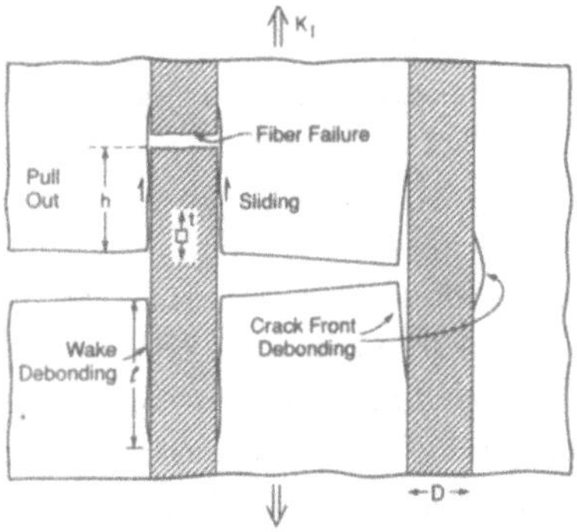

Fig. 1. A schematic of a bridging zone in a reinforced composite

TABLE I

Properties of Materials and Constituents

Material	$\alpha (C^{-1} \times 10^6)$	E(GPa)	$\mathcal{G}_c (Jm^{-2})$	$\Delta\mathcal{G}_c (Jm^{-2})$	
				Experiment	Theory
Al_2O_3	7.5	400	25 ± 5		
Si_3N_4	3.5	320	50 ± 10		
SiC	4.5	420	15 ± 5		
Amorphous Silicate Interphase	....	100	7 ± 1		
Al_2O_3/SiC		420	85 ± 15	60 ± 20	~ 80
Si_3N_4/SiC		350	90 ± 15	40 ± 15	~ 30

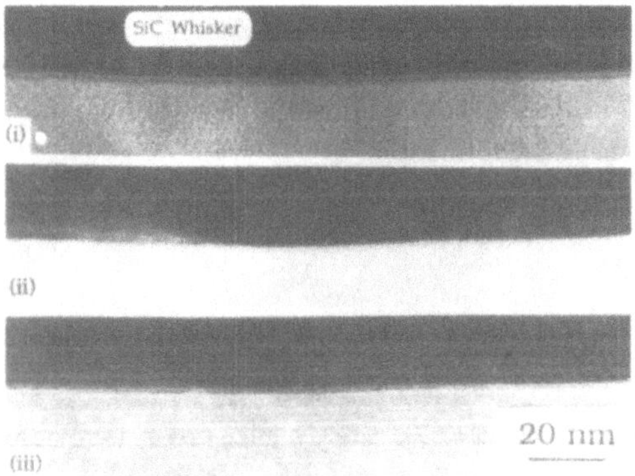

Fig. 2. Analysis of outer layer on a Tateho SiC whisker. Through-focus images at the edge of a whisker (i) $\Delta f < 0$, (ii) $\Delta f \approx 0$, (iii) $\Delta f > 0$.

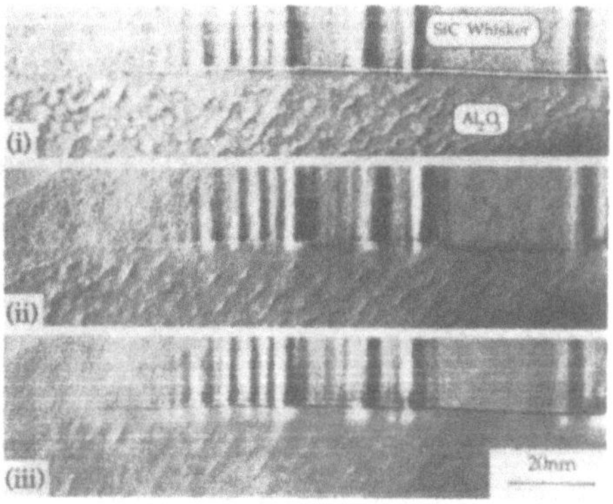

Fig.3. TEM micrograph of interface between SiC whisker and Al_2O_3 matrix (i) $\Delta f = -96$nm, (ii) $\Delta f \approx 0$, (iii) f = 96nm. The interface is parallel to the electron beam.

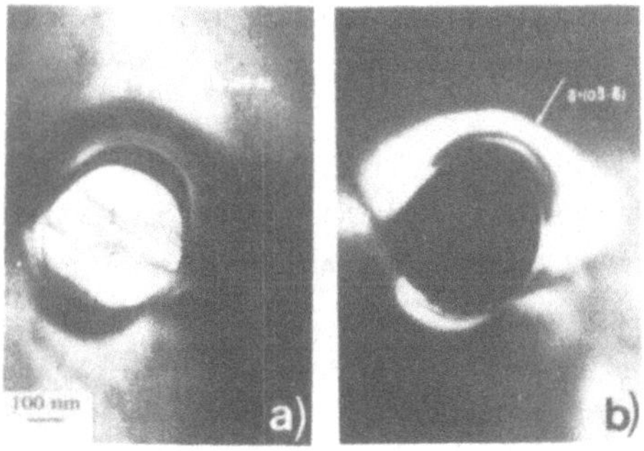

Fig. 4. Stress contours around whiskers: dynamical imaging conditions a) Bright field and b) Dark field images (courtesy of E. Bischoff)

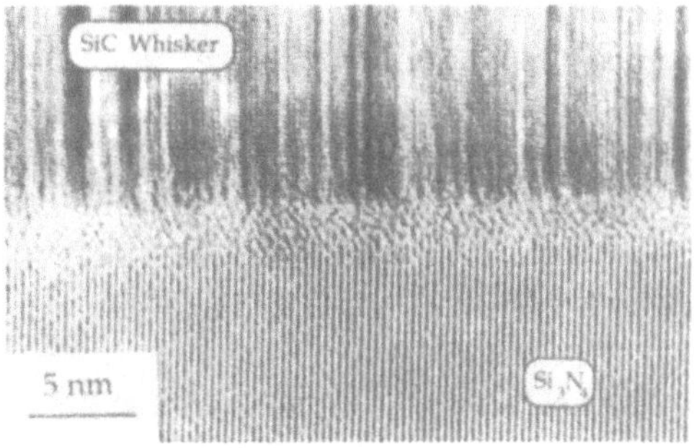

Fig. 5. High-resolution image of the amorphous layer between a matrix grain and a SiC whisker

4. SOME RELEVANT MECHANICS

The correlation between microstructural studies and mechanical properties requires micromechanical modeling. The crack propagation is modeled for an ideal composite composed of parallel whiskers which possess identical distances and are embedded in the ceramic matrix.

4.1 INTERFACE DEBONDING

Cracks on bimaterial interfaces are characterized by a strain energy release rate, $\mathcal{G}$, and a phase angle of loading, ψ.[19,20] The latter is a measure of the mode mixity, such that $\psi = 0$ refers to a pure crack opening mode and $\psi = \pi/2$ represents an interface crack subject to crack surface shear. Specifically, debonding along the interface occurs when $\mathcal{G}$ reaches the interface fracture energy Γ_i at the relevant phase angle, ψ. For a matrix crack subject to mode I loading, the incidence of *initial debonding*, rather than cracking into the whisker, is found to be governed by the ratio of interface to whisker fracture energies Γ_i/Γ_f and the whisker orientation (Fig. 8).[20] For the present composites, values of Γ_i/Γ_f needed to debond whiskers normal to the crack plane should be either < 1/3 for Si_3N_4/SiC or < 1/4 for Al_2O_3/SiC. Note that this prediction is *independent* of the residual stress.

Growth of the initial debond along the interface is influenced by additional variables, such as the residual stress and the whisker radius. Debond growth is most prevalent in the crack wake.[21,22] Analysis of wake debonding for interfaces under *residual tension* [21] indicates that debond growth in the wake occurs when the stress t on the whisker reaches a critical value t^*, given by,

$$t^* \approx 2.2E_f e_T \tag{1}$$

where E_f is Young's modulus for the fiber and e_T is the misfit strain. When $t > t^*$, the debond propagates unstably up the interface.

Residual compression results in different behavior. In this case, $\mathcal{G}$ is strongly influenced by the friction coefficient μ along the previously debonded interface.[22] The basic debonding features indicate that again, a threshold stress must be exceeded before debonds can further propagate in the wake. However, for stresses in excess of the threshold, debonding occurs <u>stably</u> to an extent determined by the friction coefficient μ and the residual strain, until $t \rightarrow E_f e_T/\nu$ (with ν being Poisson's ratio for the fiber), whereupon the interface separates and further debonding occurs unstably. The debonding behavior is thus sufficiently complex that prediction of trends with e_T, Γ_i and other material variables is unlikely to be instructive. Instead, the preceding results may be used as the mechanics background needed to facilitate interpretation of observed trends in debonding, as elaborated below.

4.2 TOUGHENING

The steady-state toughening $\Delta\mathcal{G}_c$ imparted by whiskers can be considered to have <u>four</u> essential contributions, as elaborated by Campbell et al.[7]. These contributions can be insightfully expressed in the simple form (Fig. 9);

$$\Delta\mathcal{G}_c/f\,d \approx S^2/E - Ee_T^2 + 4(\Gamma_i/R)/(1-f) + (\tau/d)\sum_i(h_i^2/R) \tag{2}$$

where d is the debond length, R the whisker radius, f the volume fraction and S the whisker "strength." The first term S^2/E is simply the *strain energy stored* in the whisker *over the debonded length* on both sides of the matrix crack, before the whisker fails ($S^2/2E$ on either side of the crack)[7]. This strain energy *dissipates* as acoustic waves and thus contributes *positively* to the toughening. The second term is the ***residual strain energy*** in each composite element ***within the debond length***, as governed by the misfit strain e_T. This strain energy is ***lost*** from the system when the fiber fails and thus, ***detracts*** from the toughness, independently of the sign of e_T. The third term is the energy needed to create the ***debond fracture surface***, with Γ_i being the fracture energy per unit area. This term must be ***positive***. The last term is the ***pull-out contribution***, with h_i being the pull-out length and τ the sliding resistance of the debonded interface. This term is again positive, because heat is generated by the frictional sliding at the interfaces. The above result is all inclusive and has not been specifically presented elsewhere.‡ Furthermore, it deviates from certain of the other results reported in the literature for each of the individual terms, principally in the magnitude of the constants of proportionality. It is believed that Eqn. (2) is the simplest possible result that is also physically consistent with the mechanisms involved.

Optimization of toughness based on these terms is discussed in Section 6. Presently, it is noted that all contributions scale with the debond length d, *indicating that large d is desirable*. However, it is also recognized that d should have a functional dependence,

$$d/R = F(\Gamma_i, e_T, S) \tag{3}$$

where the function F has yet to be determined.

The present experimental results are examined using Eqn. (2) to assess the various contributions to toughness, based on direct measurements of h, d, R, f. Thereafter, implications for high toughness are discussed.

5. COMPARISON BETWEEN THEORY AND EXPERIMENT

Initial comparison between theory and experiment can be achieved based on measured values of the debond length and using Eqn. (2) for the toughness. The relative magnitudes of the four contributions to toughness from elastic bridging, residual strain, debond surface energy and pull-out can then be readily assessed, using the material properties listed in Table I and using $d/R \approx 2-6$ for Al_2O_3 and 1–3 for Si_3N_4 with $f \approx 0.2$ and $R \approx 0.25\mu m$ for both composites. The *pull-out* contribution is negligible since only a small fraction of whiskers exhibit this phenomenon. The reduction in toughness from the ***residual strain***, $Ee_T^2 fd$, ranges between 0.3 Jm^{-2} for Si_3N_4/SiC to 2 Jm^{-2} for Al_2O_3/SiC and is thus of negligible importance (in part, accounting for the similar toughening levels measured for both composites). The toughening contribution from the *elastic strain energy* stored in the whiskers up to failure, S^2fd/E, is dominated by the choice of whisker "strength." The appropriate choice for S is unclear because the "gauge length" is small and because the whiskers may be subject to degradation upon composite processing. For initial purposes, it is *assumed* that S is in the range 4–8GPa, whereupon this component of toughening is at most 45 Jm^{-2} for Al_2O_3/SiC and 25 Jm^{-2} for Si_3N_4/SiC. Finally, the *debonding energy* contribution is estimated by noting that amorphous silicates have a fracture energy, $\Gamma_i \approx 6-8$ Jm^{-2}. The toughening caused by debonding is then of order 50 Jm^{-2} for Al_2O_3/SiC and 25 Jm^{-2} for Si_3N_4 /SiC. The energy needed to create the debond surface and the strain energy dissipated from elastic bridging thus appear to provide similar contributions to toughness. Furthermore, the toughness level provided by the combination of both processes is comparable to the measured values (Table I).

‡ The results are strictly applicable to reinforcements *normal to the crack plane*. Whiskers inclined to the crack are more likely to fail by bending and thus, result in smaller pull-out toughening.

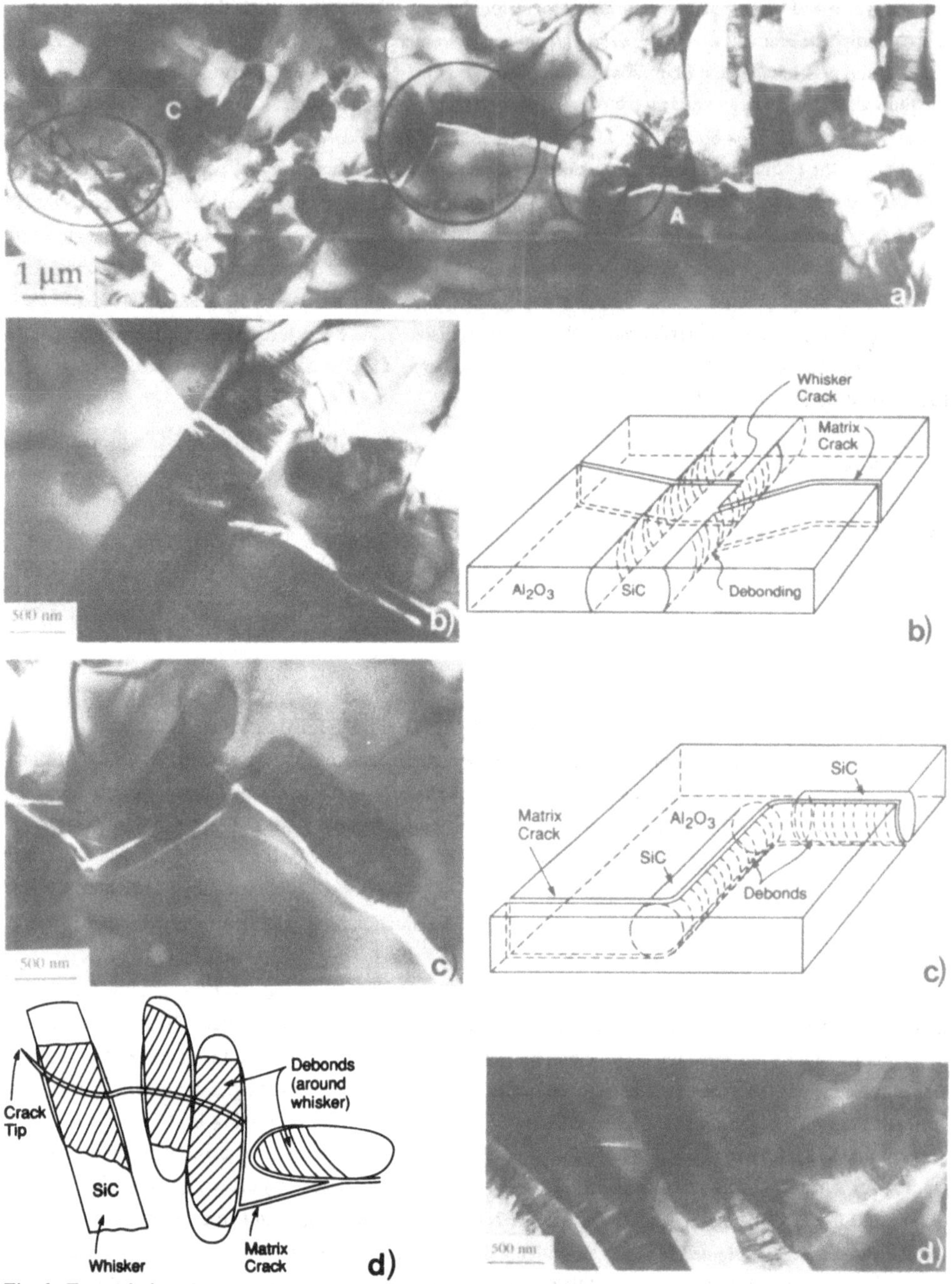

Fig.6. Transmission electron microscopy studies of the crack tip region in thick foils of Al_2O_3/SiC. A comprehensive visualization of debonding and cracking can only be achieved by imaging under different orientations by extensive tilting. Consequently, to assist in summarizing the behavior, a schematic drawing is shown in addition to *one* TEM micrograph of each crack/whisker interaction:

a) Overview of crack tip region
b) Region A with schematic 3-dimensional drawing
c) Region B with schematic 3-dimensional drawing
d) Region C bridging zone with schematic overview. The debonded interface zones between whiskers and matrix are hatched. The whiskers are not fractured.

6. IMPLICATIONS AND CONCLUSIONS

The preceding experiments and calculations firstly indicate that *non-aligned*, inclined whiskers typically fail by bending and do not provide a pull-out contribution to toughening.* Consequently, composites with *randomly oriented whiskers cannot normally be expected to exhibit high toughness.* Subject to this limitation, useful toughness increases are still possible, as governed by an optimum combination of bridging and debonding. Since residual strain is invariably detrimental, *matched thermal expansions are desirable. Enhanced debonding* is also desirable, but then the relative contributions to toughness from bridging and from the debonded surfaces requires further elaboration. The debond length is expected to scale directly with the whisker radius, to increase with increase in S and to depend inversely on Γ_i (Eqn. 3). Consequently, d/R and Γ_i are coupled in such a way that the contribution to $\Delta \mathcal{G}_c$ from the *energy of the debonded surfaces* is expected to be weakly dependent on Γ_i and *insensitive to whisker radius*, but should *increase* as the *whisker strength* increases. A corresponding assessment of the *bridging contribution* yields very different conclusions. In this case, the direct dependence on debond length suggests that this contribution should *increase* appreciably as either Γ_i *decreases* or the *whisker radius increases* and should become the dominant contribution to toughness for small Γ_i and large R, provided that the *"strength" S is also high.* Indeed, it is important to note that careful experiments[6] indicate a systematic dependence of $\Delta \mathcal{G}_c$ on R, confirming an important contribution of elastic bridging to toughening. However, the whisker "strength" is not related in simple form to the uniaxial fracture strength of the whiskers. Further research is needed to understand the operative relationships.

The factors which govern the *debond length* merit brief additional consideration. While small Γ_i and large S clearly enhance debonding, such variables as residual strain, surface roughness and friction coefficient could be important. The available calculations[7] indicate that, when the non-dimensional residual strain parameter Q $(= S\nu/E_f|e_T|)$ is much larger than unity, the Poisson contraction of the whisker is large compared with the misfit displacements. Then, residual strain is unimportant: instead, the amplitude of the roughness on the debond surface dominates debonding. Conversely, when $Q < 1$, residual strain effects dominate debonding in the sense that positive e_T (interface tension) enhances debonding and vice versa. For the present materials, the inverse trend with residual strain (*large* debond lengths for Al_2O_3/SiC in which e_T is *negative*) suggests that roughness effects and interface characteristics are more important in debonding than the residual strain. Indeed, consistent with this implication, Q is larger than unity. Whisker roughness effects would thus appear to merit further investigation.

7. ACKNOWLEDGEMENT

Fruitful discussions with Drs. E. Bischoff and B.J. Dalgleish are gratefully acknowledged. The work was supported financially in part by the Defense Advanced Research Projects Agency through the University Research Initative Program at USBC under ONR Contract N 00014-86-K-0753.

8. REFERENCES

[1] M. Rühle, B. J. Dalgleish and A. G. Evans, Scripta Metall. **21** (1987) 681–686.
[2] P. F. Becher and G. C. Wei, Comm. of the Am. Ceram. Soc. **67** (1984) C267–269.
[3] P. Becher, C. H. Hsueh, P. Angelini and T. N. Tiegs, Mtls. Sci. Eng. **A107** (1989) 257–261.
[4] K. T. Faber and A. G. Evans, Acta Metall. **31** (1983) 565–576.
[5] D. B. Marshall and A. G. Evans, Materials Forum **11** (1988) 304-312.
[6] P. F. Becher, C. H. Hsueh, P. Angelini and T. N. Tiegs, J. Am. Ceram. Soc. **71** (1988) 1050.

---

* An exception may be graphite whiskers which can sustain very high bending strains and have debonded interfaces with a low sliding resistance.

[7] G.H. Campbell, M. Rühle, B.J. Dalgleish and A.G. Evans, J. Am. Ceram. Soc., in print.
[8] L. Reimer, Transmission Electron Microscopy, Springer Berlin (1986).
[9] J.C.H. Spence, Experimental High-Resolution Electron Microscopy, 2nd ed. Oxford University Press, New York and Oxford (1988).
[10] D.C. Joy, A.D. Romig, Jr. and J.I. Goldstein (edts.), Principles of Analytical Electron Microscopy, Plenum Press, New York and London (1986).
[11] R.F. Egerton, Electron Energy Loss Spectroscopy in the Electron Microscope, Plenum Press, New York and London (1986).
[12] S. R. Nutt, J. Am. Ceram. Soc. **67** (1984) 428–431.
[13] V. K. Sarin and M. Rühle, Composites **18** (1987) 129–134.
[14] D. R. Clarke, Ultramicroscopy **4** (1979) 33–44.
[15] J. R. Porter, F. F. Lange and A. H. Chokshi, Am. Ceram. Soc. Bull. **66** (1987) 343-347.
[16] M. Wilkens and M. Rühle, Phys. Stat. Sol.(b) **49** (1972) 749–760.
[17] P. Angelini, W. Mader and P.F. Becher, Mat. Res. Soc. Symp. Proc. **78** (1987) 241
[18] W. Braue, R. W. Carpenter and D. J. Smith, Proc. 46th Ann. Meeting EMSA (ed. G. W. Bailey), San Francisco Press, Inc. (1988) pp. 734–735.
[19] J. R. Rice, J. Appl. Mech. **110** (1988) 98–103.
[20] M. Y. He and J. W. Hutchinson, J. Appl. Mech. **111** (1989) 270–278.
[21] P. G. Charalambides and A. G. Evans, J. Am. Ceram. Soc. **72** (1989) 746–753.
[22] L. S. Sigl and A. G. Evans, Mechanics of Materials, in press.

Fig.7. Transmission electron micrograph of a multiply-fractured whisker well into the crack wake

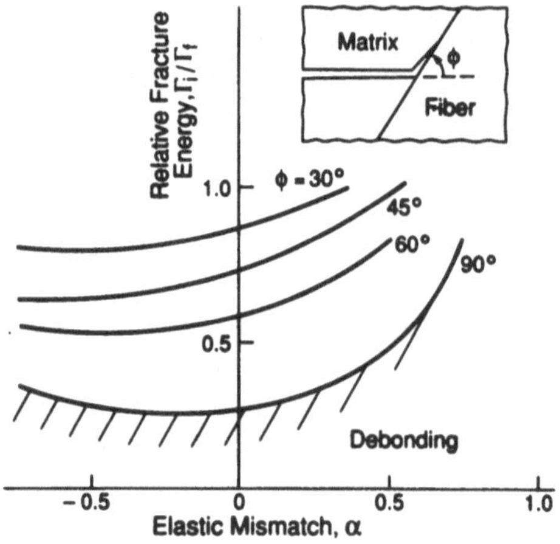

Fig.8. A debond diagram: the dependency of fracture energy on elastic mismatch and whisker orientation

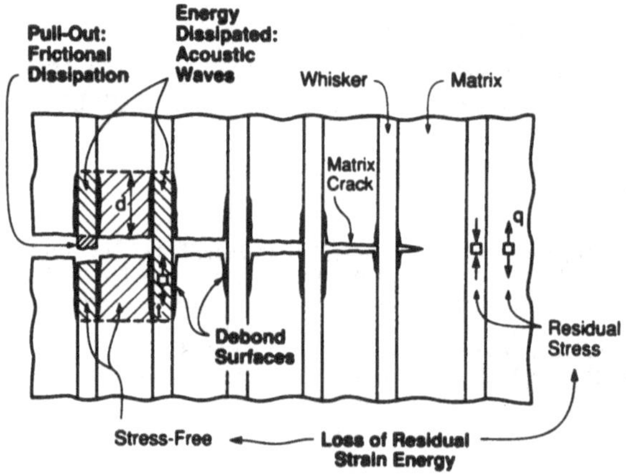

Fig.9. A schematic representation of the four mechanisms that contribute to toughening by whiskers

THE DEFECT STRUCTURE OF SiC AND Si_3N_4 CERAMICS

M. Bartsch, U. Messerschmidt, F. Appel, and P. Werner

Institute of Solid State Physics and Electron Microscopy
Academy of Sciences of the G.D.R.
Weinberg 2, Halle/Saale, 4010

ABSTRACT

The structure of pressureless sintered SiC and Si_3N_4 ceramics was investigated using high voltage and high resolution electron microscopy. The defects mainly occurring were characterized using the standard methods of diffraction contrast and high resolution techniques. With regard to the mechanical properties special emphasis was put on the detection of thin intergranular phases.

INTRODUCTION

Among other advanced ceramics silicon carbide and silicon nitride are considered to be two of the more promising materials for high - temperature applications. The exceptional mechanical properties at high temperatures result from the covalent bonding, which provides a large lattice resistance. Thus, high-temperature creep has mostly been interpreted in terms of diffusional atomic flow rather than in terms of dislocation glide. In addition, certain ways of producing the materials may form thin intergranular glassy phases of a thickness less than 500 Å. At elevated temperatures these films soften with sudden loss of trength [1,2]. Thus, the brittle-ductile transition is determined by complex defect processes depending on the applied production technique. The investigation of the microstructure may therefore give information on the mechanisms limiting the mechanical strength. In this regard, the defect structures of SiC and Si_3N_4 ceramics were studied using high voltage and high resolution electron microscopy.

EXPERIMENTAL DETAILS

SiC samples pressureless sintered from α-SiC powder with B-based additives were supplied by the WTK Meißen (G.D.R.). The pressureless sintered Si_3N_4 samples with sintering aids of Y_2O_3, Al_2O_3, and AlN were produced by the Central Institute of Inorganic Chemistry of the Academy of Sciences of the G.D.R.

Thin foils were prepared from mechanically polished sheets by ion-thinning in the standard manner (8 kV-argon ions, angle of incidence about 10°). The Si_3N_4 foils were coated with a thin carbon layer to prevent electrostatic charging.

The Jeol microscopes 100C, 4000 EX, and 750 were used operating at 100, 400, and 1000 kV, respectively.

DISLOCATIONS AND STACKING FAULTS IN SiC

In the SiC samples investigated 6H polytypes are predominant, in rare cases, however, 4H polytypes and cubic structures were found.

A number of dislocations have been analyzed as Shockley partials with a Burgers vector of $b=(a/3)\langle 10\bar{1}0\rangle$, adopting the extinction criterion $(\mathbf{gb})=0$ or 1/3 [3]. $\mathbf{g}$ is the reciprocal lattice vector used for bright-field imaging. The grown-in dislocations are usually arbitrarily curved and do not show prefered orientations.

The planar defects being predominant in the SiC samples are stacking faults. The result is in agreement with the observations made on other SiC materials [4 to 6]. Fig. 1 shows stacking faults in a large grain, as an example. A characterization of the defects using the extinction criterion $\varphi=0$, $\pm 2\pi$, ..., with $\varphi=2\pi(\mathbf{rg})$ could not uniquely be performed. $\mathbf{r}$ and φ are the displacement vector and the phase shift of the stacking faults, respectively. The reasons for this are that a variety of possible phase shifts may be related to the defects and that the $\mathbf{g}$-vectors necessary are hardly exactly to meet, since very close reflex distances occur in the diffraction patterns with a zone axis lying in the basal plane. The stacking faults were directly identified using high resolu-

Fig. 1 Stacking faults in a large SiC grain.
Acceleration voltage 1000 kV

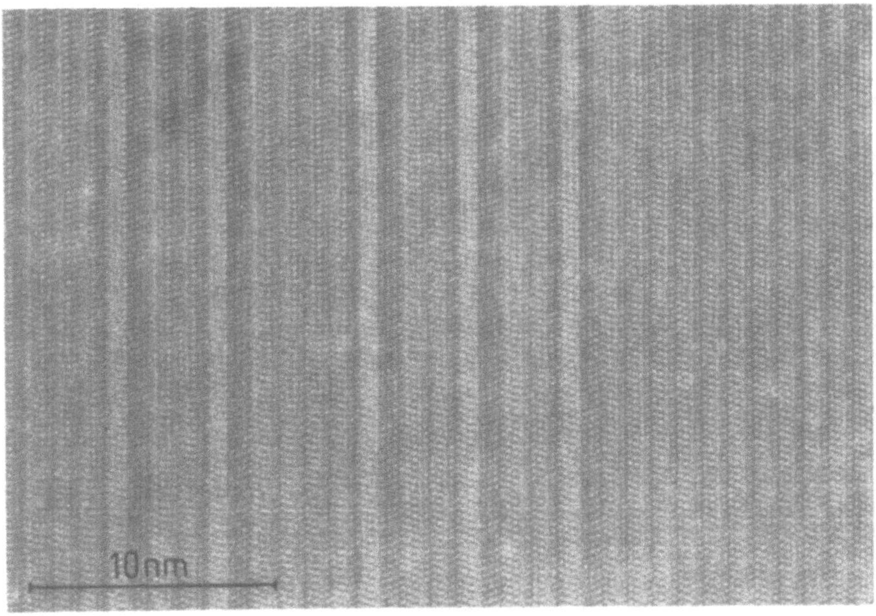

Fig.2 Structure pattern of SiC 4H showing a number of stacking faults. Foil orientation <11$\bar{2}$0>, the (0006) lattice planes with a distance of 2.51 Å are imaged. Acceleration voltage 400 kV

tion electron microscopy. Fig. 2 shows faulty-stacked basal planes in a SiC grain mainly having the 4H structure. High densities of stacking faults are usually observed in transition regions between different polytypes.

INTERGRANULAR PHASES

The detection of thin intergranular films by electron microscopy was discussed by Clarke [7]. Since the particular methods have their own limitations and drawbacks, in the present study lattice fringe imaging was supplemented by conventional bright-field and dark-field techniques. To identify intergranular phases by lattice fringe imaging at least one set of lattice planes was imaged in each adjacent grain. Thus, both grains have to be in strongly diffracting orientations. Most important for the interpretation is, however, that the grain boundary is additionally imaged edge-on. Even for small deviations from this condition the intergranular film may become obscured [7]. Fig. 3 shows a grain boundary in SiC imaged by this technique. For both grains the direction of the electron beam is <11$\bar{2}$0>. The (0001) fringe spacing is 2.51 Å. The lattice fringes start directly from the grain boundary indicating that there is no glassy phase. The result was confirmed by numerous examples of other grain boundary orientations.

The main disadvantage of lattice fringe imaging is that only small

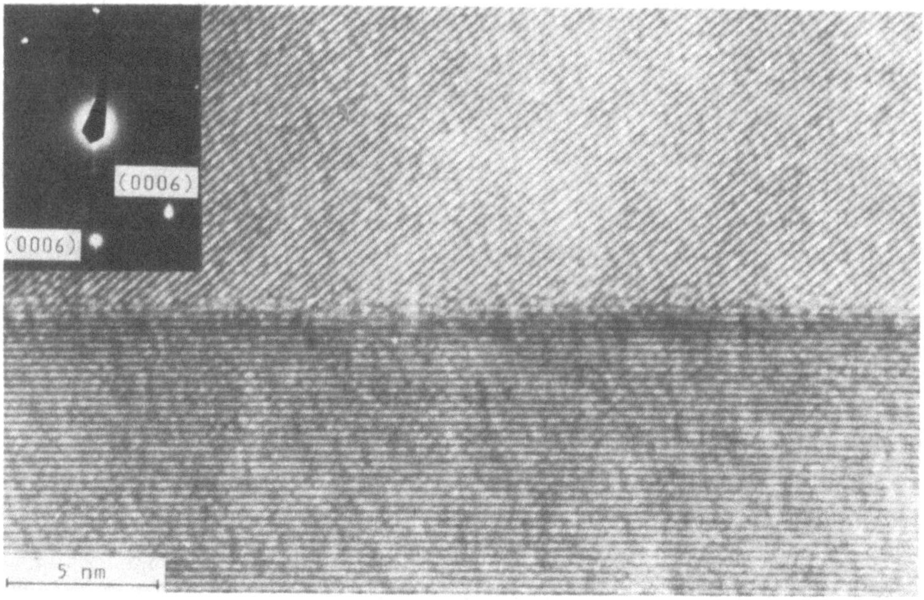

Fig.3 Grain boundary in SiC imaged by the (ØØØ6) lattice fringes in the adjacent grains. Foil orientation for both grains <11$\bar{2}$Ø>. Acceleration voltage 1ØØ kV

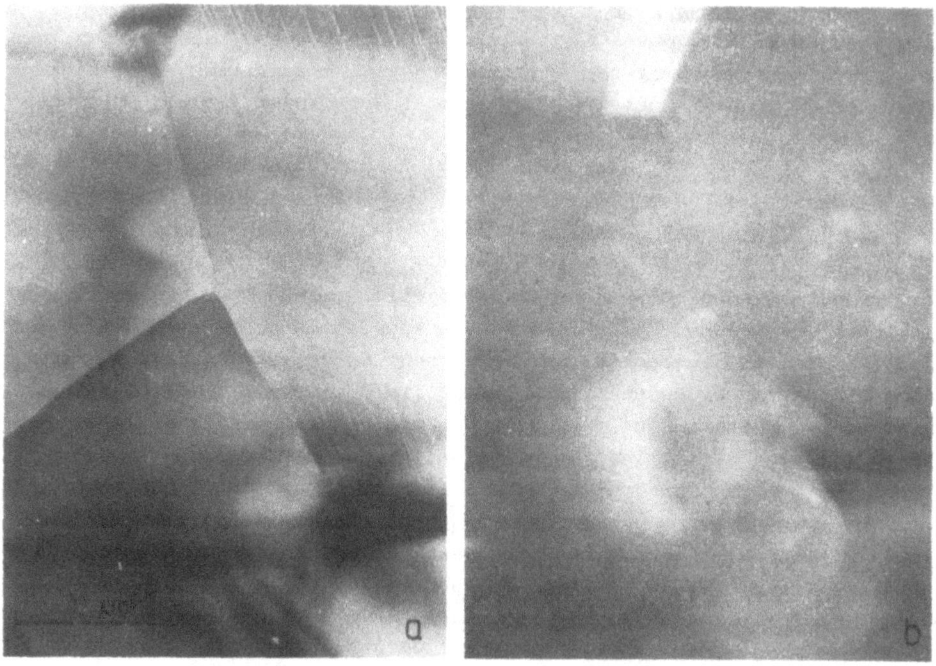

Fig. 4 Grain boundary triple node in SiC; a) bright-field image, b) diffuse dark-field image. Acceleration voltage 1ØØØ kV

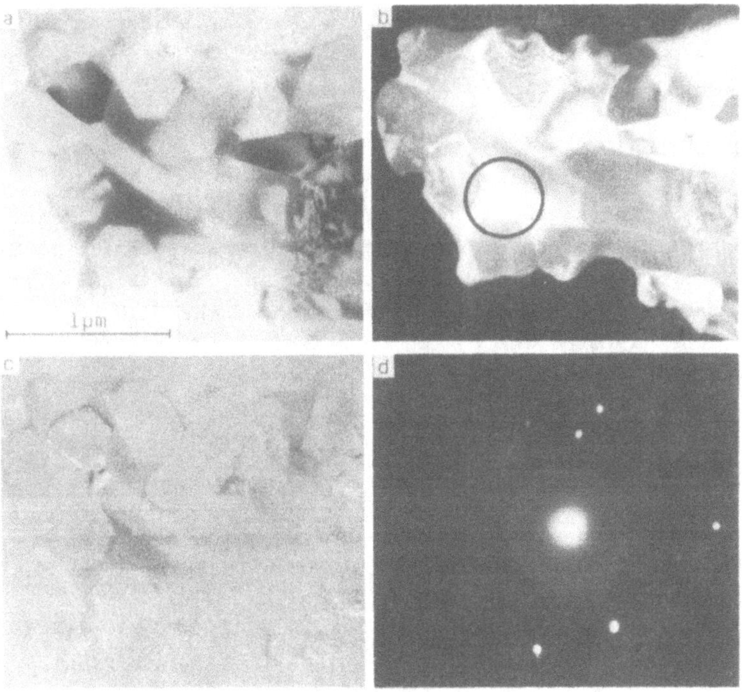

Fig. 5 Grain boundary structure in Si_3N_4; a) bright-field image, b) diffuse -dark field image, c) bright-field image formed without objective aperture, d) diffraction pattern of the region encircled in b). Acceleration voltage 1000 kV

areas can be examined at the same time. To get information on a larger scale, the other methods mentioned above were applied. In addition to the conventional bright-field technique, images were formed without objective aperture using the different electron absorption of the grains and the intergranular film. To image the structure in the diffuse dark field, the objective aperture was placed over the diffraction ring, which corresponds to the presumed amorphous phase [7]. The aperture should lie away from any diffraction spots originating from the grains. Fig. 4 demonstrates a typical example of a grain boundary node in SiC imaged in the conventional bright-field and diffuse dark-field modes. The grain boundaries always form sharp triple nodes, which confirms the absence of glassy phases. In Fig. 5 a grain boundary in Si_3N_4 is analyzed similarly. In the conventional bright-field image (Fig. 5a) there are large regions between the grains the contrast of which is nearly independent of the sample orientation. The same regions become bright in the diffuse dark-field mode (Fig. 5b). If the boundary is imaged in the bright-field mode without the objective aperture, the contrast phenomenon is just opposite (Fig. 5c). Fig. 5d shows the diffraction pattern of the region encircled in Fig. 5b. The existence of intergranular glassy phases in Si_3N_4 therefore seems to be clearly proved. Intergranular films in Si_3N_4 were also evidenced by Clarke and Thomas [8,9].

CONCLUSIONS

The high-temperature strength of the SiC and Si_3N_4 ceramics investigated is probably determined by different processes. While in SiC intergranular phases have not been detected they occur in Si_3N_4. At elevated temperatures a remarkable loss of strength of the Si_3N_4 ceramic should therefore be expected due to the by weakening of the interganular films allowing grain boundary sliding.

REFERENCES

1 F.F. Lange and J.L. Ishoe, In: Ceramics for High Performance Applications, J.J. Burke, A.E. Gorum and R.N. Katz (Eds.), Brook Hill, Chestnut Hill, MA 1973

2 P.M. Sargent and M.F. Ashby, Scripta Met. 17 (1983) 951-57

3 P.B. Hirsch, A. Howie, R.B. Nicholson, R.B. Pashley, and M.J. Whelan, In: Electron Microscopy of Thin Crystals, Butherwords, London 1965

4 K. Maeda, K. Suzuki, S. Fujita, M. Ichara and S. Hyodo, Phil. Mag. A 57 (1988) 573

5 J.Van Landuyt and S. Amelinckx, Mat. Res. Bull. 6 (1971) 613

6 A.H. Heuer, V. Lou, L. Ogbugji, and T.E. Mitchel, J. Microsc. Spectrosc. Electron. 2 (1977) 475

7 D.R. Clarke, Ultramicroscopy 4 (1979) 33

8 D.R. Clarke and G. Thomas, J. Am. Ceram. Soc. 6Ø (1977) 491

9 D.R. Clarke and G. Thomas, J. Am. Ceram. Soc. 61 (1977) 114

ACKNOWLEDGEMENT

The authors are much indebted to the WTK Meißen and to the Central Institute of Inorganic Chemistry for providing the ceramic materials.

MICROSTRUCTURAL PHENOMENA OF TiC-CONTAINING CERAMICS STUDIED BY HIGH-VOLTAGE AND HIGH-RESOLUTION ELECTRON MICROSCOPY

E. Pippel, J. Woltersdorf

Institut für Festkörperphysik und Elektronenmikroskopie
der Akademie der Wissenschaften der DDR, 4050 Halle
Weinberg 2, Postfach 250

ABSTRACT

By adding of TiC to a high performance ceramic material a composite can be produced combining the good electrical conductivity of TiC with the high strength and wear resistance of the matrix. A successful construction of such composites demands a detailed knowledge of the structure-property relations down to the atomic level. In this lecture, some results of electron microscopic investigations of the microstructure of TiC/Al_2O_3 - ceramics are reported with special view on the structure of grain boundaries and segregations, and the characterization of graphite phases.

INTRODUCTION

In principle, the improved properties of polycrystalline ceramic materials are manifested in their microstructure: Especially, grain size distributions, grain and phase boundaries, segregations of doping elements and microflaws or phase transformation phenomena play an important role. Therefore, a successful fabrication of high-performance ceramic materials requires the detailed knowledge of the structure-property relations down to the atomic level. As a powerful tool for microstructural investigations transmission electron microscopy (TEM) can be used - including high-voltage (HVEM) and high resolution (HREM) methods. Reviews of the possibilities and limitations of TEM in ceramic research are given in /1, 2/ and in special sections of recent conference proceedings of electron microscopy.

In the last years, TiC-containing ceramics have increasingly gained importance: Despite the almost covalent character of TiC and some other transition metal carbides their electrical conductivity resembles that of metals. By adding TiC to a high-performance alumina ceramic, for example, a composite can be produced combining the good electrical conductivity of TiC with a high strength and wear resistance of the matrix. The conduction mechanism in such a composite material can, in principle, be explained by two different microstructural features: On the one hand exceeding a certain content the TiC-grains coagulate and the resulting framework drastically increases the conductivity /3/. In addition, the carbon excess always present in commercial TiC powders may form thin graphite layers along the grain boundaries of the non-conducting matrix which also improve the electrical conductivity /4/.

In this lecture we present microstructural phenomena of compact sintered Al_2O_3/TiC - ceramics as obtained from HVEM and HREM images. Besides the distribution of individual grains, microflaws and dislocations, the investigations include the structure of grain boundaries and segregations with special respect to crystalline graphite phases.

EXPERIMENTAL

TEM specimens were prepared from different commercial materials as well as from pressureless sintered Al_2O_3/(30-40 weight %)TiC samples, provided by Dr. A. Krell (Zentralinstitut für Festkörperphysik und Werkstofforschung, Akademie der Wissenschaften der D.D.R., Dresden). The thinning to electron transparency was carried out by standard techniques including cutting of thin ($<$ 150 µm) slices, mechanical grinding and finally ion milling (Ar; 6 kV). In spite of the relatively good electrical conductivity of the Al_2O_3/TiC-ceramic, appreciable charging arises in the electron microscope. Probably, the three dimensional TiC-framework, effecting the conductivity of the material, becomes incomplete in very thin specimen areas ($<$ 25 nm). Therefore, the specimens had to be coated by an additional carbon layer of a thickness between 5 nm and 15 nm. For the HVEM investigations, we used a JEOL-JEM 1000 electron microscope operated at 1000 kV. HREM images were generally taken in the tilted illumination mode employing a JEM 100 C electron microscope operated at 100 kV.

RESULTS AND DISCUSSION

Liquid Phase Sinter Process

In general, Al_2O_3 and TiC-powders sinter to nearly full dense (99 %) compacts by diffusion-induced solid phase sintering. On the other hand, small contamination contents of silicon oxide can lead to a partial liquid phase sintering. Although we never found an amorphous grain boundary phase in this material, Fig. 1 shows some wave-shaped grain boundaries. Such features indicate a liquid phase occuring during early stages of the sinter process. Above all, wave-shaped grain boundaries were observed in specimens containing a small amount of TiH_2 in the starting powder. As pointed out by Cuttler et al. /5/, TiH_2 decomposes in order to form a transient liquid phase of metallic titanium which aids the sinter process. The decomposition of TiH_2 at high temperatures leads to the formation of amorphous titanium, in a first step. This behaviour is clearly shown in Fig. 2 where the TiC-grains in incompletely sintered parts of the specimen are frequently covered by an amorphous layer. The formation of TiC within titanium is possible as carbon easily diffuses from TiC into the surrounding Ti-layer. Starting with very small particles this process continues up to the entire Ti has converted into TiC in the way that the interface between TiC and Ti successively moves towards the Ti region /6/. The result will be an enlarged TiC grain with a certain carbon deficit.

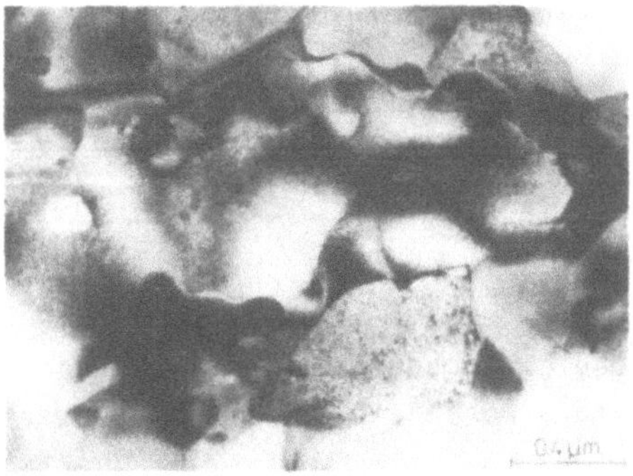

Fig. 1 Wave-shaped grain boundaries after partial liquid phase sintering

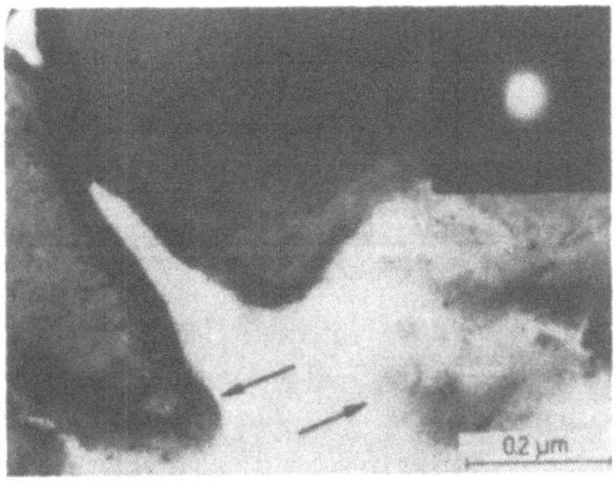

Fig. 2 Incompletely sintered region in Al_2O_3/TiC-ceramic

Dislocation Processes

Near room temperature the cleavage behaviour of TiC resembles that of ionic crystals whereas at elevated temperatures a metal like gliding is observed. The brittle/ductile transition temperature of polycrystalline TiC amounts to 800 °C, approximately. Therefore, thermal stresses can be relaxed by the generation of dislocations during cooling from the sinter temperature. Further on, it is well known that a general carbon deficit in TiC leads to a carbon to titanium ratio smaller than one which strongly influences the mechanical properties: A slight increase of the C-content (i.e., a lower vacancy number) drastically increases the micro-hardness and decreases the dislocation density obtained in deformation tests /7/. As it can be seen in Fig. 3 the dislocations sometimes are concentrated near the border of large TiC grains indicating a carbon deficit in these regions. The high dislocation density, respectively the high carbon deficit, should promote again the sinter process of TiC containing ceramics.

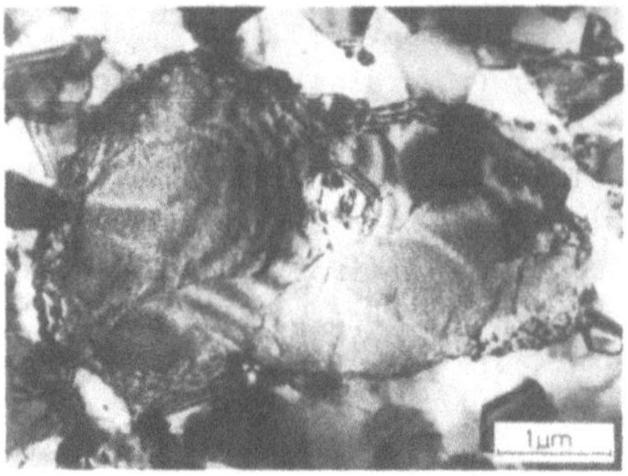

Fig. 3 Dislocations concentrated near the border of a large TiC-grain

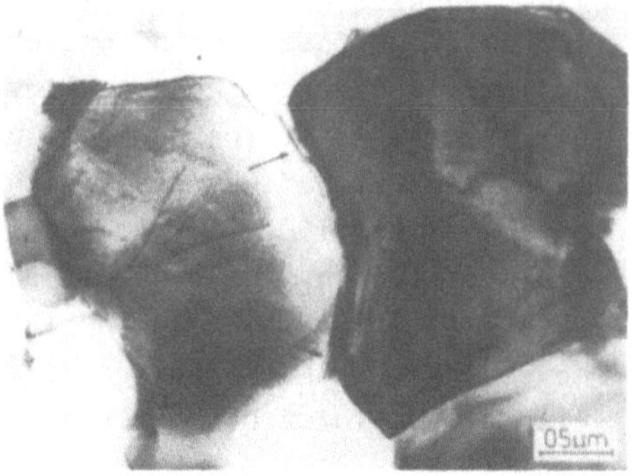

Fig. 4 Grain boundary segregations between two TiC-grains

Grain Boundaries and Segregations

Even in thin specimens we observed transcrystalline fracture of TiC particles which points to relative strong grain boundary interactions between TiC and Al_2O_3. This induced us to a more detailed grain boundary investigation in this composite material.

Just as expected, a number of segregations are observed in the grain boundaries as it can be seen in Fig. 4. This figure also demonstrates that the precipitates are mainly located at or between TiC-grains characterizing them as reaction products. Their small dimensions prevent an exact determination of the crystallographic nature by usual electron diffraction methods.

In most cases, however, the grain boundaries in Al_2O_3/TiC are completely free of any segregations or phases. Moreover, the lattice fringes of neighbouring grains propagate exactly up to the boundary. In case of overlapping grains, cf. Fig. 5, regular moiré-patterns exclude a second crystalline phase. This figure also demonstrates that the high resolution technique demands the grain boundary to be located exactly in an end-on position so as to yield information on boundary phases. It is very hard to bring such a position into line with the demand for imaging lattice planes in both adjoining grains simultaneously.

An example of a grain boundary in a position parallel to the electron beam is shown in the high resolution image, Fig. 6. As can be seen on the low magnified inset this boundary is placed between an intragranular TiC grain and the Al_2O_3 matrix. Particularly, Fig. 6 demonstrates a glassy grain boundary phase as it is typical for the case of other high performance ceramic materials almost being absent in Al_2O_3/TiC composites. Possibly, the silicon oxide contaminations which lead

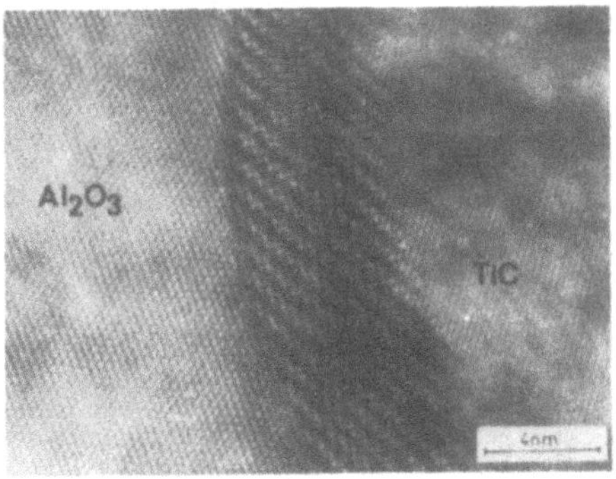

Fig. 5 Lattice fringes near an overlapping interface

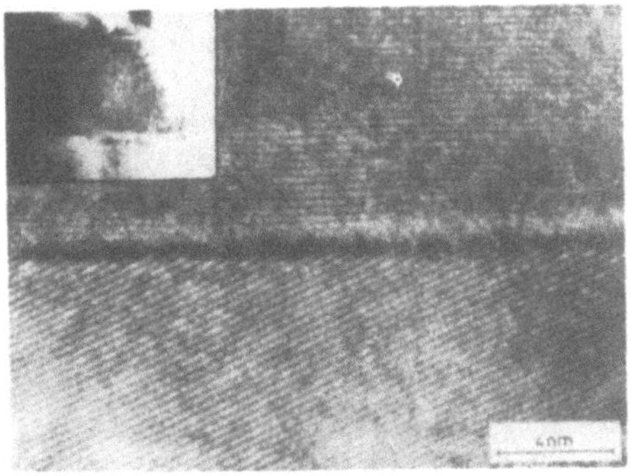

Fig. 6 Grain boundary of an intragranular TiC-particle in an Al_2O_3-matrix

to glassy phases in ZTA ceramics /2/ diffuse into TiC, undergo chemical reactions or they are concentrated at a few places and crystallized, eventually.

Graphite Inclusions

The graphite excess being always present in commercial TiC powders can be incorporated into the polycrystalline ceramic in form of both thin strips along grain boundaries and larger aggregates in triple junctions. The sinter temperature of about 1800 °C should be sufficient for a graphitization process as pointed out in /8/.

A high magnification image of a graphite agglomeration is shown in Fig. 7. Employing the lattice fringe technique small graphite fibres of only 2 - 5 nm thickness can be distinguished. Taking into consideration the extremely small interaction forces (which are in the order of metallic bonding) between the hexagonal basic layers in graphite the strong curvatures of the fibres become plausible without dislocations being necessary. According to this, the electron diffraction spots shown on the inset form a slightly distorted ring pattern reflecting the irregular arrangement of the crystalline fibres.

In contrast to the above shown agglomerates thin graphite layers were mostly observed propagating exactly parallel to the adjacent TiC or Al_2O_3 grains. An example is demonstrated in Fig. 8. The epitaxially oriented interface between TiC and C on the left gives no hint to any chemical attack, a possible irregularity rather extends for 2 or 3 lattice planes only. This is also confirmed by the electron diffraction spots of corresponding TiC and graphite containing streaks perpendicular to the interface in question.

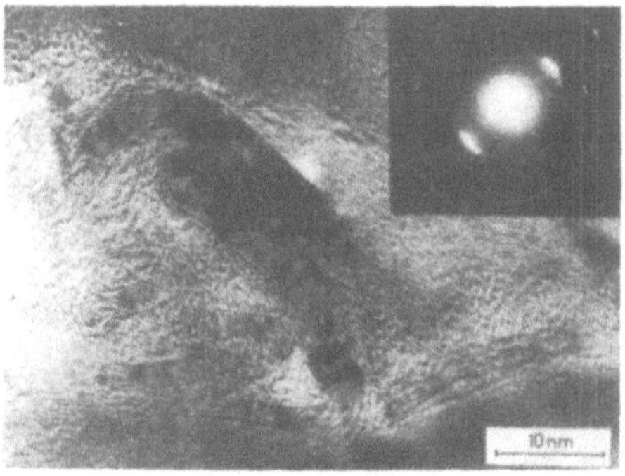

Fig. 7 Lattice fringes of strongly curved graphite fibres

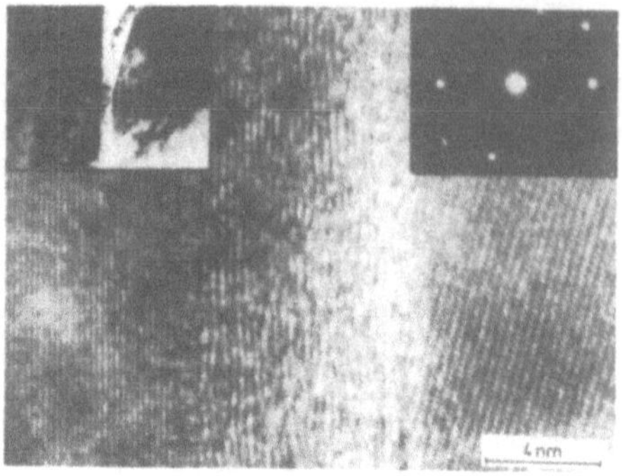

Fig. 8 Graphite fibre parallel to a TiC-grain boundary

REFERENCES

/1/ a) Science and Technology of Zirconia II, Adv. Ceram. 12 (1984)
b) Science and Technology of Zirconia III, Adv. Ceram. 24 (1988)

/2/ E. Pippel, J. Woltersdorf, Philos. Mag. A 56, 595 (1987)

/3/ R.B. Gerkila, T.Y. Tien, J. Am. Ceram. Soc. 48, 22 (1965)

/4/ N. Yamamoto et al., Wirtschaftspatent DE-2643 131, C04B 35/00, 29.1.87

/5/ R.A. Cuttler et al., Mat. Sci. Engineering A 105/106, 183-192 (1988)

/6/ C.J. Quinn, D.L. Kohlstedt, J. Mater. Sci. 19, 1242 (1984)

/7/ R.F. Davis, In: Materials Science Research, Vol. 18, R.E. Tressler, R.C. Bradt, Plenum Press, New York 1984, p. 97

/8/ A. Oberlin, Carbon, 22, 521 (1984)

HVEM AND HREM OBSERVATION OF THE MARTENSITIC TRANSFORMATION PROCESS IN ZTA AND Y-PSZ CERAMICS

J. Woltersdorf, E. Pippel

Institut für Festkörperphysik und Elektronenmikroskopie der Akademie der Wissenschaften der DDR, 4010 Halle, PF 250

ABSTRACT

Using high voltage (HVEM) and high resolution electron microscopy (HREM) some microstructural phenomena of the martensitic t $\rightarrow$ m (tetragonal to monoclinic) phase transformation process in ZTA (zirconia toughened alumina) and in Y-PSZ (yttria-partially stabilized zircónia) were observed, including the role of interface defects, the initial steps of transformation twin formation, back-transformation effects during in-situ heating experiments and influence of stabilizing additives.

INTRODUCTION

The use of the martensitic t $\rightarrow$ m phase transformation of zirconia particles which are dispersed or precipitated in a ceramic matrix enables a drastical improvement of the mechanical properties of the ceramics and leads to the production of high-tech materials /1/.

The transformation toughening and strengthening are due to the shape deformation and the anisotropic volume increase of about 3 ... 5 % (depending on the transformation temperature) on transformation of the confined tetragonal zirconia particles. Therefore, around the monoclinic particles formed which are invariably twinned (the twinning shear reduces the transformation shear), macrocrack shielding processes can be initiated. For this purpose the t $\rightarrow$ m transformation process which in ZrO_2 bulk material occurs at about 950 ^{o}C must be suppressed, except in the stress field of a propagating crack /2-8/. Such a procedure becomes possible as the transformation for different reasons is governed by a lot of parameters: The kinetics and the morphology of the t $\rightarrow$ m transformation depend on the strain energy which is caused by the shear displacements of the atoms. Since the martensitic transformation is a phase transition of the first kind it follows that parent and product phase (i.e. t and m phase) must co-exist at an intermediate stage. Therefore, a t/m interface occurs and a nucleation is necessary to initiate the transformation. After the process has started the interface moves through the grain with nearly the speed of sound, thus transforming the whole particle. It follows from this thermodynamical considerations that the start of the transformation can be controlled by the strain distribution inside the metastable t-ZrO_2 particle (which results from the thermal expansion mismatch between the inclusions and the matrix and the thermal expansion anisotropy as well as from the surrounding stress field, especially in front of the tip of a propagating crack) and by the nucleation conditions in a complex manner. Thus, the transformation process depends e.g. on the size and the shape of the particle, on the localisation within the structure (especially intra or intergranular with respect to the matrix grains), on special defect configurations in the vicinity and on the local chemical potential.

Generally, formation and growth processes of a new phase are governed by interface phenomena /9-12/, and therefore - in addition to the above mentioned parameters - the structure of the interface between the matrix and a confined particle will be of great importance for the transformation kinetics, particularly as the special atomic geometry of this interface will decisively influence the nucleation conditions. However, the details of the nucleation are not well understood. In all cases the nucleation is stress-assisted and probably it can be described best by nonclassical nucleation models particularly the localized soft mode model /13/. A discussion of classical heterogeneous and especially barrierless nucleation in the martensitic t $\rightarrow$ m transformation and details of different nonclassical models can be found in /14-20/, whereas the papers /21,22/ are helpful in providing a critical overview on this intricate subject.

In the following we present some HVEM and HREM observations of structural phenomena which are related to the t $\rightarrow$ m transformation process in ZTA and in Y-PSZ. For experimental details, preparation conditions and imaging techniques used we refer to /23,24/

RESULTS AND DISCUSSION

Some typical structural features of ZTA are shown in Fig. 1. The facet edges of the intergranular particles may act as nucleation sites because of their high curvature and respective stress concentrations. The interface between an intragranular tetragonal ZrO_2 particle and the surrounding Al_2O_3 matrix grain is shown in Fig. 2. In this special case, the (200) lattice planes of t-ZrO_2 (upper part) and the (110) planes of Al_2O_3 having distances of 0.245 nm and 0.237 nm, respectively, propagate exactly parallel and almost perpendicular to the interface (cf. the inserted optical diffraction pattern). If the relation $m = (d_1-d_2)/d_1$ (d_1 and d_2 are the lattice plane distances of $Al_2O_3$and ZrO_2, respectively, in question) is employed, the lattice-plane misfit m is evaluated to be 6.7 %. Consequently, a complete geometrical fit demands an additional (110) Al_2O_3 plane inserted at every fifteenth plane. Dislocation core-like structures (arrowed regions) as well as the corresponding strain field diffraction contrast being exactly separated can be identified in the figure. Of course, these 'additional' Al_2O_3 planes are not 'inserted' by generation, glide or climb mechanisms as present in thin bicrystal films; they are rather a simple

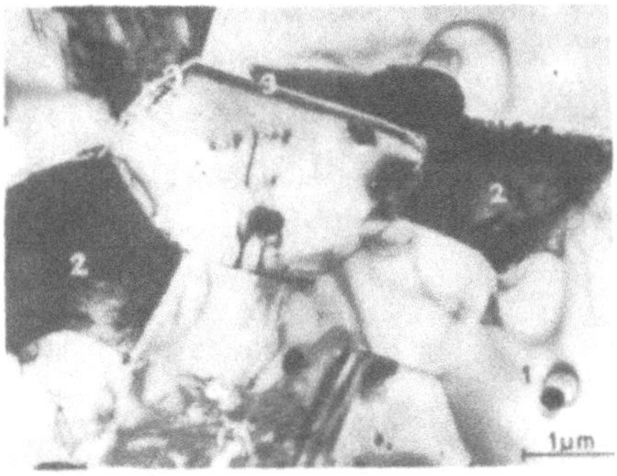

Fig. 1 HVEM image of ZTA, showing t-ZrO_2 grains (1), m-ZrO_2 grains (2), microcracks (3) and dislocation arrays (4)

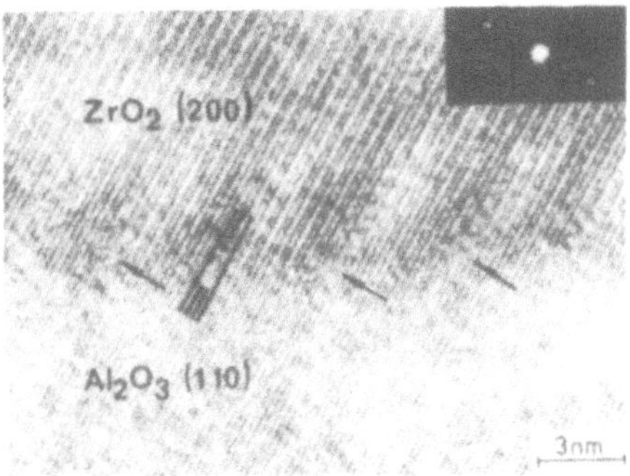

Fig. 2 Misfit dislocations in the interface between an intragranular t-ZrO_2 particle and Al_2O_3

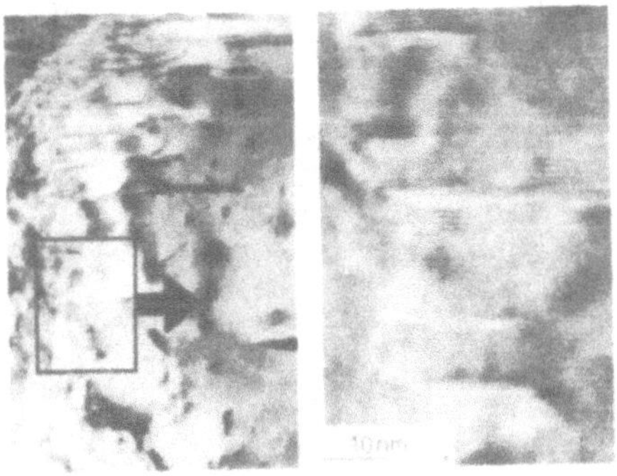

Fig. 3 Initial stages of micro-twin formation during the t → m transformation in ZTA

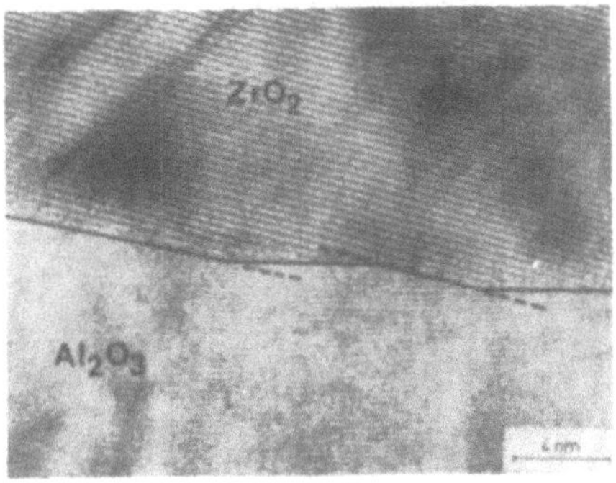

Fig. 4 Faceted phase boundary between Al_2O_3 and ZrO_2 in ZTA

geometrical result of the intergrowth process of ZrO_2 and Al_2O_3. Nevertheless, the lattice accommodation is performed.

It is evident that any energetically favourable lattice site is a preferred centre for nucleation. According to the remarks given in the introduction and following /25/ where the importance of crystal defects, especially dislocations, as an initiating centre of transformation in ZrO_2 is demonstrated, we assume that misfit dislocations of the type observed in Fig. 2 can govern the transformation behaviour of ZrO_2. Until now, it has not been possible for us to decide whether the transformation is influenced by the local strain field of a singular dislocation or by the cooperative action of the interface-dislocation arrangement. In any case, however, special interface phenomena control the transformation behaviour as it is revealed by Fig. 3. In this figure the generation of microtwins accompanying the t → m transformation of an intragranular ZrO_2 particle is shown just starting at its border line. Formation and growth of these transformation twin lamellae spread out over the whole grain thus transforming this region. The lattice fringe image in the right part illustrates the inner structure of the micro-twins as well as their embedment in the zirconia lattice.

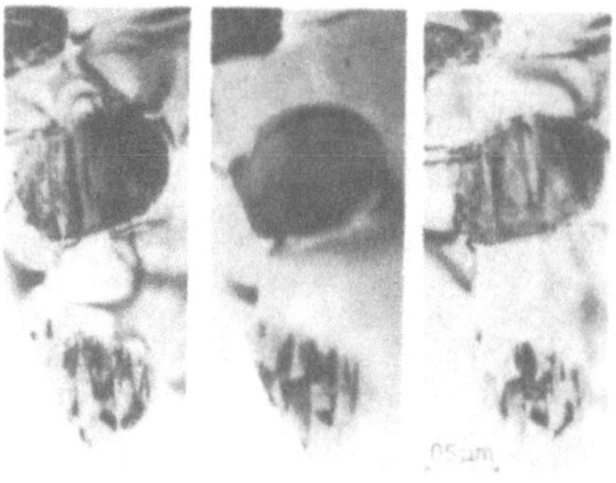

Fig. 5 In-situ observed m → t → m transformation cycles of intragranular ZrO_2 particles in ZTA, T = a) RT, b) 1200 °C, c) RT

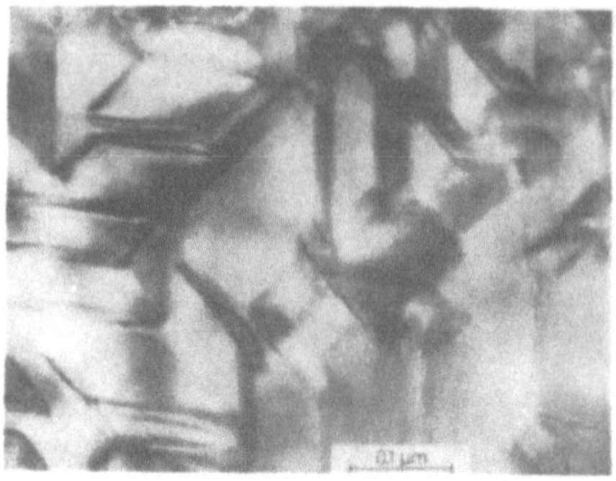

Fig. 6 Mosaic-like twin pattern ((100) and {110} twin planes) in a zirconia particle in ZTA after t → m transformation

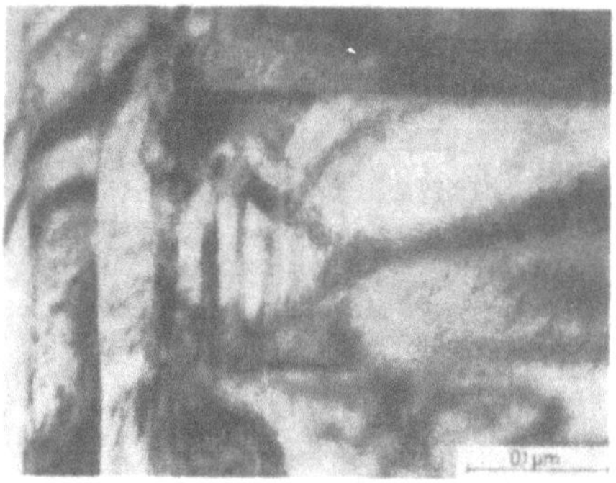

Fig. 7 Monoclinic zirconia with twin structures containing a triangle twinned on a finer scale

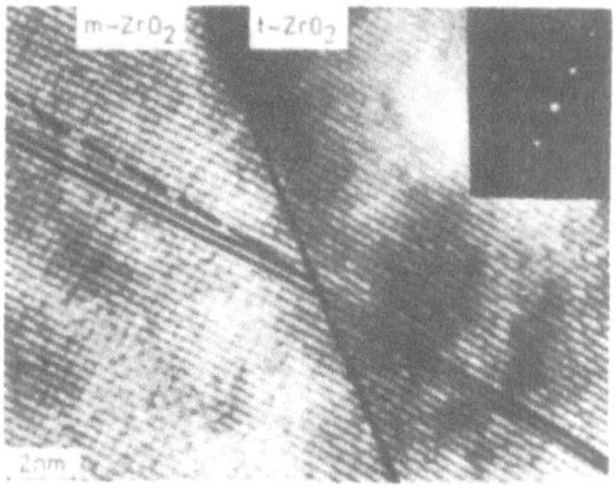

Fig. 8 {111} lattice fringes crossing the complete coherent t/m interface in a Y-PSZ grain

Sometimes the transformation of intragranular zirconia particles is accompanied by the occurrence of lattice periodicities of a few nanometers leading to a micro-faceting of the interface. Fig. 4 gives an example where the faceting is significantly marked by the arrangement of the atomic planes. Possibly such a faceting in an atomic scale - particularly if connected with defect configurations - can affect stress concentrations which are sufficient to set off the nucleation.

Fig. 5 gives an example of in-situ heating experiments in the high voltage electron microscope. Subsequent complete transformation cycles between 1300 °C and room temperature could be observed using a special heating device. After m → t back-transformation which occurs successively between 1100 and 1300 °C the twinning structures disappear and strain contours are reduced correspondingly. Moreover, it can be recognized that after the next t → m transformation the twins can occur at other positions than before due to changed nucleation conditions.

The process of strain relaxation during the phase transformation leads to a great variety of twinning structure formations as both the (100) and the (110) planes are twinning planes in m-ZrO_2 and - in addition - the latter ones again enable different types of twins because of the different length of the diagonals in the monoclinic elementary cell. The resulting diversity of twin orientations and the corresponding structural interaction of the different twin lamellae leading to mosaic-like patterns is shown in Fig. 6. Sometimes triangular configurations of parallel oriented twins can be observed which show a twinning on a finer scale as presented in Fig. 7. Such structures could give evidence of some orthorhombic oriented early stages of the phase transition /cf. 26,27/, i.e. the larger base lamellae initially were orthorhombic and during the further phase transition they transformed into the monoclinic modification forming twins as well as further and correspondingly smaller orthorhombic triangular configurations. Such an early orthorhombic stage would present a gradual transition from the tetragonal to the monoclinic symmetry: Firstly the lattice parameters are changed with the axes remaining orthogonal and then, in a next step, the direction of the c-axis is changed, affecting the transition from the orthorhombic to the monoclinic symmetry.

In case of partially stabilized zirconia where the transformation process is influenced by additives like magnesia or yttria (by which the chemical potential and thus also the driving force for nucleation is changed) the arrangement of twin lamellae is more irregular then in the foregoing cases. A t/m-interface can occur within one and the same zirconia particle separating the transformed and the non-transformed region in a partially transformed macrograin. Such a transformation behaviour may be due to an inhomogeneous distribution of the yttria stabilizer leading to local differences in the chemical potential. In addition, the complicated arrangement of self-stresses, particularly in the vicinity of larger agglomerates, may be responsible for the incomplete transformation process. The arrangement of the atomic planes in the interfacial region of such a t/m phase boundary is demonstrated in Fig. 8. It is clearly shown that the interface between the transformed and the non-transformed zirconia regions is completely coherent; only a corresponding change of direction occurs if the atomic planes cross the interface. The series of operating reflexes represented in the inset is in perpendicular orientation to the atomic planes and the streak formation is perpendicular to the boundary according to the theory of electron diffraction. The streak formation indicates a nearly atomically sharp interphase boundary.

Thanks are due to Dr. A. Krell, ZFW Dresden, and to Prof. Michalowsky. BA Freiberg, for providing the ceramic materials and to our colleague D. Hoehl for assistance in the in-situ-experiments.

REFERENCES

1 a) Science and Technology of Zirconia, Adv. Ceram. 3 (1981)
b) Science and Technology of Zirconia II, Adv. Ceram 12 (1984)
c) Proc. 2nd U.S. - Australian - West German Transf. Tough. Workshop, J. Am Ceram. Soc., 69 (3), (1986)
d) Science and Technology of Zirconia III, Adv. Ceram. 24 (1988)
2 A.G. Evans, A.H. Heuer, J. Amer. Ceram. Soc. 63, 241 (1980)
3 W.M. Kriven, Adv. Ceram. 3, 168 (1981)
4 W.M. Kriven, Adv. Ceram. 12, 64 (1985)
5 W. Kreher, W. Pompe, J. Mater. Sci., 16, 694 (1981)
6 W. Pompe, W. Kreher, Adv. Ceram 12, 283-292 (1984)
7 A.H. Heuer, N. Claussen, W.M. Kriven, M. Rühle, J. Amer. Ceram. Soc. 65, 642 (1982)
8 S. Schmauder, W. Mader, M. Rühle, Adv. Ceram. 12, 251 (1984)
9 J. Woltersdorf, Thin Solid Films 85, 241 (1981)
10 E. Pippel, phys. stat. sol. (a) 67, 443 (1981)
11 J. Woltersdorf, Appl. Surf. Sci. 11/12, 495 (1982)
12 J.H. van der Merwe, J. Woltersdorf, W.A. Jesser, Mater. Sci. Engineering 81, 1 (1986)
13 P.C. Clapp, phys. stat. sol. (b) 57, 561 (1973)
14 I.-Wei Chen, Y.-H. Chiao, Acta met. 33, 1827 (1985)
15 G.B. Olson, M. Cohen, Proc. Int. Conf. Solid-Solid-Phase Transf., Carnegie-Mellon Univ., Pittsburgh,PA (1981)
16 G. Olson, M. Cohen, J.de Physique, Coll.C4, supp.12,43,75 (1982)
17 G.B. Olson, M. Cohen,in: Dislocations in Solids, Vol. 7, F.R.N. Nabarro, ed; Elsevier Science Publishers, 1986, Chapter 37, Sect. 4
18 T. Suzuki, M. Wuttig, Acta met. 23, 1069 (1975)
19 G. Guénin, P.F. Gobin, Proc. Int. Conf. on Martensitic Transformations, ICOMAT 79 (Mass.) M.I.T., 1979, 532
20 G. Guénin, P. Gobin, J.de Physique, Coll.C4, suppl.12,43,57 (1982)
21 M. Rühle, A.H. Heuer, Adv. Ceram. 12, 14 (1984)
22 A.H. Heuer, M. Rühle, Acta met. 33, 2101 (1985)
23 E. Pippel, J. Woltersdorf, Philos. Mag. A 56, 595 (1987)
24 J. Woltersdorf, Dissertation B, Univ. Halle, 1989
25 I.-Wei Chen, Y.-H. Chiao, Adv. Ceram. 12, 33 (1984)
26 W. Wunderlich, M. Rühle, in: High Tech Ceramics, ed. by P.Vincencini, Elsevier Science Publishers B.V., Amsterdam, 1987, 1219
27 V. Dravid, M. Notis, C. Lyman, J.Amer.Ceram.Soc., 71, C-219 (1988)

5. HIGH TEMPERATURE DEFORMATION AND RECOVERY

ON THE COMPOSITE MODEL OF DISLOCATION STRUCTURE AS APPLIED TO HIGH TEMPERATURE CREEP IN METALS

A. Orlová, F. Dobeš and J. Čadek
Institute of Physical Metallurgy, Czechoslovak Academy of Sciences,
616 62 Brno, Czechoslovakia

ABSTRACT

The iso-strain rate composite model of dislocation structure is outlined and applied to the steady state creep of subgrain forming metals and solid solution alloys. The steady state values of the local stresses acting in soft and hard regions as well as the volume fraction of hard regions are calculated and the strain transient dip test technique to measure the internal stress in creep is simulated. Correlations of the measured effective and internal stresses, respectively, with the square root of free dislocation density are compared. Some methods to estimate the volume fraction of hard regions on the basis of TEM are outlined. Finally, the stress dependence of steady state creep rate is briefly discussed.

1. INTRODUCTION

For metals and solid solution alloys exhibiting Metal Class creep behaviour the process of heterogeneization of spatial dislocation distribution during primary creep is characteristic. This process, which ultimately leads to cell or subgrain structure formation, has been described in considerable detail by many authors (e.g. refs. 1-4).

Taking into account the heterogeneous dislocation distribution, the model based on the idea that a single phase system can be considered as a "composite" consisting of regions strongly differing in dislocation densities was proposed by Nix and Ilschner |2| to interprete high temperature creep in metals. A similar composite concept has been widely applied to account for various other phenomena of plastic behaviour in metals (e.g. refs. 5, 6) with remarkable success. In the present paper, the steady state creep of metals and solid solution alloys exhibiting Metal Class creep behaviour is considered.

1.1. The Composite Model

Since dislocation densities are high in cell walls, while in cell interiors they are usually at least an order of magnitude lower, the shear stresses acting in cell walls, τ_w, are higher, while the shear stresses τ_c in the cell interiors are lower than the applied stress τ_a. Hence, the cell walls represent hard, while the cells represent soft regions of dislocation structure. The condition of mechanical equilibrium in a composite under consideration is expressed as

$$\sigma = \sigma_c f_c + \sigma_w f_w \ ; \qquad (1)$$

f_c and f_w are the volume fractions of soft and hard regions, respectively, $f_c + f_w = 1$. The local stresses in soft and hard regions are coupled through long range internal stresses, $\sigma_{iB} = \sigma - \sigma_c$ and $\sigma_{iF} = \sigma_w - \sigma$, respectively. These stresses are interrelated through Albenga's law: $f_c \sigma_{iB} = f_w \sigma_{iF}$. The local stresses estimated from measured radii of curvature of bowed-out dislocation segments |7, 8| in the nearest vicinity of subboundaries - in hard regions - are several times higher than applied stress, while the local stresses in subgrain interiors - soft regions - represent only a fraction of the applied stress.

Usually however, such parameters of dislocation structure generated in creep of polycrystals as the dislocation density in subgrains, ρ_s, the subgrain size, L, and the subgrain misorientation, θ, are measured (e.g. Ref. 9). Knowing L and θ and assuming regular substructure and regular dislocation structure of subboundaries, the density of dislocations stored in subboundaries, ρ_{sB}, can be estimated. The total dislocation density is then

$$\rho = \rho_s + \rho_{sB} \, . \qquad (2)$$

Since $\rho_s = \rho_c f_c$ and $\rho_{sB} = \rho_w f_w$, the total dislocation density can be also expressed as

$$\rho = \rho_c f_c + \rho_w f_w \, . \qquad (3)$$

1.2. The Dip Test Technique

The internal stress in creep, σ_i, measured by the strain transient dip test technique |10| is defined as the residual stress $\sigma - \Delta\sigma_{crit}$, where $\Delta\sigma_{crit}$ is the instant reduction of applied stress σ which results in zero creep rate as measured immediately after the stress reduction. The applied stress decrement $\Delta\sigma_{crit}$ is identified with the effective stress σ^*. Thus, the applied stress

$$\sigma = \sigma_i + \sigma^* \, . \qquad (4)$$

The composite model of dislocation structure applied to high temperature creep of metals and solid solution alloys exhibiting Metal Class creep behaviour is quite a natural challenge to a sound interpretation of measured internal stress in terms of inhomogeneous distribution of dislocations and/or the local stresses. In the present paper, the iso-strain rate composite model will be outlined and applied to estimate the local stresses σ_c and σ_w acting in soft and hard regions of dislocation structure, respectively, and to simulate the strain transient dip test technique. Possible relations of dislocation densities ρ_c and ρ_w to the long range internal stresses σ_{iB} and σ_{iF}, respectively, will be considered. Further, some methods to estimate the volume fraction of hard regions, f_w, based on TEM data will be outlined. Finally, a possibility to interprete power-law recovery creep in terms of measured internal stress σ_i

will be discussed in relation to the requirements of the composite model of dislocation structure.

2. THE ISO-STRAIN RATE MODEL AND THE MEASURED INTERNAL STRESS

The iso-strain rate model |2, 11|, which can be represented by the rheological diagram |12| in Fig. 1, is based on the compatibility condition. Since the total strain is

$$\varepsilon = \varepsilon_c + \sigma_c/E = \varepsilon_w + \sigma_w/E \ , \tag{5}$$

where E is the Young modulus, the total strain rate is

$$\dot{\varepsilon} = \dot{\varepsilon}_c + \dot{\sigma}_c/E = \dot{\varepsilon}_w + \dot{\sigma}_w/E \ . \tag{6}$$

Assuming the volume fraction f_w not changing with time, for the steady state creep at constant applied stress the relation

$$\dot{\sigma}_c f_c + \dot{\sigma}_w f_w = 0 \tag{7}$$

follows from eqn. (1). The above assumption about time independence of f_w is certainly fulfilled in the steady state creep and almost certainly immediately after a sudden applied stress reduction in the course of creep.

Combining eqns. (1), (6) and (7) the following relation for the total strain rate $\dot{\varepsilon}$ is obtained

$$\dot{\varepsilon} = \dot{\varepsilon}_c f_c + \dot{\varepsilon}_w f_w \ . \tag{8}$$

In what follows, the equations describing temperature and stress dependence of the strain rates $\dot{\varepsilon}_c$ and $\dot{\varepsilon}_w$ as derived by Nix and Ilschner |2|,

$$\dot{\varepsilon}_c = A_c \cdot (\sigma/G)^2 \cdot \exp\left[-\Delta F/kT\right] \cdot \sinh\left[(\Delta F/kT)(\sigma_c/\sigma)\right] \tag{9}$$

and

$$\dot{\varepsilon}_w = A_w \cdot (D_L b \sigma/kT) \cdot (\sigma_w/G)^3 \ , \tag{10}$$

respectively, will be used. In eqn.(9), ΔF is the Helmholtz free energy associated with dislocation motion past local obstacles, $\Delta F = C_i G b^3$, C_i is a constant |13|: Equation (10) should hold for FCC metals and solid solution alloys in which the dislocations are split.

Solving numerically eqns. (1), (9) and (10) the steady state values of σ_c, σ_w and f_w are obtained for the applied stress given. For a Fe-21 Cr - 37 Ni solid solution alloy |14|, the results are presented in Table 1. At any given temperature, the σ_c/σ and σ_w/σ ratios increase, while the fraction f_w decreases with increasing applied stress.

A sudden reduction of the applied stress in the course of creep is certainly associated with the change of dislocation velocity, while immediately after the stress reduction the dislocation structure can be considered practically identical with that just before the stress reduction. Hence, eqns.(9) and (10) can be re-written to describe the strain rates $\dot{\varepsilon}_c$ and $\dot{\varepsilon}_w$ immediately after the stress reduction |11| and the critical applied stress reduction $\Delta\sigma = \Delta\sigma_{crit}$ can be calculated. As already mentioned, the

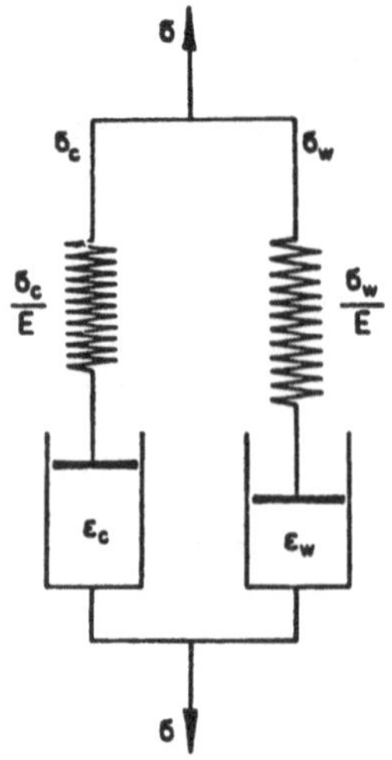

Fig. 1: A rheological diagram of the iso-strain rate composite model |12|

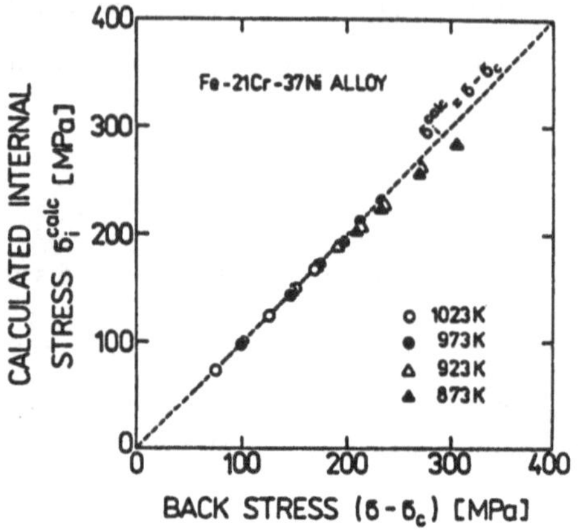

Fig. 2: Calculated internal stress, σ_i^{calc}, plotted against the back stress $(\sigma - \sigma_c)$ in soft regions

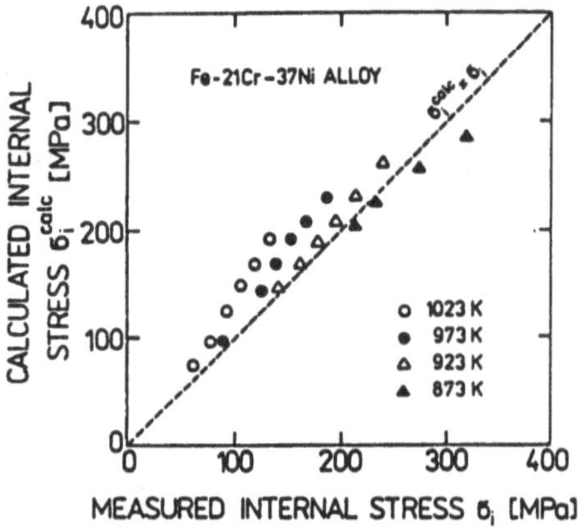

Fig. 3: Calculated internal stress, σ_i^{calc}, plotted against the internal stress, σ_i, measured by the transient dip test technique in steady state creep

residual stress $\sigma - \Delta\sigma_{crit}$ represents the internal stress as measured by the strain transient dip test technique, the calculated value of internal stress is denoted σ_i^{calc}. For the Fe-21Cr-37Ni alloy, σ_i^{calc} is plotted against the back stress $\sigma - \sigma_c$ acting in soft regions in Fig. 2. It can be seen that the stress σ_i^{calc} is approximately equal to the back stress $\sigma - \sigma_c$. Thus, the internal stress σ_i^{calc} represents the level of the long range internal back stress acting in soft regions at any given applied stress and temperature.

Table 1: Local stresses σ_c and σ_w acting in soft and hard regions, respectively, and volume fraction of hard regions, f_w, for a Fe-21Cr-37Ni alloy in steady state creep at various temperatures and applied stresses

T|K|	σ |MPa|	σ_c |MPa|	σ_w |MPa|	f_w
873	225	15.57	1 525.5	0.139
	250	18.13	1.637.5	0.143
	300	30.00	2 370.0	0.115
	350	44.45	3 353.0	0.092
923	150	2.48	739.5	0.200
	175	5.08	962.5	0.177
	200	8.68	1 210.0	0.159
	225	12.65	1 451.3	0.148
	250	15.50	1 600.0	0.148
	300	28.23	2 364.0	0.116
973	100	0.35	414.0	0.241
	150	3.90	957.0	0.153
	175	4.81	1 030.8	0.166
	200	5.20	1 952.0	0.186
	225	13.25	1 606.5	0.133
	250	16.03	1 750.0	0.135
1023	75	0.08	275.3	0.272
	100	0.21	384.0	0.260
	125	0.16	471.3	0.260
	150	0.94	630.0	0.237
	175	5.36	1 160.3	0.147
	200	6.04	1 208.0	0.161

In Fig. 3, the calculated internal stress σ_i^{calc} is plotted against the measured internal internal stress σ_i. From the plot it follows that the long range internal back stress predicted on the basis of the above iso-strain rate model correlates rather satisfactorily with the measured internal stress. For the temperatures of 973 K and 1023 K the values of σ_i^{calc} are higher than those of σ_i . Nevertheless, the approximate equality of the calculated internal stress σ_i^{calc} to the back stress $\sigma - \sigma_c$ acting in soft regions (Fig. 2), together with the results shown in Fig. 3, support the earlier suggestion |15| that the internal stress measured by the strain transient dip test technique represents the level of long range back stresses acting in the soft regions of dislocation structure generated in high temperature creep.

3. LOCAL STRESSES AND THE LOCAL DISLOCATION DENSITIES

The relation between the total dislocation density ρ and the dislocation densities in soft and hard regions, ρ_c and ρ_w, respectively, eqn.(3), can be obtained combining the condition of mechanical equilibrium, eqn. (1), with the relations

$$\sigma_c = \alpha_c Gb\sqrt{\rho_c} \quad , \tag{11}$$

$$\sigma_w = \alpha_w Gb\sqrt{\rho_w} \tag{12}$$

and

$$\sigma = 2\alpha\sqrt{f_c f_w} \cdot Gb\sqrt{\rho} \tag{13}$$

assuming $\alpha_c = \alpha_w = \alpha$. Equation (13) was derived by Mughrabi |16| for the composite under consideration.

We have shown that the level of the internal back stress $\sigma_{iB} = \sigma - \sigma_c$ is well represented by the internal stress σ_i as measured by the strain transient dip test technique. Thus, the effective stress $\sigma^* = \sigma - \sigma_i$ can be identified with the local stress σ_c. Accordingly, we should expect that

$$\sigma^* = \alpha^* Gb\sqrt{\rho_c} \quad , \tag{14}$$

where α^* is a constant. Identifying, for simplicity, ρ_c with the density of free dislocations ρ_s , values of $\tau^* = \sigma^*/2$ are plotted against $Gb\sqrt{\rho_s}$ for a number of metals and cell subgrain forming alloys in Fig. 4. From the figure it follows that a constant τ_0^* must be added to the right hand side of eqn. (14):

$$\tau^* = \alpha^* Gb\sqrt{\rho_s} + \tau_0^* \ . \tag{15}$$

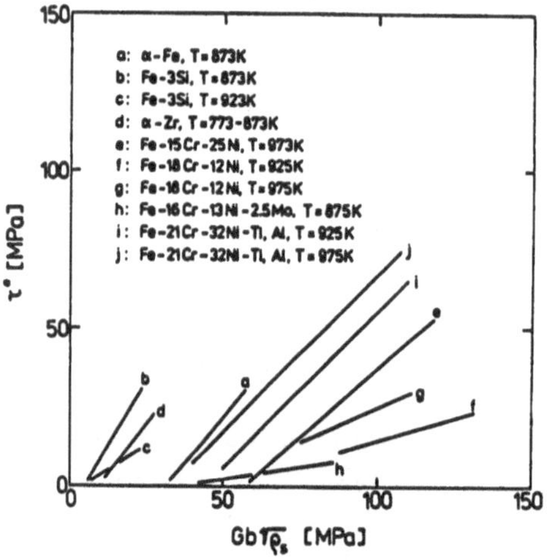

Fig. 4: Correlation of measured effective stress $\tau^* = \sigma^*/2$ with dislocation density ρ_s , eqn.(15)

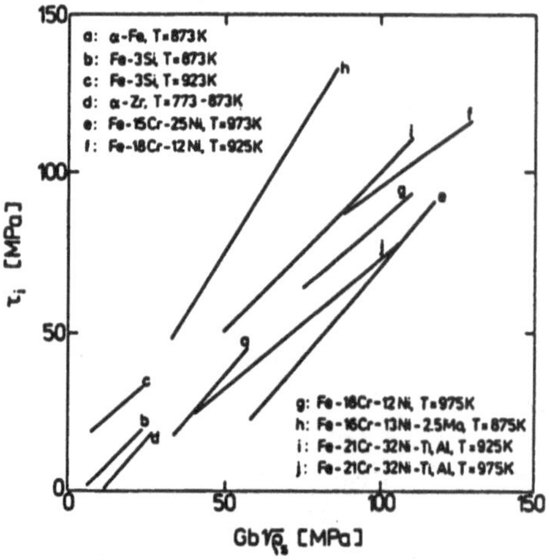

Fig. 5: Correlation of measured internal stress $\tau_i = \sigma_i/2$ with dislocation density ρ_s , eqn. (16)

The quantity τ_0^* is mostly negative and its physical meaning is not clear. Values of the constant α^* vary in a relatively broad range. Hence, the correlation between τ^* and $\sqrt{\rho_s}$ cannot be considered satisfactory.

The measured effective stress was interpreted (e.g. ref. 17) as that part of the applied stress which really acts on dislocations and does work during their glide motion. The measured internal stress was interpreted as the stress which the dislocations encounter during their glide motion and which is created by other dislocations present. The above definitions, together with the unsatisfactory correlation of τ^* with $\sqrt{\rho_s}$, lead us to an attempt to correlate with the dislocation density ρ_s just the measured internal stress σ_i.

Relations between $\tau_i = \sigma_i/2$ and $\sqrt{\rho_s}$ for the same metals as in Fig. 4 are shown in Fig. 5. Again, we can write:

$$\tau_i = \alpha_i \, Gb\sqrt{\rho_s} + \tau_{i0} \, . \qquad (16)$$

The additive constant τ_{i0}, if it differs from zero at all, is negative, but can be interpreted in quite an acceptable way |9|. The average value of the constant α_i is close to one. A comparison of results described by means of eqns. (15) and (16) clearly shows that the dislocation density ρ_s is more probably related to τ_i than to τ^*.

4. SOME METHODS TO ESTIMATE VOLUME FRACTION OF HARD REGIONS

With respect to the importance of a reliable estimate of the volume fraction of hard regions some methods of this estimate |18|, namely the methods involving the parameters of dislocation structure measurable by means of TEM, are outlined in the present section.

Let us approximate a subgrain by a cube of the edge length L, measured as the mean distance between the centers of neighbouring subgrains. Then, if the volume fraction of hard regions is f_w, the thickness of a subboundary is

$$h = L\,[1 + \sqrt[3]{(f_w - 1)}\,] \, . \qquad (17)$$

This relation is rather insensitive to the shape of equiaxed subgrains. Under appropriate assumptions of subboundary structure |18| the dislocation density in hard regions is given by the relation

$$\rho_w = \rho_{sb}/f_w = 4\sqrt{3} \cdot \theta/(b \cdot L \cdot f_w) \, , \qquad (18)$$

where θ is the mean subgrain misorientation.

Of the several methods proposed |18| to estimate the volume fraction of hard regions, only two will be outlined in this section.

<u>Method (a)</u> Mughrabi |16| proposed for hard regions the following relation

$$h\sqrt{\rho_w} = K_1 \, , \qquad (19)$$

where K_1 is a constant. Using eqns. (19), (17) and (18) the volume fraction f_w can be calculated numerically. For low volume fractions f_w, i.e. for $L \gg h$, an approximate relation

$$f_w \cong 3\sqrt{3} \cdot K_1^2 \cdot b/(4\theta L) \qquad (20)$$

is obtained. Note that the relation between f_w and K_1 is not linear. The result of f_w estimate for aluminium |19| and K_1 = 4 |16| is shown in Fig. 6. The values of f_w range from 0.05 to 0.10 and thus seem quite reasonable. The fraction f_w increases slightly with increasing applied stress.

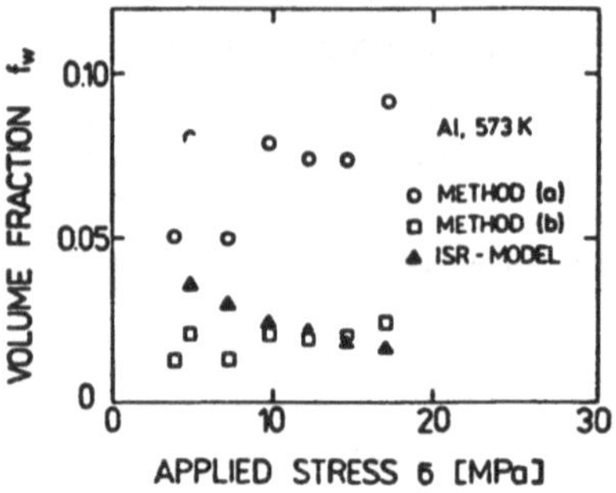

Fig. 6: Estimation of volume fraction of hard regions from substructure characteristics - methods (a), eqn. (20) (K_1 = 4) and (b), eqn. (22) (K_2 = 1) - and iso-strain rate (ISR) model (Section 2) in the steady state creep.

Method (b). In the composite model, the hard region represents a layer adjoining a sub-boundary in which a significantly higher stress acts than in the subgrain center. The range of the elastic stress field of low-angle boundary is proportional to the mean spacing between dislocations forming this boundary. Assume that the distance at which a significant stress field of the boundary acts (the half thickness of a hard region) is the K_2 multiple of the mean spacing λ between dislocations; K_2 is a constant. Then the following relation is obtained

$$f_w = 1 - [1 - K_2 b \sqrt{3} / (\Theta L)]^3 . \tag{21}$$

For low values of f_w the approximate relation

$$f_w \simeq 3\sqrt{3} \cdot K_2 b / (\Theta L) \tag{22}$$

holds. If $K_1^2/4$ is set for the constant K_2, eqn. (22) becomes identical with eqn. (20). Hence, for K_2 = 4, the fraction f_w calculated for aluminium |18| is equal to that following from the application of the method (a). Values of f_w for aluminium and K_2 = 1 are shown in Fig. 6.

While the method (a) leads to the relation $f_w \sim K_1^2$, the method (b) leads to a linear relation between f_w and K_2. Therefore, in the latter method the choice of the constant K_2 is not as critical as in the former one. However, a physical basis to choose the value of the constant K_2 is not available at present.

The ISR model based calculation of f_w in aluminium by the method applied in Section 2 for a Fe-21Cr-37Ni alloy provides values of f_w that can be considered acceptable as compared with those obtained using the methods (a) and (b) (Fig. 6). However, the calculation leads to f_w decreasing with increasing applied stress.

At the present time, there does not seem to be any possibility to measure the volume fraction of hard regions developed during creep directly. Consequently, it can be hardly decided which of the outlined methods provides realistic results.

5. ON THE APPLIED STRESS DEPENDENCE OF STEADY STATE CREEP RATE

The temperature and stress dependence of steady state creep rate $\dot{\varepsilon}_s$ at temperatures above 0.5 T_m (where T_m is the melting temperature) can be described by the well known phenomenological equation

$$\dot{\varepsilon}_s b^2 kT/D_L G \Omega = A_n (\sigma/G)^n \qquad (23)$$

where D_L is the lattice diffusion coefficient, A_n a dimensionless constant, Ω the atomic volume, k Boltzmann constant and n the applied stress exponent. The value of the stress exponent n is typically 5 for pure metals and solid solution alloys exhibiting Metal Class creep behaviour.

However, as early as 1975, Weertman |20| showed that if all "unnatural" or ad hoc assumptions are rejected, all the theoretical models of creep lead to the "natural" third power-law, i.e. $n = 3$. Needless to emphasize that full explanation of the discrepancy between "natural" and "experimental" or phenomenological stress dependence of creep rate would contribute to understanding of dislocation mechanisms in creep in quite a principal way. The idea that the plastic deformation of crystalline solids takes place under the action of a stress represented by the difference between the applied stress σ and a back stress σ_B may seem promising. The back stress σ_B may depend on temperature and, especially, on applied stress. According to this idea the steady state creep rate can be described by the equation

$$\dot{\varepsilon}_s b^2 kT/D_L G \Omega = A_m [(\sigma/G) - (\sigma_B/G)]^m , \qquad (24)$$

where, again, A_m is a dimensionless constant and m the stress exponent the value of which generally differs from that of the phenomenological applied stress exponent n in eqn. (23).

Rather frequently (e.g. refs. 21, 22) the back stress has been identified with the internal stress measured by the strain transient dip test technique. This approach was also applied to the steady state creep in a Fe-21Cr-37Ni alloy at temperatures 873 K to 1023 K |15|.

Considering along with Argon and Takeuchi |23| that the natural third power-law can be modified by introducing internal stress σ_i, i.e. simply replacing applied stress σ by the difference $\sigma - \sigma_i$, we obtain an equation of the same form as eqn. (24) with $\sigma_B = \sigma_i$ and $m = 3$; $A_m \equiv A_3$. Comparing it with eqn. (23) we obtain the relation

$$\sigma_i/G = \sigma/G - C \cdot (\sigma/G)^\beta , \qquad (25)$$

where $C = (A_n/A_3)^{\frac{1}{3}}$ and $\beta = n/3$ are constants. The experimental data for a Fe-21Cr-37 Ni alloy in $(\sigma_i/G),(\sigma/G)$ coordinates can be fitted by eqn. (25) in which $\beta = 1.8$ ($n = 5.4$ |14|) if the temperature dependence of measured internal stress is neglected |15| The normalized steady state creep rate $\dot{\varepsilon}_s b^2 kT/D_L G\Omega$ plotted against $(\sigma/G) - (\sigma_i/G) = C \cdot (\sigma/G)^\beta$ obeys the "natural" power law, since the stress exponent equals 3.01 ± 0.05 |15|, Fig. 7.

One would expect, in full accordance with contemporary models (e.g. refs. 24, 25) that it would be dislocation processes in hard regions, not those of soft regions, that would govern high temperature recovery creep of metals and subgrain forming solid solution alloys. Then a question arises how this expectation can be reconciled with the fact that the creep can be satisfactorily interpreted, under certain conditions at least in a phenomenological way in terms of measured internal stress, i.e. the level of long range internal back stress acting in the soft regions.

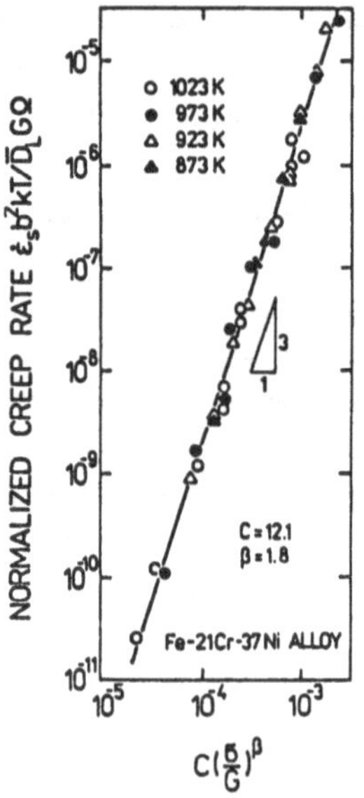

Fig. 7: Normalized steady state creep rate plotted against $(\sigma - \sigma_i)/G = C\,(\sigma/G)^{\beta}$ where $\beta = n/3 = 1.8$ |15|.

The deformation processes in the hard and soft regions act in parallel. In the limit of recovery control (for high enough temperatures) we can write

$$\dot{\varepsilon}_s \simeq \dot{\varepsilon}_w = F_w\,(\sigma_{iF} + \sigma)\,, \tag{26}$$

where F_w represents the functional form describing the deformation processes in hard regions |13|. Undoubtedly, it is possible to assume that deformation in hard regions is controlled by recovery. Thus, the applied stress in the "natural" third power-law creep equation should be replaced by the local stress σ_w acting there as it was, in fact, done by Nix and Ilschner |2| (see eqn. (10)), but not by the difference $(\sigma - \sigma_{iB})$. The relation

$$\sigma_{iF} = \sigma_{iB}\,(f_c/f_w) \tag{27}$$

then can be used to estimate σ_{iF} assuming $\sigma_{iB} \simeq \sigma_i$ and knowing f_w.

Under the conditions of negligible recovery rate the creep rate can be described as

$$\dot{\varepsilon}_s \simeq \dot{\varepsilon}_c = F_c\,(\sigma - \sigma_{iB}) \tag{28}$$

where the function F_c represents a law describing the deformation process in soft regions |13| in which dislocations can surmount the relatively weak obstacles. Accordingly F_c should be replaced by an appropriate rate law for thermally activated dislocation glide. The natural power-law creep equation cannot be used.

Generally, of course, the creep rate is contributed by both the deformation processes in hard and the deformation processes in soft regions. This was taken into account in our iso-strain rate model calculations.

6. CONCLUDING REMARKS

1. The calculated internal back stress σ_i^{calc} is affected by the value of ΔF rather significantly, however its unexpected temperature dependence cannot be fully accounted for by the value of C_i (which was chosen rather arbitrarily) and by the temperature dependence of ΔF. The assumption of homogeneously distributed "square" obstacles |26| to dislocation glide in the soft regions cannot be considered quite realistic. Nevertheless, the above rather good correlation between the measured and calculated internal stresses is certainly encouraging. In fact, quite a tolerable relation between σ_i^{calc} and σ_i was found also for aluminium |11|.

2. Experimental estimates of local stresses acting in subgrains of a crept Al-11Zn alloy show that σ_c varies considerably with the distance from the subgrain center towards the subboundary |8|. Hence, it seems that our calculation of the local stress σ_c provides "effective" value of these stresses. To obtain better agreement between calculated σ_c and σ_c measured by means of TEM the contribution to σ_c of moving screw dislocation density within cells or subgrains should be most probably taken into account in a way similar to that applied recently by Mughrabi |27| for the dislocation wall structure of persistent slip bands. Also the calculation of the fraction f_w provides an "effective" fraction of hard regions which is not easily correlatable with the fraction of cell walls or subboundaries as estimated on the basis of dislocation structure data obtained by TEM.

3. In the case of high stacking fault energy FCC metals and subgrain forming alloys as well as in the case of BCC metals and solid solution alloys the subgrain boundaries represented by plane arrangements of dislocations usually form in creep at high enough temperatures. It is considered |28| that, in this case, the local stresses originate from subboundary flexing under the action of applied stress. Even then, however, σ_w is assumed |18| to be related to the density of dislocations stored in subboundaries, ρ_{sb}, as

$$\sigma_w = \alpha_w G b \sqrt{\rho_{sb}/f_w} \,. \tag{29}$$

This relation is by no means self-understood and should be addressed in further composite model development.

4. The recovery creep in metals and subgrain forming solid solution alloys can be rather satisfactorily interpreted in terms of the difference of applied and measured internal stress only if the exponent β in eqn. (25) is equal to $n/3$, which, however, is not always the case. Of course, this interpretation, as a phenomenological one, contributes little to better understanding of high temperature creep. The measured internal stress represents a level of internal back stresses acting in soft regions,

however, dislocation processes that govern temperature and stress dependence of high temperature creep rate are most likely those operating in hard regions, where high internal forward stresses act. The application of eqn. (24) with $m = 3$ to describe steady state creep is thus inconsistent with the iso-strain rate composite model as applied to power-law recovery creep.

5. Despite of still not fully satisfactory state of development of the iso-strain rate composite model of recovery creep this model seems promising as to better understanding of high temperature creep in metals and cell or subgrain forming solid solution alloys, i.e. alloys exhibiting Metal Class creep behaviour.

REFERENCES

|1| S. Takeuchi and A.S. Argon, J. Mater. Sci. 11 (1976) 1542.

|2| W.D. Nix and B. Ilschner, In: Proc. 5th Intern.Conf. on Strength of Metals and Alloys, P. Haasen, V.Gerold and G. Kostorz (Eds), Pergamon, Oxford 1980, p. 1503.

|3| H. Oikawa and T.G. Langdon, In: Creep Behaviour of Crystalline Solids, B. Wilshire and R.W. Evans (Eds), Pineridge Press, Swansea 1985, p. 33.

|4| A. Orlová and J. Čadek, Mater. Sci. Eng. 77 (1986), 1.

|5| H. Mughrabi, In: Proc. 5th Intern. Conf. on Strength of Metals and Alloys, P. Haasen, V. Gerold and G. Kostorz (Eds.), Pergamon, Oxford 1980, p. 1615.

|6| F.B. Prinz and A.S. Argon, Acta Metall. 32 (1984) 1021.

|7| H. Mughrabi, Acta Metall. 31 (1983) 1367.

|8| M.A. Morris and J.L. Martin, Acta Metall. 32 (1984) 549; 32 (1984), 1609.

|9| A. Orlová, D.Sc. Thesis, Institute of Physical Metallurgy, Czechoslovak Academy of Sciences, Brno 1989.

|10| C.N. Ahlquist and W.D. Nix, Scr. Metall. 3 (1969) 679.

|11| F. Dobeš, Czech. J. Phys. B38 (1988) 524.

|12| W. Blum, S. Vogler, M. Biberger and A.K. Mukherjee, Mater. Sci. Eng. A 112 (1989) 93.

|13| W.D. Nix, J.C.Gibeling and K.P. Fuchs, In: Mechanical Testing for Deformation Model Development, R.W. Rohde and J.C. Svearengen (Eds.) ASTM STP 765, American Society for Testing and Materials, 1982, p. 301.

|14| E. Hajná, M. Pahutová and J. Čadek, Metallic Mater. 23 (1985) 174.

|15| J. Čadek, Mater. Sci. Eng. 94 (1987) 79.

|16| H. Mughrabi, Mater. Sci. Eng. 85 (1987) 15.

|17| I. Saxl and F. Kroupa, Phys. Stat. Solidi (a) 11 (1972) 167.

|18| F. Dobeš and A. Orlová, submitted to Mater. Sci. Eng.

|19| A. Orlová, Z. Tobolová and J. Čadek, Philos. Mag. 26 (1972) 1263.

|20| J. Weertman, In: Proc. J.E. Dorn Symp. on Rate Processes in Plastic Deformation and Materials, J.C.M. Li and A.K. Mukherjee (Eds.), American Society for Metals, Metals Park, OH, 1975, p. 315.

|21| T. Hostinský and J. Čadek, Philos. Mag. 31 (1975) 1177.

|22| K. Milička, D.Sc. Thesis, Institute of Physical Metallurgy, Czechoslovak Academy of Sciences, Brno 1985.

|23| A.S. Argon and S. Takeuchi, Acta Metall. 29 (1981) 1877.

|24| W.D. Nix, J.C. Gibeling and D.A. Hughes, Metall. Trans. A, 16A (1985) 2215.

|25| J. Weertman, In: Proc. 2nd Intern. Conf. Creep and Fracture of Engineering Materials and Structures, B. Wilshire and D.R.J. Owen (Eds.), Pineridge Press, Swansea 1984, p. 1.

|26| H.J. Frost and M.F. Ashby, Deformation Mechanisms Maps, Pergamon, Oxford 1982, p. 21.

|27| H. Mughrabi, Phys. Stat. Solidi (a) 104 (1987) 107.

|28| J.C. Gibeling and W.D. Nix, Acta Metall. 28 (1980) 1743.

INVESTIGATION OF RECRYSTALLIZATION TEXTURE DEVELOPMENT BY HVEM

P.-J. Wilbrandt

Institut für Metallphysik, Universität Göttingen,
Hospitalstr. 3-5, D-3400 Göttingen, Federal Republic of Germany

ABSTRACT

The mechanisms of recrystallization texture development were investigated by annealing thin specimens in the HVEM. The specimens were prepared from deformed single crystals of copper, aluminium and dilute copper phosphorus alloys. The orientations of the first recrystallized grains coincide with those of the deformed microstructure. New orientations are exclusively formed by annealing twinning. The decrease of interfacial energy by the orientation change is regarded as the parameter controlling twin formation. Information about the importance of the new orientations for texture formation was obtained by additional recrystallization experiments in bulk material. The grain boundary mobility turned out to be the second essential parameter influencing recrystallization texture development.

INTRODUCTION

Annealing of a deformed metal or alloy specimen often changes the initial deformation texture to a completely different recrystallization texture [1]. This phenomenon is observed in geological materials, too [2]. Many efforts were made to identify and to understand the mechanisms controlling recrystallization texture development. Two different kinds of investigations are described in literature. Those of the first kind analyse the textures of the deformed and recrystallized microstructure and interpret the relations between the two textures [3]. By local measurements the investigations of the second kind try to get information about the processes involved. Electron microscopy is frequently used because the high resolution of this method allows a detailed analysis of the microstructure. High voltage electron microscopy (HVEM) additionally provides the possibility to investigate relatively thick specimens (10μm in Al, 2μm in Cu) [4]. Thus the recrystallization process can be studied "in situ" by annealing specimens in the microscope. Thermal grooving does not restrict grain boundary motion in specimens of this thickness [5].

An essential disadvantage of electron microscopic investigations is the limited field of view. Consequently, the detection of specimen areas being representative for the whole specimen is difficult. Moreover, a deformation microstructure (DM) with a complex morphology makes the interpretation of recrystallization experiments difficult. In the case of our experiments these problems were solved by using specimens of tensile deformed single crystals. The crystals were prepared from copper, aluminium and diluted copper phosphorus alloys. The investigation of different materials allows conclusions on the influence of stacking fault energy (SFE) and impurity content. Information about the significance of the HVEM-results and the late stages of the recrystallization process was obtained by additional annealing experiments with bulk specimens.

PREPARATION OF THE DEFORMATION MICROSTRUCTURE

<100>- and <110>-oriented crystals of high purity copper (99.9998% Cu) were grown by the Bridgman technique in graphite moulds. The same method was used to prepare <110>-oriented single crystals of less pure aluminium (99.998% Al) and of diluted copper phosphorus alloys (Cu-0.03at% P, Cu-0.09at% P and Cu-0.7at% P). Tensile deformation of the <100>-oriented crystals up to 20% elongation causes multiple slip on different slip systems of type {111}<110>. This creates an extremely homogeneous DM consisting of dislocation-free subgrains. All subgrains have the same orientation as the undeformed crystal. The dislocations are concentrated in the walls separating the subgrains.

With the exception of the Cu-0.7at% P alloy the DM of all tensile deformed <110>-oriented crystals (80% elongation) consists of clearly separated regions of two different orientations. Depending on their orientation they are named as "matrix" (M) or "band" (B). The border between M and B is considered to be a high angle grain boundary (HAGB) because no evidence for a continuous transition from one orientation to the other was found. Both regions were deformed by slip on different {111}-planes each with an effective <211>-slip direction (coplanar double slip). Therefore, the deformation induced lattice rotations occurred about one common <110>-axis perpendicular to the tensile direction.

<110>-oriented single crystals of the Cu-0.7at% P alloy deform at 77K by single slip. Deformation twinning occurs if the elongation exceeds 85%. The final DM consists of matrix regions interspersed by layers of many

thin deformation twins. The change in microstructure in comparison with the low P alloys indicates a change in SFE with P content, not just an interaction of the solute atoms with the moving defects. Experiments were made with crystals deformed up to 120% elongation.

HVEM RESULTS ON RECRYSTALLIZED GRAINS

The HVEM micrographs in the present section are selected examples that illustrate the different processes during recrystallization. More HVEM observations were published elsewhere [6]. If not stated otherwise, the examples are regarded as representative for all materials investigated.

The micrograph in fig. 1a elucidates the beginning of the recrystallization process. It was obtained in an Al-specimen and shows a small recrystallized grain completely surrounded by the DM. The circles mark the positions where orientations were determined by selected area diffraction. As can be seen in figs. 1b and 1c the new grain with the "matrix related" orientation formed just at the HAGB between M and B.

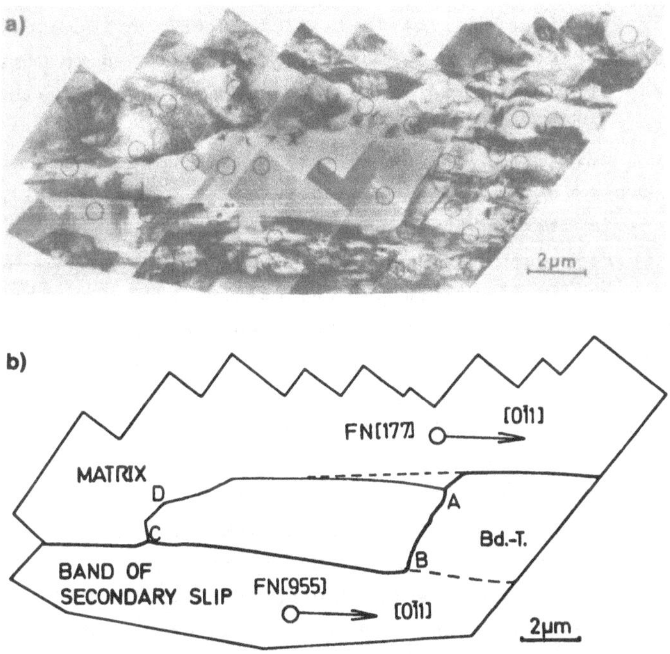

Fig. 1a,b Recrystallized grain in an Al-specimen after annealing for 10 min at 200°C and 10 min at 300°C (with identified microstructure)

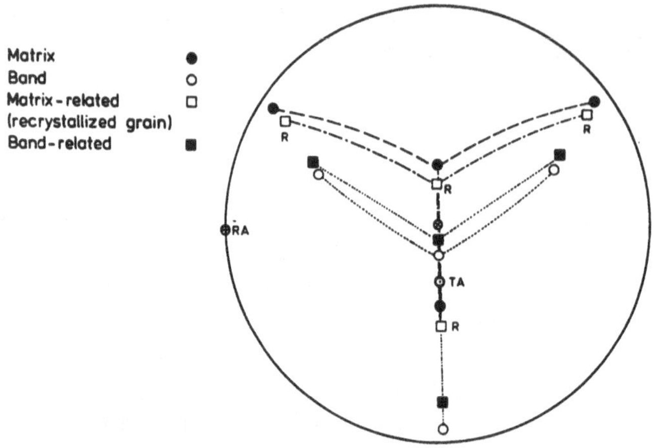

Fig. 1c {111}-pole figure for fig. 1a

Generally, ledges in the HAGB's were found to be the most preferable sites for the beginning of recrystallization. The "matrix related" orientation was also detected in this specimen region prior to annealing. Thus in our DM no new orientations are generated in the beginning of the recrystallization process.

The formation of the small recrystallized grain can be explained simply by movement of a pre-existing HAGB [7]. According to Bailey [8] a segment of a HAGB can move if its length is larger than $2\gamma/E_v$ (γ: total interfacial energy, E_v: stored deformation energy in the volume). As pointed out by Huang and Form [9] the classical nucleation models do not describe the formation of the first recrystallized grains because, altogether, no increase in free energy is needed to form the initial strain-free site.

Recrystallization twinning turned out to be the only process which creates new orientations. Even in aluminium with its high SFE this process occurs as demonstrated by the HVEM micrograph in fig. 2a. A large recrystallized grain grows into a region of the DM which, despite of a small region in the middle, has matrix orientation (see figs. 2b and 2c). Twinning on two different {111}-planes changes the orientation from R1 to R2 and R3, respectively. The slight orientation variations of the adjacent matrix regions are regarded as reason for this effect. The orientation relationship (OR) R2-M2 can be characterized by a 40° rotation about a common <111>-axis (near the $\Sigma 7$ coincidence orientation relationship). In the following, the description of the OR's by coincidence orientation relationships is preferred in order to simplify the comparison of the results. A near Σ7 OR has been found by Liebmann et al. [10] to favour rapid grain growth in Al. Also in our Al-specimen the grain R2 grew rapidly during further annealing. Yet the influence of grains with such an OR to the DM on further texture formation must not be overestimated because they can change their orientation by further twinning. In our specimen the OR R1-M can be approximated by Σ7, too, but nevertheless the new orientation R3 was formed.

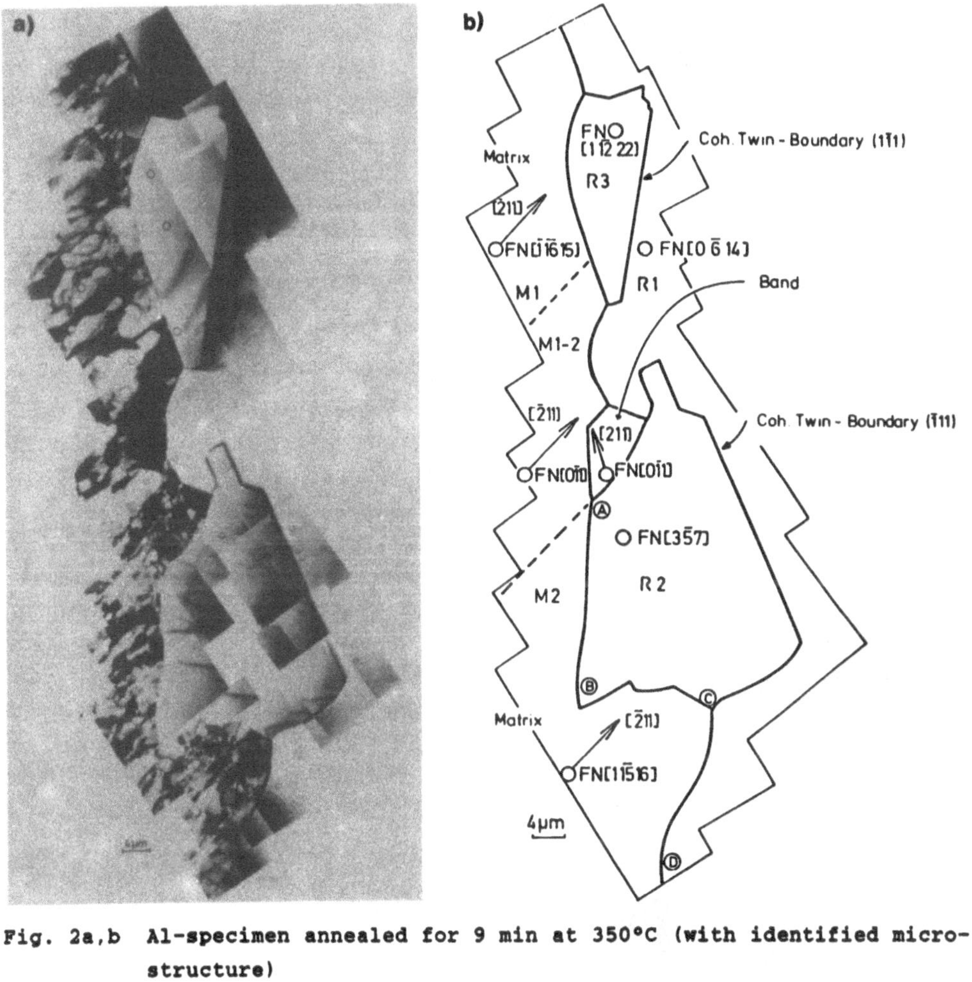

Fig. 2a,b Al-specimen annealed for 9 min at 350°C (with identified microstructure)

Generally, twinning can not be repeated at random on the four available {111}-planes because this would generate an almost "grey" texture after 6 generations [11]. Certain twinning sequences should be

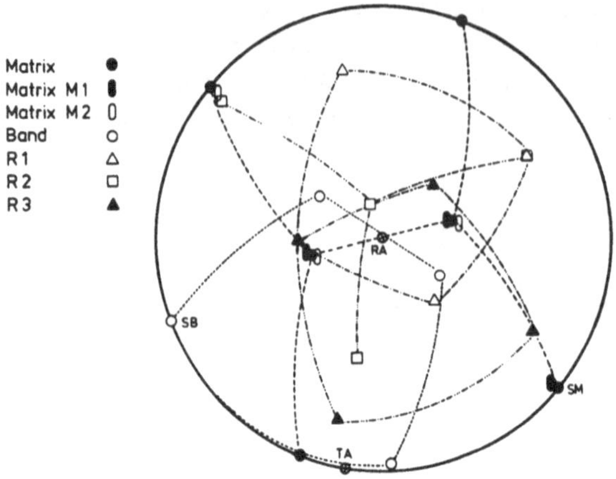

Fig. 2c {111}-pole figure for fig. 2a

preferred. The grain boundary energy is regarded as the parameter controlling the path of twin formation. Fig. 3 shows a grain in a Cu-0.03at% P-alloy which grows from the right-hand side and has twinned several times. Finally, the orientation R2 was formed which has a near Σ11 OR to the matrix. The relation R2-M across straight {311} and {332}-boundaries (segments CD, EF and FG, respectively) is interesting: On the basis of data of Hasson [12] both interfaces are predicted to be of low energy.

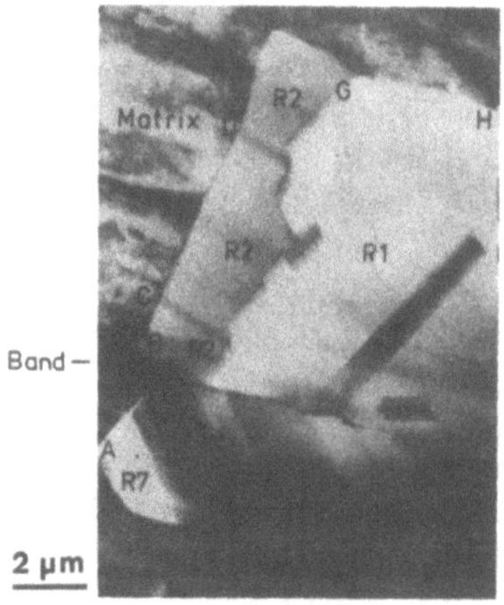

Fig. 3 Recrystallized grain in Cu-0.03at% P after annealing for 20 min at 500°C

The recrystallized grain in fig. 4 is a another proof of the influence of the GB-energy on annealing twin formation. Again the observation was made in a Cu-0.03at% P alloy. The OR between the grain and the whole surrounding DM can be approximated by $\Sigma 11$. There is evidently no incentive for further twinning. Especially the straight GB segments 4-5 and 6-7 (parallel to {311}-planes) as well as 1-2 and 3-4 (parallel to {322}-planes) could again be identified as low energy interfaces.

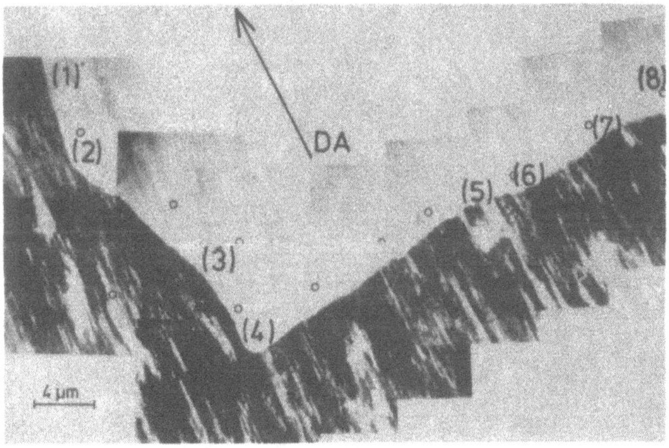

Fig. 4 Recrystallized grain without twins in Cu-0.03at% P after 20 min anneal at 360°C

The influence of a larger phosphorus content on the recrystallization of copper phosphorus alloys is demonstrated by the HVEM micrograph in fig. 5a which was obtained in a Cu-0.7at% P specimen. Here the recrystallized grains have exclusively orientations of the DM (see figs. 5b and 5c). The specimen has recrystallized by movement of low-angle GB's. Such a behaviour differs completely from that of the low alloyed material. It reveals a relatively high mobility of low-angle grain boundaries (LAGB's) in the high alloyed materials.

TEXTURE MEASUREMENTS AND BECK EXPERIMENTS

The HVEM observations described in the previous section give information only about the processes occurring on the microscale. Information about the preferred orientations at the end of primary recrystallization were obtained by additional X-ray texture measurements. Fig. 6a is an idealized {111}-pole figure of the DM of an Al specimen, fig. 6b shows the

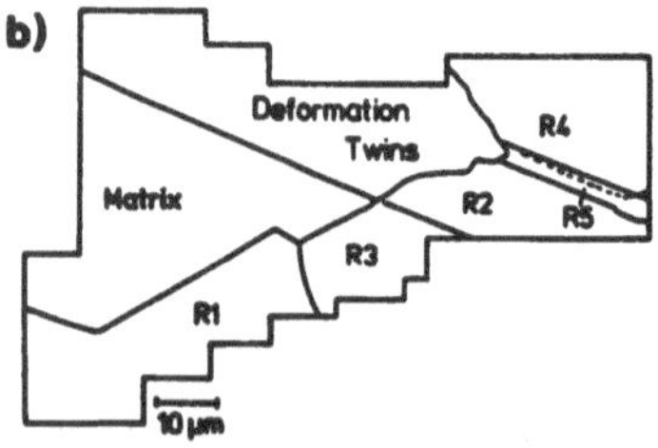

Fig. 5a,b Recrystallized grain in Cu-0.7at% P after annealing for 10 min at 600°C (with idendified microstructure)

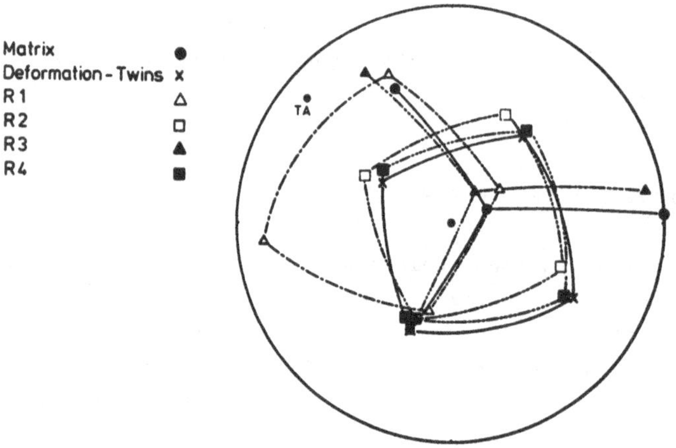

Fig. 5c {111}-pole figure for fig. 5a

corresponding texture. The OR's between the DM and the idealized orientations of the texture can be characterized by coincidence relations with Σ-values between 13 and 25. Such a preferred occurrence of orientations with high Σ-values was not found in the HVEM specimens of this material. Similar results were obtained for the copper specimens. The textures of the copper phosphorus specimens, however, also show orientations with a low Σ value. Their frequency of occurrence increases with increasing phosphorus content. The extreme case is the texture of the Cu-0.7at% P alloy. With one exception, the orientations of the recrystallization texture have $\Sigma 1, \Sigma 3$ and $\Sigma 9$ OR's to the DM. The differences between the texture measurement results and the HVEM observations indicate that more information about grain growth is needed.

a)

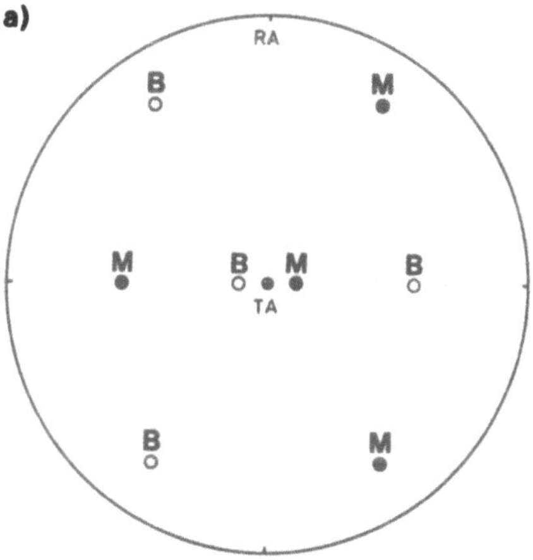

b)

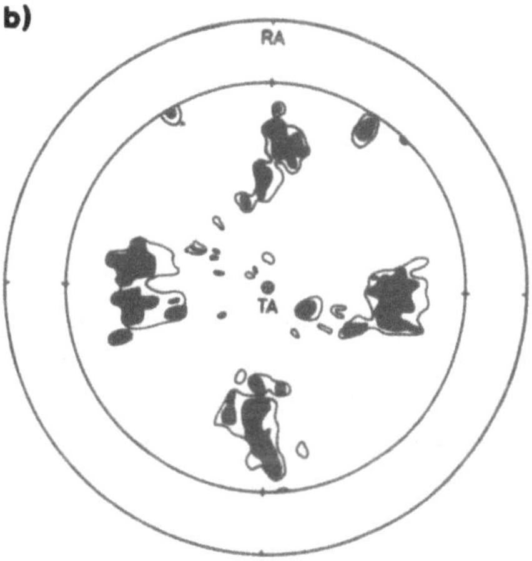

Fig. 6a Idealized {111}-pole figure of the DM of the Al-specimen

Fig. 6b {111}-pole figure with the texture of the recrystallized Al-specimen (annealed for 15 min at 350°C)

A suitable experiment for grain growth studies has been proposed by Beck [13]. We abraded specimens of tensile deformed <100>-oriented Cu single crystals at one end. During subsequent annealing the recrystallization process started just in the abraded specimen region. As the recrystallization nuclei are randomly oriented the recrystallization microstructure is dominated by grains which have grown fastest. Because of the extreme homogeneity of the DM the OR's of these grains and, therefore, the orientation relations for rapid grain growth can be determined with great accuracy.

A typical example of the recrystallization microstructure in these specimens is presented in fig. 7. Only grains with a diameter of more than 2mm were investigated. 85% of these grains have OR's to the DM wich can be idealized by Σ13a, Σ19b and Σ25b coincidence relations.

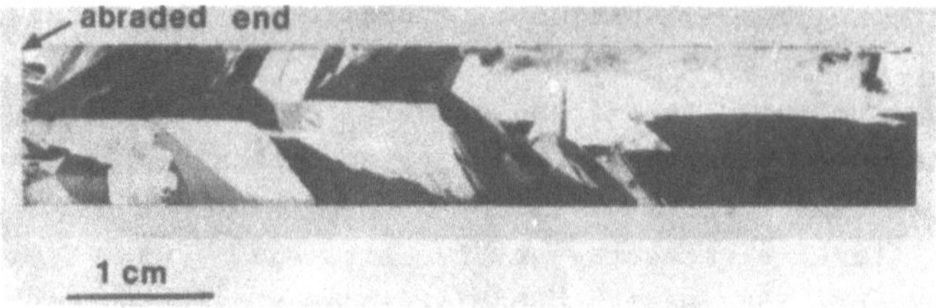

Fig. 7 Microstructure of a recrystallized specimen annealed for 2 h at 480°C. The specimen was prepared from a tensile deformed {100}-oriented Cu single crystal

CONCLUSIONS FOR THE EXPLANATION OF TEXTURE DEVELOPMENT

The old theories of recrystallization texture development try to explain the phenomenon either as a result of oriented nucleation [e.g. 14] or as a result of oriented growth [e.g. 15]. The results summarized in the preceeding sections reveal that the basic ideas of both concepts are equally needed to explain recrystallization texture development. Additionally, annealing twinning was identified as the only process which creates new orientations. These results are the basis for our three-stage-model for recrystallization texture development.

The nucleation of small recrystallized grains is the first stage. A recrystallization nucleus is defined as a dislocation free region which is at least partly separated from the DM by a mobile grain boundary and has a sufficient size to grow during further annealing. Several mechanisms leading to the formation of such regions are proposed in literature [16]. In our HVEM specimens nuclei are formed by the movement of preexisting HAGB's. The mobile grain boundaries account for the growth capability of the nuclei. Growth is only possible if the resulting increase in free energy is compensated by the release in volume energy. Again it must be emphasized that these recrystallization nuclei are not formed by reversible thermal fluctuations. Nearly all TEM and HVEM observations showed that the nuclei have orientations of the DM. The only exception is the nucleation by formation of recovery twins observed by Huber and Hatherly [17].

In the second stage of the texture development new orientations are created by annealing twinning. As indicated by the HVEM investigations in Al this process can occur in a metal with a high SFE. Annealing twinning in metals and alloys with medium and low SFE is a well-known phenomenon [18]. As random twinning over 6 generations would lead to a 'grey' texture there must be a reason for the preference of certain sequences. The gain in interfacial energy by the orientation change is considered as the only mechanism controlling twin chain development. This is evidenced by our HVEM observations of grains with special low energy boundaries whose orientations are created by twinning.

In the last stage a growth selection among different recrystallized grains occurs. Yet contrary to the old concept only orientations created by annealing twinning and orientations of the DM are involved. The Beck experiments in pure Cu indicate that $\Sigma13a$, $\Sigma19b$ and $\Sigma25b$ OR's favor rapid growth. Foreign atoms impede general boundaries more than some special boundaries [19]. An extreme example for the influence of impurities is the recrystallization behaviour of the Cu-0.7at% P alloy. Here a preferred growth of grains with LAGB's was observed. These boundaries are rather immobile in pure metals.

There are other factors than the OR's which determine grain growth. First, the anisotropy of the dislocation arrangement in the DM was found to determine the preferential growth of the new grains [20]. Second, the GB structure has bo be taken into account because there exist appreciable differences between the mobility of tilt and twist boundaries [21]. The influence of alloying on the GB mobility was already mentioned.

On the whole, this model for recrystallization texture development has the advantage of taking into consideration all experimental results described in literature. The role of the different processes involved could be understood only by systematic HVEM-investigations. Yet, there still exist many open questions which need further crucial experiments.

ACKNOWLEDGEMENTS

The author thanks Prof. Dr. P. Haasen, Dr. A. Berger, Dr. F. Ernst and U. Klement for many helpful discussions. All investigations of the Göttingen recrystallization group were financially supported by the Deutsche Forschungsgemeinschaft. The aid of the Max-Planck-Institut für Metallforschung (HVEM investigations) and of the Institut für Werkstoffe der TU Braunschweig (texture measurements) is gratefully acknowledged.

REFERENCES

1 G. Wassermann and J. Grewen, Texturen metallischer Werkstoffe, Springer-Verlag, Berlin-Göttingen-Heidelberg 1962

2 H.-R. Wenk, Preferred Orientation in Deformed Metals and Rocks, Academic Press, Orlando 1985

3 J. Hirsch, Habilitation Thesis, Aachen 1988

4 T. Imura and H. Saka, Memoires of the Faculty of Engineering, Nagoya University, 28 (1976) 54-112

5 W. Roberts and B. Lehtinen, Phil. Mag. 26 (1972) 1153-66, see also W. Roberts and B. Lehtinen, Phil. Mag. 29 (1974) 1431-3

6 A. Berger, P.-J. Wilbrandt, F. Ernst, U. Klement and P. Haasen, Progr. in Mat. Sci. 32 (1988) 1-95

7 A. Berger, P.-J. Wilbrandt and P. Haasen, Acta Met. 31 (1983) 1433-43

8 J.E. Bailey, Phil. Mag. 5 (1960) 833-42

9 K.-P. Huang and W. Form, Z. Metallkde 80 (1989) 686-91

10 B. Liebmann, K. Lücke and G. Masing, Z. Metallkde 47 (1956) 57-63

11 P.-J. Wilbrandt, phys. stat. sol. (a) 61 (1980) 411-8

12 G. Hasson, M. Biscondi, P. Lagarde, J. Levy and C. Goux, In: The Nature and Behavior of Grain Boundaries, H. Hu (Ed.), Plenum Press, New York, London 1972 p. 3-40

13 P.A. Beck, Acta Met. 1 (1953) 230-4

14 I.L. Dillamore, H. Katoh, Met. Sci. 8 (1974) 73-83

15 K. Lücke, Can. Met. Q. 13 (1974) 261-74

16 R.D. Doherty In: Recrystallization of Metallic Materials, F. Haeszner (Ed.), Dr. Riederer Verlag, Stuttgart 1978, p. 32-62

17 J. Huber, M. Hatherly, Met. Sci. 13 (1979) 665-9

18 P.-J. Wilbrandt, In: Proceedings of the Eighth International Conference on Textures of Materials, J.S. Kallend and G. Gottstein (Eds.), The Metallurgical Society, Warrendale 1988, p. 573-84

19 K.T. Aust and J.W. Rutter, Trans AIME 215 (1959) 119-27

20 G. Czjzek and F. Haeszner, Z. Metallkde 51 (1960) 567-72

21 G. Gottstein, H.C. Murmann, G. Renner, C. Simpson and K. Lücke, In: Proceedings of the Fifth International Conference on Textures of Materials, G. Gottstein and K. Lücke (Eds.), Springer-Verlag, Berlin-Heidelberg-New York 1978, p. 521-31

LOMER-COTTRELL DISLOCATIONS AFTER HIGH TEMPERATURE DEFORMATION OF [112] COPPER SINGLE CRYSTALS

P. ANONGBA, J.L. MARTIN
Institut de Génie Atomique, Ecole Polytechnique Fédérale de Lausanne,
PHB Ecublens, 1015 Lausanne, Switzerland

ABSTRACT

[112] Copper single crystals have been deformed up to stage IV at temperatures of 923 K and 1145 K and a shear strain rate of $2.10^{-2}s^{-1}$ (temperature regime 2). A unique aspect of the substructure consists of long edge segments in Lomer-Cottrell orientation. A formation mechanism is proposed in which the screw portions of {100} loops are more mobile than the edges.

INTRODUCTION

The shear stress-shear strain curves $\tau(\gamma)$ and the associated hardening rates $\theta \equiv d\tau/d\gamma$ of [112] Copper single crystals have been recently studied [1] [2]. Above room temperature, in addition to the well known stages II and III which are generally observed in f.c.c. single crystals, they exhibit three additional stages, named IV, V and VI. As the temperature increases, three temperature regimes 1, 2 and 3 are successively observed, corresponding to particular properties of the hardening stages and typical microstructure configurations. In particular, in regime 2, slip line observations show that several non-compact slip systems are activated in stage IV [1][2].

In the present study the geometrical features of dislocations gliding on {100} planes are investigated in stage IV in temperature regime 2. The possible mechanisms of their formation and migration are given.

EXPERIMENTAL TECHNIQUES

The [112] single crystals were deformed under the conditions given in Tab. 1. Slip traces were observed using an optical microscope. The dislocation substructure observations were performed with a Hitachi H 700 H microscope operating at 200 kV. Thin foils were prepared by cutting discs 0.35 mm thick parallel to $(0\bar{1}\bar{1})$ and (112) respectively for specimens deformed at 1145 K and 923 K by spark machining. The discs were then electropolished by the jet technique using a solution 25% HNO_3 - 75% CH_3OH at - 11°C and 12 V.

The Burgers vector $\vec{b}$ of the dislocations was determined using at least two different diffraction vectors satisfying the invisibility principle $\vec{g} \cdot \vec{b} = 0$. In some cases, the dislocation may be visible even if $\vec{g} \cdot \vec{b} = 0$; for a pure edge dislocation, the residual contrast is symmetrical about the dislocation [3] (see e.g. Fig. 1(i) below). The glide plane of the dislocation was generally determined by $\vec{b}$ and the direction defined by two points on the dislocation. Some difficulties were encountered during dislocation observations due to their

Table 1: Deformation conditions for the specimens used in this study (regime 2 of temperature)

Temperature [K]	Imposed shear strain rate $[s^{-1}]$	Initial flow shear stress [MPa]	Highest shear stress achieved [MPa]	Hardening stage achieved
923	2×10^{-2}	4	14	IV
1145	2×10^{-2}	3.5	7	IV

instability with respect to the foil surfaces.

RESULTS

Transmission Electron Microscope (TEM) Observations after deformation at T = 1145 K, $\dot{\gamma} = 2.10^{-2}s^{-1}$.

Fig.1(i) shows a dislocation (A,B) reacting with other dislocations at several points such as a, b and c. In Fig. 1(ii), taken in $0\bar{2}2$ reflection, the image of dislocation (A,B) vanishes; it has also been observed to vanish for $11\bar{1}$ and $1\bar{1}1$ reflections, therefore its Burgers vector is 1/2 [011]. The foil orientation is presented in Fig. 2. A careful examination of dislocation (A,B), Fig. 1(i), reveals that it consists of portions which are perpendicular to [100] and connected by shorter portions perpendicular to $[11\bar{1}]$ (in J) or $[1\bar{1}1]$ (in I). At point J, the short portion is 83 nm long. The portions of dislocation (A,B), perpendicular to [100], Fig. 1(i), are parallel to $[01\bar{1}]$, Fig. 1(iii). Therefore, their direction is $[01\bar{1}]$ and they are edge dislocations with (100) as glide planes i.e. in Lomer-Cottrell orientation. The dislocation (A,B) is lying in several distinct planes: parallel (100) planes, (the {100} segments), parallel $(11\bar{1})$ planes and $(1\bar{1}1)$ planes (the {111} segments).

At points I and J, Fig. 1(i), the {111} portions of dislocation (A,B) are respectively in the direction $\vec{I}$ = $[21\bar{1}]$ and $\vec{J} = [2\bar{1}1]$ (Fig. 2): Therefore, the complete dislocation (A,B) Fig. 1, is on average in the foil plane i.e. $(0\bar{1}\bar{1})$. Fig. 1(iv) gives a schematic representation of various reactions: the Burgers vector and the glide plane are indicated for each reacting dislocation. At points, a, b and c, dislocations with $1/2\ [1\bar{1}0]$ as Burgers vector react with dislocation (A,B): these dislocations glide in the (111) planes but cross slip onto the $(11\bar{1})$ planes at the nodes and take a nearly screw character, Fig. 1(i). Such reactions stabilize the dislocation (A,B) in the foil. Some of them have obviously occured at {111} portions along AB (see e.g. in b in Fig. 1(i)).

Fig. 3 shows a dislocation loop having two main segments in Lomer-Cottrell orientation with 1/2 [011] as Burgers vector and lying in (100) planes. Its portions J_1 and J_2 are parallel to [101], (Fig. 3), therefore glide on $(11\bar{1})$ planes.

Fig. 4 shows the substructure within a subgrain; it mainly consists of Lomer-Cottrell dislocation segments having $\vec{b} = 1/2\ [011]$ as Burgers vector and lying in (100) planes. Therefore dislocation such as (A,B) are very numerous under such deformation conditions. The relative density of the Lomer-Cottrell segments in the subgrains is about 30%. However they are not observed for other deformation stages and temperature regimes.

TEM Observations after deformation at 923K, $\dot{\gamma} = 2.10^{-2}s^{-1}$.

Lomer-Cottrell dislocations are also observed. Their Burgers vector (Fig. 5) is 1/2[101] and their glide plane is (010). The {111} portions lie in the $(\bar{1}11)$ and $(11\bar{1})$ planes respectively (see i.e. I and J). The relative density of the {100} edge portions in the subgrains is about 10%. The {111} segments are longer than at 1145K. They can be as long as 130 nm.

Slip Line Observations.

The slip systems are as follows [1] [2] :

• [011] $(1\bar{1}1)$ or [101] $(\bar{1}11)$, the traces of which are parallel to [110] on the $(1\bar{1}0)$ face. They are observed in all temperature regimes for most of the hardening stages.

• Various non compact systems of the type [011] $(X\bar{1}1)$ and [101] $(\bar{1}X1)$, with X = 2, 3, 5... . They have been identified from their traces on the $(1\bar{1}0)$ and $(11\bar{1})$ faces. They are typical of stage IV in regime 2 [1] [2].

• [011](100) or [101](010), the traces of which are parallel to [001] on the $(1\bar{1}0)$ face of the specimen. The corresponding traces are composed of few straight short segments. An increasing importance of these segments with increasing temperature is not clearly observed. They are also typical of stage IV in regime 2.

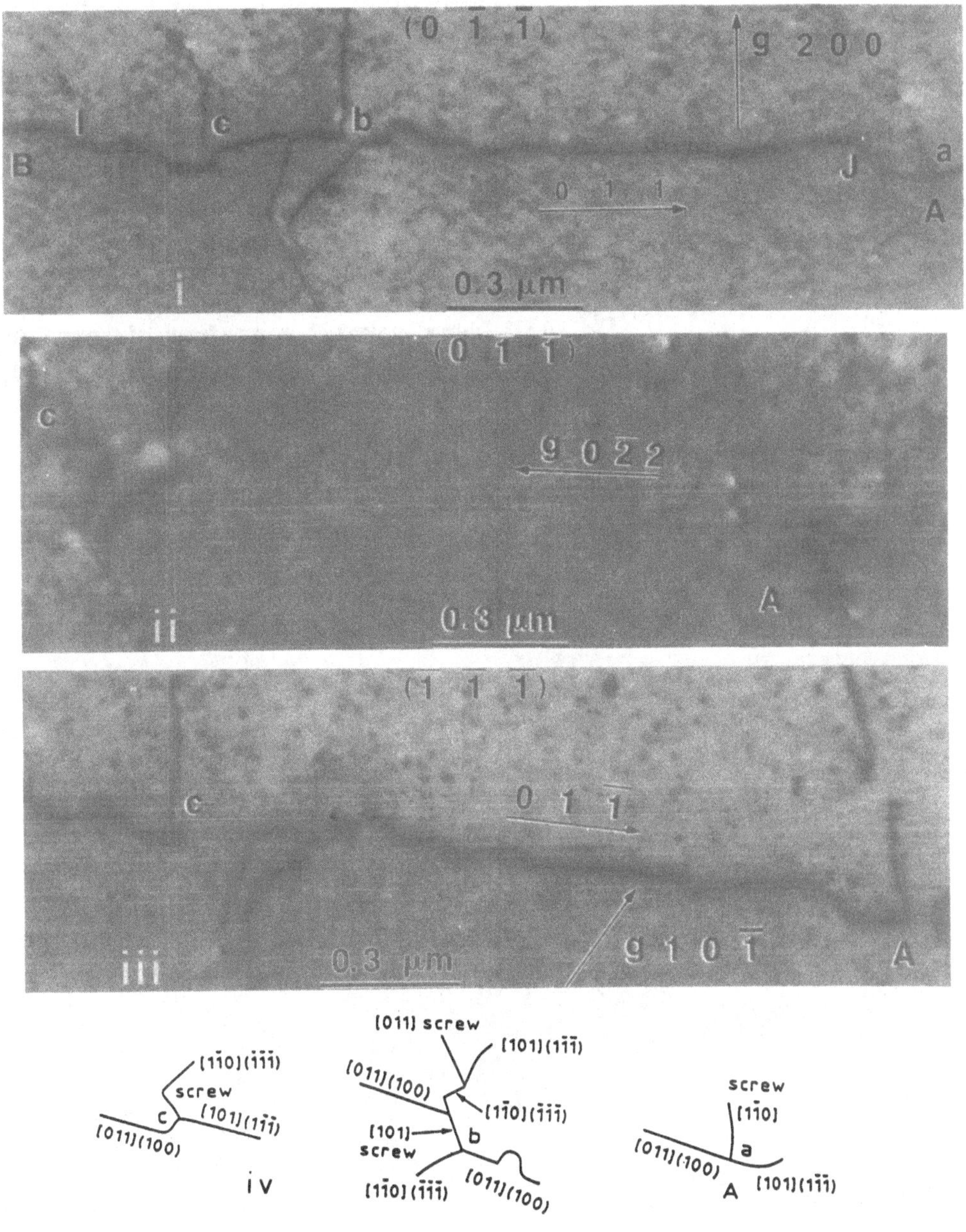

Fig. 1 Typical dislocation configuration after deformation at T = 1145 K and $\dot{\gamma} = 2.10^{-2}s^{-1}$. $(0\bar{1}\bar{1})$ foil. (i) The dislsocation (A,B) displays a symmetrical image. $\vec{g} = [\bar{2}00]$ and $\vec{g} \cdot \vec{b} = 0$. The corresponding segments are in Lomer-Cottrell orientation . (ii) The image of the dislocation (A,B) vanishes. $\vec{g} = [0\bar{2}2]$. (iii) $(\bar{1}\bar{1}\bar{1})$ orientation. The dislocation (A,B) direction is close to $[0\bar{1}1]$. (iv) Schematic representation of various reactions along dislocation (A,B). In (i) to (iii) the beam directions are indicated.

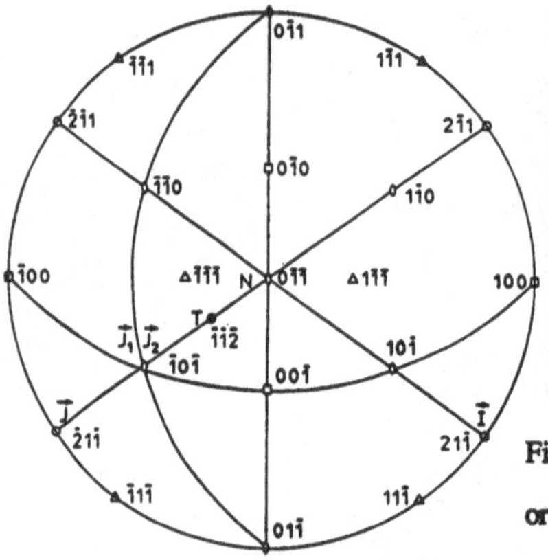

Fig. 2 Crystallographic orientation of specimens and thin foils. T = [112]: tensile axis. N = [01̄1̄]: Thin foil normal for specimens deformed at T = 1145 K.

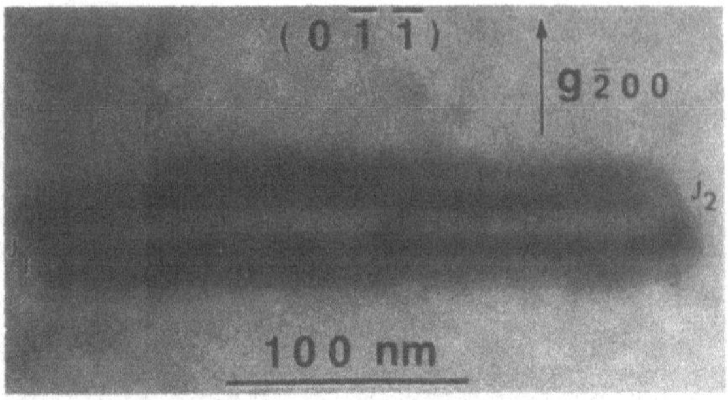

Fig. 3 Dislocation loop with two main segments in Lomer-Cottrell orientation $\vec{b}$ = 1/2 [011]. (01̄1̄) foil. T = 1145 K. $\dot{\gamma} = 2.10^{-2}s^{-1}$.

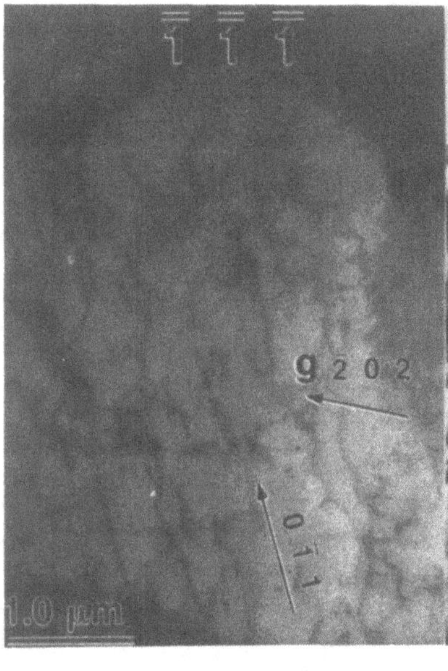

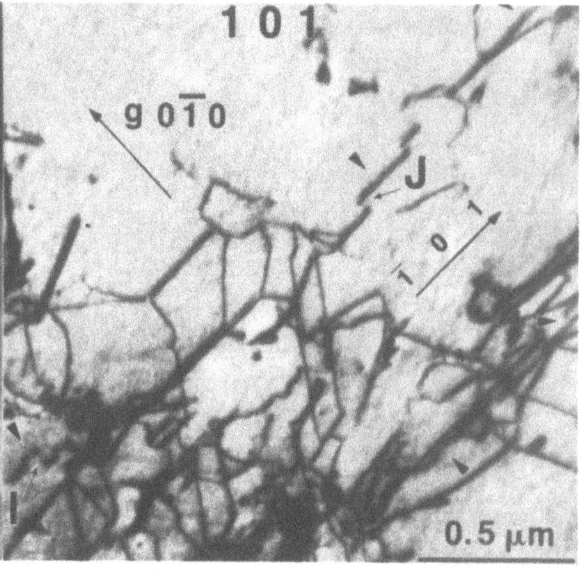

Fig. 4 Typical substructure in a specimen deformed to stage IV in temperature regime 2 (T = 1145 K, $\dot{\gamma}$ = $2.10^{-2}s^{-1}$). The straight dislocation segments in the subgrain which are parallel to [01̄1] are in Lomer-Cottrell orientation on (100) planes with $\vec{b}$ = 1/2[011].

Fig. 5 Typical dislocation substructure within subgrains in stage IV (regime 2). (112) foil. (T = 923 K, $\gamma = 2.10^{-2}s^{-1}$). Some straight dislocation segments (►) parallel to [101̄] are in Lomer-Cottrell orientation.

DISCUSSION

The Lomer-Cottrell dislocations which have been previously reported in the litterature correspond to deformation conditions below or at room temperature (see e.g. [4] to [8]). The highest density of long segments were observed after deformation at very low temperatures (near liquid helium) [4]. These dislocations originate at the intersection of active {111} planes, according to the original idea of Lomer and Cottrell. The deformation conditions of the present study are completely different. Therefore the formation mechanism of the dislocations which are described above, has to be discussed.

The model that we propose refers to a mechanism of (001) glide which was derived from experimental observations of non compact slip in aluminium after deformation above 200°C [9][10]. In this kink pair mechanism, the screw dislocation moves along the {001} plane and dissociates at each dense row of this plane along one out of the two available {111} planes. We assume that the same mechanism holds in copper, with some differences which are connected to the lower stacking fault energy. As the temperature rises, the density of {100} kinks on screw dislocations increases, thus stabilizing them in the {100} plane. Unlike in aluminium, as the screw segment of a loop moves on {100}, the edge segments split into a Lomer-Cottrell dissociation. This is in agreement with the long edge segments which were described above. These edge segments have a low mobility at 923 and 1145 K if any, as compared to the screws which are never observed in the subgrains. As they glide along {100} the latter can eventually cross slip onto a {111} type plane then cross slip back onto a {100} type plane parallel to the previous one. This accounts for the {111} segments of the observed dislocations, which connect {100} segments. The respective densities of the {111} and {100} segments along the dislocation depend of the relative stability of the screw segments on {100} and {111}. At lower temperatures the screws spend less time on {100} and more time on {111}. This is in agreement with a higher relative density of Lomer-Cottrell segments and shorter {111} segments at 1145 K than at 923 K. Fig. 6 is a sketch of this mechanism.

The formation of the prismatic dislocation loops similar to the one in Fig. 3 is a consequence of the cross slip of the screws between {001} and {111} planes. The mechanism is probably similar to the one observed during in situ experiments on prismatic glide in Magnesium for the "closed loop" multiplication process [11]. Fig. 7 shows a schematic representation of the process. A screw dislocation moving on {100} separates into two segments at an obstacle (Fig. 7a). One segment glides further on {100} while the second one cross slips onto {111} (Fig. 7b). Later on, the first segment cross slips onto {111} and the second one back to {100} (Fig. 7c).

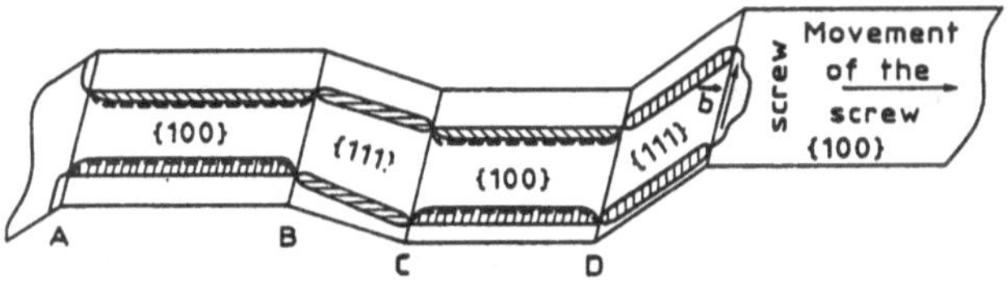

Fig 6 Formation mechanism of long Lomer-Cottrell segments on {100} type planes in temperature regime 2.

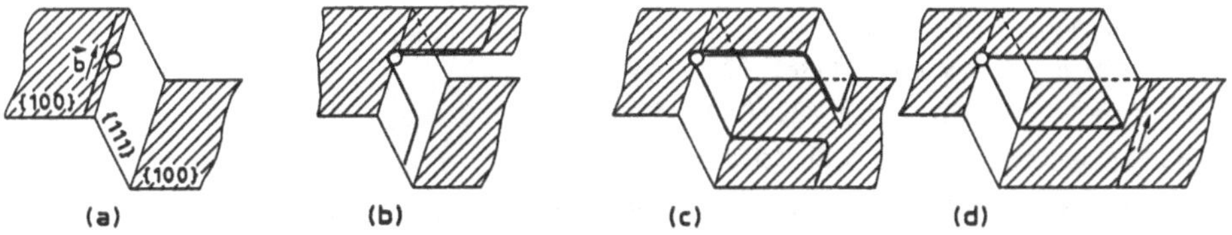

Fig 7 Schematic representation of the formation of prismatic loops similar to the one on Fig.3. a) to d) various cross slip events forming the prismatic loop (see text).

The two segments combine into a screw dislocation, leaving the prismatic loop behind. The loop segments in {100} split into the Lomer-Cottrell dissociation so that they lie along < 110 > directions.

As already mentioned, the mobility of the edges on {100} is probably low since a high mobility would correspond to more abundant {100} slip traces which are not clearly observed. The friction on the edges is due to the splitting geometry. One can imagine how this friction can decrease with the help of thermal activation by three possible mechanisms which are represented on Fig.8. From a) to c), the corresponding activation energies decrease.

Isolated Lomer-Cottrell dislocations with some similar features, have been reported in Copper samples after room temperature deformation and an alternate mechanism of formation was proposed [7][8]: this particular mechanism cannot account for the present observations since it predicts only one glide plane for the {111} segments.

Work is in progress to better characterize the edge segments on {100}: splitting geometry through weak beam experiments, mobility through in situ experiments.

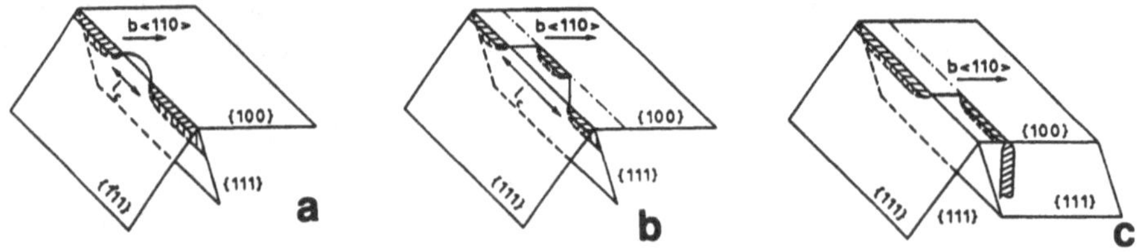

Fig 8 Possible mechanisms for edge dislocation motion in {100} planes.a) recombination over a critical length l_c. b) propagation of kink pairs along edges. c) single kink mechanism. From a) to c) the corresponding activation energy decreases.

ACKNOWLEDGEMENTS

The authors wish to thank J. Bonneville and J. Douin for fruitful discussions. The financial help of Fonds National Suisse is greatfully acknowledged.

REFERENCES

1 P. Anongba, J. Bonneville and J.L. Martin (1988), in "Strength of Metals and Alloys", vol.1, Pergamon Press, New York, p. 265
2 P. Anongba, PhD Thesis, EPF Lausanne, December 1989
3 P.B. Hirsch, A. Howie, R.B. Nicholson, D.W. Pashley and M.J. Whelan, Electron Microscopy of Thin Crystals, Butterworths, London (1967), p. 178
4 S.J. Basinski and Z.S. Basinski, in: Dislocations in Solids, Vol. 4, F.R.N. Nabarro (Ed.), North-Holland, Amsterdam 1979, p. 261.
5 U. Essmann, Phys.stat.sol. 12 (1965) 723
6 J.W. Steeds, Proc. Roy. Soc. A 292 (1966) 343
7 H.P. Karnthaler, Phil. Mag. A, 38 (1978) 141
8 A. Korner and H.P. Karnthaler, Phil. Mag. A 44 (1981) 275
9 M. Carrard and J.L. Martin, Phil. Mag. A 56 (1987) 391
10 M. Carrard and J.L. Martin, Phil. Mag. A 58 (1988) 491
11 A. Couret and D. Caillard, Acta Met 33 (1985) 1447

DUCTILITY IMPROVEMENT IN COPPER BASE ALLOYS AT ELEVATED TEMPERATURES

M. Kanno

Department of Materials Science, Faculty of Engineering,
The University of Tokyo, 7-3-1 Hongoh, Bunkyo-ku, Tokyo 113, Japan

Abstract

Phosphor bronzes with high tin content are difficult to hot-work. In addition, several copper base alloys show a ductility trough at intermediate temperatures. This paper describes improvement in hot ductility in some copper base alloys caused by a trace amount of additional elements. The ductility improvement observed at temperatures above 873K in a Cu-8mass%Sn alloy is attributable to an increase in grain boundary strength probably brought about by the inhibition of intergranular oxidation. On the other hand, the improvement appearing at intermediate temperatures in several alloys is ascribed to the reduction in the the amount of segregation of sulfur impurity at grain boundaries.

Introduction

In general, metals and alloys show high deformability at elevated temperatures and that is why many engineering materials are produced through hot working. However, there are some materials which show poor hot workability. For example, it is known that hot working is very difficult in phosphor bronzes with high tin content but the reason for the poor hot workability has not been made clear. Therefore, phosphor bronze materials are usually made through cold working and intermediate annealings. This fact is one of the cause of high production costs of these materials.

On the other hand, some copper base alloys show a ductility trough at intermediate temperatures and this is called an intermediate-temperature embrittlement(1,2). There is a great possibility that this embrittlement in these alloys is related to edge crackings appearing in hot-worked plates when the working temperature is lowered. Nevertheless, few attempts have been made to improve the ductility at intermediate temperatures, because the trough is considered to be inherent in copper base alloys.

This paper is a summary of some studies which have been made by the present author's group to improve hot ductility of copper base alloys(3,4,5,6). It will be shown here that a Cu-Sn alloy becomes hot-workable by adding trace elements. Trace additions are also successful to improve ductility at intermediate temperatures for some copper-base alloys precipitation-hardened as well as solution-hardened.

Experimental

As to hot-shortness of phosphor bronzes, different deoxidizing elements such as B,

Mg and P were singly added to a Cu-8mass%Sn binary alloy for about 0.1mol%. Since these alloys were made by vacuum melting and casting, an analytical value of the Mg content became smaller than the above nominal value because of vaporization of Mg.

In relation to the intermediate-temperature embrittlement, the following binary and ternary alloys(with additions) were used: a Cu-8mass%Sn alloy added B, Y, Ce, La or Ca of about 0.1mol%; a Cu-30mass%Zn alloy added Y of about 0.1mass%; a Cu-0.7mass%Cr alloy added Zr of about 0.06mass%.

All the alloy ingots were homogenized, hot- or cold-worked, and finally cold-swaged to rods having diameter of about 8mm. Each rod was heat-treated at 973-1073K to control grain size, and in the case rods of age-hardenable Cu-Cr alloys, they were solution-treated (grain size control was made simultaneously) at 1243K, quenched and aged at 748K to reach maximum hardness. Then, these rods were machined to round tensile specimens with a gauge length of 10 mm and a diameter of 4 mm.

Tensile test was carried out in vacuum or in argon at temperatures ranging from 293 to 1073K by use of Instron type machine. The initial strain rate was varied over a range of 3.3×10^{-3} to $2.0\times10^{2}s^{-1}$, where the rate of $3.3\times10^{-3}s^{-1}$ was mainly adopted. For especially high strain rates of over 10 s^{-1}, a thermorester was used. The ductility was assessed by measuring reduction in area(RA) at fracture. Metallographic observations were made by means of optical and transmission electron microscopy on specimens quenched immediately after testing. SEM observations and Auger electron microscopy(AES) were performed on fracture surface. In addition, dispersed particles were analyzed by means of energy dispersive X-ray spectroscopy(EDS).

Results and discussion

In connection with hot-shortness of phosphor bronzes, hot ductility of Cu-8mass%Sn alloys with different deoxidizing elements was first examined, since it has been known that phosphor exerts deleterious effect on hot ductility of some ferrous materials (7). The ductility of the binary and P-bearing alloys was as high as about 70% at room temperature, but it becomes markedly low at elevated temperatures as shown in Fig.1. A commercially available phosphor bronze also showed almost the same low ductility as that of the P-bearing alloy in Fig. 1. Therefore, it can be said that P addition causes no specially deleterious effect on hot ductility of a Cu-Sn alloy. On the other hand, different behavior of hot ductility is observed in the case of B and Mg additions: though both alloys show an intermediate-temperature embrittlement being similar to that of the binary alloy, ductility at high temperatures is increased by the additions.

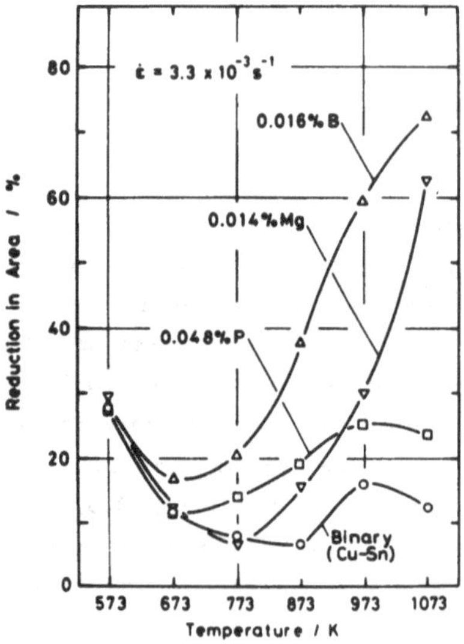

Fig.1 Hot ductility of Cu-8mass %Sn-X alloys.

The effect is more marked in the case of B addition and the value as high as about 70% is obtained at 1073K by the addition. The B-bearing alloy also showed high ductility of over 60% at this temperature even when strain rate was increased up to $2.0 \times 10^{2} s^{-1}$.

Fig. 2 shows optical micrographs of specimens after fracture. Both the binary and P-bearing alloy exhibit intergranular fracture at intermediate and higher temperatures, while B-bearing and Mg-bearing alloys, transgranular fracture at 1073K, being accompanied by pronounced dynamic recrystallization. The ductility increase in the latter alloys at 1083K might be attributable to dynamic recrystallization, since it is reported that dynamic recrystallization brings about good ductility at high temperatures in a Cu-Ni alloy(9). However, static recrystallization behavior was shown to be the same between the former and the latter alloys, and moreover, dynamic recrystallization is also observed in the binary and P-bearing alloys to a lesser extent. Therefore, marked dynamic recrystallization which occurred in the latter alloys is not a cause for the high ductility in the latter alloys but a result which was brought about by suppression of intergranular fracture in these alloys.

(a)
(b)
(c)

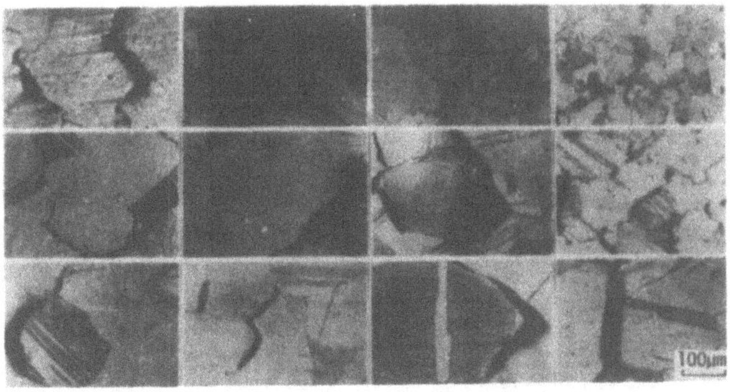

Binary 0.048%P 0.014%Mg 0.016%B

Fig.2 Micrographs in cross section of specimens fractured at the temperature:(a) 1073K, (b) 873K, (c) 673K.

Optical micrographs of cross sections of specimens with different amount of strain showed that intergranular fracture started from the specimen surface. This implies that grain boundary strength is lowered by the reaction between the specimen and heating atmospheres. On the basis of this idea, an oxidation experiment was carried out. Four kinds of specimens were annealed at 1073K in air and cross sections were observed by optical microscope and SEM. Oxide films formed on specimens were much thinner in B- and Mg-bearing alloys than in the binary and P-bearing alloys. In addition to this, structures under oxide films were different in these alloys as shown in Fig. 3. A lot of particles are present in the binary and phosphor-bearing alloys near specimen surface and they were revealed to be tin oxide particles which were formed by internal oxidation. No particles of tin oxide, however, were found in boron- and magnesium-bearing alloy. This means that internal as well as external oxidation of a tin bronze is extremely suppressed by B and Mg addition. Consequently, it can be concluded that internal oxidation through grain boundary is almost completely suppressed in B- and Mg-bearing alloys leading to the increase in grain boundary strength.

Fig. 4 shows the variations of hot ductility of a tin bronze caused by additional elements which have strong affinity to oxygen. It is shown that Y and Ce(La) improves ductility at high temperatures above 873K. The cause of this improvement in hot ductility is considered to be the same as that in the case of B and Mg addition. The addition of Ga results in no improvement at high temperatures and this may be due to

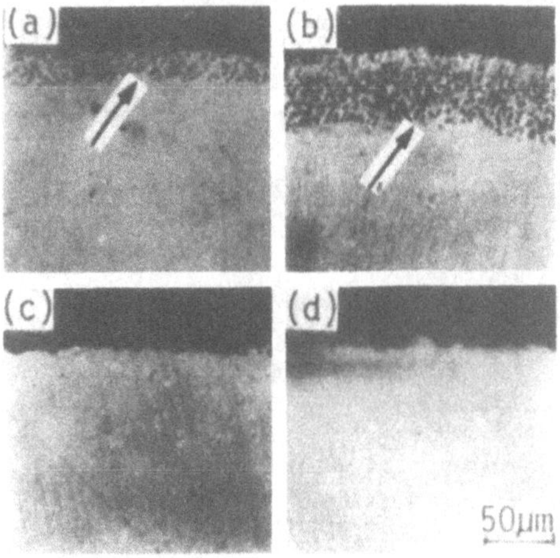

Fig.3 micrographs in cross section of specimens after removing oxide film. Annealed at 1073K for 36ks in air. (a) the binary, (b) P-bearing alloy, (c) B-bearing, (d) Mg-bearing alloy..

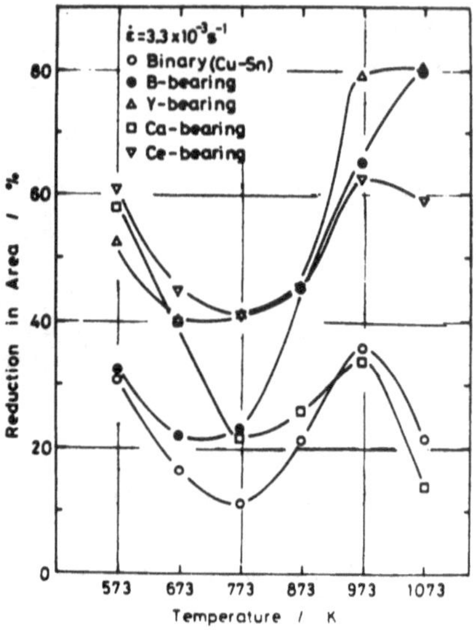

Fig.4 Hot ductility of Cu-8mass%Sn-X alloys.

the fact that Ca and Sn form compounds with low melting points. On the other hand, another kind of improvement in hot ductility is observed at intermediate temperatures. It is shown in Fig.5 that the intermediate temperature embrittlement in the binary and B-bearing alloys were caused by intergranular fracture, while Y, Ce and Ca changed fracture mode from intergranular to transgranular.

In relation to intergranular fracture of alloys, the following factors have been considered to be significant: (a) grain boundary sliding(10), (b) stress concentration at grain boundaries caused by coarse slip bands(11), (c) preferential deformation in the interior of PFZ(precipitate free zone)(12), (d) fracture of grain boundary precipitates or decohesion of interface of grain boundary precipitates and matrix(13), (e) strength decrease of grain boundaries on account of grain boundary segregation of trace impurities(14). All these factors must be taken into consideration to make clear the cause of the ductility improvement at intermediate temperatures. Needless to say,

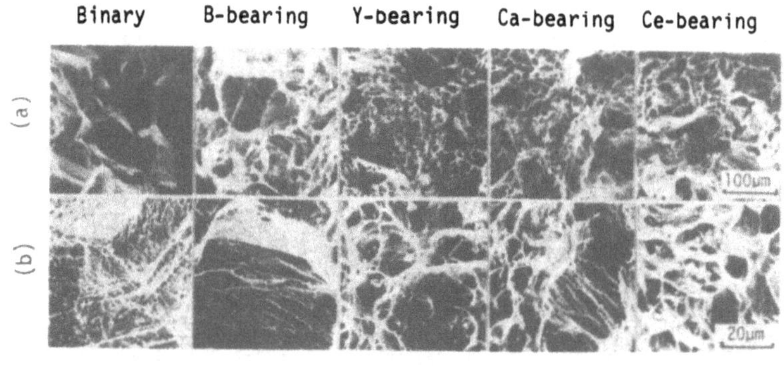

Fig.5 SEM micrographs of fracture surface of specimen tested at 673K (refer to Fig.4). (a) low magnification, (b) high magnigication.

however, factors(c) and (d) which concern with the intergranular fracture of age-hardened alloys have no relation to the mid-temperature embrittlement of solid-solution hardened alloys.

Concerning the factor (a), surface observations were made by use of optical microscope and SEM, but no grain boundary sliding was observed when tested at least at intermediate temperatures. As to the factor (b), TEM revealed that planar distribution of dislocations and coarse slip bands were observed by OM and SEM in the binary and boron-bearing alloys. Thus, there must be a intense stress concentration at the grain boundaries in these alloys. In the Y-, Ca- and Ce-bearing alloys, however, nearly the same kind of coarse slip bands were observed. It means that only the former alloys exhibit intergranular fracture under the same degree of stress concentration at the grain boundaries. This suggests that the factor (e) is correlated with the intergranular fracture of the former alloys.

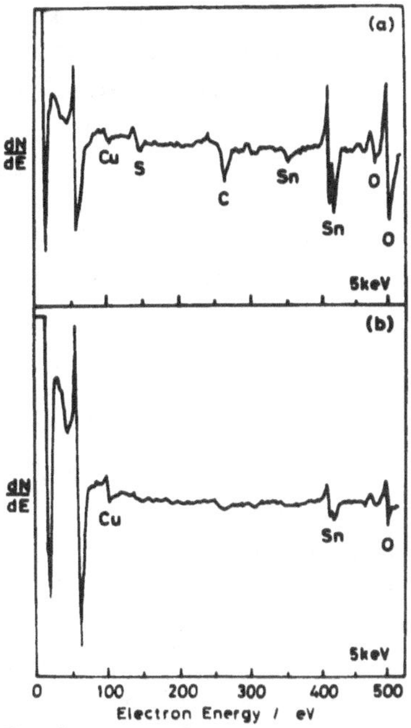

Fig.6 Auger electron spectra taken from interfeanular fracture surface of the Cu-8mass%Sn binary alloy: (a) before Ar-spattering, (b) after Ar-spattering.

In order to make clear what kind of impurities are segregated to the grain boundaries, Auger electron spectroscopy was made on intergranular fracture surface of the binary alloy(Fig.6). Although the content of each impurity in the binary alloy is below 10 molppm, sulfur is detected from the fracture surface. Here, the presence of carbon and oxygen is due to contamination of the fracture surface. Thus, it is obvious that sulfur is significantly segregated to the grain boundaries in the binary alloy and it becomes extremely important to know whether sulfur segregation is reduced, for example, by the Y addition. It was impossible, however, to evaluate the amount of sulfur segregation on grain boundaries in the Y-bearing alloy, since the alloy almost shows transgranular fracture.

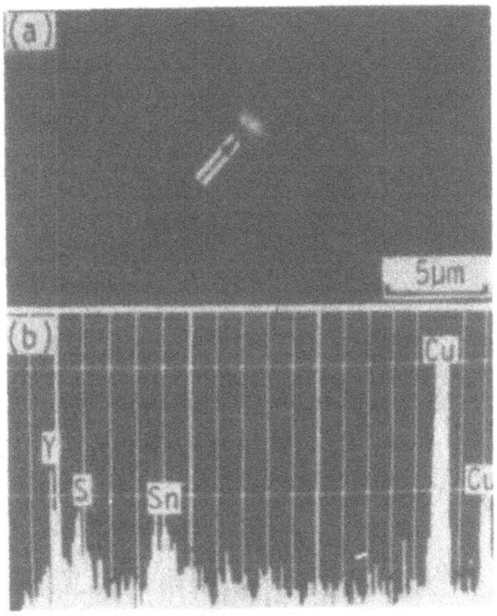

Fig.7 SEM micrograph of dispersed particles appearing in a Cu-Sn-Y alloy(a) and the result of EDS analysis on the particle(b).

If yttrium addition results in reduction of sulfur segregation, it may be brought about by sulfide formation. There were particles in the Y-bearing alloy which were not present in the binary alloy (it was possible to observe these particles by use of SEM only after removing abrasive particles from specimen surface by electropolishing). EDS analysis reveals that the particle is a yttrium sulfide

(Fig.7). It means that the concentration of sulfur at the grain boundaries must be lowered to a great extent by Y addition compared to the binary alloy on account of sulfide formation. Thus, it is concluded that the improvement in ductility at mid-temperatures is ascribed to the reduction in the amount of segregation of impurity sulfur at grain boundaries.

In addition, the present authors have shown that trace additions are also successful to improve ductility at mid-temperatures in other copper base alloys: an yttrium addition in a Cu-30mass%Zn alloy and a zirconium addition in a Cu-0.7mass%Cr alloy (5,6). The cause for the improvement was shown to be the same as above. Finally, it must be pointed out that the effect of trace impurity of sulfur on the other properties of copper base alloys must be examined closely from now on.

Summary

The ductility of a Cu-8mass%Sn at temperatures above 873K is revealed to be increased by a single-addition of a small amount of boron and magnesium. The effect is more pronounced in the case of boron addition and the alloy showed good ductility as high as over 60% which is high enough for the alloy to show good hot-workability. The similar increase in hot-ductility is attained by a single addition of yttrium and cerium(lanthanum). These phenomena are attributable to the inhibition of preferential oxidation along grain boundaries. The small addition of yttrium, cerium(lanthanum) and calcium to the tin bronze also improved the ductility at intermediate-temperatures and the same kind of ductility improvement is observed both in Cu-30mass%Zn-(0.004-0.010)mass%Y and in Cu-0.7mass%Cr-0.05mass%Zr alloys. This improvement at mid-temperatures is ascribed to the increase in grain boundary strength caused by the decrease in the amount of sulfur segregation at grain boundaries on account of sulfide formation.

References

1 C.C. Koch and A.R. Troiano, Trans. ASM, 57 (1964) 519-26.
2 M.J. Saarivirta and P.P. Tabenblat, Trans. Met. Soc. AIME, 218(1960) 935-9 3 M.
3 Kanno and N. Shimodaira, T. Ohsako and H. Suzuki, Trans. Japan Inst. Met., 28 (1987) 73-80.
4 M. Kanno and N. Shimodaira, Trans. JIM, 28 (1987) 742-8.
5 M. Kanno and N. Shimodaira, Scripta Met., 21 (1987) 1487-92.
6 M. Kanno, Z. Metallkunde, 79 (1988) 684-8.
7 C.L. Briant and S.K. Banerji, Metall. Trans., 10A (1979) 123-4
8 M.K. Tseng, D.Y. Lee and H.L. Marcus, Mat. Sci. Eng., 60 (1983) 73-77.
9 R.W. Evans and F.L. Jones, Met. Technol., 5 (1978) 1-6.
10 R.D. Gifkins, Act. Metall., 4 (1956) 98-9.
11 J.D. Evensen, N. Ryum and J.D. Embury, Mat. Sci. Eng., 18 (1975) 221-9.
12 E. Hornbogen and H. Kreye, J. Mat. Sci., 17 (1982) 9979-88.
13 A.K. Vasudeven, E.A. Ludwiczak, S.F.Baumann, P.R. Howell, R.D. Doherty and M.M. Kersker, Mat. Sci. and Technol., 2 (1986) 1205-9.
14 A. Joshi and D.F. Stein, J. Inst. Met., 99 (1971) 178-181.

RETARDATION OF INTERGRANULAR CREEP FRACTURE BY APPLICATION OF HIGH HYDROSTATIC PRESSURE

V. Sklenička[+], V.I. Betekhtin[++], J. Čadek[+], A.G. Kadomtsev[++], J. Svoboda[+], K. Kuchařová[+], A. Dlouhý[+], A.I. Petrov[++]

[+]Czechoslovak Academy of Sciences, Institute of Physical Metallurgy, Žižkova 22, 616 62 Brno, Czechoslovakia

[++]Academy of Sciences of the USSR, A.F. Ioffe Physico-Technical Institute, Politekhnicheskaya str. 26, Leningrad 194 021, USSR

ABSTRACT

Intergranular creep damage development in copper is examined with particular reference to the potential recovery effect of a high hydrostatic pressure on creep cavitation. It is shown that a short term application of the hydrostatic pressure of 1 GPa at room temperature to copper specimens crept at various stages of creep can substantially reduce the amount of creep cavitation due to "mechanical" closuring of the heavily cavitated areas at grain boundaries. This effect can result in a significant increase in creep life.

INTRODUCTION

Materials used for high temperature components are subject to long term deterioration of integrity due to the accumulation of creep cavities at grain boundaries. Rejuvenation treatment procedures to remove creep damage and hence effectively regenerate deteriorated material would therefore be of value for the extension of the overall life of components.

In recent years, the possibility of recovering service induced creep cavitation by reheat treatments (diffusional sintering of cavities) /1/ or by hot isostatic pressing (superimposed hydrostatic pressure is necessary to achieve sintering in acceptably short period) /2/ has gradually turned into a reality /3/. In contrast, there has to date been little consideration of cavity removing by the application of a high hydrostatic pressure at room temperature.

The aim of this paper was to check the possibility of applying high hydrostatic pressure to retard intergranular creep cavitation in copper

EXPERIMENTAL MATERIAL AND PROCEDURE

The material used for investigation was copper of 99.99% purity. Flat tensile specimens with a gauge length of 50 mm and a cross-section of 9.0x3.2 mm^2 were annealed in a sealed evacuated silica tube for 10 h at 1073 K. The average linear intercept grain size obtained from this heat treatment was 0.15 mm.

Constant-stress uniaxial creep tests were conducted in purified dried hydrogen at a temperature 773 K and a stress 12.5 MPa up to various predetermined fractions t/t_f of the fracture life t_f. Typically, creep tests were interrupted at $0.2t_f$, $0.5t_f$ and $0.7t_f$, specimens were cooled to room temperature and subsequently subjected to a hydrostatic pressure of 1 GPa for 600 s in a PTI-type piston cylinder apparatus /4/ in silicon oil. After this procedure some of pre-crept specimens were pulled to creep fracture.

Standard transmission electron microscopy (TEM) and scanning electron microscopy (SEM) were employed in assessing the effect of a hydrostatic pressure on both the interrupted and fractured creep specimens. In addition, to examine creep cavitation development in the unfractured specimens the special fractographic technique /5/ was used to create premature brittle intergranular failures (the "artificial" fracture surfaces) in these specimens.

EXPERIMENTAL RESULTS AND DISCUSSION

Creep Data

Examples of creep curves for specimens pulled to creep fracture without pressuring are shown in Fig. 1a. It can be seen that primary creep stage is replaced by the stage of steady state in a relatively short time. The steady state region is not well-established (frequently it reduces to a point). The tertiary stage is the dominant stage with respec' to both the fraction of time to fracture and the fraction of the total creep strain. No substantial changes in the shape of creep curves and the resulting creep lifes were found for creep specimens pressured prior to creep expeosures (Fig. 1b).

At first sight, quite different situation follows from Fig. 1c which shows creep curves of the interrupted specimens (at time $0.2t_f$, $0.5t_f$ and $0.7t_f$, respectively) and subjected to a hydrostatic pressure prior to recommencing the creep test. Comparing Fig. 1a with Fig. 1c, it appears that the resulting times to fracture of specimens pressured at

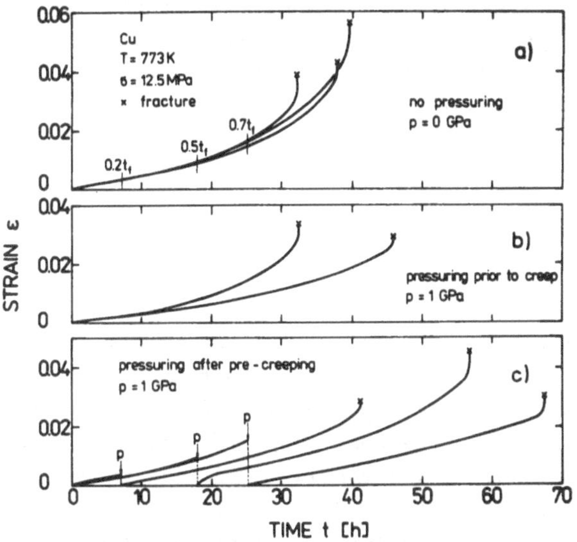

Fig. 1a,b,c Strain/time plots showing the effect of pressuring on the creep life

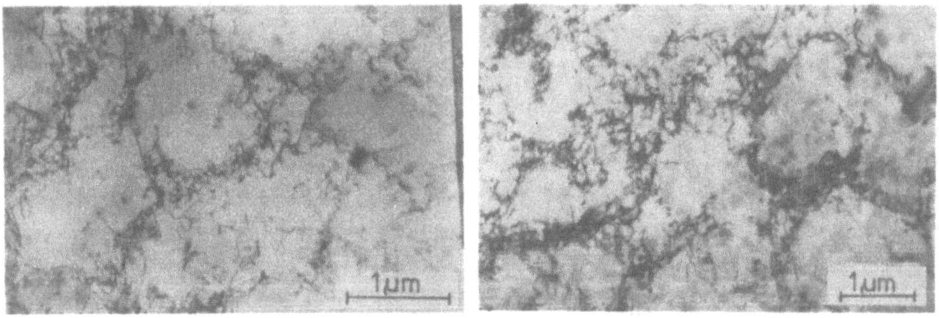

Fig. 2a,b TEM micrographs of dislocation structures in specimen crept to normalized time $0.5t_f$

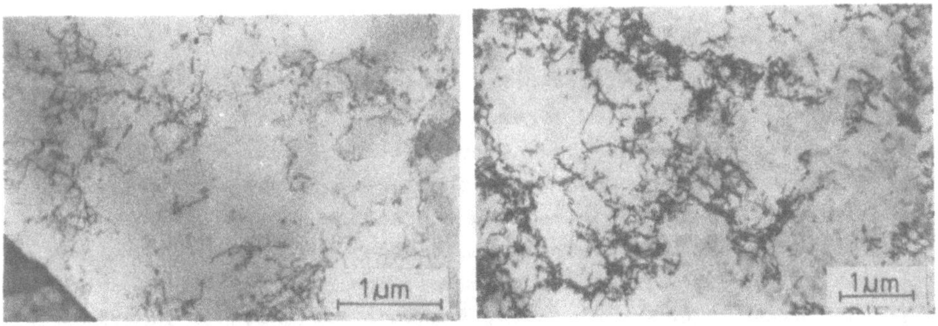

Fig. 3a,b TEM micrographs of dislocation structures in specimen crept to normalized time $0.5t_f$ and then pressured

$0.5t_f$ and $0.7t_f$ considerably increase. On the other hand, there appears to be little beneficial effect of a hydrostatic pressure on creep life in case of specimen pressured after a previous creep exposure up to $0.2t_f$. There is, however, some ambiguity in evaluating this effect as the times to fracture of specimens without pressuring are little scattered (Fig. 1a) about an average value of t_f.

TEM and SEM Results

TEM was used to examine the substructural activities associated with both creep and pressuring. Independently of the specimen history two dislocation configurations were seen: (i) sets of loosely knitted tangles of dislocations (Fig. 2a),(ii) occasionaly observed cellular substructure (Fig. 2b). Thus, considering the spatial variation in dislocation substructure of individual specimens it does not seem likely that the application of a hydrostatic pressure can be associated with the sub stantial difference in the resulting substructure. To check on this possibility in more detail, the substructure of a specimen which has been crept up to $0.5t_f$ was investigated (Fig. 2). Then a piece of this specimen was cut from the gauge length and subsequently pressured. Micrographs illustrating the substructure of this piece are shown in Fig. 3. It is immediately apparent that the resulting dislocation configurations in Fig. 3 closely resemble the creep substructure (Fig. 2).

SEM was used to characterize the features of cavitation structure. Under the creep conditions of temperature and stress given, the SEM analysis clearly revealed that the pure intergranular creep fracture occurs via extensive cavitation. Fig. 4a shows an SEM image of the fracture surface on the specimen pulled to creep fracture without pressuring. The SEM micrograph of the creep fractured specimen which was exposed to a hydrostatic pressure at $0.7t_f$ is shown in Fig. 4b. Both fracture surfaces consisted of heavily cavitated grain boundary areas, separated by ridges that appeared to be ruptured ligaments. No significant difference between the details of cavitation is evident when Fig. 4a and Fig. 4b are compared. This result suggests that the level of critical cavitation da mage corresponding to creep fracture state does not reflect the effect of superimposed hydrostatic pressure to creep loading.

However, the fact that approximately the same value of the critical cavitation damage characterizes the fracture state cannot immediately result in the same creep life. Clearly, the time to fracture reflects the rates at which the damage processes proceed. Then the time to fracture is determined by the incubation time for cavity nucleation and the

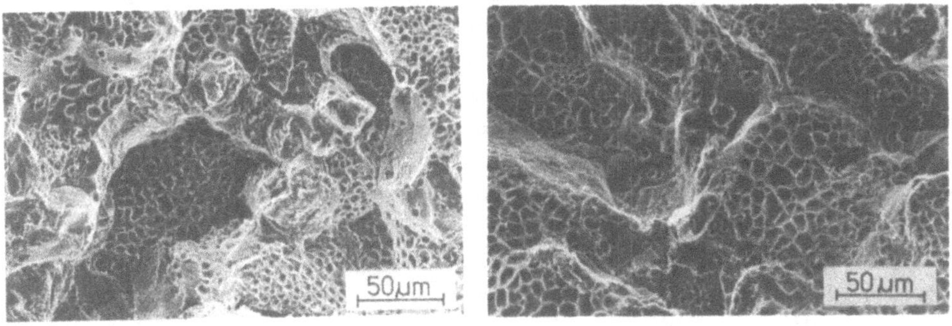

Fig. 4a,b SEM micrographs of creep fracture surface: (a) no pressuring, (b) pressuring at normalized time $0.7t_f$

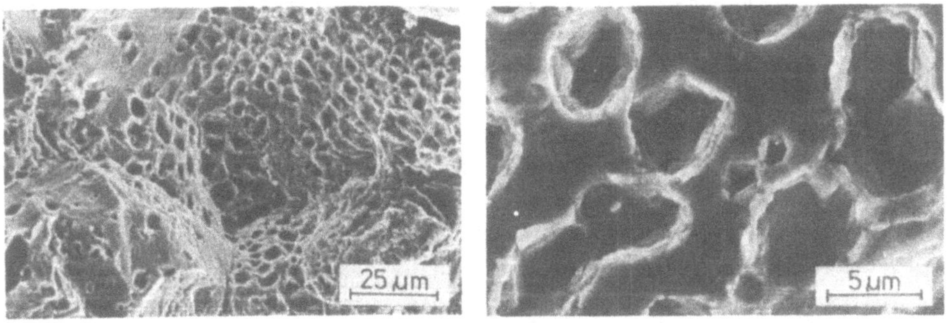

Fig. 5a,b SEM micrographs of intergranular creep cavitation. Specimen crept to normalized time $0.5t_f$

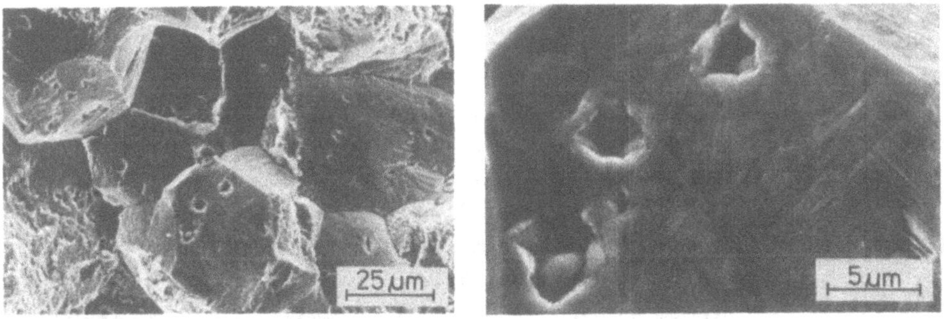

Fig. 6a,b SEM micrographs of intergranular creep cavitation. Specimen crept to normalized time $0.5t_f$ and then pressured time necessary for the cavity nuclei to grow and reach the critical level of damage /6/. Consequently, by analysing cavity development one can get an idea of the resulting effect.

SEM analysis of the cavitation morphology in the course of creep has been made using the special fractographic technique /5/. This analysis revealed that for creep exposure time $t \approx 1/3t_f$ single (isolated) and bicavities prevail, later on the chains of cavities are formed; at $t \approx 2/3t_f$ flat clusters of cavities and areal microcracks can be observed. The same procedure was applied to the interrupted creep specimens at predetermined fractions t/t_f and subsequently subjected to a hydrostatic pressure. SEM analysis of the structure of creep cavitation provided clear evidence for "mechanical" closuring of creep cavitation due to the application of a hydrostatic pressure. However, the effect of the reduction in cavitation depends critically on the amount and morphology of intergranular damage. Comparing the level of cavitation on an artificially created "as-fractured" surface of the specimen crept to $0.5t_f$ (Fig. 5) with the significantly lower level observed in the same specimen subsequently exposed to a hydrostatic pressure (Fig. 6) suggests that pressuring may primarily enhance closuring of the extensive damaged regions at grain boundaries (i.e. areas of advanced coalescence of cavities). In a separate study /7/ we have observed that a hydrostatic pressure cannot significantly retard or remove small and isolated cavities.

Therefore, the results of this study lead us to conclude that a significant increase in creep life can be reached only if a hydrostatic pressure is applied after rather advanced intergranular creep cavitation had developed. This conclusion is consistent with the results of the creep tests (Fig. 1).

REFERENCES

1 H.E. Evans, In: Mechanisms of Creep Fracture, Elsevier Appl. Sci. Publishers Ltd., London, 1984, pp. 257-277

2 W.J. McCall, Combustion 42 (1971) 27-33

3 A.K. Koul and R. Castillo, Metall. Trans. 19A (1988) 2049-2066

4 A.I. Petrov, A.M. Kartashov, V.N. Azhimuratov and V.I. Betekhtin, Physical-Chemical Mechanics of Materials 3 (1984) 109-114

5 J. Svoboda and V. Sklenička, Metallic Materials 24 (1986) 489-503

6 J. Čadek, In: Creep in Metallic Materials, Elsevier Appl. Sci. Publ. Ltd., Amsterdam, 1988, pp. 330-337

7 V. Sklenička, K. Kuchařová, J. Čadek, V.I. Betekhtin, A.G. Kadomtsev, A.I. Petrov, Metallic Materials, 27 (1989) 504-518

6. FATIGUE

HETEROGENEOUS PLASTIC DEFORMATION IN FATIGUED PURE FCC METALS

C. Holste, K. Mecke, W. Kleinert

Sektion Physik, Pädagogische Hochschule Dresden

ABSTRACT

The heterogeneity of plastic deformation at early fatigue stages of cyclically strained nickel is considered on different structure levels. The development of the typical dislocation arrangement during cycling is reflected in a structure map and is shown to become more transparent, when the statistics of deformation processes are taken into account. Finally, the degree of heterogeneity and the existence and order of magnitude of long-range stresses are discussed.

INTRODUCTION

Phenomena of inhomogeneous plastic deformation in cyclically strained crystalline materials are of great practical interest because in most cases they are directly correlated to crack initiation and propagation processes. Today the idea is widely accepted that the spatially inhomogeneous dislocation distribution arising during fatigue loading should involve heterogeneous plastic deformation on different structure levels in single-phase metals, too (1,2,3). The study of basic features of inhomogeneous slip in cyclically deformed fcc metals may be a helpful key to understand the defect structure evolution and crack initiation in more complex metallic systems.

Although the heterogeneity of cyclic plasticity is generally recognized to be true and is considered in several composite models (2,4,5), the degree of heterogeneity and its experimental proof as well as the connection between spatially statistical deformation processes and structure development have to be examined in more detail. In the present paper a brief survey is given concerning investigations in this field on nickel polycrystals and monocrystals oriented for single slip. The amplitude range considered is restricted to the regimes A and B (4,6) of the cyclic stress strain curve (CSSC) of monocrystals. Since for nickel qualitatively the same processes of cyclic plasticity proceed as for copper (6), for example, the results discussed may be representative of fcc metals.

SCALE LEVELS

It is well-known that the dislocations in cyclically strained metals are frequently inhomogeneously distributed in the deformed specimens. A more detailed analysis of typical dislocation structures reveals that this inhomogeneous distribution exists at different scale levels. At a microscopic level, which is characterized by dimensions of the order of the mean distance of individual dislocations, the distance, segment length and dipole width of screw and edge dislocations vary randomly. At a higher scale level alternating volume regions of high and low local dislocation density can be distinguished (1,2). The distance between the dislocation-dense (DRs) or the width of the dislocation-poor regions (PRs) in glide direction are of the order of 1 μm in fcc metals. Dimensions of this order have been termed as mesoscopic dimensions by HOLSTE and BURMEISTER (5). According to the different orientations of the individual grains in a polycrystal the mesoscopic dislocation structure changes from grain to grain at a polycrystalline level. Finally, the dislocation arrangement may locally vary at a macroscopic level. A typical example is the so-called matrix-PSB composite in cyclically strained monocrystals oriented for single slip (4,6). Another example in this context is the observation that the dislocation arrangement in the surface layer of a cyclically deformed specimen often differs significantly from the microstructure in the bulk of the sample.

The marked inhomogeneous dislocation arrangement produces a local stress distribution at the different scale levels. The origin of local stress changes may be quite differently comprehended, depending on the scale level considered. At a microscopic scale level stress changes are caused by the stress field of individual dislocations which can be seen in a TEM micrograph. At higher scale levels a continuum mechanical consideration is more suitable. At these levels local flow stresses can be defined, abstracting from the behaviour of individual dislocations. The spectrum of local flow stresses is correlated to a frequency distribution of plastic strains and consequently to a spectrum of residual stresses at the different scale levels. In the loaded state of the fatigued specimen the local resolved shear stress τ_{loc} at any point is given by the sum of the macroscopic flow stress τ and the local internal stress τ_{INT} which may be split up additively into the residual macroscopic stress $\Delta\tau_{mac}$, the residual polycrystalline stress $\Delta\tau_{pc}$, the residual mesoscopic stress $\Delta\tau_{mez}$ and the contribution from individual dislocations $\Delta\tau_{mic}$.

DISLOCATION STRUCTURE MAP

Based on the work of MECKE et al. (4), MECKE and BLOCHWITZ (7) and

KLEINERT (8) in the case of nickel crystals oriented for single slip, it will be possible to characterize the development of the dislocation configuration during cycling in a structure map (cf. Fig. 1).

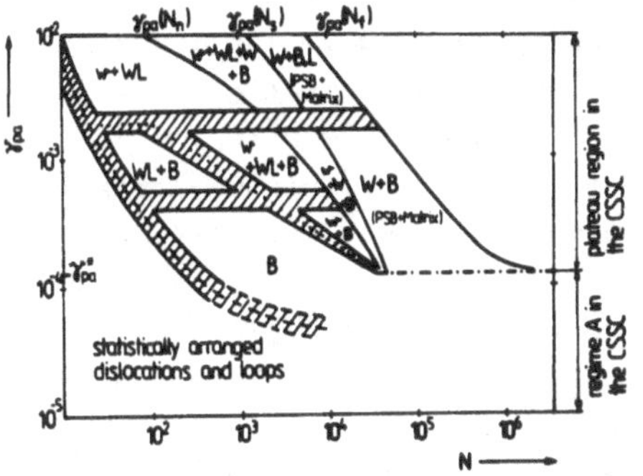

Fig. 1 Dislocation structure map for Ni single crystals cyclically strained at room temperature

At small amplitudes we find at larger cycle numbers N the well-known dipolar bundle or vein structure (B), for example cf. Fig. 7 in (7). For larger amplitudes and smaller cycle numbers a structure type is typical which will be termed "wall-like structure" (WL). The wall-like DRs in this configuration appear in the $(1\bar{2}1)$ section randomly arranged, they can be elongated over a longer distances but not in specially preferred directions and have rather irregular borderlines (Fig. 7a in (6)). A striking feature of the dislocation arrangement, particularly in the higher cycle region, is the wall structure (W). The walls of this structure type are often very regularly arranged in slip direction (cf. Figs. 8c, 6c, 6d in (6)). If the alignment of walls in glide direction extends over macroscopic distances the wall configuration defined here is identical with the PSB-structure (e.g. 1,2). The individual walls (w) can be observed as single structure elements in the dislocation arrangement (cf. Figs. 6b, 7c in (6)). Finally, a special type of bundle structure, the labyrinth structure (L) should be mentioned (cf. Fig. 9 in (6)).

In the structure map shown in Fig. 1 each fatigue stage characterized by typical dislocation arrangements is represented by a corresponding field bounded by scatter bands. Suitable γ_{pa}-N borderlines turned out to be the MANSON-COFFIN line $\gamma_{pa}(N_f)$, the plot of γ_{pa} versus the cycle number up to the saturation stage $\gamma_{pa}(N_S)$, the γ_{pa}-dependence of the cycle number N_n at which first PSBs appear, the plot of γ_{pa} versus the cycle number at which first individual walls are observed (dotted line) and the γ_{pa}-dependence of the cycle number at which the first regularly arranged bundles are found (dashed line).

The main characteristics and tendencies in the γ_{pa}-N space of the map can be summarized as follows. Firstly, the larger the strain amplitude the more pronounced will be the formation of wall-like and wall structures. Note that at high amplitudes individual walls appear already at early stages of fatigue. In the saturation region the volume fraction of the wall structure will be the larger the greater the strain amplitude is chosen, and tends to one for $\gamma_{pa} \approx 10^{-2}$. At small amplitudes only the bundle structure is observed. The occurrence of bundle structure at larger amplitudes at later fatigue stages can easily be understood if the fact is considered that with the nucleation of PSBs the strain amplitude in the matrix diminishes drastically. Secondly, the most striking point is the fact that the different structure types can generally develop side by side during cycling. This becomes especially evident at medium amplitudes ($\gamma_{pa} \approx 2 \cdot 10^{-3}$) where all three structure types coexist, for example cf. Fig. 6c in (6). Thirdly, the typical ladder-like wall structure forms step by step, starting with single walls and growing up to the typical wall arrangement.

DEFORMATION MECHANISMS AND STRUCTURE DEVELOPMENT

Considering the deformation mechanisms and the development of PSB-matrix composite in the high amplitude regime of the structure map (plateau range of CSSC, cf. Fig. 1) besides individual dislocation mechanisms statistical processes at a mesoscopic level have to be taken into account. During cycling the plastic strain in this case is concentrated more and more in the PRs. There may exist regions with fairly high plastic deformation not only in the wall-like but also in the bundle structure. Looking at the structure map it can be seen that large plastic strain is accompanied with a strong tendency to wall formation. Starting from microscopic dislocation mechanisms and from the statistics of mesoscopic deformation processes KLEINERT (8) has explained the observed connection between local plastic strain and wall formation. At larger local strain amplitudes ($\gamma_{pa} > \gamma_{pa}^{*}$) in a given PR the mean distance of moving dislocations becomes small enough to enable edge dislocation segments, due to interaction processes, to bow out overcritically in groups from a DR and, in connection with long-range screw dislocation movement, to form edge dislocation walls at the periphery of the opposite DR.

Once a local wall configuration has arisen the wall formation can spread along the slip direction supported by the stress field of the local wall configuration. Altogether, the structure development during cycling at larger amplitudes is characterized by two competing processes. On the one hand, the structure at the border of the DRs is "hardening" due to segment-length shortening and other processes, on the other hand the plastic strain in the PRs increases locally due to the enlargement of the DR-volume fraction. In each mesoscopic structure region characterized by its local athermal flow stress $\tilde{\tau}_{loc}$ there exists a critical inter-action distance and due to that a critical strain amplitude γ_{pa}^{cr} above which collective dislocation movement (edge-wall formation) becomes possible.

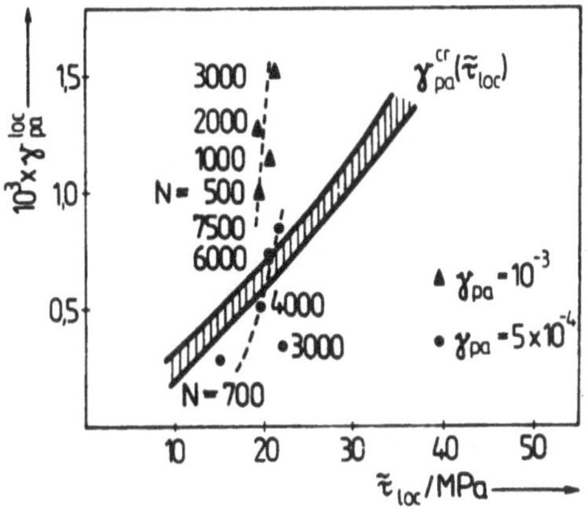

Fig. 2 Connection between critical strain γ_{pa}^{cr} and local stress τ_{loc}, illustrating the condition for wall-formation

Considering the local dislocation processes KLEINERT (8) has calculated a $\gamma_{pa}(\tilde{\tau}_{loc})$-scattering band which represents in a $\gamma_{pa}-\tilde{\tau}_{loc}$ space a border line for wall formation (cf. Fig. 2).

Information about the amount and frequency distribution of local flow stresses τ_{loc} and strain amplitudes γ_{pa}^{loc} at a mesoscopic level can be obtained by evaluating mechanical hysteresis loops which characterize different fatigue stages (3,6,8). Knowing the stress and strain distribution density functions $\varphi(\tau_{loc})$ and $h(\gamma_{pa}^{loc})$, determined from a hysteresis loop, it may be decided whether the wall formation at the fatigue stage considered is favoured or not. In Fig. 2 for different fatigue stages the points are marked which correspond to the value pairs: maximum local strain on the one hand and corresponding minimum athermal local stress on the other hand. It can be seen that at the larger strain amplitude wall nucleation is possible already for low cycle numbers N whereas at the lower amplitude wall formation can start not before N = 6000, in good agreement with TEM observations (Fig. 1).

In Figs. 3b and 3c the density functions $\varphi(\tau_{loc})$ and $h(\gamma_{pa}^{loc})$, respectively, are shown for the wall structure in a Ni monocrystal. The function $\varphi(\tau_{loc})$ evaluated from a loop branch ought to be well correlated to suitable distributions of mesoscopic structure parameters determined from TEM micrographs. Fig. 3d represents the frequency density distribution of the channel widths $s_{TEM}(d)$ found out by evaluating nearly 500 channels in a wall structure as well as the width distribution $s_{HS}(d)$ calculated from $\varphi(\tau_{loc})$. In the latter calcula-

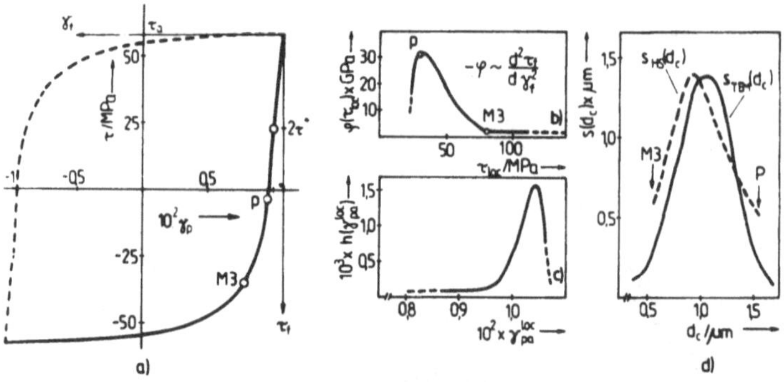

Fig. 3 Parameters and distribution density functions of a hysteresis loop reflecting the behaviour of wall structure in a nickel monocrystal

tion the athermal flow stress, $\tilde{\tau}_{loc} = \tau_{loc} - \tau^*_{loc}$, has been assumed to be given by the channel width d_c through $\tilde{\tau}_{loc} = \alpha Gb/d_c$, with α = 1.7 (from line tension consideration for bowing-out screw dislocations), the shear modulus G = 75 GPa and the absolute value of the BURGERs vector $b = 2.5 \cdot 10^{-10}$ m. According to the COTTRELL-STOKES law for the thermal or friction stress $\tau^*_{loc} = c\tau_{loc}$ was supposed. The constant c was set equal to the proportion of macroscopic loop values $\tau^*/\tau_a = 0.3$ (cf. Fig. 3a). Fig. 3d shows a fairly good agreement between the two distribution functions supporting the idea (3) that at P the first channels begin to deform plastically at a mesoscopic level and already at M3 all channels take part in the plastic deformation. At M3 in channels with the lowest flow stress the plastic strain equals to $\gamma_{pa} = 1 \cdot 10^{-3}$.

LONG-RANGE INTERNAL STRESSES AND DEGREE OF HETEROGENEITY

Whether long-range internal stresses actually arise and which order of magnitude is present at a given structure level has to be proved experimentally. There are several experimental results which point directly and indirectly to the existence of such stresses.

On the basis of a simple composite model, HOLSTE and BURMEISTER (5) have shown that the mean square residual stress changes in a characteristical way during one loading cycle. All stress-dependent physical quantities measured at different strains γ_p within the cycle should reflect this behaviour. Actually, this was found for coercive-force measurements (e.g. 9,14). The typical cyclic change of residual stresses occurs irrespectively of the scale level at which the stresses vary. To obtain in a special experiment information concerning the kind of stresses responsible for the observed cyclic variation a detailed analysis is necessary.

At a mesoscopic structure level most direct measurements of internal stresses are possible by evaluating of the curvature of dislocation segments in a dislocation distribution pinned by neutron irradiation in the stress-applied or unloaded state (10). The segment curvature is directly correlated to the acting athermal local stress $\tilde{\tau}_{loc}$. Investigations on copper (16) and nickel (11) monocrystals have shown that there exists a characteristical stress distribution in the PSB-wall structure. After unloading from the tension loop tip many parts of the PRs (channels) are under pressure and the walls are strained in tension. In nickel, on the average, in the channel centre for the residual mesoscopic stress $\Delta\tau^{c}_{mez}$ = -25 MPa was found and in the walls $\Delta\tau^{w}_{mez} \approx 100$ MPa could be estimated. These residual mesoscopic stresses result from the determined athermal local stresses if a mean value $\bar{\tau}^{*}_{loc}$ = 15 MPa is assumed for the thermal stress in the channels and walls. The mean value $\bar{\tau}^{*}_{loc}$ follows from the hysteresis loop evaluation.

A study of the shift of X-ray diffraction line profiles in dependence of the X-ray measuring direction makes it possible to determine all parts of internal stresses mentioned in the first section of this paper. In (12) a procedure is discussed which allows to split up the total line shift into portions originating from stresses at different scale levels. The experimental results hitherto obtained for nickel can be summarized for the loaded state as follows. In polycrystalline nickel cycled at the low amplitude $\gamma_{pa} = 5\cdot10^{-4}$ up to $N = N_S$ (τ_a = 80 MPa) for an averaged slip system $\Delta\tau_{mac} \approx -30$ MPa, $\Delta\tau_{pc} \approx 0$ and in the matrix structure $\Delta\tau^{PR}_{mez} \approx +15$ MPa have been found (13). The positive value of $\Delta\tau_{mez}$ in the PRs seems to be reasonable, if it is taken into account that in the matrix bundle structure the overwhelming part of macroscopic plastic deformation is realized in the bundles (8). For a larger amplitude $\gamma_{pa} = 3\cdot10^{-3}$ (cell structure, τ_a = 110 MPa) $\Delta\tau_{mac} \approx 0$, $\Delta\tau_{pc} \approx 0$ and $\Delta\tau^{PR}_{mez} = -20$ MPa have been measured (14). In Ni monocrystals only data are available for the low amplitude $\gamma_{pa} = 5\cdot10^{-4}$ (23). In the saturation region ($N=N_S$) $\Delta\tau_{mac}$ was found to be different from zero and negative. At $N = N_n$, $\Delta\tau^{PR}_{mez} = -10$ MPa, and at $N = N_S$, $\Delta\tau^{PR}_{mez} = +5$ MPa have been determined. The positive mesoscopic stress in the PRs at N_S reflects the deformation behaviour of the matrix bundle structure mentioned above.

The experimental results indicate considerable long-range stresses at a mesoscopic structure level. These stresses may be determined using composite models (1,2,5) but they should also result from the consideration of the dislocation arrangement.

In Fig. 4 the distribution of mesoscopic internal stresses in a wall structure of typical PSBs in Ni monocrystals is represented. The stress field was calculated on the basis of the specific dislocation configuration shown in Fig. 4a. Corresponding to the discussion in the foregoing section it is assumed that during the plastic deformation of the channels (c) edge dislocations arise at the periphery of the walls (w) which cause the stress field in the wall structure. Assuming micro-plastic deformation (local displacement of dipole edge dislocations) of the walls(with the heigth h) up to the maximum strain γ_{pa} in the wall structure the number n of interface dislocations at the wall border is given by $n = \gamma_{pa} \cdot h/b$. In computer-aided calculations twenty dipolar walls have been considered because for small wall numbers the stress field in a given channel was shown to depend considerably on the number of walls. The stress distribution along the x-axis represented in Fig. 4b results after averaging along the y-axis and is in fairly good agreement with the distribution found from TEM evaluations, if the thermal stress of 15 MPa is taken into account. It should be mentioned

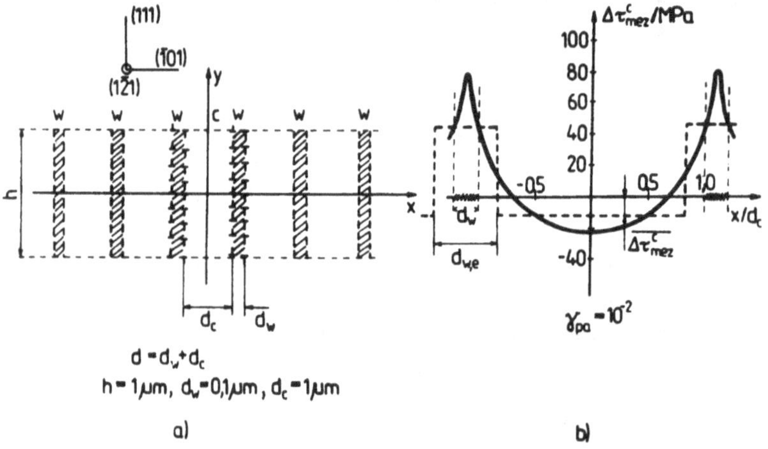

Fig. 4 Configuration of interface dislocations and stress field in the wall structure of PSBs (unloaded state)

that BURMEISTER and HECKER (16) have deduced an analytical expression for the stress field of a configuration quite similar to that in Fig. 4a. Considering the results represented in this paper it may be concluded that a microplastic deformation of walls through the entire cycle would be possible, contrary to the statement of MUGHRABI (10) who has shown for Cu crystals that the channels should yield plastically long before the maximum cycle strain is reached.

However, the numerical evaluation of TEM micrographs (10,11) as well as the quantitative stress estimation are burdened with considerable uncertainties. These concern among others the line tension approximation for the dislocation curvature evaluation (10,16), the geometrical parameters of the wall configuration and in our case the amount of thermal stress. The strong spatial dependence of the residual stress in a channel in x- and y-direction points out that composite models which start from accommodation factors Γ and from averaged stresses in TEM-observable structure elements can describe the local stress situation only roughly (16). Starting from an effective volume fraction of walls $f_e = d_{we}/d = 0.18$ (Fig. 4b) we obtain with $\overline{\Delta\tau^c_{mez}} = -12$ MPa in our case $\Gamma \approx 0.07$, using equation (15) in (2), for example. From the present experimental and theoretical results it can not clearly be decided, whether the walls are strained microplastically up to the PSB-deformation amplitude $\gamma_{pa} = 10^{-2}$ or begin to yield plastically (e.g. at M3 in Fig. 3) as soon as all channels take part in plastic deformation. Further experimental and theoretical work will have to be carried out in this field.

REFERENCES

1 H. Mughrabi, Rev. Phys. Appl. 23 (1987) 369

2 O.B. Pedersen, Proc. Int. Coll. "Basic mechanisms in fatigue of metals", Brno 1988, p 169-83

3 C. Holste, W. Kleinert and W. Fischer, Cryst. Res. Technol. 22 (1987) 419-27, 553-58

4 H. Mughrabi, Acta metall. 31(1983) 1367-79

5 C. Holste and H.-J. Burmeister, phys. stat. sol. (a) 57 (1980) 269-80

6 K. Mecke, C. Blochwitz and U. Kremling, Cryst. Res. Technol. 17 (1982), 1557-70

7 K. Mecke and C. Blochwitz, Cryst. Res. Techn. 17 (1982) 743-58

8 W.Kleinert, Thesis (B), Päd. Hochschule Dresden, 1988

9 H.-D.Schumann, Thesis (B), Päd. Hochschule Dresden, 1983

10 H. Mughrabi, phys. stat. sol. A104 (1987) 107

11 K. Kirsten, Diplomarbeit Päd. Hochschule Dresden, 1989

12 G.K. Schmidt, K. Morgenstern, E. Thiele and E. Wällnitz, Z. Metallkde 78 (1987) 851

13 C. Blochwitz and E. Thiele, to be published

14 J. Brechbühl, K. Morgenstern, H.-D. Schumann, E. Thiele, W. Tirschler, A. Tosch and E. Wällnitz, Proc. 10th Int.Coll. "Mechanical fatigue of metals", Dresden 1989, p.12-9

15 G.K. Schmidt and E. Thiele, Proc. Int.Coll. "Basic mechanisms in fatigue of metals", Brno 1988, p.51-5

16 H.-J. Burmeister and M. Hecker, Proc. 10th Int.Coll. "Mechanical fatigue of metals", Dresden 1989, p.20-5

DISLOCATION STRUCTURES IN NICKEL SINGLE CRYSTALS FATIGUED AT 78 K

K. Mecke, W. Kleinert and M. Hübner
Sektion Physik Pädagogische Hochschule Dresden, G.D.R.

ABSTRACT

The development of the dislocation arrangement in nickel single crystals during cyclic deformation at 78 K was investigated in dependence on the cycle number and on the plastic strain amplitude. During cyclic deformation different types of dislocation arrangements of mainly wall-like character develop up to the saturation stage over the entire specimen volume. The investigations of the surface suggest a non-homogeneous plastic deformation.

INTRODUCTION

In the last ten years the dislocation arrangements in f.c.c. metals fatigued at room temperature have been investigated intensively. The results obtained are the same for copper and nickel.

With regard to fatigue test at low temperature investigations on copper have so far been made by BASINSKI [1] and PEDERSEN [2] only. The dislocation structure on copper single crystals cycled to saturation at 78 K contains both vein-like structures and wall structures with the same volume fraction as at room temperature fatigue. The occupied volume of PSBs is a function of the plastic strain amplitude, but it is approximately independent of temperature. The ladder spacing decreases sharply and the saturation stress amplitude increases with decreasing test temperature.

In fact, there are differences concerning the mesoscopic dislocation arrangements in copper and nickel after fatiguing at 78 K. After fatigue to saturation at 78 K a wall-like structure was mainly observed, but after fatigue at the same plastic strain amplitude and at room temperature there was 4 % wall structure only.

EXPERIMENTAL METHODS

In the experiments concerning the dependence of the dislocation structure on the cycle number and the dependence of the dislocation structure on the plastic strain amplitudes, too, we used a special shape of our specimens (Fig. 1). The specimens were oriented for single slip so that the small side plane was a (121)-plane.

After fatiguing a desired number of cycles a thin slice of the rspecimen was cut for TEM investigations.

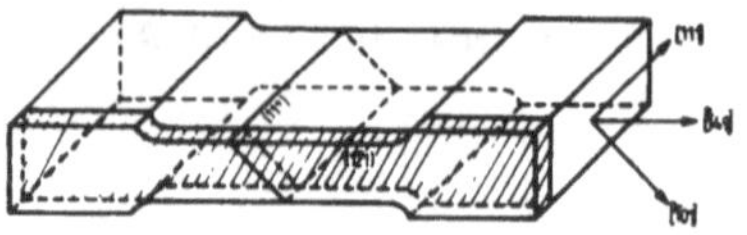

Fig. 1 Special shape and orientation of the specimens used in the present work

After cutting the slice and polishing the surfaces the fatigue process could be continued with the same specimen.

Two experiments were made: The first experiment was made for investigations of the dislocation structure in dependence on the cycle number. A nickel single crystal was fatigued in push pull mode at a constant plastic strain amplitude of 2.5×10^{-4}. TEM investigations were carried out after fatiguing a desired number of cycles. The second experiment was made for investigations of the dislocation structure in the stabilized states of mechanical properties in dependence on the plastic strain amplitude. A nickel single crystal was fatigued in succession at different plastic strain amplitudes. At the saturation stages the fatigue process was interrupted and the single crystal was cut for TEM investigations. Some other crystals were fatigued from the virginal state to the saturation state. A view of all experiments is given in Fig. 2.

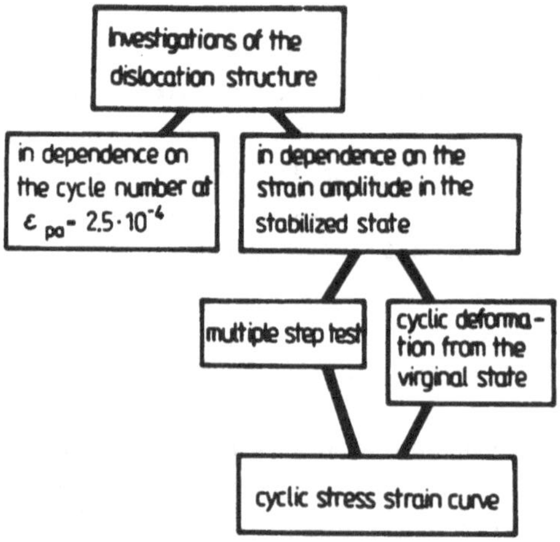

Fig. 2 A view of the experiments

The test temperature of 78 K was obtained by direct immersion of the specimen in liquid nitrogen.

EXPERIMENTAL RESULTS

The dislocation structure in dependence on the cycle number (for further details see [4]):

After the first 500 cycles the dislocation structure is very similar to the structure developed during fatigue at room temperature under otherwise similar test conditions (Fig. 3a). There are dislocation-poor and dislocation-dense regions which are irregular in shape and arrangement. Fig.3b represents the dislocation structure in the range of the stabilization of the mechanical properties. A wall-like structure has developed more and more in the largest part of the specimen volume. The structure of the small strips elongated in [101]-direction reminds of the matrix structure after room temperature fatigue. The mean distance of the dislocation-dense regions has decreased down to about 0.6 μm.

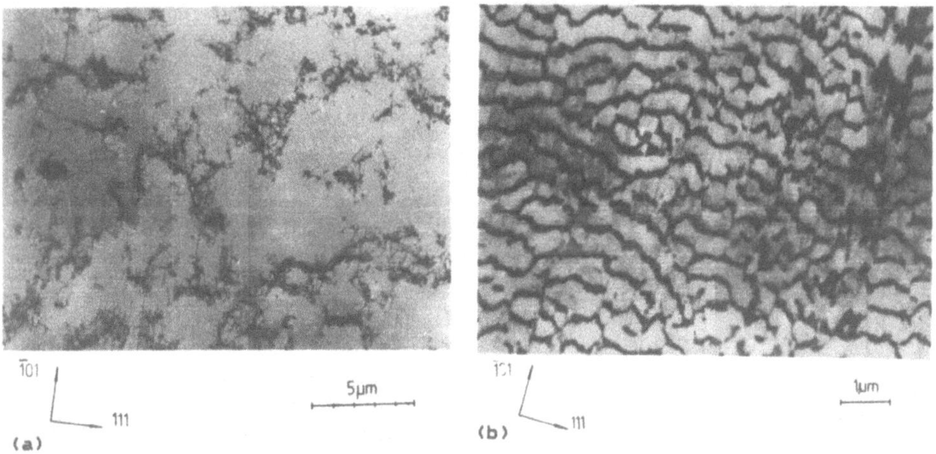

Fig. 3 Dislocation structure during cyclic deformation at 78 K (a) after 500 cycles and (b) at the saturation stage with the plastic strain amplitude 0.00025

The dislocation structure in dependence on the plastic strain amplitude:

Concerning the dislocation arrangement there are no qualitative differences between the results of multiple step test and single step test.

At the saturation stage after cyclic deformation at the smallest plastic strain amplitude of 6.5×10^{-5} there is a uniform dislocation arrangement without differences in the whole specimen consisting of poor and dense regions in an irregular arrangement (Fig. 4a).

After cyclic deformation at the plastic strain amplitude of 1×10^{-4} the dislocation dense regions are still irregularly arranged (Fig. 4b). The density of screw dislocations varies locally. Fig. 4b shows a region with a small screw dislocation density.

Fig. 3b shows the dislocation structure after fatiguing at the amplitude 2.5×10^{-4}. The higher the strain amplitude the better will be the regular arrangement of the dislocation-dense regions. The dislocation arrangement develops differently in different places. This development continues at the next higher strain amplitude.

After fatigue at the amplitude 5×10^{-4} a uniform dislocation arrangement was not observed in the whole specimen as shown in Figs. 4c and 4d. A wall-like structure was found in large parts of the specimen volume (Fig. 4c). The strips can be observed in [101]-direction as well. But regions with structures which are similar to the matrix structures after room temperature fatigue have also been observed (Fig. 4d). The boundaries between such different types of structures are soft. The mean distance between the dense regions is 0.5 μm.

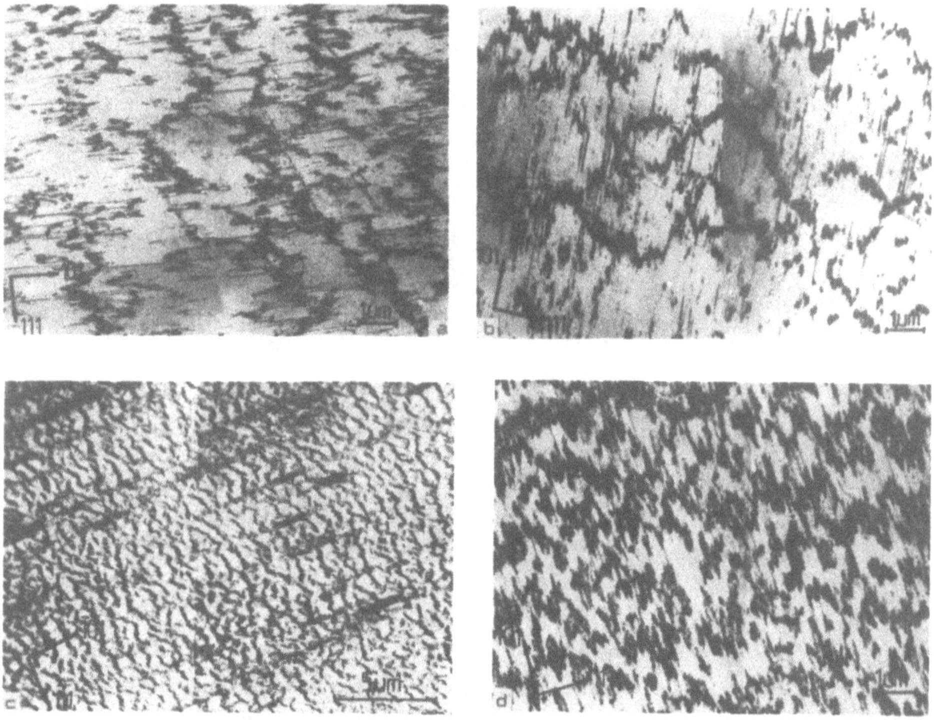

Fig. 4 Dislocation structure at the saturation stage after cyclic deformation at different plastic strain amplitudes (a) 0.000065; (b) 0.0001; (c),(d) 0.0005

CONCLUSIONS

At room temperature fatigue there are significant differences between the dislocation structure in the matrix and in PSB regions. These pronounced differences were not observed after fatigue at 78 K. Especially after fatigue at high strain amplitudes, we observed, however, several types of dislocation arrangements with soft boundaries.

Using a special specimen shape it was possible to investigate simultaneously the development of the dislocation structure in bulk material as well as the slip markings on the surface of one and the same specimen at several fatigue stages.

The results of our surface observations suggest the need to distinguish between volume regions with high slip activity (PSB and transient active slip bands) and volume regions with a small slip activity, i. e. regions without slip markings. We believe there is a correspondence between the different structure types and the different types of slip markings.

References

1. Z.S. Basinski, A.S. Korbel and S.J. Basinski, Acta Metall. 28 (1980) 191-207.
2. O.B. Pedersen and L.L. Lisiecki, Proc. 8th Int. Conf. on the Strength of Metals and Alloys, Vol.1, Tampere 1988, Pergamon Press, Oxford 1988, p.719-724.
3. K. Mecke, J. Bretschneider, C. Holste and W. Kleinert, Cryst. Res. Technol. 21 (1986) K135-K138.
4. K. Mecke, W. Kleinert and M. Hübner, Mater. Sci. Eng. A 111 (1989) L3 - L7.

7. FRACTURE

APPLICATIONS OF ELECTRON MICROSCOPY TO MICROMECHANISMS OF BRITTLE FRACTURE

P. Bowen*, C.A. Hippsley+ and J.F. Knott[x]

*School of Metallurgy and Materials, University of Birmingham, Birmingham, U.K.
+Materials Physics and Metallurgy Division, AEA Technology, Harwell, Oxon, U.K.
[x]Department of Materials Science and Metallurgy, University of Cambridge, Cambridge, U.K.

ABSTRACT

The paper describes the contribution of electron microscopy in elucidating micromechanisms of brittle fracture under monotonic loading at both low temperatures (catastrophic failure) and elevated temperatures (sub-critical crack growth) in ferritic steels. Examples are drawn from pressure vessel steels and weld-metals. In particular close attention is given to both micromechanisms of transgranular cleavage fracture and intergranular fracture. The latter mechanisms involve the segregation of phosphorus and sulphur. The uses of scanning electron microscopy, transmission electron microscopy and scanning Auger microscopy are all considered.

INTRODUCTION

It is now accepted that macroscopic engineering properties such as strength and toughness are controlled by events occurring on the microstructural size scale. Modelling of strength in terms of dislocation movement is well established, but toughness is much less understood at this fundamental level. Catastrophic failure derives from extreme microscopic fracture events which may occur over a size range from sub-micron to sub-millimetre. A general challenge is thus to predict engineering toughness using micromechanistic models based on observed or assumed behaviour at this microstructural level.

Ideally, microstructural features that control toughness can be identified by direct observation and microanalysis in situ on fracture surfaces at failure initiation sites. However, in the vast majority of studies such features cannot be identified positively and must therefore be deduced. These deductions are often based on measured changes in toughness that result from controlled changes in microstructure. Such changes may be produced by controlled heat-treatment and/or variation in alloy composition. This approach may be difficult in practice because microstructural features often cannot be varied independently e.g. in mild steel cooled directly after austenitising, the grain boundary carbide size varies in a regular manner with the ferrite grain size if the carbon content is constant (1). Moreover, the measured macroscopic toughness may not relate unambiguously to a single microstructural feature e.g. the fracture toughness (K_{IC}) not only involves a combination of local tensile stress and a microstructural distance, but is also observed usually to be an inverse function of yield stress. It is evident that both approaches rely upon the determination of fracture micromechanisms from accurate microstructural, fractographic and microanalytical information. This paper reviews how such micromechanisms can be identified for a number of brittle fracture phenomena using Scanning Electron Microscopy (SEM), Transmission and Scanning Transmission Electron Microscopy (TEM and STEM) and the Scanning Auger Microprobe (SAM) and their associated microanalytical facilities, namely energy dispersive spectroscopy (EDS) and Auger Electron Spectroscopy (AES). It is a personal review based on the work with which the authors have been associated jointly over the past decade.

TRANSGRANULAR CLEAVAGE FRACTURE

Cleavage Fracture in A533B Nuclear Pressure Vessel Steel

For a single chemical composition (0.25 C, 1.51 Mn, 0.63 Ni, 0.53 Mo, wt.%) changes in heat-treatment have produced equivalent fracture toughness values at -140°C and +150°C, i.e. a temperature shift of approximately 300°C (2). In this present paper it is sufficient to consider three coarse grained conditions (prior austenite grain size of ≈130 μm produced by austenitising at 1250°C for 1 hr) heat-treated to give (i) an autotempered martensite (by oil-quenching from 1250°C); (ii) a predominantly mixed lower and upper bainite (fluidised bed transformation at 420°C) and (iii) a predominantly upper bainite (transfer to holding furnace at 400°C). 0.2% proof stresses, fracture toughness values and carbide sizes are given in Fig. 1, from which the

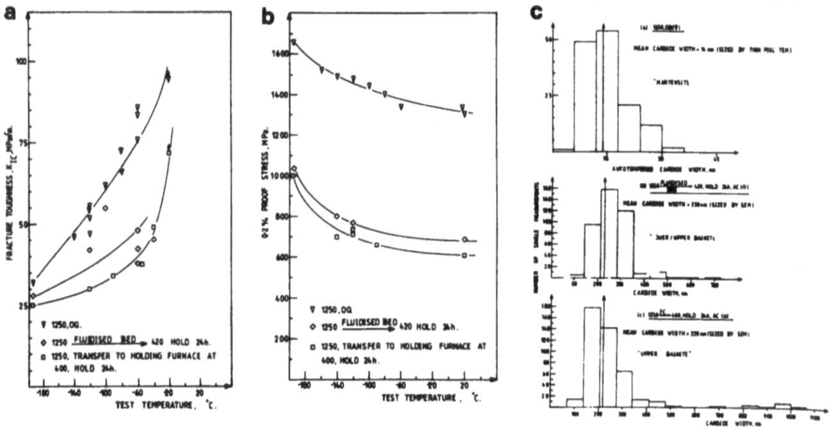

Figure 1(a) Fracture toughness values, (b) proof stresses, (c) carbide sizes for A533B.

importance of carbide size is clearly indicated since the martensitic condition possesses greater cleavage fracture resistance in spite of its vastly superior strength. On closer inspection, however, it is noted that significant differences in carbide size distribution are measured using thin-foil TEM compared with SEM - see Fig. 2. It might be expected that the improved resolution of TEM would account for many smaller carbides being measured, but it is important to observe that the coarser carbides are also missing from the size distribution. This is not thought to be due to sampling alone (since nominally identical "areas" have been examined (for the quenched and tempered condition) in Fig. 2 and the distributions are qualitatively similar), but because the foil was too thin (≈300 nm maximum) to support the larger carbides. This illustrates a problem in fracture micromodelling where often only the coarsest carbide sizes are considered (3, 4).

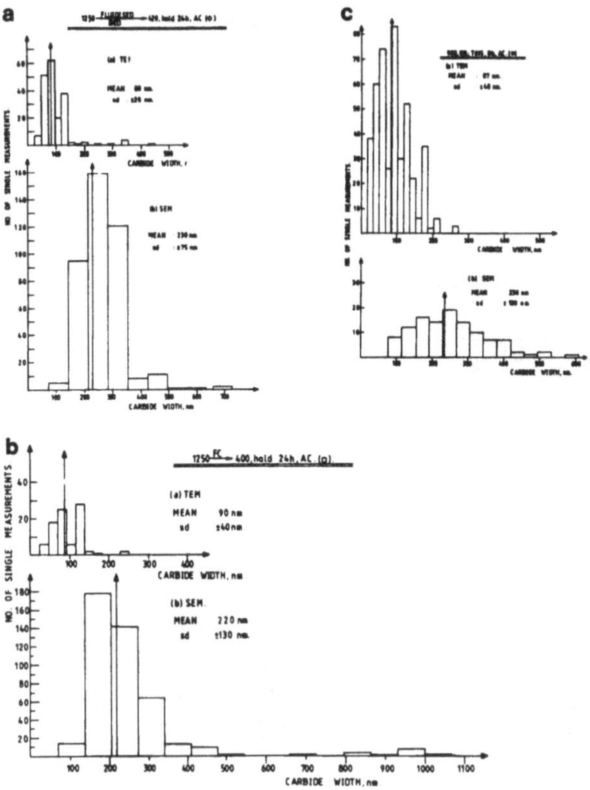

Figure 2 A comparison of carbide size distributions measured using SEM and TEM for A533B. In Figure 2(c) nominally identical "areas" have been examined by both techniques

The present results are more consistent with carbides sized using SEM. Local cleavage fracture stress measurements are shown in Tab. 1 for the entire range of microstructures evaluated (3), and

Table 1 Summary of coarsest observed carbide width (diameter), carbide shape and σ_F^* values observed for the full range of microstructures considered (3)

Coarsest observed carbide width (diameter)-SEM, nm	Carbide shape	σ_F^* MPa
30[a]	Plate	3700
36[a]	Plate	3700
110[a]	Plate	3300
340	Plate	2200
350	Plate	1900
450	Plate	1700
450	Spheroidal	3200
600	Spheroidal	3300
720	Plate	1900
1000	Plate	1700
1100	Spheroidal	1200

[a] Sized by TEM (thin-foil)

from which the shape of the carbides can also be deduced to be important: spheroidal carbides are less damaging than plate carbides of thickness equivalent to the diameter of the spheroid. In conclusion considerable care may be necessary to quantify such trends in microstructure. Further considerations are detailed elsewhere (3,4,22).

Cleavage Fracture in Weld-metals

Earlier studies (5,6) in this area on C-Mn weld-metals have indicated that the fracture of a single non-metallic (silicate) inclusion of greater than 0.9 μm in diameter can cause the catastrophic failure of a macroscopic testpiece, see Fig. 3. Moreover, in-situ EDS has identified inclusions at

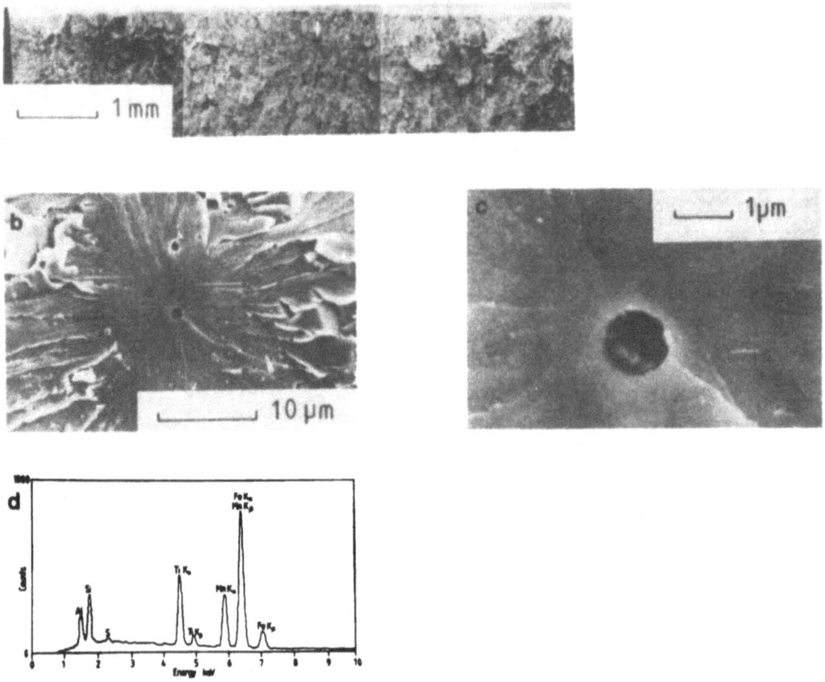

Figure 3(a), (b), (c) Details of cleavage fracture initiation site; (d) qualitative X-ray microanalysis of inclusion located at centre of initiation site.

these unique fracture initiation sites. They contain Mn, Al, Si and occasionally Ca. As shown in Fig. 4(a), local values of tensile stress at observed cleavage initiation sites (as estimated by finite element methods) appear to be in excellent agreement with a modified Griffith relationship (and yield an effective surface energy, γ_p, of 7 Jm^{-2}) with the microcrack size equated to the initiating inclusion diameter. This work has confirmed the crucial importance of critical tensile stress for cleavage since most sites are observed close to positions of known peak tensile stress. However, recent work on 2.25Cr-1Mo weld-metals (2.26Cr, 1.05Mo, 0.08C, wt.%) has considered 130 sites of local bursts of cleavage (under fatigue loading) and has indicated not only the importance of large inclusions (≥ 1 μm) but also their location in large ferrite grains (≥ 10 μm), see Fig. 4(b), (c) and (d). The microstructure local to the microcrack nuclei must also be taken into account therefore in any quantitative micromodelling of cleavage fracture (in weld-metals). To control average cleavage fracture resistance the use of cleaner electrodes and/or welding

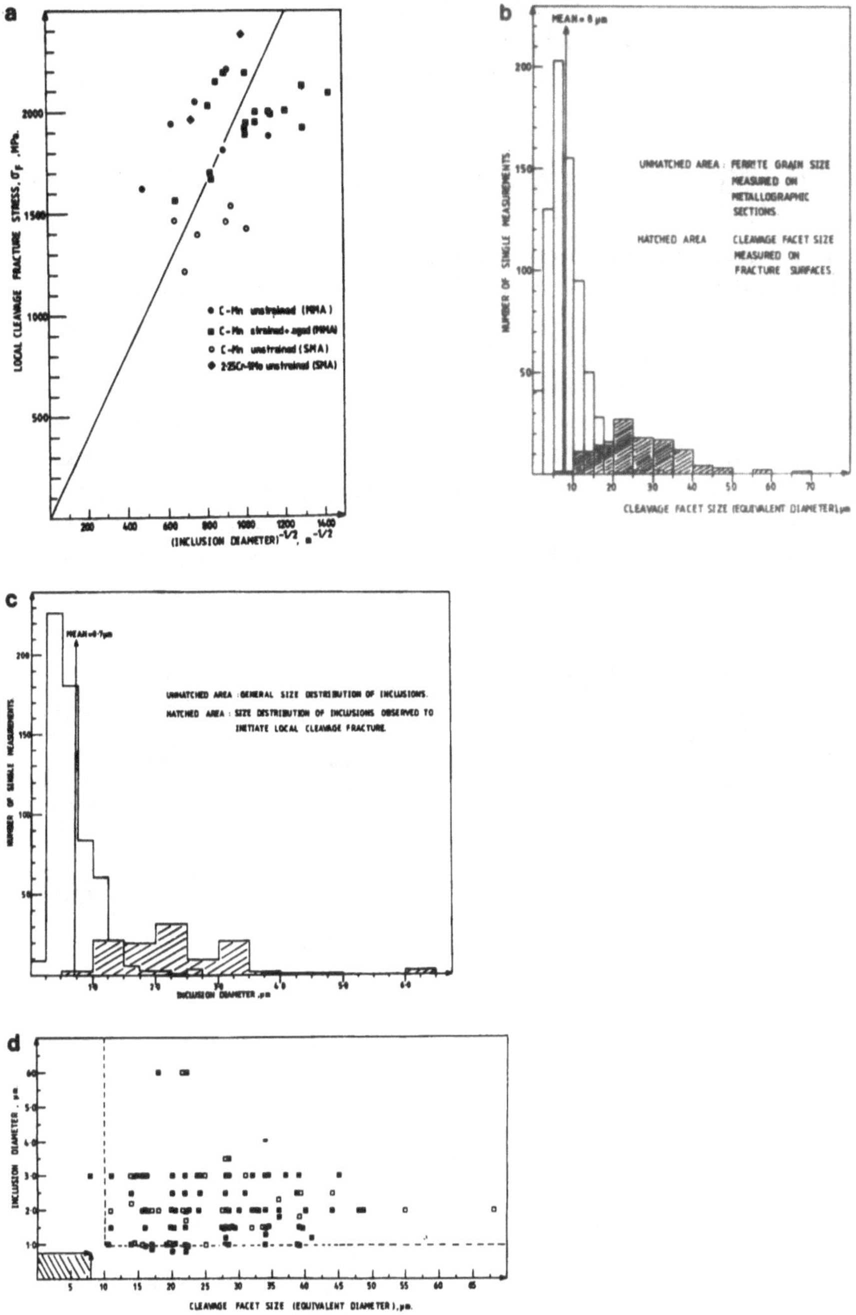

Figure 4(a) Relationship between local tensile stress and initiating inclusion size (5,6); (b) importance of large inclusions, (c) importance of large ferrite grains (d) combined local cleavage initiation sites (2.25Cr-1Mo weld metals).

practice(7) (to reduce inclusion sizes), and/or a reduction in pro-eutectoid ferrite content (8) may be suggested. For modelling purposes, the 2.25Cr-1Mo weld-metals were heat-treated subsequent to welding to reduce the ferrite content to zero (by reaustenitising and quenching to produce a martensitic/lower bainitic microstrostructure). In this way it was possible to achieve plastic collapse (maximum load crack-opening displacement values, δ_m) at -60°C, compared with unstable cleavage fracture (unstable crack opening displacement values, δ_u) at +20°C in the as-welded condition(8).

ONE STEP TEMPER EMBRITTLEMENT (OSTE) IN A533B NUCLEAR PRESSURE VESSEL STEEL

The mechanism of OSTE in quenched and tempered martensites has attracted much attention over the past few decades (9-11), and it is now attributed most generally to the conjoint action of the precipitation of cementite and the presence of impurity elements (notably phosphorus) which lower the cohesion of the prior austenite grain boundaries and promote intergranular fracture. The advent of AES and SAM has elucidated this phenomenon by direct in situ observations and analysis of grain boundaries. The study reported here (12) combines careful heat-treatment, detailed fractography, AES and fracture toughness measurements for a coarse-grained autotempered martensitic condition in A533B. Heat treatment was performed to hold both prior austenite grain size (130 µm) and matrix strength constant while varying the degree of phosphorus segregation in austenite. To achieve this, two austenitising treatments were used and both involved holding at 1250°C for 1 hour followed by either: oil quenching (OQ) to room temperature; or

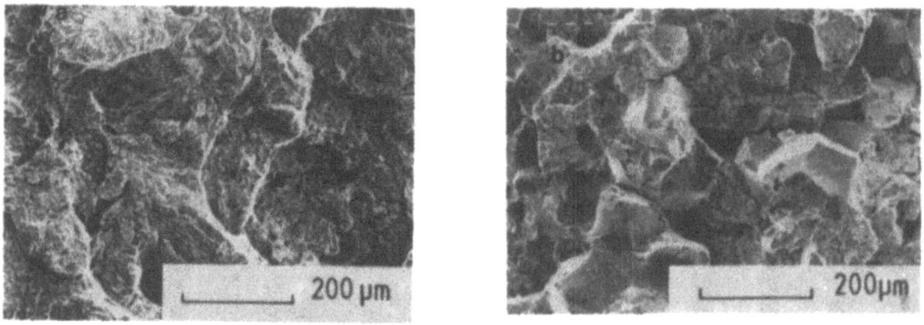

Figure 5(a) Transgranular quasi-cleavage OQ(RT) condition, (b) predominantly intergranular fracture SQ(290) condition.

furnace cooling to 900°C, holding for 1 hour, then oil-quenching (SQ) to room temperature. This latter treatment increases the driving force for phosphorus segregation to austenite grain boundaries. Tempering was carried out additionally at 290°C for 1 hour in some testpieces and coarsened the mean carbide width from 14 to 38 nm, while matrix strength did not decrease substantially (12). As a result of step-quenching and tempering SQ (290), the fracture mechanism changes from predominantly transgranular cleavage to intergranular decohesion - Fig. 5. A summary of results is presented in Tab. 2. It is clear from the Table that reductions in fracture toughness from 58 to 50 $MPam^{1/2}$ can occur due to segregation of phosphorus alone [compare OQ(RT) with SQ(RT)]; or due to carbide coarsening alone, K_{IC} decreases from 58 to 46 $MPam^{1/2}$, [compare OQ(RT) with OQ(290)]. The largest reductions in fracture toughness from 58 to 40

Table 2 Summary of (OSTE) brittle fracture behaviour at -120°C. (12).

Condition	K_{IC} MPam$^{1/2}$	% I.G. Fracture	AUGER PHR pct(P_{120}/Fe_{703}) Bulk Analyses x	sd
OQ(RT)	58	1	1.7	±0.3
OQ(290)	48	9	1.9	±1.2
SQ(RT)	52	23	2.7	±0.5
SQ(290)	43	63	3.3	±1.4

MPam$^{1/2}$ are observed however, when both increased phosphorus segregation and carbide coarsening occur [compare OQ(RT) with SQ(290)]. These results are consistent with the proposed micromechanisms of OSTE. In the present context attention is drawn to the fact that despite an increase from 23 to 63% in the area fraction of intergranular fracture observed on tempering the SQ condition, little change in the average peak height ratios (PHR %) are noted, and this is not statistically significant at the 95% confidence level. This is considered to be due to sampling equivalent numbers of individual grain boundaries for both SQ(RT) and SQ(290) conditions (the spatial resolution of the microanalysis area is of the order of 2 μm x 2 μm under the beam conditions employed), whereas it is impossible to identify the most heavily segregated boundaries (within the additional 40% of facets (63-23%)) produced on tempering (12). The authors have used the above results to support the mechanism of carbide rejection of phosphorus in OSTE when prior levels of phosphorus segregation are high.

HIGH TEMPERATURE BRITTLE INTERGRANULAR CRACK GROWTH (HTBIGF)

A number of recent papers (13-19) have considered several aspects of high temperature intergranular crack growth in high strength 2.25Cr-1 Mo- steels under static loading conditions. Such stress-relief-cracking or reheat cracking of thick-section welded joints has long been recognised in the pressure-vessel industry (20), and its engineering significance has now been reduced greatly through modified welding practice and compositional control. Nevertheless, the underlying mechanism of crack-growth is of considerable scientific interest. It occurs over the temperature range 300-600°C, and intriguingly subsequent fracture at low temperatures (e.g. -196°C) of a testpiece unloaded from such temperatures does not produce significant intergranular fracture. It is clear, therefore, that the mechanism of sub-critical, stable, intergranular cracking can occur only under the combined action of stress and temperature and is a dynamic mechanism of crack growth.

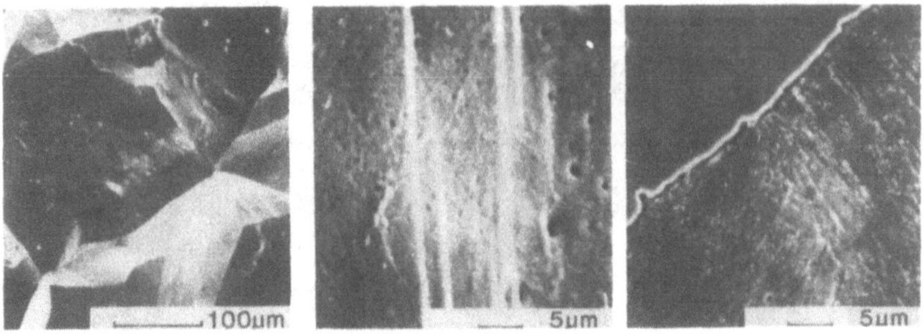

Figure 6 Step-wise crack propagation mechanism in HTBIGF.

Significant advances in understanding the fundamental mechanisms of crack growth have been made through the use of high resolution SEM and SAM. Although at low magnifications the brittle intergranular facets appear to be featureless, at high magnifications (using a low working distance geometry scanning electron microscope) they are seen to consist of smooth facets but with ridges perpendicular to the crack growth direction (see Fig. 6). The observation of such ridges of plastically deformed material supports strongly the step-wise crack propagation mechanism suggested on theoretical grounds (13,18,19). In addition, these discrete jumps across a large intergranular facet enable the position of the high temperature crack-tip to be located precisely using the method shown in Fig. 7. Low temperature brittle fracture is achieved using a fracture stage located within the high vacuum of the Auger system, and the position of the crack-tip may be defined by the change in oxygen profile (since the high-temperature crack surfaces will be contaminated with oxygen and carbon). From the Figure it is seen that a peak in sulphur concentration is observed just ahead of the high-temperature crack-tip. Several studies of this type have unequivocally confirmed the dynamic segregation of sulphur to be responsible for HTBIGF (13,15-19), using both Auger line profiles and Auger maps, examples of which are also given in Fig. 7. Such evidence is complemented by step-quenching experiments (identical in concept to those reported earlier in the study of OSTE) where as the final austenitising tempeature prior to quenching is lowered then the degree of phosphorus segregation at austenite grain-boundaries is increased but the amount of sulphur in solid-solution (and the distribution of manganese sulphide inclusions) available for subsequent dynamic segregation under stess is reduced markedly (16).
The precise source of the sulphur is still to be resolved completely, but the reduced crack growth rates measured under equivalent stress intensities at high temperature (500°C) subsequent to the final austenitising temperature being lowered is consistent with the controlling influence of sulphur

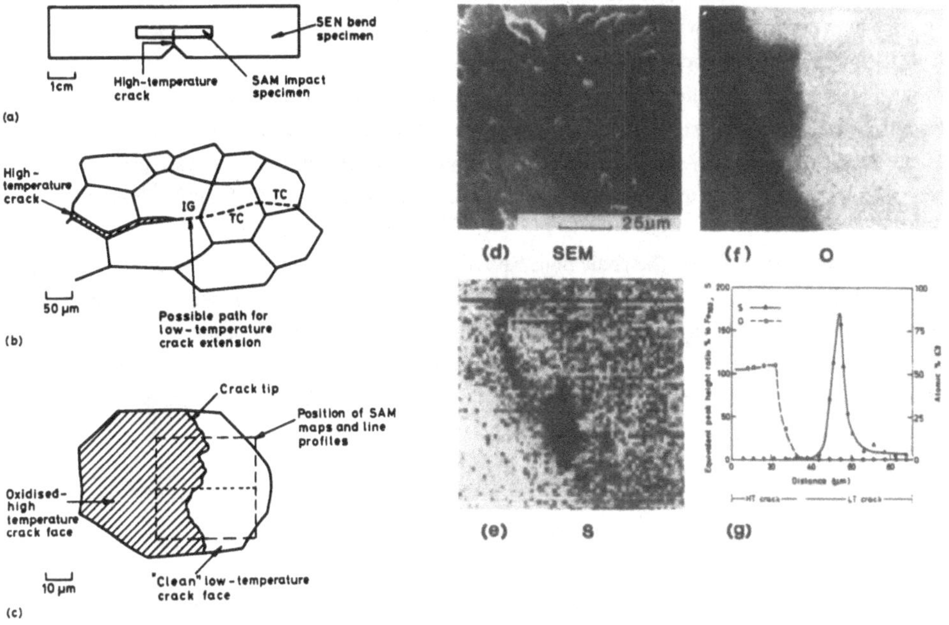

Figure 7 Identification of sulphur as embrittling element in HTBIGF.

alone(16).

Finally, it is appropriate to consider the state of the art measurement of grain boundary segregation by EDS using a dedicated scanning transmission electron microscope with a field emission electron source (FEG-STEM), and capable of producing a focused probe of ~1 nm

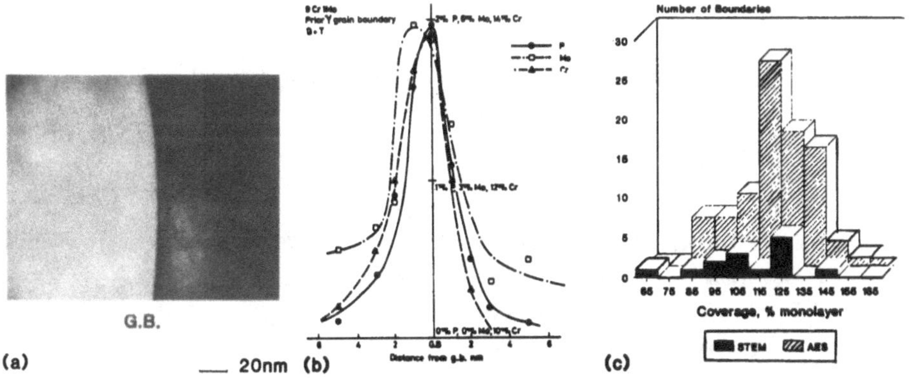

Figure 8(a), (b). EDS profile obtained from precipitate free grain boundary in unembrittled 9Cr-1Mo, (c) comparison of P segregation measured by AES and STEM-EDS for an embrittled 9Cr steel.

diameter. Such a technique is essential for segregation studies if the boundaries cannot be exposed by low temperature fracture within the Auger system, and its high inherent resolution enables clean (i.e. precipitate free) grain boundaries to be detected for analysis, Fig 8(a). Care is required to ensure that measured concentration profiles are not affected by grain boundary curvature, departure from the ideal orientation (probe at normal incidence) and "beam-broadening" in thicker foils. Monte Carlo calculations have been performed to assist in the quantification of data and to predict how some of the above factors influence the profiles measured (21). An EDS profile from a 9Cr-1Mo (wt.%) ferritic steel is shown in Fig. 8(b), note that these boundaries could not be exposed for analysis by AES. The predictions have also been tested by measuring the extent of grain boundary segregation in 9 Cr (wt.%) ferritic steel doped with 0.06 (wt.%), where comparisons using AES were possible, see Figure 8(c). The good correlation obtained suggests that the combination of FEG-STEM with EDS can be used with confidence to measure grain boundary segregation in materials that cannot be examined readily by surface analysis methods.

CONCLUSIONS

This paper has endeavoured to illustrate the wide range of electron microscope techniques that can promote an understanding of micromechanisms of brittle fracture under monotonic loading in ferritic steels. In particular it has emphasised the control of macroscopic fracture toughness by both microstructural features and segregation of tramp impurity elements.

ACKNOWLEDGEMENTS

Much of the work presented here was supported, at least in part, by the UKAEA Underlying Fracture Studies Research Programme over the last decade. Thanks are due to I.A. Vatter and J.M. Titchmarsh for permission to incorporate Figure 8, and to J.H. Tweed and D.E. McRobie for permission to incorporate Figure 2 into this publication.

REFERENCES

1. D.A. Curry and J.F. Knott, Metals Sci. 12 (1978) 511.
2. P. Bowen, "Effects of microstructure on toughness in pressure vessel steel", Ph.D. Thesis, University of Cambridge, November 1984.
3. P. Bowen, S.G. Druce and J.F. Knott, Acta Metall. 34 (1986) 1121-1131.
4. P. Bowen, S.G. Druce and J.F. Knott, Acta Metall. 35 (1986) 1135-1746.
5. J.H. Tweed and J.F. Knott, Metals Sci. 17 (1983) 45.
6. D.E. McRobie and J.F. Knott, Mater. Sci. Tech. 1 (1985) 357.
7. P. Judson and D. McKewon, Weld. Inst. Autumn Conference (1982) 3.1-3.8.
8. P. Bowen and J.F. Knott, in preparation.
9. C.L. Briant and S.K. Banerji, Int. Metall. Rev. 4 (1978) 164.
10. C.L. Briant and S.K. Banerji, Metall. Trans. 10A (1979) 123.
11. J.R. Rellick and C.J. McMahon, Jr., Metall. Trans. 5 (1974) 2439.
12. P. Bowen, C.A. Hippsley and J.F. Knott, Acta Metall. 32 (1984) 637-647.
13. C.A. Hippsley, H. Rauh and R. Bullough, Acta Metall. 32 (1984) 1381.
14. J. Shin and C.J. McMahon Jr., Acta Metall, 32 (1984) 1535.
15. C.A. Hippsley, Mater. Sci. Technol. 1 (1985) 475.

16. P. Bowen and C.A. Hippsley, Acta Metall. 36 (1988) 425-439.
17. C.A. Hippsley, Acta Metall. 35 (1987) 2399-2416.
18. H. Rauh, C.A. Hippsley and R. Bullough, Acta Metall. 37 (1989) 269-279.
19. C.A. Hippsley, D. Buttle and C.B. Scruby, Acta Metall. 37 (1988) 441.
20. J.C. Ritter and R. McPherson, Metal. Tech. 1 (1974) 506.
21. I.A. Vatter and J.M. Titchmarsh, Ultramicroscopy 28 (1989) 236-239.
22. G.P. Gibson, M. Capel and S.G. Druce, European Symposium on Elastic-Plastic Fracture Mechanics: Elements of Defect Assessment, Freiburg, FRG, 9-12 October 1989.

CRACK INITIATION BY SHEARING OF KINK BANDS

G. Kästner

Institute of Solid State Physics and Electron Microscopy of the Academy of Sciences of the German Democrat. Republ., POB 250, Halle/S. DDR 4010

ABSTRACT

After briefly classifying certain modes of crack initiation in crystalline solids, the paper describes a mechanism for initiating cracks by kink bands which are strained by shear of the main slip system.

INTRODUCTION

The fracture behaviour of crystalline solids is usually explained by statistical models. This treatment, however, rarely considers details of crack nucleation /e.g. 1/ known from numerous microscopic investigations /2 to 17/. They are classified in Table 1. It considers active processes of slip or of deformation twinning which initiate cracks by interacting with certain heterogeneities, among them also local, active slip or twinning may occur. This table reveals some gaps. One of them is filled by the topic of the present paper. It is concerned with strains and cracks developing at kink bands which become inclined (tilted) in a field of plastic shear.

GLIDE POLYGONIZATION AND FORMATION OF KINK BANDS

Plastic deformation of single or of polycrystals often proceeds by single slip within limited regions. The surrounding material exerts constraints, e.g. bending or local shear orthogonal to the slip plane. This can lead to glide polygonization. Examples are mobile bend zones /3/, kink planes and kink bands, each of which are composed of edge dislocation walls. If these walls become tilted during proceeding shear, accommodation strains arise parallel to the wall. These strains can lead to local deformation and cracking as will be shown below.

Experimentally, the development of long-range stresses during deformation can easily be observed by optical birefringence in transparent crystals /3, 16/. In the present work, NaCl single crystals were strained up to 60% shear /17/ according to known geometric principles /18/. The resulting stress-strain curve is shown in Fig. 1 at S and S2. It indicates extremely low work hardening typical of single slip. Stronger hardening could be attributed to residual bending due to insufficient

Table 1 Modes of initiating cracks in single and polycrystals by an "active element" (deformation by slip or twinning) which interacts with a heterogeneity. References /2 to 15/

active element				interacting heterogeneity
slip plane	slip band (Luders band)	homogeneous strain	twin lamella	
Zener 1949 Petch 1954 Stroh 1955		cracking of interfaces and of second phases		rigid barrier
Johnston, Stokes & Li 1962			Marcinkowski& Lipsitt 1962	brittle grain
Orowan 1954 Gilman 1954				deformable grain
		Marcinkowski & Lipsitt 1962	Rose 1868 Cahn 1954-55 Argon & Orowan 1964	twin lamella
Cottrell 1958				single slip plane
	Stokes, Johnston, & Li 1961 Argon & Orowan 1964 Briggs, Clarke & Tattersall 1964			slip band
		present work		kink band, deformation band
				grip constraint
			Hull 1960	matrix constraint

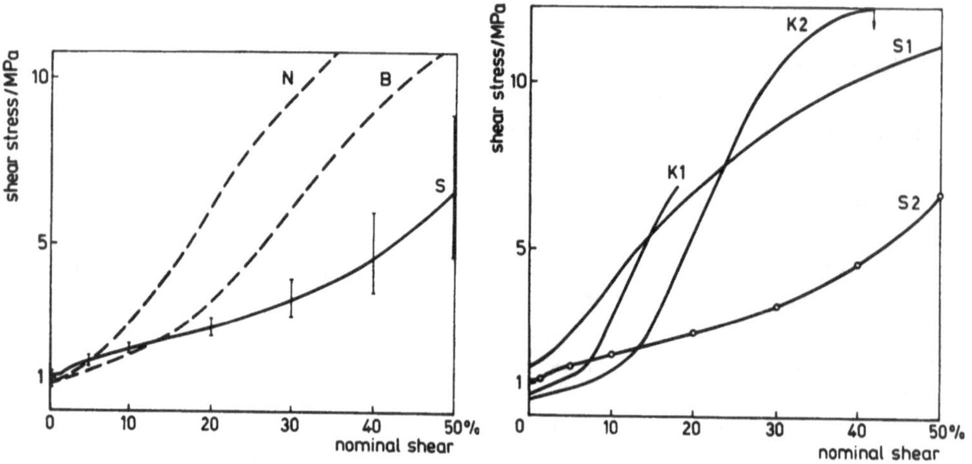

Fig. 1 Typical stress-strain curves of NaCl single crystals. Shear tests (S, S1, S2) compared to compression (K1, K2)

gripping (curve B) or to blocking of the main slip by orthogonal slip (curve N). Such features may explain why curve S1 obtained by /18/ differs from S2 in spite of equal geometries of the shear tests. Similar curves K1 and K2 represent typical compression tests.

To obtain S2 in a reproducible way, the specimens have to be carefully cleaved and annealed (in vacuo). Any damage to the surface has to be avoided.

The geometry of slip has been examined by optical and electron microscopy. For the latter purpose, the shear tests were performed in vacuo where evaporation of gold allows to decorate slip steps of atomic height. The beginning slip is confined to slip bands /16, 17/, similar to observations on MgO /2, 3/ and to Lüders bands. The slip bands grow in width and combine until the whole free gauge length shears quite homogeneously above 1.3% strain. Up to this stage, hardening is extremely low.

During further single slip, two mechanisms of forming kink bands were observed. A typical sequence is shown in Fig. 2. The drawing illustrates the slip geometry: Horizontal, parallel lines represent the main slip planes of a specimen glued at an angle of 45° between grips (top and bottom line). They are driven horizontally along the arrows. Semi-circles indicate residual torques exerted at the specimen. They relax plastically by glide polygonization in the sense of the dislocation walls indicated

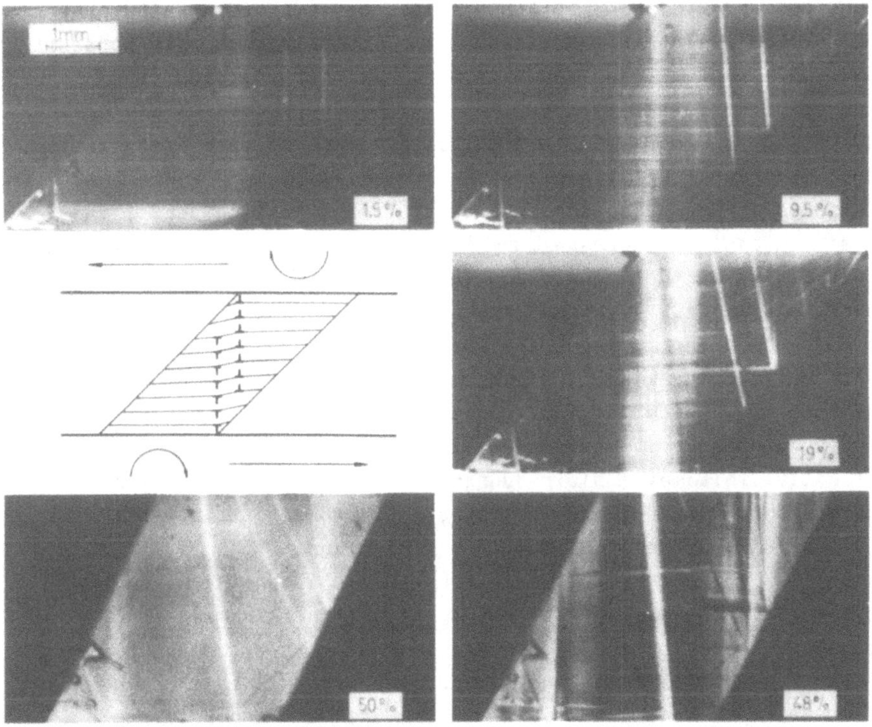

Fig. 2 Shear test of a NaCl single crystal, optically observed between crossed polarizers set // ⟨100⟩ (⟨110⟩ at 50%)

schematically. Each wall corresponds to a mobile bend zone visible as a bright diffuse band at 19% shear. At 48%, both zones have combined to form a narrow strong kink band in the sense of the drawing.

In a second, similar mechanism known from /2/, a kink band is formed at the site of an inactive slip band which (orthogonally) intersects the main slip system. Two examples of narrow, orthogonal slip bands are seen in Fig. 2 at 1.5% strain. They are subjected to a shear stress locally the same as in the main slip system, for known reasons of local stress equilibrium. However, these orthogonal bands usually become inactive: Their first growth in length stops within the grips, hence their widening (mainly by cross slip) remains restricted. This can be closely compared to the slip-band behaviour at bending /16/. Moreover, these bands become interpenetrated and hardened /2/ by the main slip bands which widen freely as they do not stick in the grips. As a result, the inactive orthogonal bands represent geometrically a kink state /2/. They remain fixed within the material, therefore they become tilted during shear as imaged in the sequence of Fig. 2. At 50% strain, also the bright central band is seen tilted but to a smaller degree since it has been formed later.

ACCOMMODATION STRAINS AND CRACKING

If a single kink plane (low angle tilt boundary, represented by an edge dislocation wall) has a position mirror-symmetric to both adjacent lattices, it is free of long-range strains. If, however, the kink plane is tilted off from this symmetry, the lattice coherency is violated, hence so-called accommodation strains arise /19/. They are essentially orthogonal to the active slip planes and tensile within a kink band /2/. If the band is tilted by an angle α due to the operating shear, the tensile strain rapidly increases, by the factor $1/\cos\alpha$ in the simplest case. This stress is the main reason for the bright birefringence of the kink bands seen in Fig. 2, except the stage at 50% shear where the polarizers have been set to image the shear stress. The latter is seen to be enhanced within the kink bands where the material is hardened as known from the locally enhanced dislocation density./2, 22/.

At proceeding shear, the tensile stress within a kink band deforms it plastically. This may explain "unpredicted slip" within kink bands as early observed on Al /20, 21/. In the case of NaCl, plastic tension can occur by slip on 4 systems oblique to the band and to the image plane of Fig. 2. On this surface, a groove is thus formed along the kink band. This is clearly shown by Fig. 3b and 3c. At 3a, the kink state of a band is directly indicated by the displacement of a crossing cleavage step S.

A further result of interpenetrating slip within the kink band may be the formation of vacancies. This may specifically account for the known drop of bulk density on deforming crystals as often observed.

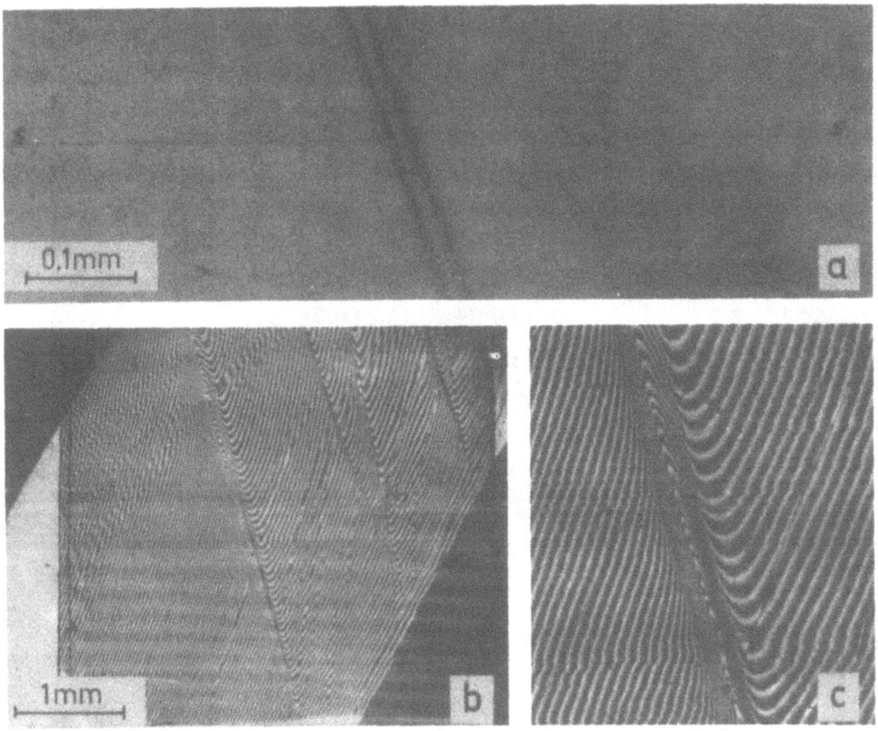

Fig. 3 Surface grooves along kink bands of a NaCl crystal sheared as in Fig. 2. Optical micrographs; b,c by Tolansky interferometry

A final result of extreme tension in a kink band is cracking. An example of a sequence similar to Fig. 2 is shown in Fig. 4. At 30%, the bend zones originating from the necks between sample and grip combine to form a bright band. It has appreciably tilted at 40%. New bands not tilted exhibit compressive stress which apparently balances the tension. At 43%, a crack has originated at the center of the central band. This crack runs nearly perpendicular to the maximum tension stress resulting from this band and from the applied stress. The crack has stopped, probably by the compressive bands mentioned above. New compressive and tensile bands arise, as expected, from the center and the tip, resp., of the crack. - The lower left image of Fig. 4 indicates the redistribution of shear stresses.

Apparently, imaging of internal stresses is an informative tool.

ACKNOWLEDGEMENTS

The author is indebted to Prof. Dr. H. Bethge and Prof. Dr. V. Schmidt for supporting this work.

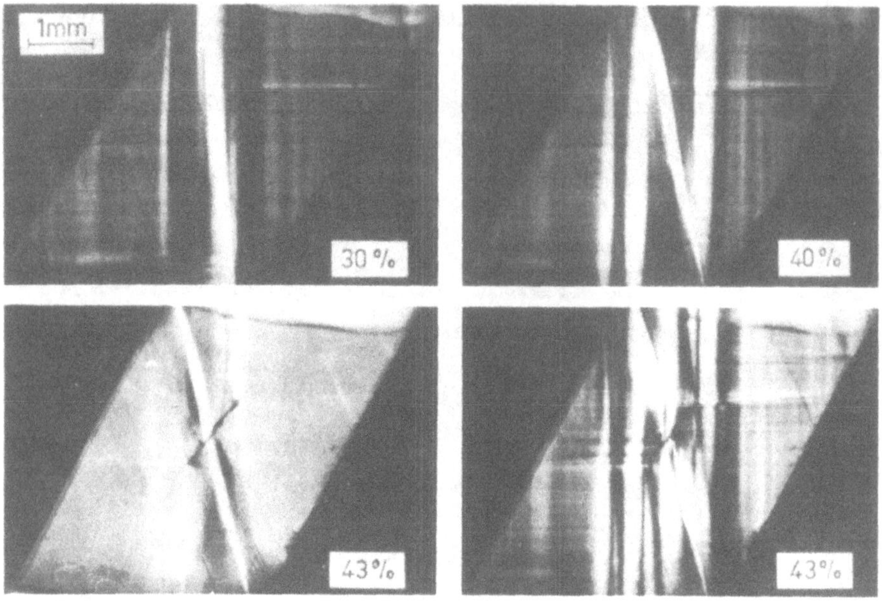

Fig. 4 Crack initiation during shear of a NaCl crystal, observed optically between crossed polarizers as in Fig. 2

REFERENCES

1 R.J. Stokes, In: Fracture of Metals, Polymers, and Glasses, Proc. 4th Symp. Fundamental Phenomena in Materials Science, L.J. Bonis (Ed.), Plenum Press, New York 1967, p. 151

2 A.S. Argon and E. Orowan, Phil. Mag. 9 (1964) 1023

3 A. Briggs, F.J.P. Clarke and H.G. Tattersall, Phil.Mag. 9 (1964) 1041

4 R.W. Cahn, J. Inst. Metals 83 (1954-55) 493

5 A.H. Cottrell, Trans. Met. Soc. AIME 212 (1958) 192

6 J.J. Gilman, J. Metals (New York) 6 (1954) 621

7 D. Hull, Acta Met. 8 (1960) 11

8 T.L. Johnston, R.J. Stokes and C.H. Li, Phil. Mag. 7 (1962) 23

9 M.J. Marcinkowski and H.A. Lipsitt, Acta Met. 10 (1962) 95

10 E. Orowan, In: Dislocations in Metals, M. Cohen (Ed.), Am. Soc. Mining and Eng., 1954, p. 191

11 N.J. Petch, Progr. Met. Phys. 5 (1954) 1

12 G. Rose, Abh. Preuss. Akad. d. Wiss. (1868) 57

13 R.J. Stokes, T.L. Johnston and C.H. Li, Phil. Mag. 6 (1961) 9

14 A.N. Stroh, Proc. Roy. Soc. A 232 (1955) 548

15 C. Zener, Fracturing of Metals, Amer. Soc. Metals, Clevel. 1949, p. 3

16 G. Kästner, phys. stat. solidi 36 (1969) 261

17 G. Kästner, Thesis, Universität Halle/Saale, Physik. Inst., 1969

18 L. Graf, D. Mayer u. E. Müller, Z. Metallk. 56 (1965) 750

19 M.J. Stowell, Phil. Mag. 7 (1962) 677

20 H. Lange and K. Lücke, Z. Metallk. 44 (1953) 183 and 514

21 P. Haasen and G. Leibfried, Fortschr. Phys. 2 (1954) 73

22 A.S. Argon, A.K. Nigam and G.E. Padawer, Phil. Mag. 25 (1972) 1095

INVESTIGATION OF DEFORMATION AND FRACTURE OF HETEROPHASE NICKEL ALLOYS AT ROOM TEMPERATURE BY in situ HIGH VOLTAGE ELECTRON MICROSCOPY

M.M. Kantor, L.M. Olshansky, I.L. Svetlov, L.P. Sorokina

Baikov Institute of Metallurgy, Acad. Sci. of U.S.S.R.
Leninsky Prospect 49, Moscow 117911 U.S.S.R.

ABSTRACT

The initial stages of deformation and fracture in two heterophase Ni-base alloys have been studied by an analysis of crystallographic characteristics of slip traces and the growth behaviour of microcracks during in situ deformation experiments in the HVEM. It has been found that in the early stages the deformation-induced fracture and the nucleation of microcracks occurred only in the most plastic structural constituent.

EXPERIMENTS AND RESULTS

The data available on micromechanisms of fracture of heterophase metallic materials are scarce and contradictory. In particular, V.M. Finkel states in his monograph [1] that "commercially pure two-phase materials are characterized by a trivial mechanism of the nucleation of fracture: by the appearance of microcracks in the brittle component". In some cases, this point of view has been supported by experiments. For instance, in a study of the fracture mechanism of a two-phase eutectic alloy consisting of a nickel-base solid solution and single crystalline tantalum carbide fibres, fracture was observed to start by cleavage in the {001} planes of the TaC fibres [2]. On the other hand, the results of electron microscope studies of the location of microcracks in foils of other heterophase alloys suggest that the microcracks can initially appear and propagate in the more plastic component of an alloy [3, 4].

The present report, based on the results of in situ experiments in a high-voltage electron microscope, describes the initial processes leading to fracture of heterophase nickel alloys during deformation at room temperature.

Two alloys have been investigated. One of them was a heat-resisting nickel-chromium base alloy doped with aluminium. It consisted of two structural constituents: the nickel-base γ solid solution and dispersed equi-axed cubic particles of an ordered γ' phase with an f.c.c. crystal lattice. The size of the particles was 300-800 nm. Since the lattice parameters of the γ and γ' phase in this alloy differ only slightly and the size of γ' particles is of the same order as the foil thickness, the Ashby-Brown contrast is not observed but δ-bands appear on the {100} interfaces between the γ and γ' phase under particular diffraction conditions [5]. They are the evidence of a very slight coherent deformation between the γ and γ' phases. For that reason, the stress field around the γ' phase particles does not noticeably contribute to the strength of the alloy. In such cases, the resistance to plastic flow can be caused by the destruction of the superstructure [6].

The other alloy was a eutectic nickel-chromium base heterophase alloy with additions of aluminium, niobium and carbon. It was prepared by the method of directional crystallization. The alloy contains three constituents: the γ and γ' phases which in their morphology and chemical composition are similar to the structural components of the former alloy. The third phase consists of long, well-faceted crystals of niobium carbide which have a rod-like shape with plane faces [7].

Specimens for in situ experiments were prepared by electrolytic polishing in an electrolyte composed of 250 ml methanol, 150 ml butyl alcohol, and 25 ml $HClO_4$. The specimens were deformed in tension at room temperature in a standard attachment to the high-voltage electron microscope JEM-1000. The electron-microscopical studies were carried out at an accelerating voltage of 1000 kV.

Under an applied load, dislocation slip was observed within the solid solution in both alloys, which resulted in an increase of the local dislocation density up to a value in the order of $10^9 cm^{-2}$. In the eutectic alloy, misfit dislocations at the matrix-fibre interfaces were observed along with dislocations in the γ phase. With increasing deformation, the dislocation density at the matrix-fibre interfaces was increasing. With a further increase of the applied stress, numerous slip traces were formed in the γ solid solution. In some cases, the arrangement of slip traces indicated the action of at least two slip systems (Fig. 1). This corresponds to the conclusion drawn by Bauer, Lyles and Wilsdorf that at least two and often a greater number of slip systems are active in the fracture zone [8]. The analysis of the slip traces has shown that the slip takes place in planes which refer to the {111}<110> systems characteristic of f.c.c. metals and that it is associated with a local change of the orientation of the crystallite being studied.

It should be mentioned here that we never observed indications of slip in the ordered γ' phase described recently [9]. For that reason, we assume that the plastic deformation in our experiments is localized in interlayers of the γ solid solution.

The nucleation of microcracks in the two-phase alloy also takes place only in volumes of the solid solution. During the microcrack propagation, narrow regions with a weaker diffraction contrast appear along the crack edges (Fig. 2). They appear due to a change of the diffraction conditions as a consequence of the plastic deformation of the zones of the material in the course of the crack propagation. For that reason, the width of these regions can be regarded as a measure of

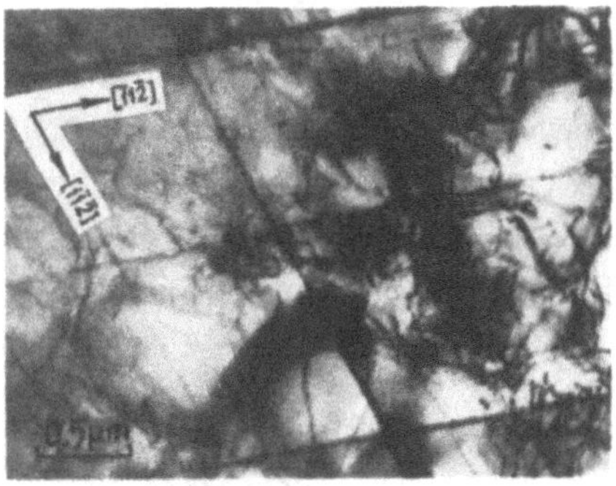

Fig. 1 A system of two conjugate slip bands in the matrix of a heterophase eutectic nickel-base alloy. The plane of the foil is (110)

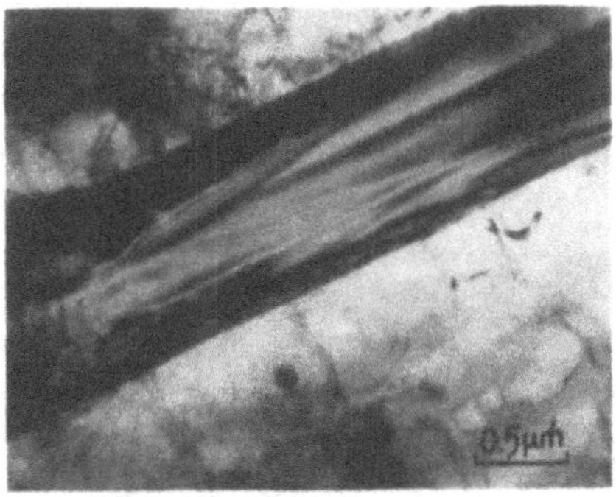

Fig. 2 Changes of diffraction contrast at the edges of a crack due to plastic deformation of the material

the local plasticity of the material just before fracture [10]. The nature of these processes is sometimes explained by a thinning of the crack edges [10], though the probable orientation change of these regions with the propagating microcracks cannot be excluded [10].

Fig. 3 shows the plastic zone in front of the tip of a crack in a two-phase alloy. The crack extends over more than 15 um and is formed by numerous dislocations distributed relatively uniformly. It therefore follows that the crack propagation is associated with intensive multiplication and motion of dislocations in front of the crack tip. The motion of a crack is preceded by a continuous thinning of the regions at its tip, which is evidenced by a redistribution of dislocations in their slip planes and an increase of the number of dislocations as the slip lines move away from the tip.

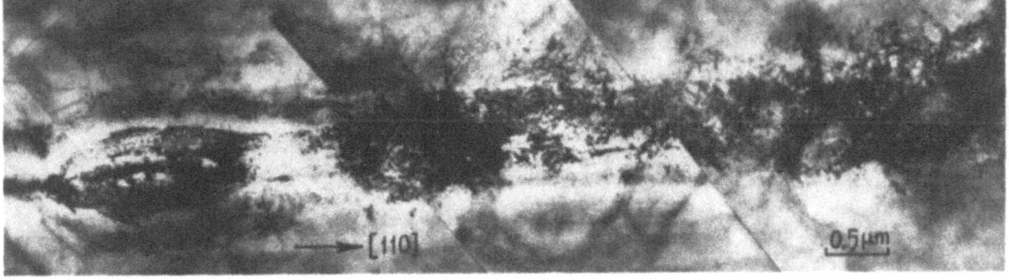

Fig. 3 Plastic zone in front of a crack tip in two-phase nickel-base alloy. The plane of the foil is (001)

The location of the nucleation sites of microcracks in the eutectic alloy depends on the orientation of the axis of the applied load relative to the growth direction of the rod-like NbC crystals [4]. In specimens with the load applied in the growth direction of the NbC crystals, microcracks appear and propagate in interlayers of the γ phase. In specimens where the load axis is perpendicular to the growth direction of the NbC crystals, the nucleation of microcracks takes place in the matrix-fibre interface (Fig. 4).

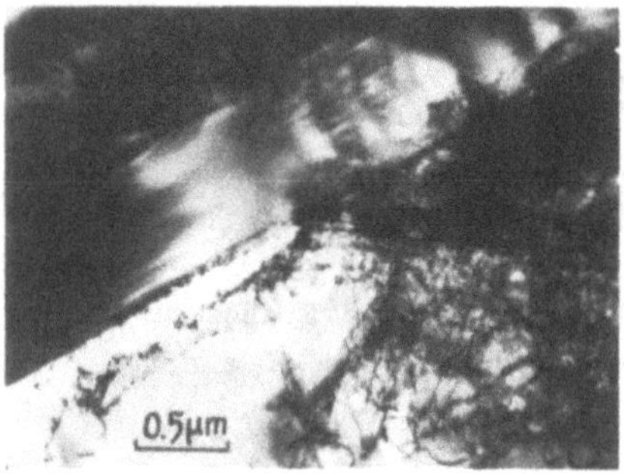

Fig. 4 Nucleation of a microcrack in the matrix-fibre interface after load application

Further increasing the load causes the cracks to grow further and acquire a saw-tooth shape with their edges parallel to the slip traces (Fig. 5). The last observation suggests that microcracks propagate along the slip planes. The changes in the direction of the propagation of microcracks with the formation of a large number of segments within the same crystallite were observed during the fracture of various materials [4, 10], and various explanations of this phenomenon have been proposed. In the case considered, it can be assumed that the tooth-like shape of microcracks is the result of dislocation slip in two crystallographically equivalent slip systems of the γ solid solution. Another cause of the tooth-like shape may be associated with the fact that microcracks, as they propagate only in the slip planes of the γ phase, change their direction of propagation each time they encounter a γ' particle.

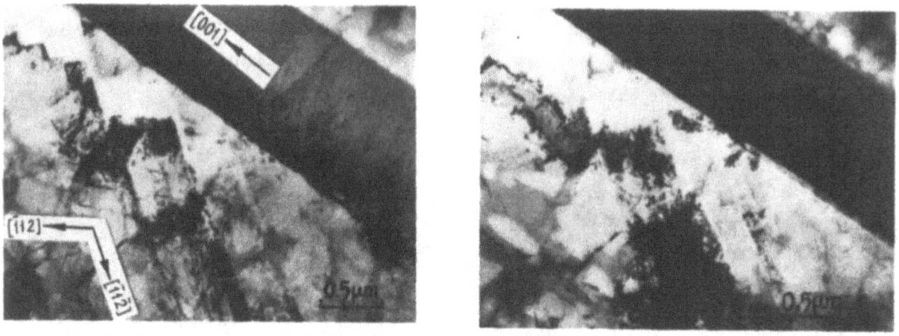

Fig. 5 Propagation of a microcrack in a heterophase eutectic alloy during an increase of load

During an increase of the deformation a microcrack propagates by widening of the existing segments and formation of new ones. It should

be noted that a microcrack mainly extends in a direction at 45° to the tensile axis, which is the evidence that the fracture takes place under the action of shear stresses and leads to rupture along the slip planes.

Thus, the results of investigating the elementary processes during in situ straining experiments inside a HVEM lead to the conclusion that, though the alloys investigated consist of a number of structural constituents, the plastic deformation prior to fracture and the nucleation of microcracks take place only in the most plastic component, the γ solid solution. This conclusion agrees with the localization of fracture in the most ductile component of a two-phase chromium-nickel alloy the macroscopic fracture of which is brittle [3]. Thus, it has been demonstrated that the concept of the initiation of microcracks in a brittle constituent as the only mechanism of fracture nucleation is probably not valid in heterophase alloys. It should also be recognized that the discussion of the mechanism of fracture of such alloys only on the basis of a comparing the plastic properties of the structural components is quite insufficient, since it does not take into consideration the role of other factors important to this process.

Straining experiments inside the column of an HVEM are well suited to diagnose the mechanical behaviour of materials, including heterophase alloys, as they allow an identification of the elementary processes of plasticity and fracture.

REFERENCES

1 V.M. Finkel, Physics of Fracture, Metallurgizdat, Moscow 1970 (in Russian)

2 J.F. Stohr and R. Valle, Phil. Mag. 32 (1975) 43

3 A.V. Perepelkin, G.F. Sartshan, S.A. Firstov, and G.G. Kurdjumova, Fiz. Metallov i Metallovedenie 48 (1979) 588

4 M.M. Kantor, L.M. Olshansky, I.L. Svetlov, and L.P. Sorokina, Fiz. i Khim. obrabotki metallov N 2 (1983) 132

5 L.P. Sorokina and M.P. Usikov, Fiz. Metallov i Metallovedenie 32 (1971) 281

6 H. Gleiter and H. Hornbogen, Mater. Sci. Eng. 2 (1968) 285

7 S.T. Kishkin, I.L. Svetlov, L.P. Sorokina, and M.P. Usikov, Fiz. Metallov i Metallovedenie 53 (1982) 1127

8 R.W. Bauer, R.L. Lyles and H.G.F. Wilsdorf, Z. Metallk. 63 (1972) 525

9 N. Clement, D. Caillard, P. Lourse, and A. Coujou, Scripta Metall. 23 (1989) 563

10 U.H. Lindborg and B.L. Averbach, Acta Metall. 14 (1966) 1583

EM INVESTIGATIONS ON CRACKS IN LITHIUM NIOBATE CRYSTALS

L. Malicskó

Research Laboratory for Crystal Physics, Hungarian Academy of Sciences,
H - 1502 Budapest, 112. P.O.Box 132, Hungary

ABSTRACT

The origin and the propagation of cracks were studied in lithium niobate single crystals by different electron microscopic methods including the in-situ high-resolution one as well. It was established that the cracks are mainly nucleated at inclusions. Their propagation occurs in jumps by the formation of subsequent new crack nuclei before the actual crack tips. The splitting of the crystal lattice happens through intermediate amorphization processes in nano-regions.

INTRODUCTION

Pure and doped lithium niobate /LN/ single crystals are widely used in the micro- and optoelectronics as special supports and/or optically active construction elements. The grown boules of single crystals can randomly contain cracks reducing the quality and frequently causing further problems during the shape-processing /1-3/.

The present work reports some results concerning the origin and the propagation of cracks investigated in LN crystals by different electron microscopic methods including also the high-resolution one previously rarely used for such purposes /4,5/.

EXPERIMENTAL

The pure and metal ion doped LN single crystals investigated were grown from melts by Czochralski technique described already /2,6/. From the crystal boules regions containing cracks were optical microscopically /OM/ selected and then prepared into /001/ plates and electron transparent discs /3,7,8/. The polished plates were investigated in scanning electron microscopes /SEM/ equipped with energy dispersive X-ray spectrometers /EDS/, and the thinned discs in conventional, high-voltage /HV/ and high-resolution /HR/ transmission electron microscopes /TEM/.

The as-grown cracks in the TEM samples served as pre-cracks for the in-situ study of the propagation processes. The propagation of the pre-cracks was simply activated by thermal stresses produced in the samples by the suitable regulation of the electron beam. In this way also the conditions for the HRTEM imaging of the /110/ type lattice planes with spacing of 0.257 nm could simultaneously be assured.

RESULTS AND DISCUSSION

The crack trace lines on the /001/ surfaces of the hexagonal LN samples, are crystallographically expected to be parallel to the /110/ type lattice planes /2,9/. But according to the OM and SEM investigations the trace lines of the extended cracks often show direction deviations. This fact points to the presence of local inhomogeneities influencing the crack propagations. The observed cracks were almost always initiated at inclusions /cavities and/or foreign particles/. The danger of crack formations was found to be increased in the presence of certain dopants or impurities, resp. /3,7/.

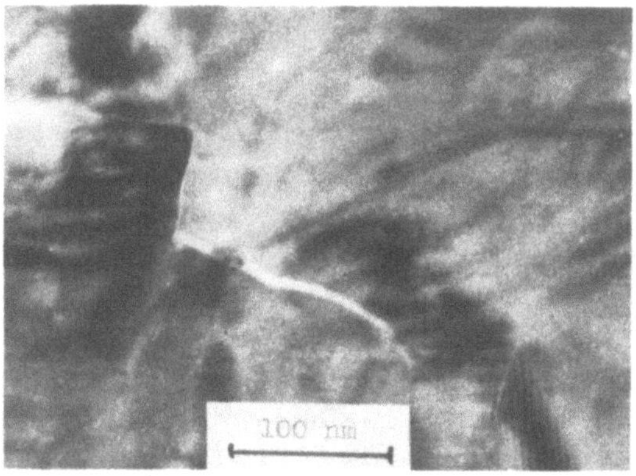

Fig. 1 TEM bright field image taken at $\bar{B}$ nearly /001/ and showing a kinked microcrack along the boundaries of small misfitted grains /see the moiré patterns/ in a Ti-doped LN sample.

Constitutional supercooling states randomly developing in microscopic scales at the growth fronts of the crystals can lead to the formation of microregions which contain then a lot of small, strongly disturbed and misfitted matrix and/or foreign grains initiating microcracks /Fig. 1/. As seen, the extension of the microcracks formed at such grains seems to be initially determined by the courses of the grain boundaries.

Dislocations and/or twins connected apparently with the extended cracks, could also be observed in the surroundings of crack tips. The twins shown in Fig. 2 have probably stopped the motion of the crack.

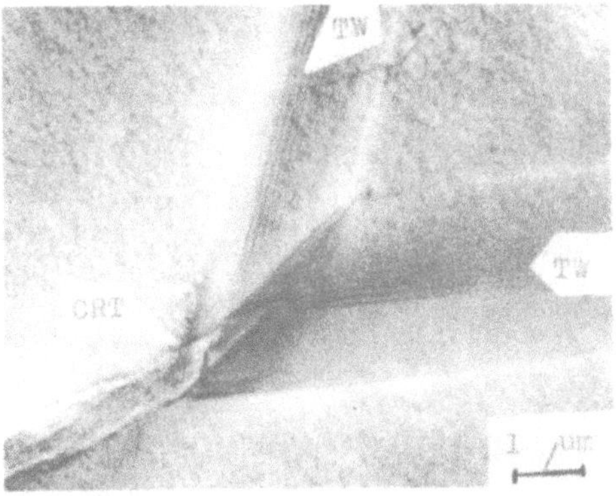

Fig. 2 HVTEM bright field image taken at a nearly two beam condition with $\bar{g}$=110 on a strongly Ti-doped LN sample, showing twins /TW/ at a crack tip /CRT/.

Before the actual crack tips of the moving cracks, several 100 nm long and some nm broad trace stripes can be found. These stripes consisting of the rows of alternating crack nuclei and "compact bridges",

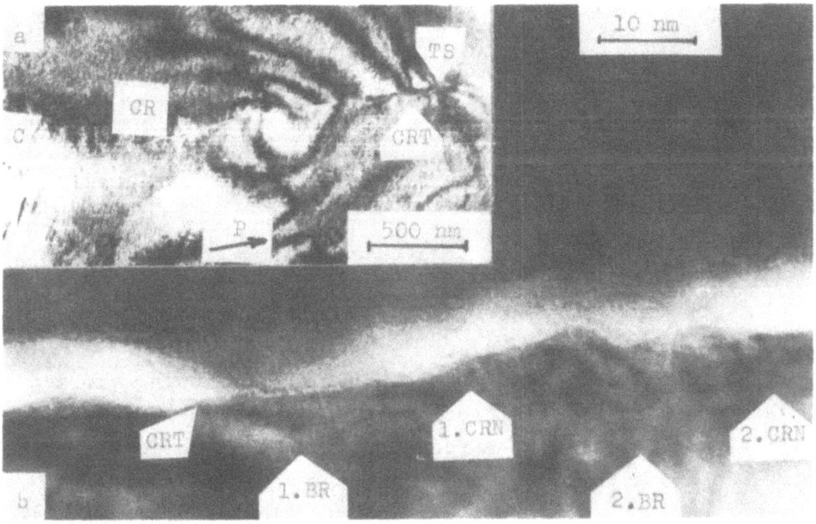

Fig. 3 TEM bright field images taken at $\bar{B}$ near /001/ on a crack in a strongly Ti-doped LN sample. a - The general view of the crack /CR/ initiated at a cavity /C/, showing deviations from the main propagation direction /P/ and a trace stripe /TS/ before the actual crack tip /CRT/. b - The magnified detail of the surroundings of the crack tip with the trace stripe consisting of small compact bridges /BR/ and lens-like crack nuclei /CRN/.

are apparently prepared for the extension of the cracks /Fig. 3/. The curvatures in the course of the crack /Fig. 3a/ and the presence of the crack nuclei in the trace stripe /Fig. 3b/ point to that the crack propagation happens in jumps influenced by local micro-conditions /e.g. inhomogeneities/.

On the compact bridges between the neighbouring crack nuclei the different states of the strongly stressed crystal material could be studied in the nano-scales by using a HRTEM. On the registered HRTEM images the percental local cross-contractions of the /110/ type lattice planes were measured by OM and considered as characteristic quantities for the stressed states.

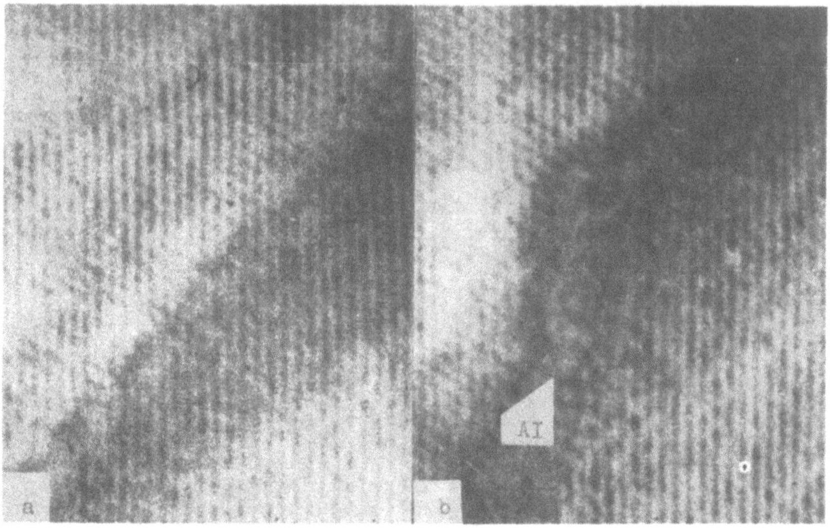

Fig. 4 HRTEM images of the /110/ type lattice planes in two compact bridges of different deformation states: a - before the splitting of the lattice planes, and b - after the formation of an amorphous interlayer /AI/. Spaçing: 0.257 nms.

During the splitting of the bridges the following average deformation states could be well distinguished /Fig. 4/. At average cross-contractions of /1.5±0.2/ percent, lattice plane drifts appear and accumulate along certain narrow contrast lines /Fig. 4a/. At average cross-contractions of /3.6±1.1/ percent, single lattice planes begin to split. Then, at cross-contractions of /2.6±1.2/ percent decreased already, an amorphous interlayer appears between the ends of the lattice planes already split. At further decreasing cross-contractions down to /1.6±0.7/ percent the interlayer thickness increases /Fig. 4b/. Finally reaching a thickness of a few nms the interlayer also splits. Probably same splitting processes of the lattice planes take place during the crack nucleations at the actually weakest nano-regions of the stressed crystal lattice.

SUMMARY

The cracks are nucleated first of all at inclusions and extend subsequently during the growth and/or other treatments /e.g. shape-processing/ as far as they will be stopped at other defects. The crack propagation happens in jumps by the formation of lens-like crack nuclei before the actual crack tip under the influence of the local lattice inhomogeneities of the crystals. The splitting of the LN crystal lattice occurs through local amorphization processes observed in detail.

Acknowledgements. The author thanks to the Directorate of the IFE der AdW der DDR in Halle for enabling SEM, HVTEM and HRTEM investigations, and to Dr.R. Scholz and Mrs.I. Perczel for their valuable helps. The work was supported by the Hungarian Research Foundation /AKA/.

REFERENCES

1 A. Räuber, In: Current Topics in Material Sciences, Vol. I, E. Kaldis /Ed./, North-Holl.P.Co., Amsterdam 1978, p. 481-602
2 K. Polgár, L. Jeszenszky, K. Raksányi and E. Hartmann, Acta Phys. Hung. 47 /1979/ 125-32
3 L. Malicskó, I. Cravero and Th. Krajewski, Cryst.Res. and Technol. 19 /1984/ 999-1006
4 Sh. Wen, J.El.Micr. Technique 7 /1987/ 301-12
5 Strength of Metals and Alloys, P.O. Kettunen, T.K. Lepistö and M.E. Lehtonen /Eds./, Pergamon Press, Oxford 1988.
6 F. Schmidt and R. Voszka, Cryst.Res. and Technol. 16 /1981/ K127-29
7 L. Malicskó, Acta Phys. Hung. 61 /1987/ 219-22
8 L. Malicskó, Á. Barna and R. Scholz, Cryst.Res. and Technol. 23 /1988/ 223-9
9 A.W. Vere, J.Mat.Sci. 3 /1968/ 617-21
10 B.A. Bilby and R. Bullough, Phil.Mag. 45 /1954/ 631-46

HVEM INVESTIGATIONS OF MICROPROCESSES OF DEFORMATION AND FRACTURE IN POLYMERS

G.H. Michler

Scientific research and Coordination Centre, Kombinat VEB Chemische Werke Buna, 4212 Schkopau, and Institute of Solid State Physics and Electron Microscopy of the Academy of Sciences of the GDR, 4010 Halle/S.

ABSTRACT

High-voltage electron microscopy (HVEM) was used to study the micromechanical processes of deformation directly on thin deformed samples of amorphous and rubber-modified polymers. In these polymers the microprocesses are closely connected with the initiation and formation of crazes. Craze formation and its effects on the fracture toughness are discussed in dependence on the type and morphology of polymers.

INTRODUCTION

Today, a fundamental problem in polymer research as well as in materials science in general is the determination of correlations between structure and properties. In polymers structures of most different sizes are distinguished (including the molecular structure of the macromolecules, the type of arrangement of neighbouring macromolecules, and a great number of supermolecular structures). Recent studies showed that the mechanical properties can often be related directly to the supermolecular structure (the morphology). Electron microscope investigations have essentially promoted the knowledge of the morphology of polymers. This is due to the progress in instrumentation but mainly in the preparation techniques, particularly in ultramicrotomy. In addition to the morphology, the micromechanical processes of deformation and fracture can advantageously be studied by electron microscopy.

In comparison to inorganic materials, the direct EM investigation of polymers involves three general problems:

1. Preparing thin specimens from bulk polymers requires special techniques, since the traditional methods for inorganic materials are not suitable.
2. Polymers (as organic substances) are particularly sensitive to electron beam irradiation.
3. The contrast between structural details is very poor, since polymers usually consist of the same light elements (C, H, O, ...).

The HVEM implies several advantages in overcoming these difficulties /1/.

Depending on their morphology polymers can be divided into several groups: amorphous polymers, semicrystalline polymers (consisting of phases in crystalline and amorphous states), polymer-polymer combinations (consisting of two or more different polymer components), and composites (including particle-filled and fibre-reinforced polymers). Rubber-modified polymers (and other polymer blends having a morphology of a dispersely distributed rubber phase in a hard-polymer matrix) possess favourable mechanical properties and reveal micromechanical processes of plasticity very interesting from the physical point of view. These processes are often related to the formation of so-called crazes, which are highly localized plastic deformation zones in polymers. Some results of studying the formation of crazes in dependence on the morphology of amorphous, glassy and high-impact polymers are given in this paper.

INVESTIGATION TECHNIQUE AND POLYMERIC MATERIALS

From bulk polymeric materials 0.5 to 5 μm thick sections were prepared using an ultramicrotome. The sections were uniaxially deformed in a microtensile device and investigated in a 1000 kV high-voltage electron microscope (HVEM), enabling not only the morphology to be determined but also the micromechanical processes (formation of crazes), which are directly influenced by the local morphology. A more detailed investigation of the morphology has been performed by transmission electron microscopy using different preparation techniques, which are discussed more thoroughly in /1-3/.

The samples investigated are amorphous, glass-like or ductile polymers (polystyrene (PS), polymethylmethacrylate (PMMA), styrene-acrylonitrile-copolymers (SAN), polyvinyl-chloride (PVC), polycarbonate (PC) and rubber-modified polymers on the basis of PS and SAN with different rubber phases, particularly HIPS (high-impact polystyrene), ABS (acrylonitrile-butadiene-styrene), and ACS (acrylonitrile-chlorinated polyethylene-styrene). Besides commercial types several model materials were used to find out to which extent the formation of crazes depends on several structural parameters, including the size and size distribution of the rubber particles.

FORMATION OF CRAZES IN AMORPHOUS POLYMERS

In general, crazes are small, long zones of highly plastically deformed polymeric material with their longest extension perpendicular to the loading direction. They have relatively sharp boundaries to the undeformed or less deformed material. The micrographs in Fig. 1 show a

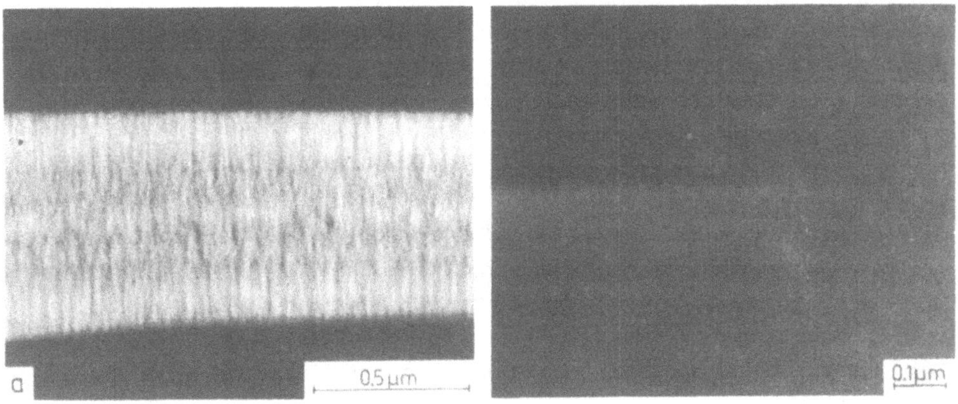

Fig. 1 Craze in PS: a) interior structure with fibrils, b) domains in the pre-craze ahead of the craze tip (deformation direction vertical)

craze in PS. Inside the craze (a) the amorphous polymeric material is highly plastically deformed yielding fibrils having a deformation of about 100 % to 200 % /4/. The propagation direction (the longitudinal direction) of the crazes is orthogonal to the deformation direction. In PS, the crazes are up to several 100 μm long and up to a few μm thick. The transformation zone of the craze tip is called "pre-craze", consisting of bright domains of 20 to 50 nm in diameter (b). These domains are slightly plastically deformed and act as craze nuclei.

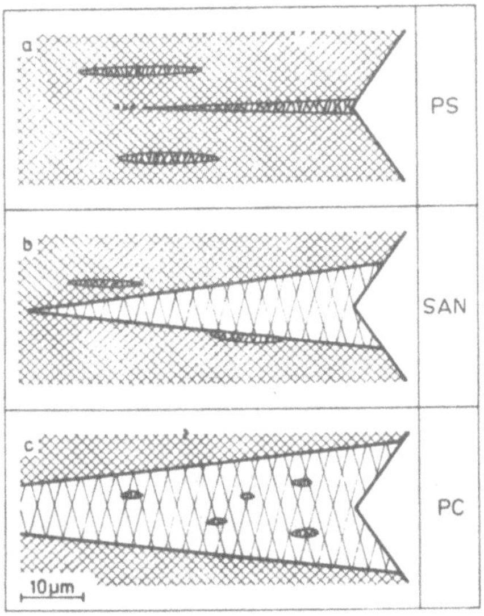

Fig. 2 Scheme of some characteristics (size and interior structure) of crazes in different amorphous polymers

The length and the thickness of the crazes may vary considerably in each sample, but characteristic ranges of thicknesses exist for the different types of polymers, which is schematically shown in Fig. 2 for crazes in PS, SAN, and PC, respectively. The interior structure of crazes may also be very different: crazes in PS show an interior structure of plastically stretched fibrils, crazes in PC consist of more homogeneously (on a supermolecular scale) deformed material, and in SAN samples often both types of crazes coexist.

The craze nuclei in the pre-craze were found to have typical minimum diameters d^{min}, correlating with the minimum thickness

g^{min} of the boundary layers of the craze. These minimum values are temperature-dependent material constants and characterize the length of the release of a local stress concentration by shifting macromolecular segments. This results in a minimum craze thickness D^{min}, defined by

$$D^{min} \approx 2\, g^{min} \approx d^{min} \qquad (1)$$

In more ductile polymers such as PVC there is a competition between crazes and the other type of local plastic bands, i.e. shear bands. Fig. 3 shows diffuse microshear bands in the stress-field ahead of a crack (a) with small crazes at the crossing points of the shear bands (b). Further details of the structure of crazes in amorphous polymers are given in /4, 5/.

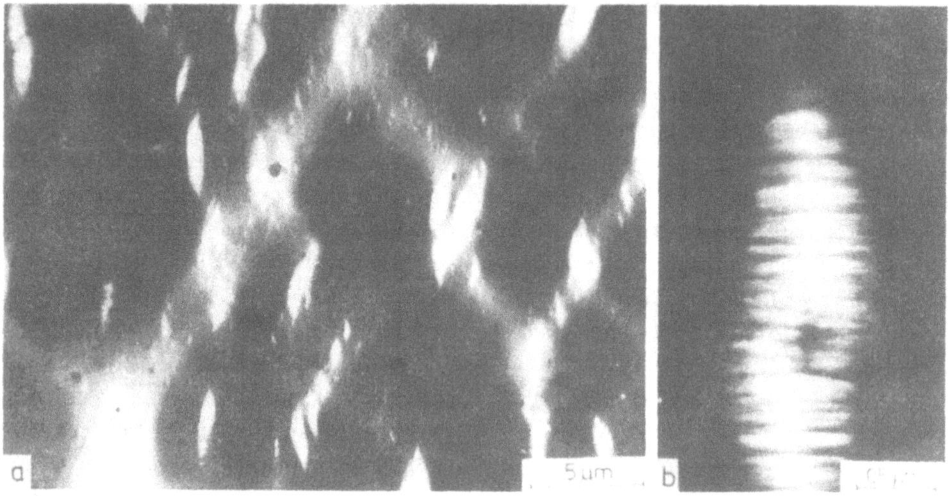

Fig. 3 Coexistence of microshear bands and fibrillated crazes in PVC-S

CRAZES IN HIGH-IMPACT POLYMERS

Basic Processes

Direct HVEM investigations of deformed samples of HIPS and related polymers show micromechanical processes of plasticity, which have been arranged in three stages /2, 6/:

Stage 1 - craze initiation: Each rubber particle acts as a stress concentrator and initiates the transformation of the surrounding material into crazes; see Fig. 4 (mechanism of multiple crazing, first deduced by BUCKNALL /7, 8/).

Stage 2 - superposition effect: For a rubber content of more than

15 vol.-% the average distance between particles is sufficiently low to superimpose the stress-fields of the neighbouring particles, favouring the formation of broad crazes and broad, long craze bands.
Stage 3 - crack propagation: Crack propagation starts by rupture of the crazes, but it is hindered by the limited length of the crazes and by the weak rubber particles.

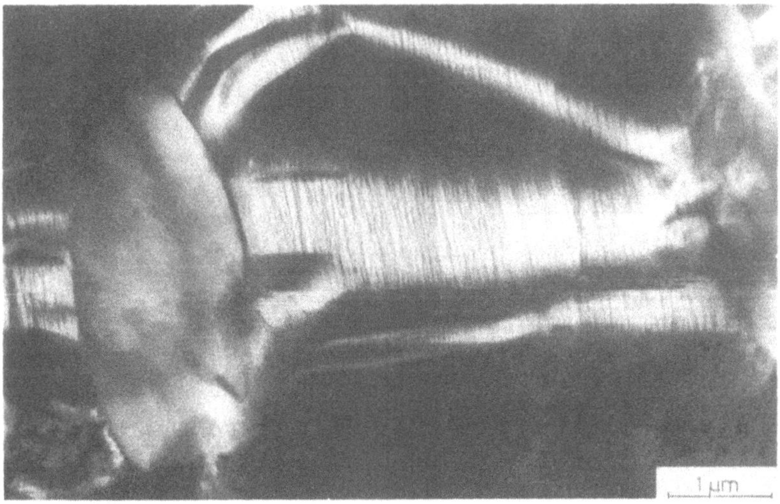

Fig. 4 Initiation of crazes at rubber particles in HIPS (1 µm thick section, vertical deformation direction, HVEM)

Influence of the Diameter of the Rubber Particles

An applied external stress causes local stress concentrations at the equator region of the weak rubber particles. The size D_Z of this stress concentration zone is nearly the same as the particle radius $D_K/2$

$$D_Z \approx D_K \;/\; 2 \qquad (2)$$

In this zone crazes are preferentially initiated ("craze initiation zone"). The size of the zone D_Z and hence the diameter D_K of the rubber particles define the thickness of the crazes. The smallest possible craze thickness together with eqs. (1) and (2) yield the criterion of the minimum effective diameter of rubber particles

$$D_K^{min} \approx 4 \cdot g^{min} \qquad (3)$$

For PS and SAN the minimum thickness of the boundary layer was determined to be $g^{min} \approx$ 10 nm, yielding a minimum effective particle diameter of $D_K^{min} \approx$ 40 nm. The craze initiation is most effective, if the size D_Z of the zone of stress concentration correlates with the characteristic thicknesses of the crazes, leading to a criterion of the optimum particle diameter D_K^{opt}. Particles of diameters essentially larger than the optimum diameter are less effective for the craze initiation for two reasons: (i) Since the zone of stress concentration is much larger than the optimum thickness of crazes, only some thinner crazes are initiated having a total craze volume smaller than the volume of an imaging thicker craze (of a thickness equal to D_Z). (ii) Larger particles imply larger distances between the particles (if the rubber volume content is constant), giving rise to longer cracks after the rupturing of the crazes. In PS and SAN critical crack lengths are 10...20 µm /12/, promoting a crack propagation before an intensive craze formation starts.

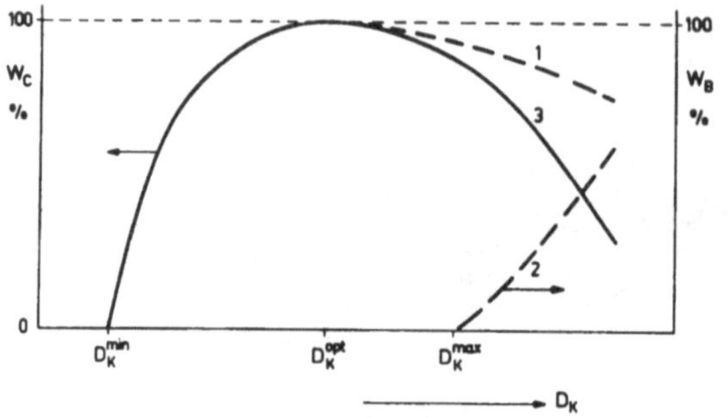

Fig. 5 Effectiveness curves for high-impact polymers as a function of the particle diameter
curve 1 - probability W_C of the differently large particles to initiate crazes
curve 2 - probability W_B of a premature crack propagation
curve 3 - resulting probability of the craze formation

These results are schematically shown in Fig. 5 as effectiveness curves depending on the particle diameter. Critical diameters of the particles are D_K^{min} (minimum critical diameter for the craze initiation), D_K^{opt} (optimum diameter for the craze initiation), and D_K^{max} (maximum diameter without premature crack propagation).

Influence of the Frequency Distribution of Rubber Particle Diameters

Curve 3 in Fig. 5 marks the effectiveness of the differently large rubber particles to produce crazes; it also holds true of a frequency distribution of particle diameters. Under stress, a polymer system having a wide particle diameter distribution first produces crazes at the largest particles but only with a reduced effectiveness. Solely particles having diameters of about the optimum size initiate crazes of a maximum volume content. The wider the diameter distribution is the greater is the number of particles with reduced effectiveness. Therefore, an optimum diameter distribution should be relatively narrow over D_K^{opt}, i.e., the frequency distribution of particle diameters should quickly drop to zero for large particles.

Influence of the Interparticle Distance

When the influence of the particle diameter on crazing in Fig. 5 was discussed, the role of the average interparticle distance A was already indicated. For a constant rubber volume content of about 20 % the correlation between particle diameter D_K and distance A is roughly given by $\bar{A} \sim \bar{D}$. According to this, the entire curve of Fig. 5 can be considered in dependence not on the diameter but on the distance A: For the smallest, ineffective particles of diameters $D_K < D_K^{min}$ the interparticle distance A is too small to realize a plastic yielding of the small matrix strands between the particles. If $D_K \approx D_K^{opt}$, the distance A correlates with the optimum thickness of crazes, indicating optimum yielding conditions. The dangerous role of larger distances A (to initiate premature crack formation) was mentioned above. That means that in particle-modified polymer systems a high, local plasticity can only be realized, if the size of the polymeric parts between the particles lies within an optimum range, i.e. if there are volumes having an optimum diameter A^{opt}.

The role of the interparticle distance in blends of a pseudo-ductile matrix with rubber was also stated by WU /13, 14/. He found a sharp brittle/ductile transition when the average interparticle distance is at a critical value.

The physical significance of such an optimum size of highly deformable polymeric parts in a heterogeneous system is not yet well understood. There may be a correlation with the concept of the entanglement network of the macromolecules (as in /15/).

REFERENCES

1 G.H. Michler, Kristall u. Technik 14 (1979) 1357-63
2 G.H. Michler, Polymer 27 (1986) 323-8
3 G.H. Michler, Ultramicroscopy 15 (1984) 81-100
4 G.H. Michler, Colloid & Polymer Sci. 263 (1985) 462-74
5 G.H. Michler, Colloid & Polymer Sci. 267 (1989) 377-88
6 G.H. Michler, Plaste u. Kautschuk 26 (1979) 497-501, 680-4
7 C.B. Bucknall, R.R. Smith, Polymer 6 (1965) 437-46
C.B. Bucknall, Br. Plastics 40 (1967) 118-22 (Nov.)
8 C.B. Bucknall, "Toughened Plastics", Applied Science Publishers Ltd., London 1977
9 M. Dillon, M. Bevis, J. Mater. Sci. 17 (1982) 1895-902
10 G.H. Michler, K. Gruber, Plaste u. Kautschuk 23 (1976) 346-51, 496-503
11 G.H. Michler, Plaste u. Kautschuk 35 (1988) 347-50
12 C.B. Bucknall, D.C. Street, J. Appl. Polymer Sci. 12 (1968) 1311-20
13 S. Wu, Polymer 26 (1985) 1855-63
14 S. Wu, Proc. of the 32. Microsymp. Macromolecules "Polymer Blends", Prague 1989, ML 2
15 E. Donth, G.H. Michler, Colloid & Polymer Sci. 267 (1989) 557-67

MECHANISM OF ENVIRONMENT-INDUCED CRACKING

M. Möser

Academy of Sciences of the GDR, Institute of Solid State Physics and Electron Microscopy, Halle/S., G.D.R. 4050

ABSTRACT

Some media such as hydrogen, sulfides and carbides but also metals are known to be able to embrittle metals. The present paper tries to elucidate this medium-induced cracking (environmental cracking). In regions of current flow (slip bands) a fast transport of foreign atoms may occur along the dislocations in certain temperature ranges. This leads to supersaturation at sites where gliding is stopped. In form of pressure bubbles, as short-time precipitates, the medium inhibits further gliding. For hot cracking and relaxation cracking, the precipitation of the attacking medium (carbide or sulfide) along the grain boundaries represents the frozen-in state of the process of embrittlement.

INTRODUCTION

Some media such as hydrogen, sulfides and carbides but also metals are known to be able to embrittle and to crack steel and other metallic materials. The existing concepts used to explain these phenomena are mainly based on the classical form of the Griffith-equation

$$\sigma = \left\{ \frac{2\, E\, \gamma_s}{\pi\, a} \right\}^{1/2}$$

with E being Young's modulus, a the crack length and γ_s the surface energy. The medium should reduce the surface energy and therefore promote cracking. In this connection, the TROIANO-ORIANI-theory /1/ should be mentioned for hydrogen cracking and the REHBINDER-effect for liquid-metal embrittlement.

However, relative to the total fracture work, the surface energy is very small. Therefore, Orowan extended the Griffith equation by introducing the real fracture energy (γ_{eff}) as the sum of the surface energy (γ_s) and the work of plastic deformation (γ_{pl}).

In hydrogen cracking of steel, for instance, the fracture work is at least a hundred times greater than the surface energy. Accordingly, the reduction of the work of plastic deformation has to be explained rather than that of the surface energy. The latter should be confined to describe the action of wetting agents.

THE TRIBOSORPTION CONCEPT

For metals, an embrittlement is usually caused by the formation of very fine precipitates impeding the movement of dislocations. The question arises in which way a foreign medium can produce the same effect.

The medium has to enter the matrix. As medium-induced cracking may reach considerable velocities, normal diffusion can be excluded as a transport mechanism; instead, a certain kind of pumping should occur. Solely atoms but not molecules can certainly penetrate the metal lattice. Whenever the medium is a compound, the molecules have to be dissociated a priori. Embrittling media which exist in the solid state at room temperature, have their main activities in the range of their melting point or slightly below it, i.e. partially in the solid state. The latter fact points out that solely the vapour phase is responsible for the embrittlement. An embrittling effect is caused by those elements which are almost unsoluble in the matrix. But the reverse case occurs in regions of current flow (slip bands), i.e. where the solubility is high. The reason may lie in the fact that m o v i n g dislocations are small regions of very high temperatures, thus being able to carry foreign atoms.

When the movement of dislocations is locally stopped the solubility of the foreign atoms drops resulting in a strong supersaturation. Now, the foreign atoms tend to agglomerate and to rebuild the compound (recombination) as far as former compounds (hydrogen, sulfides, carbides) are concerned. The places of agglomeration (accumulation) and recombination should be provided by voids induced by the reaction of dislocations carrying the foreign atoms. The media accumulating in voids should remain in the gaseous state to form pressure bubbles as a kind of short-time precipitates inhibiting further gliding. The process of embrittlement is schematically summarized in Fig. 1

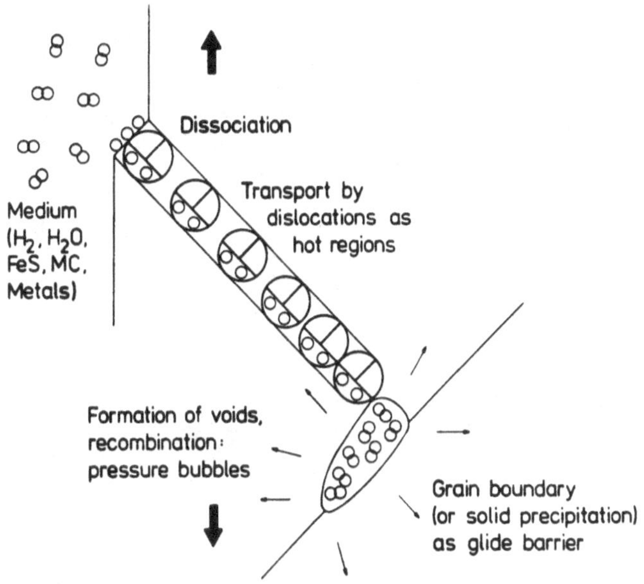

Fig. 1 Model of medium-induced embrittlement: During deformation the foreign atoms are pumped into slip-bands where they form pressure bubbles in deformation-induced voids

When the process takes place at very high temperatures ("hot cracking"), a fine dispersion of the attacking medium may be observed at the grain boundaries, since they are the places of gliding. This is particularly valid for regions only embrittled but not yet cracked. The so-called stress corrosion cracking is considered to be usually caused by hydrogen. Due to its mechanical activation the process of fast uptake and transport of foreign atoms is termed TRIBOSORPTION /2,3/.

The conclusions quoted here have mainly been drawn from failure analyses, some examples of which will be presented.

HYDROGEN CRACKING

An impressive fracture surface structure known to the welders is the so-called fisheye structure. Two of such structures are shown in Fig. 2a. Fisheyes are generated during the tensile or bending tests when welding is performed with moist electrodes etc. The crack zone is located around a larger defect in which hydrogen had accumulated. Due to their glittering appearance, fisheyes contrast with the surrounding brittle or ductile final fracture. In general, fisheyes are detected on a microscopic scale, only. The mini-fisheye shown in Fig. 2b was found beneath a weld.

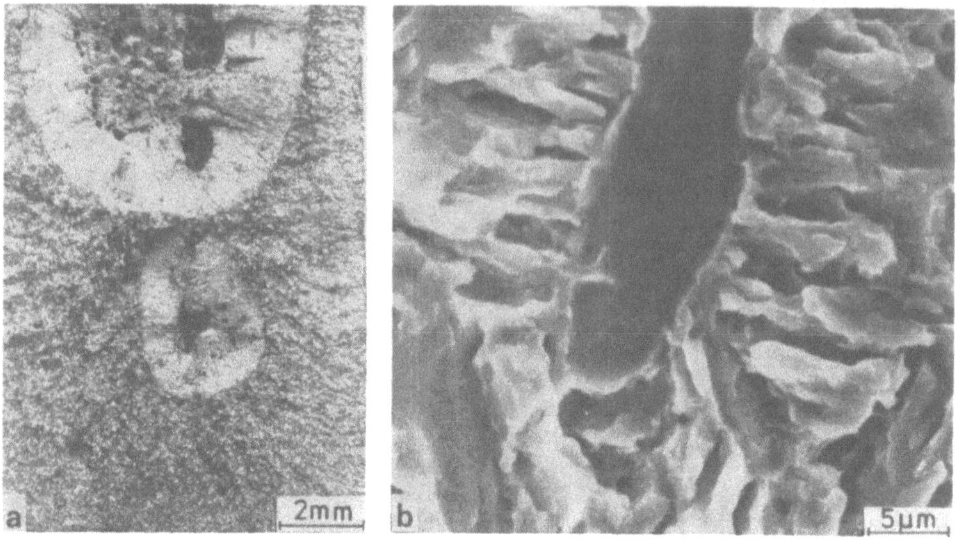

Fig. 2 Fisheyes
a) fracture surface of a welded-bent sample showing two fisheyes
b) mini-fisheye beneath a weld: a flat inclusion hole served as the crack centre.

The fracture facets of hydrogen cracks are much smaller than those of normal cleavage. Kikuta, Araki and Kuroda /4/ identified them as {110}-planes whereas cleavage follows the {100}-planes.

A strong hydrogen absorption under plastic tension as the basic process of cracking was revealed radiographically by Louthan et al./5/ in austenitic CrNi steel, Fig. 3.

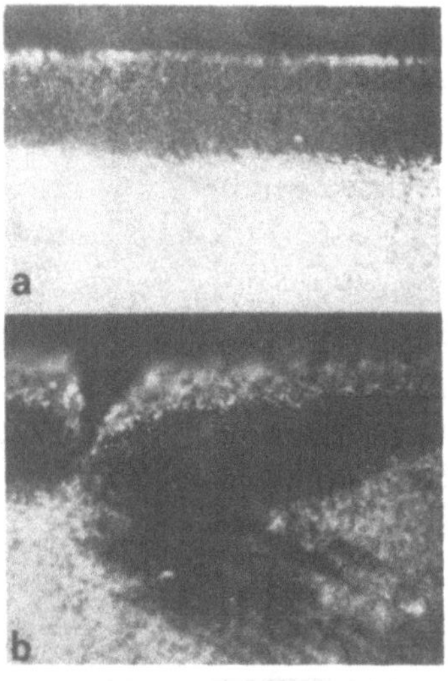

Fig. 3 Radiography of CrNi-steel penetrated by hydrogen (tritium) /5/
a) pure diffusion (sample unloaded)
b) tribosorption along slip-bands during deformation at a notch

Rotors of exhaustion turbo-superchargers are produced of a nickel alloy by casting. As carbon is nearly unsoluble in nickel, it is enriched in the melt during solidification, forming coarse (primary) carbides (see Figs. 4a,b). In order to attain a higher ductility, the parts are solution-annealed at a relatively high temperature. During subsequent air cooling extensive cracking occurred. The cracks were opened for fractographic investigations. A region of intergranular brittleness was found ahead of the cracks. In this region the primary carbides have disappeared, leaving cavities behind (Fig. 4c). The carbides now form a fine dispersion along the grain boundaries. Such particles are shown in Fig. 4d. The particles are surrounded by matrix walls (dimple walls). Thus, the carbide melt is unlikely to be dispersed simply by flowing along the grain boundaries.

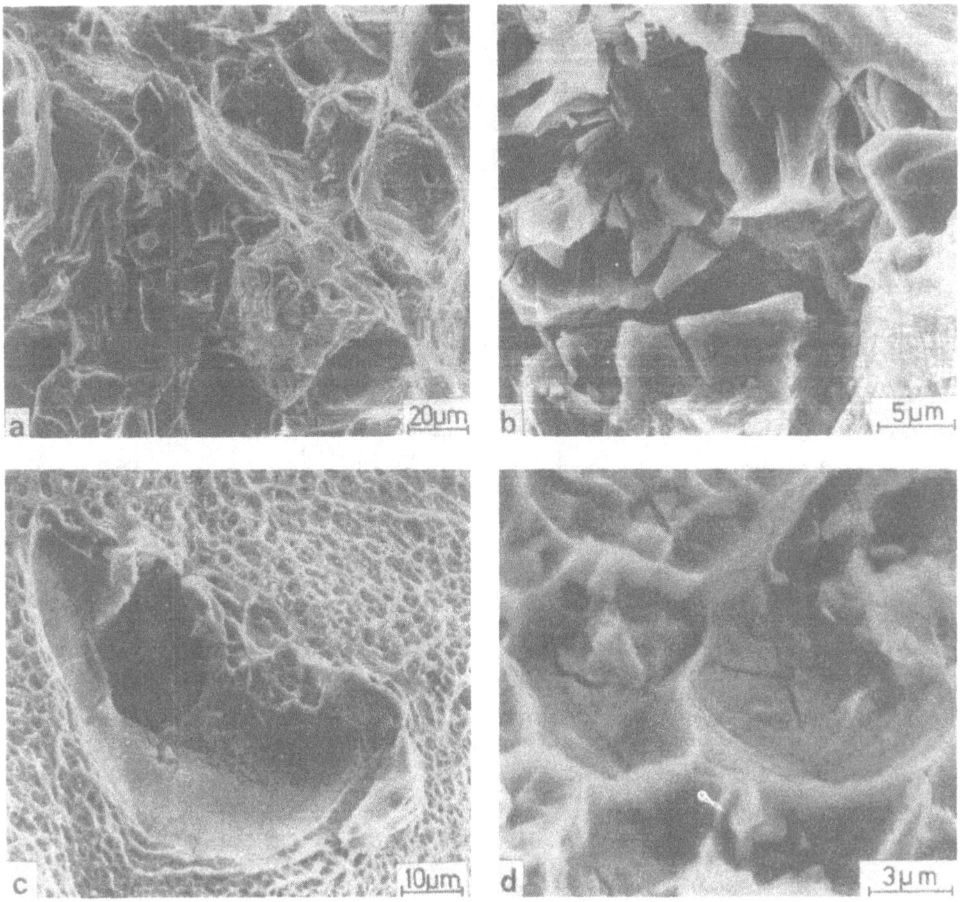

Fig. 4 Hot cracking of a nickel-alloy
a) and b) regular state: dimples with a large primary carbide
c) embrittled zone ahead of a hot-crack: grain boundary covered with fine dimples; a hole shows the former place of a primary carbide
d) fine carbide in the dimples: carbide dispersion

As the process of cracking partly remained subcritical, the distribution stage of the embrittling medium was frozen in.

HOT CRACKING OF STEEL

In structural steel sulfides are known to be responsible for hot cracking. Usually, the sulfide inclusions are large and flattened (Fig. 5a).

In the present example cracks were found in the heat-affected zone of a weld. Here again an embrittled ("overheated") region was detected in front of the cracks. Cracking occurred also intergranularly. As shown in Fig. 5b, the grain boundaries are covered with a dispersion of sulfides. Fan-like structures ("sulfide flowers") can be identified.

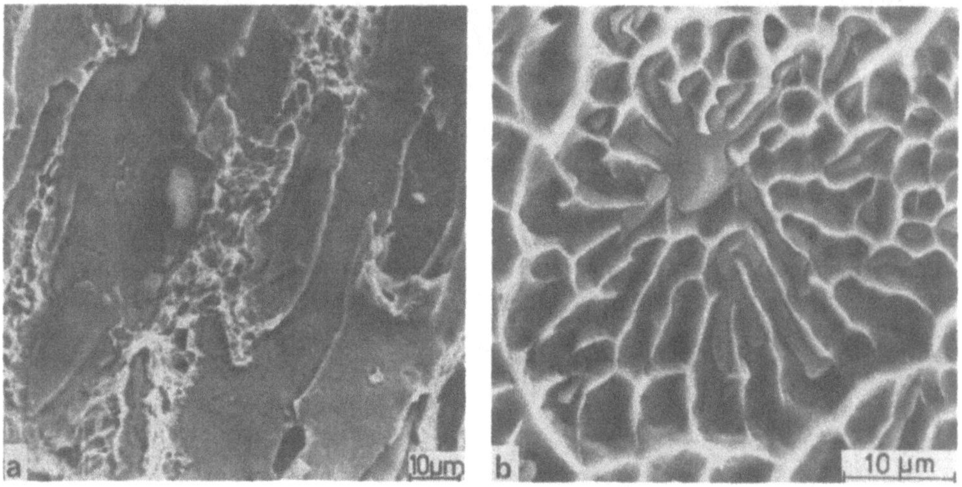

Fig. 5 Hot cracking of structural steel
a) regular state: flat sulfide inclusions
b) embrittled zone with integranular dimple fracture: sulfide dispersion

The grain boundaries were formed in the primary austenite, which does not exist at room temperature, i.e. these boundaries are only shells of impurities.

RELAXATION CRACKING

Relaxation cracking is known to occur in the heat-affected zone of welded pipes and vessels made of low-alloyed steel when they are annealed for stress-relief. An example is shown here of a hot-steam pipe having burst along the longitudinal weld some days after starting the plant. This kind of cracking is generally considered to be due to a precipitation hardening but also to overheating. Cracking occurred intergranularly; the grain boundaries show an extremely fine dispersion of sulfides (Fig.6)

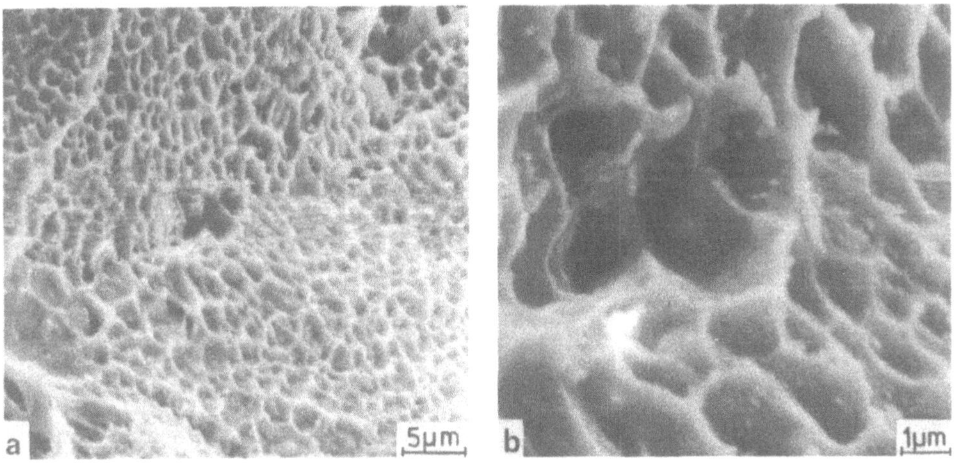

Fig. 6 Relaxation cracking
a) grain boundaries with fine dimples
b) detail: microsulfides in dimples

Finally, a historic procedure should be mentioned in connection with deformation-induced fast mass transport. The old metallurgists were not able to melt the iron ore but they had to reduce it in the solid state. Afterwards, they purified the steel blooms by forging out the sulfides. Thus, metals may be said to behave like a sponge.

REFERENCES

1. R.A. Oriani, Ber. Bunsenges. Physik Chemie 76 (1972) 848-857
2. G. Heinicke (Ed.), Tribochemistry, Akademieverlag, Berlin 1984
3. M. Möser, In: Electron Microscopy in Solid State Physics, H.Bethge, J. Heydenreich (Eds.), Elsevier, Amsterdam 1987, p.366-385
4. Y. Kikuta, T. Araki and T. Kuroda, In: Fractography in Failure Analysis, ASTM STP 645, Philadelphia 1978, p. 107-127
5. M.R. Louthan, G.R. Caskey, J.A. Donovan and D.E. Rawl, Mater. Sci. Eng. 10 (1972) 357-368

THE DEVELOPMENT OF FRACTURE UNDER UNLOADING

I. Novikov, V. Yermishkin, V. Grebneva, and E. Samoilov

Academy of Sciences of the U.S.S.R., Baikov Institute of Metallurgy, Leninsky Prospect 49, Moscow, USSR

ABSTRACT

In the present paper experimental data on the growth rate of plastic zones estimated at various conditions are discussed and interpreted.

EXPERIMENTAL RESULTS

[Ø1Ø] oriented microsamples of molybdenum single crystals were in situ deformed in the HVEM. The samples were stretched until microcracks were generated. The initiation and propagation of cracks occurred in accordance with the dislocation shearing mechanism [1]. A narrow plastic zone was formed normal to the applied load and expanded into a region being not transparent in the microscope before deformation. As soon as the crack path became localized inside the plastic zone, we changed the load by a fixed step of value ΔP_o, Fig. 1. The sequence of loading and unloading of the samples is schematically shown in Fig. 2. The velocities of the load changes were adjusted by the heat current of the wire controlling the operation of the spring loading device. Low velocities

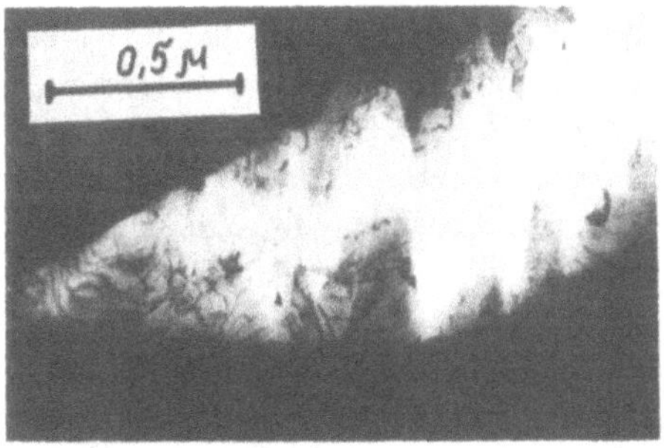

Fig. 1: Propagation of the plastic zone ahead of a crack into regions being not transparent in the electron microscope

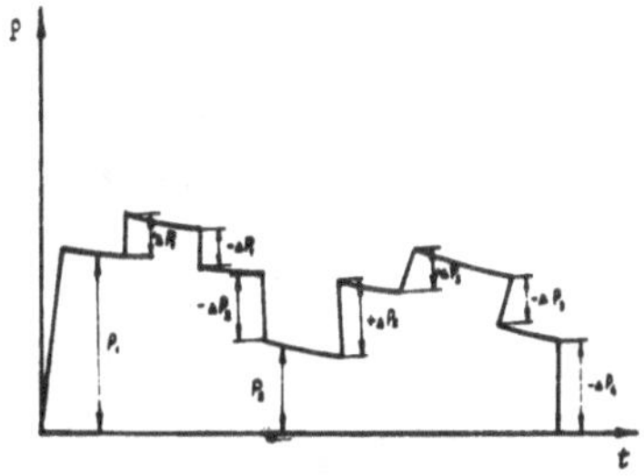

Fig. 2: The loading scheme used at the in situ straining experiments

were realized by current changes within 2Ø to 6Ø s and high ones by instantaneously changing the current The propagation of the crack and of the plastic zone was observed at an acceleration voltage of 1ØØØ kV and continuously recorded with the help of a video system. Tab. 1 contains the parameters of the load changes and the related velocities of the propagation of the plastic zone.

Table 1: Velocities of the propagation of the plastic zone at instantaneous load changes

loading condition	P[p] load level	P[p] load step	time of load change [s]	plastic zone velocity [cm/s]
loading	50	+25	1	$0.16*10^{-3}$
constant load	75	-	-	$0.36*10^{-3}$
unloading	75	-25	1	$0.95*10^{-3}$
loading	0	+50	1	$0.102*10^{-4}$
unloading	50	-50	1	$0.31*10^{-4}$
loading	0	+75	60	0
unloading	75	-75	60	$0.49*10^{-3}$
loading	0	+75	1	$0.22*10^{-3}$
unloading	75	-75	1	$1.5*10^{-3}$
loading	75	+50	10	$0.463*10^{-4}$
unloading	125	-50	10	$1.6*10^{-4}$

DISCUSSION

According to the data given in the Tab. 1 the following conclusions may be drawn:

1. The plastic zone increases at both loading and unloading. Referred to constant load steps, the propagation velocity of the plastic zone during unloading is larger than that during loading by a factor of 3 to 5.
2. The effect increases with increasing loading rate.
3. The velocity of the propagation of the plastic zone depends on both the load and the crack length.

The crack and the plastic zone propagate simultaneously. A preferred formation of plastic zones during unloading was also found in LiF [2], the results, however, were not satisfactorily explained. In our opinion it can be explained as follows. The load applied on the microspecimen creates a local stress field near the crack tip. The tensor components of this stress field can be expressed by the stress intensity factor K_σ. The emission of the dislocations from the crack tip results in the formation of a plastic zone in front of the crack tip. In this case the dislocation stress field shields the stress field of the crack, which may be described by the stress intensity factor of the effective value: $K_{ef} = K_\sigma - K_d$. K_d is the stress intensity factor related to the dislocations in the plastic zone. If K_σ is equal to K_d, dislocations are not any more emitted from the crack tip. The condition for dislocation emission may be written as $K_\sigma > K_d$. After instantaneous unloading, K_σ is zero and the dislocations arranged in front of the plastic zone are effected by stress fields of the dislocations in the plastic zone of the crack. Thus, dislocations with a Burgers vector directed away from the plastic zone may leave them. In that way the plastic zone is extended. Similarly dislocations with a opposite sign of the Burgers vector may increase the plastic zone during loading.

REFERENCES

1 I.I. Novikov, V.A. Yermishkin, V.S. Grebneva, V.G. Sharkov, and E.H. Samoilov, Fiz. Tverd. Tela 25 (1983) 2096-2104

2 K.Y. Chia, S.J. Burns, Scripta Metall. 18 (1984) 467-72

3 B.S. Majumdar, S.J. Burns, Acta Metall. 29 (1981) 579-88

THE FAILURE OF THIN SPECIMENS OF BCC AND FCC METALS AND ALLOYS

M.N. Spassky

Institute of Physical Metallurgy of the I.P. Bardin Central
Research Institute for Iron and Steel Industries, Moscow, U.S.S.R.

The deformation of thin specimens by means of a fast or slow penetration of an indentor into the rigid rim of foil discs (thin disc specimens for transmission electron microscopy) at 20°C leads to the generation of brittle and ductile cracks in them. Observation of the path of cracks and the analysis of plastic deformation at the edges of cracks in the foils of BCC and FCC metals and alloys with different initial dislocation structures (recrystallized, cold-worked crystals, martensitic steels containing up to 0.8 % carbon) have shown the following.

Brittle cracks were produced only in foils of BCC recrystallized alloys. Transcrystalline brittle cracks in Cr, Mo and Fe foils had dislocation-free edges. The traces of the edges were parallel to the traces of the $\{001\}$ planes on the surface of the foils. In microsections of foils thicker than about 100 nm, tangles of dislocations appeared near the blunt tips ($C \leqq 20 \mu m^{-1}$) of the stopped cracks with a density of $\approx 10^{10} cm^{-2}$. In microsections of foils up to a thickness of $\approx$ 30 nm, tangles of dislocations have not been observed near the tips of the stopped cracks but the foils were locally bent. Intercrystalline brittle cracks were mostly generated in deformed polycrystalline Mo foils. Near the tips of these cracks there were no tangles of dislocations.

Ductile cracks in the recrystallized foils were arising from stress concentrators. The lattice distortions along the edges of the ductile cracks were proportional to both the initial density of dislocations and to the thickness of the destroyed microsection of the foil. The spreading ductile cracks in foils of martensitic steels always started with the generation of leading microcracks in front of the tip of the main crack, which was followed by their subsequent merging after the bridges between them had broken. This was accompanied with the formation of tangles of dislocations of the same sign. Lattice distortions at the edges of the ductile cracks were growing along with the degree of complexity of the initial dislocation structure of the foil.

The conclusion may be drawn that foils deformed at 20°C retain the tendency to brittle failure shown by the bulk specimens. Ductile failure of foils became similar to that of the bulk specimens when the thickness of the foils exceeded the average distance between the elements of their structure. Otherwise the slip dislocations freely reached the surface of the foil. Apparently, the behaviour of cracks in foils of alloys with a complicated structure is the more similar to the behaviour of cracks in bulk specimens and the more distinctly it displays the peculiarities of this phenomenon, the better the foils retain the original structure of bulk specimens.

ENERGETIC ASPECTS OF THE FRACTURE OF MATERIALS UNDERGOING A BRITTLE-DUCTILE TRANSITION

A.D. VASILEV

The Institute of Problems of Materials Science,
Krjijanovskogo str., 3, Kiev 252680, U.S.S.R

ABSTRACT

SEM studies of fracture surfaces of polycrystalline materials indicate various fracture mechanisms during the brittle-ductile transition under uniaxial tension. In view of the energy for fracture, the change of the mechanisms is determined by processes being energetically favourable at the given temperature and loading rates. The general temperature dependence of the fracture toughness can be represented by a curve having maximum at the upper limit of the brittle-ductile transition.

INTRODUCTION

In materials undergoing a brittle-ductile transition, over a wide temperature range various fracture mechanisms are identified by scanning electron microscopy (SEM). Under uniaxial tension the following sequence of fracture mechanisms is generally observed with increasing temperature [1, 2 etc.]: cleavage (Fig. 1a); brittle intergranular fracture, if the grain boundaries are weak (Fig. 1b); cleavage or intergranular fracture with plastic relaxation at the crack tip (Figs. 1c and 1d); dimple fracture (Fig. 1e). Other processes observed are dimple fracture along grain boundaries, delamination combined with cleavage or dimple fracture, and, at sufficiently high temperatures, high-temperature intergranular fracture (Fig. 1f), which is the most typical process in ceramic materials.

The data published, e.g. fracture mechanism maps for various materials [3] and a number of publications on fractography, allow the nature of the change of the fracture mechanisms to be discussed.

The present study is an attempt to carry out such an analysis for various materials over a wide temperature range by comparing the energies necessary for fracture with those required for various mechanisms.

Let us now consider the influence of the temperature on the fracture toughness of different BCC metals for certain fracture mechanisms.

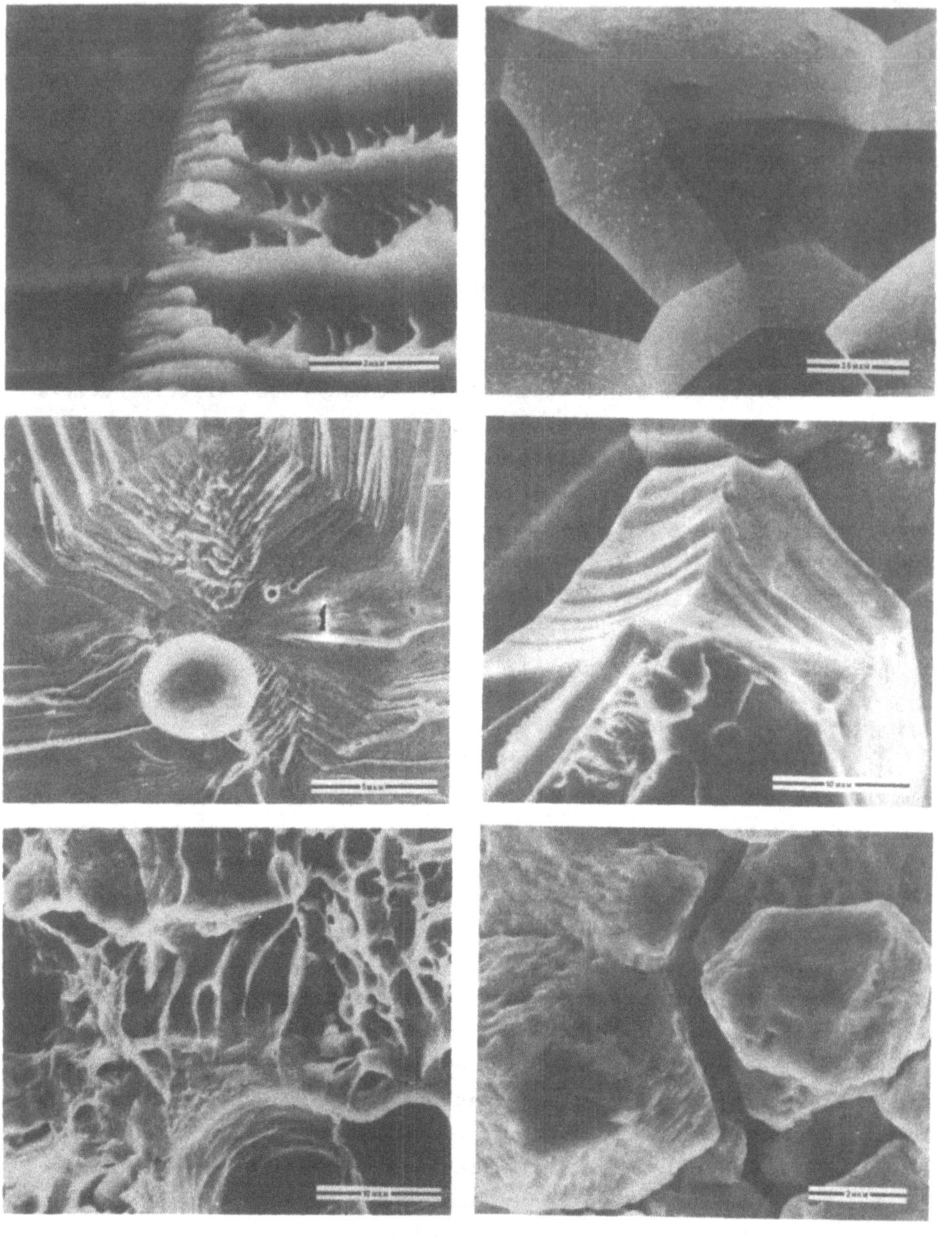

Fig. 1. SEM images of fracture surfaces indicating the following fracture mechanisms: a) cleavage; b) brittle intergranular fracture; c) cleavage with lines of periodic relaxation; d) intergranular fracture due to plastic deformation of grains; e) dimple fracture; f) high-temperature intergranular fracture.

CLEAVAGE. The greater part of studies on the influence of the temperature on the fracture toughness [1, 4] is related to the temperature dependence of the fracture toughness resulting from cleavage.

As shown in [4], the temperature dependence of the fracture toughness by cleavage γ_{cl} can be described by the dependence of the following type:

$$\gamma_{cl}(T) = \gamma_0(T) + A/\sigma_y^2(T), \tag{1}$$

(Fig. 2, curve 1), where $\gamma_0(T)$ is the true surface energy, weakly depending on the temperature; A is a constant; $\sigma_y(T)$ is the temperature-dependent part of the yield stress.

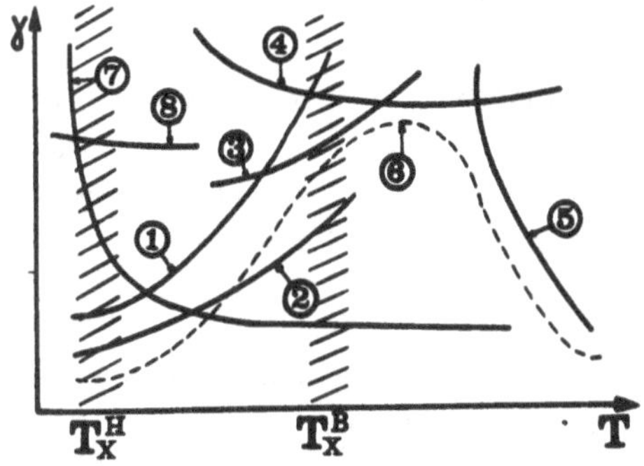

Fig. 2. Schematic representation of the temperature dependence of the fracture toughness of some BCC metals: 1 -cleavage; 2 - brittle intergranular fracture; 3 - cleavage with delamination; 4 - dimple fracture; 5 - high-temperature fracture mechanisms; 6 - resulting dependence; 7 - yield stress; 8 - fracture stress; T_x^H and T_x^B - lower and upper limits of the brittle-ductile transition.

It should be pointed out that the fracture toughness significantly increases with the temperature dependence of the yield stress considerably decreasing [4 - 7] (Fig. 2, curves 1 and 7).

BRITTLE INTERGRANULAR FRACTURE. As it follows from [6 - 8] the toughness of materials with brittle intergranular fracture increases with increasing testing temperature. Thermomechanical treatments influence the extent of the fracture toughness and its temperature dependence [7, 8]. The reasons of this difference are primarily associated with different values of the intergranular strength, which is considerably controlled by the segregation of impurities.

Intergranular fracture occurs if the tensile stress exceeds the intergranular strength σ_{ig}, at least within a certain zone of size X. The intergranular fracture toughness K_{1c}^{ig} can be written as

$$K_{1c}^{ig} = K_{ig}^{\circ} + BX\sigma_{ig}, \tag{2}$$

where K_{ig}^{o} is the toughness of a completely brittle fracture when X=Ø and B is aconstant [6].

The size X of the zone was determined by the method of finite elements. It turned out that

$$X = C\,\sigma_{ig}/\sigma_y - D\,, \tag{3}$$

where C and D are constants.

To estimate the influence of the temperature on the intergranular fracture toughness, the temperature influence on the fracture stress σ_{ig} has to be evaluated. The temperature dependence of the yield stress is well-known.

The temperature dependence of intergranular fracture stress was estimated in an impact test of sharply notched specimens. It has been found for an intergranularly fractured steel that the intergranular fracture stress is independent of the temperature over the investigated temperature range -100 to 50° C, i.e., $\sigma_{ig}(T)$ = const.

Thus, like the cleavage fracture toughness, the temperature dependence of intergranular fracture toughness is determined by the temperature dependence of the yield stress and can be expressed in a generalized form by a dependence which is inversely proportional to the yield stress (Fig. 2, curve 2).

DIMPLE FRACTURE. The temperature dependence of the dimple fracture toughness γ_d may be descibed by the following model.

Let us consider the work of the formation of a dimple fracture as the work done by the external forces yielding the stress σ which moves away the crack surfaces by a distance equal to the dimple depth $h_d(T)$. Then

$$\gamma_d(T) = \tfrac{1}{2}\,\sigma(T)\,h_d(T)\,. \tag{4}$$

Precise values of stresses at the crack tip are still unknown. Since the plastic relaxation at the crack tip limits the stresses to a level which is proportional to the yield stress σ_y, we can assume:

$$\sigma(T) \approx \sigma_y(T)\,. \tag{5}$$

From [2 et al.] it follows for a sufficiently wide temperature range

$$h_d(T) \approx const\,. \tag{6}$$

According to the considerations made above and to eqs. (5) and (6), the temperature dependence of dimple fracture toughness

$$\gamma_d(T) = \tfrac{1}{2}\,h_d \cdot \sigma_y(T) \tag{7}$$

(Fig. 1, curve 4) is mainly determined by the temperature dependence of the yield stress and decreases with increasing temperature. This phenomenon is observed in certain steels and aluminium alloys [9, 10].

HIGH-TEMPERATURE INTERGRANULAR FRACTURE. The high-temperature intergranular fracture of certain materials is caused by intergranular sliding. According to [11], in this case the fracture toughness can be described as:

$$\gamma_{HT} = \sigma^2 \pi d / 12G \quad , \tag{8}$$

where d is the grain size and G the shear modulus. At high temperatures under the applied stress the material deforms with a creep rate $\dot{\varepsilon}$ [12], with

$$\sigma = (B_0 T \dot{\varepsilon} d \exp(U_{ig}/kT))^{1/2} , \tag{9}$$

where B_0 is a constant and U_{ig} - activation energy of diffusion of the impurities along the grain boundaries.

By combining (9) and (8), we obtain

$$\gamma_{HT}(T) = C_0 T \dot{\varepsilon} d^2 \exp(U_{ig}/kT) , \tag{10}$$

where C_0 is a constant.

Accordingly, the toughness of the high-temperature intergranular fracture decreases with increasing temperature (see Fig. 2, curve 5).

According to these considerations, the change of fracture mechanisms is determined by the energy of the processes operating under the specific loading conditions. The total temperature dependence of fracture toughness is composed of the temperature dependences of individual mechanisms which are predominant in certain temperature intervals. The resulting temperature dependence of fracture toughness is represented by curve 6 in Fig. 2. The left-hand side of the maximum is determined by cleavage, whereas the right-hand part is controlled by mechanisms of coalescence of the pores and the high-temperature types of fracture.

The course of certain curves in this diagram depends on the kind of the material and of its structure. This may explain the experimental finding that the sequence of the fracture mechanisms in polycrystalline Cr and Mo is essentially different [13]. Thus, up to the temperature of the transition into the ductile state the following behaviour is observed. While in Cr the temperature dependence of the fracture toughess is only determined by cleavage, in Mo it arises from both intergranular fracture and cleavage with periodic relaxation and delamination.

In materials having weak grain boundaries, intergranular fracture directly results in dimple fracture. Then, the fracture toughness is determined by intergranular fracture, only. In this group of materials, dimple fracture is sometimes not observed so that curve 4 lies considerably higher than the point of intersection of the curves 2 and 5. In that case the low-temperature intergranular fracture is directly converted into high-temperature intergranular fracture. This is observed in ceramic materials.

The relation between the fracture toughness and the yield stress can be represented in a general form as $\gamma \sim \sigma_y^{-n}$ [14]. However, as it follows from the data given, the sign and the value of the exponent n depend on the temperature of the loading and of the operating fracture mechanism. As examples, for cleavage n = -2 and for the coalescence of pores n = 1 have to be used.

REFERENCES

1. A.D. Vasilev, Scanning electron fractography of body centered cubic (BCC) metals. SEM/1986, Vol.111, p. 917-32.
2. V.V. Rybin, High Plastic Deformations and Fracture of Metals, Metallurgija, Moscow 1986, in Russian
3. M.F. Ashby, in: Advances in Applied mechanics, vol.23, I.W. Hutchinson and T.Y. Wu (Eds.), Academic Press, London 1983, p. 117-77.
4. A.D. Vasilev, S.A. Firstov and V.I. Trefilov, Phys. Chem. of Mater. Treat., 1981, No.3, p. 100-4, in Russian
5. J.F. Knott, Fundamentals of Fracture Mechanics, Metallurgija, Moscow 1978, p. 138, in Russian
6. O.N. Romaniv, Fracture Toughness of Structural Steels, Metallurgija, Moscow 1979, in Russian
7. A.D. Vasilev, N.I. Danilenko and S.A. Firstov, in: Physics of Fracture, IPM AN UkrSSR, Kiev 1985, p. 246-8, in Russian
8. E.I. Uskov, A.V. Babak and A.P. Dyachkov, Problem of Strength, 1985, No.1, p. 40-3, in Russian
9. A.Ya. Krasovsky, Brittleness of Metals at Low Temperatures, Naukova Dumka, Kiev 1980, in Russian
10. S.I. Kishkina, Fracture Resistance of Aluminum Alloys, Metallurgija, Moscow 1981, in Russian
11. J. Mikin and N. Petch, in: Fracture of Solids, Metallurgija, Moscow 1967, p. 198-221, in Russian
12. M.V. Grabsky, Structural Superductility of Metals, Metallurgija, Moscow 1975, in Russian
13. A.D. Vasilev, V.I. Trefilov and S.A. Firstov, Ukranian Phys. J., Vol.30 (1985), No.4, p. 603-6, in Russian
14. O.N. Romaniv, Phys. Chem. Mech. Mat., 1981, No.100, p. 28-45, in Russian

MICROMECHANICAL MODELS FOR THE NUMERICAL PREDICTION OF FRACTURE RESISTANCE

B. Voss, D.-Z. Sun, W. Schmitt

Fraunhofer-Institut für Werkstoffmechanik, Wöhlerstr. 11,
D-7800 Freiburg, FRG

ABSTRACT

In order to study the correlation between microstructure, microscopical damage and macroscopical material failure two damage models were implemented into the finite element program ADINA. This paper focusses on the model that is based on the Gurson yield function and incorporates the complete micromechanical model into the constitutive relations. Quantitative metallography was applied on specimens with different amounts of deformation and damage to determine physical parameters of the model related to the material microstructure. The numerical results for the local and global behaviour of specimens were compared with experiments.

INTRODUCTION

In numerous cases ductile fracture can be characterized by global failure criteria of continuum mechanics, as the J-integral or the crack tip opening displacement CTOD. However these single parameter criteria do not take into account micromechanical aspects, local inhomogeneities or stress triaxiality. Several authors [1,2] have shown that the geometry dependence of fracture toughness (e.g. because of triaxiality) makes the application of the global fracture criteria difficult, especially for complex geometry and loading conditions.

A series of damage models, based on the concept of continuum damage mechanics have been established in order to find alternatives. Here, the ductile fracture process is described by nucleation, growth and coalescence of voids. In this work a combined numerical and experimental effort was taken to use these mathematical formulations implemented into the finite element code ADINA [3] to predict the specimen behaviour of the experiments.

MODELLING OF DUCTILE FAILURE

Ductile damage is normally described by three processes that may be active at the same time at different locations of a structure but even in parallel in close/ly neighboured regions:

Void nucleation

Under external loading after a sufficient amount of strain or stress voids nucleate e.g. by seperation of the matrix material from an inclusion or by dislocation processes at grain boundaries.

Void growth

Existing voids grow with increasing strain or stress. Growth models may include multiaxiality of the stress state and interaction of neighbouring voids.

Void coalescence

Voids of sufficient size and density will coalesce, mainly by shear fracture of the ligaments, thus forming microcracks or initiating crack extension of existing macro cracks.

THE MODIFIED GURSON MODEL

This method for ductile fracture analysis has been introduced by NEEDLEMAN and TVERGAARD [4] on the basis of the yield condition suggested by GURSON [5]. A plastic potential applicable to porous solids is given by

$$\Phi = \frac{3\sigma'_{ij}\sigma'_{ij}}{2\sigma_m^2} + 2q_1 f^* \cosh\left(\frac{\sigma_{kk}}{2\sigma_m}\right) - [\,1 - (q_1 f^*)^2\,] = 0 \qquad (1)$$

The parameter q_1 was introduced by TVERGAARD [6] to improve the prediction of the Gurson model at small f-values. f^* is a function of the void volume fraction f. For $f^* = 0$ the plastic potential (Equ. 1) is obviously identical with that of von Mises. If f^* reaches the limit $1/q_1$ the material loses its load carrying capacity because all stress components have to vanish in order to satisfy Equ. 1. f^* is a function of the current void volume f:

$$f^* = \begin{cases} f & f \le f_c \\ f_c + f_{uc}\,(f - f_c) & f \ge f_c \end{cases} \qquad (2)$$

where f_{uc}=slope factor, emphasizing the effect of voids by coalescence, for $f>f_c$=threshold value.

The development of f is given by:

$$\dot{f} = \dot{f}_{nucleation} + \dot{f}_{growth} \qquad (3)$$

with $$\dot{f}_{nucleation} = B\,(\dot{\sigma}_m + \dot{\sigma}_{kk}/3) + D\,\dot{\varepsilon}^p_m \qquad (4)$$

and $$\dot{f}_{growth} = (1 - f)\, \dot{\eta}^p_{kk} \quad . \tag{5}$$

The first part in Equ. 4 models void nucleation controlled by the maximum normal stress at the phase boundary particle/matrix and the second part models void formation caused by the equivalent plastic strain ε^p_m. Following CHU and NEEDLEMAN [7] D is described by

$$D = \frac{f_N}{s_N \sqrt{2\pi}} \exp\left[-\frac{1}{2}\left(\frac{\varepsilon^p_m - \varepsilon_N}{s_N}\right)^2\right] \tag{6}$$

with f_N=volume fraction of void forming particles, ε_N=mean value of the equivalent plastic strain for void nucleation and s_N=standard deviation of ε^p_m.

NUMERICAL AND EXPERIMENTAL INVESTIGATIONS

FE-meshes of some specimens investigated are shown in Fig. 1, cylindrical notched tensile specimens with different notch radii and a fracture mechanics specimen (C(T)) with an enlarged section of the crack tip region.

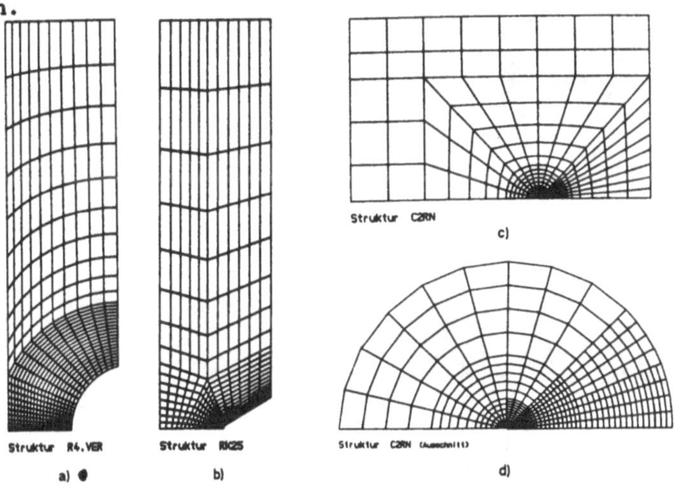

Fig. 1 FE-meshes of some specimens. (a) Axisymmetric notched specimen, notch radius 4 mm. (b) as before, notch radius 0.25 mm. (c) Compact tensile specimen (C(T)). (d) Crack tip region of (c)

MATERIAL

All experiments and calculations reported here were performed with a ferritic steel type A 710 grade A Cl.3. The chemical composition and mechanical properties are given in Tab. 1. This steel was chosen because of its homogeneity and the spherical shape of its MnS-inclusions (Fig 2).

Element	C	Mn	P	S	Ni	Cr	Mo	Cu	Si	Al	Nb
Weight-%	.06	.5	.013	.004	.88	.75	.21	1.16	.28	.03	.039

σ_y [MPa]	σ_u [MPa]	elong. [%]	RA [%]
619	707	24.1	72

Tab. 1 Chemical composition and mechanical properties of the steel A 710 grade A Cl.3

NUMERICAL RESULTS

Determination and Selection of Parameters

Fig. 2 shows a sketch of a notched specimen cut after formation of a microcrack and some examples of inclusions found in different regions. Inclusions in the elastic region of the specimen are bonded to the matrix material. Although high plastic deformations are limited to the reduced section close to the notch, up to the dash line debonded inclusions dominate.

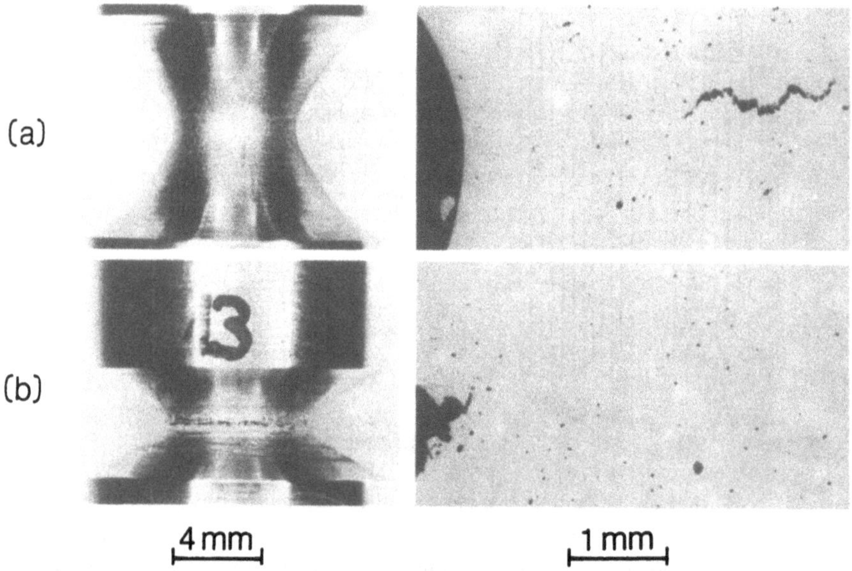

Fig. 2 Cut through a fractured notched tensile specimen. Bonded inclusions (O -symbols, upper micrograph) above the dash line, debonded inclusions below (+ -symbols, center micrograph and bottom micrograph)

Thus the inclusion volume determined by quantitative optical microscopy was used as an initial void volume f_o=0.00057 in the numerical simulation. B in Equ. 4 was set to be 0 because no additional voids could be nucleated at these inclusions. The parameter D was calculated using ε_N=0.3, s_N=0.1 (from [4]) and f_N=0.004. As in [6] q_1 was set to 1.5.

By quantitative analysis of sections of specimens which were unloaded close to final rupture a volume fraction of voids at fracture f_f=0.15 could be determined. Using the condition $f^*(f_f)=1/q_1$ from Equ. 1 this allows to calculate the slope f_{uc} in Equ. 2

$$f_{uc} = (1/q_1 - f_c) / (f_f - f_c) \tag{9}$$

as a function of the void volume fraction at onset of coalescence f_c. A good fit of the sudden load drop before final rupture of a smooth tensile bar was obtained with f_c=0.03.

Simulation of Notched Tensile Specimens

The global behaviour of tensile specimens with different notch radii could be simulated in close agreement with the experiments.

The model predicts localised damage in the center region for a notched tensile specimen with a notch radius of 4 mm and close to the notch for 0.25 mm radius. Fig. 3 shows the respective specimens unloaded before final rupture. The upper micrograph shows the localized void growth region surrounding the crack in the center of the specimen and the lower one shows first crack extension from the root of the notch, both being in good agreement with the numerical predictions, presented in more detail in [8,9].

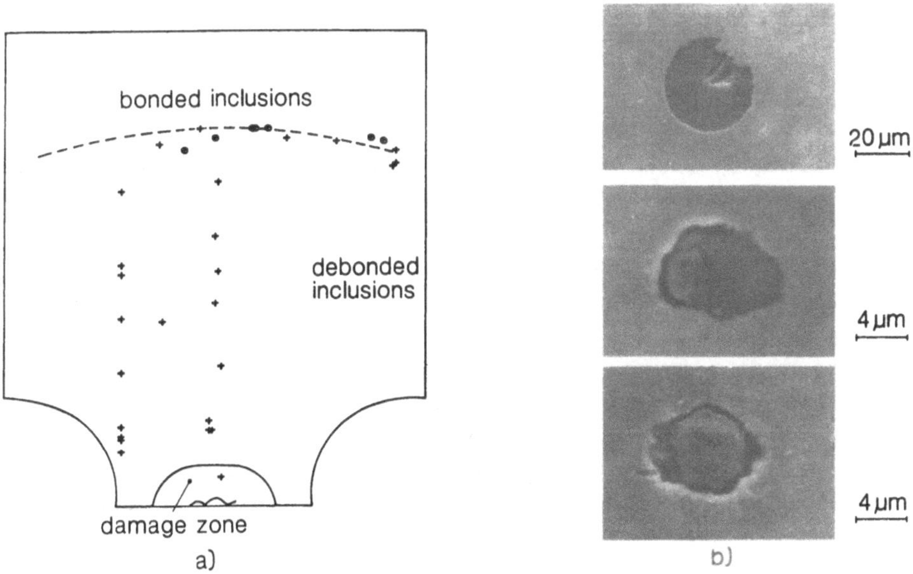

Fig. 3 Notched regions of specimens unloaded close to rupture. (a) notch radius 4 mm. (b) notch radius 0.25 mm

Simulation of Cracked Specimens

The behaviour of the cracked specimens was predicted using the same set of parameters as used for the tensile specimens. In addition it was necessary to introduce a critical length l_c [9] over which f has to exceed f_c to trigger the acceleration in Equ. 2. The mesh size has to be smaller than l_c, and l_c has to be adjusted once by optimizing the agreement between the experimental and the numeric load vs. displacement diagrams.

The results are presented as J-Δa-curves because those are often used in fracture mechanics to characterize the ductile fracture resistance of materials. Fig. 4 shows the experimental curve of a sidegrooved C(T)-specimen with B=25 mm and W=50 mm. The simulation curve resulting from the modified Gurson model is in close agreement with the experiment. The initiation and extension of the crack is obtained by the loss of load carrying capacity in elements ahead of the crack tip due to accumulation of damage.

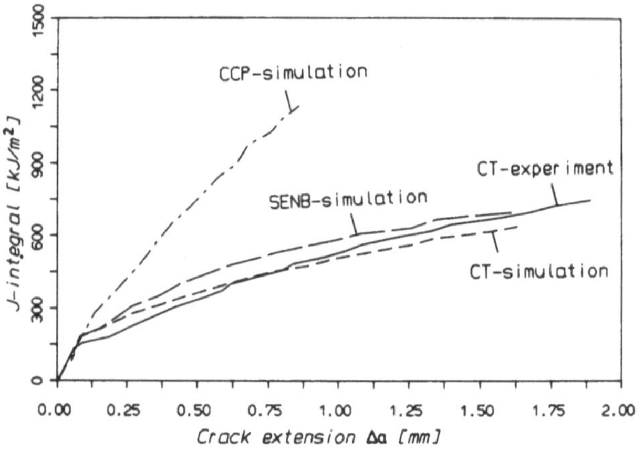

Fig 4 Experimental J-R curve of a C(T)-specimen and simulated curves for specimens with different loading conditions.

The same set of parameters and details of the crack tip meshing were used to calculate the behaviour of other specimen types in order to investigate the influence of stress state. For the bend specimen (SENB) with bend loading prevailing as in the C(T)-specimen the variation of stress state at the crack tip causes only slight changes of the R-curve. In contrast, the slope of the R-curve is significantly increased for the center cracked pannel (CCP) with pure tensile loading. This prediction is in agreement with common experience and indicates that this model covers variations of stress state as well.

CONCLUSIONS

The numerical model based on the Gurson yield function can predict the fracture behaviour of cracked specimens after fitting the material specific parameters to the results of simple tests with uncracked specimens.

It could be shown that the effect of multiaxiality is modelled by the modified Gurson model. Thus it can be expected that the prediction of the fracture behaviour of structures with complex loading conditions can be improved compared to the single or two parameter fracture criteria used now. Because of the high numeric effort and present limitations of computer power the application is still limited to basic research and laboratory specimens.

A number of material specific parameters involved in the models are measurable in principle as e.g. the initial inclusion volume fraction, the void volume fraction at different levels of plastic deformation or their size distributions. At present it is feasible to estimate some of them or to measure them with limited accuracy and to adjust some others on the basis of simple tensile experiments. This allows to minimize the effort for time consuming microscopic investigations that would be needed to get results with statistical significance.

On the other hand more sophisticated microstructural investigations may allow to identify and quantify the relevant material properties involved in the models and to improve these models. Additional experiments with different materials, varying inclusion size and macroscopic flow properties are in progress.

REFERENCES

1 D.-Z. Sun, D. Dormagen and W. Dahl, Steel Research 56 (1985), 445-9

2 D.Aurich and E. Sommer, Steel Research 59 (1988), 358-67

3 ADINA, Report ARD 87-1, ADINA R&D, Inc., Watertown, Mass., USA

4 A. Needleman and V. Tvergaard, J.Mech.Phys.Solids 32 (1984), 461-90

5 A.L. Gurson, J.Engng.Mat.Tech. 99 (1977), 2-15

6 V. Tvergaard, Int.J.Fract. 18 (1982), 237-52

7 C.C. Chu, A. Needleman, J.Engng.Mat.Techn. 102 (1980), 249-56

8 D.-Z. Sun, D. Siegele, B. Voss and W. Schmitt, Fatigue Fract.Engng. Mater.Struct. 12 (1988), 201-12

9 D.-Z. Sun, B. Voss and W.Schmitt, presented at Europ.Symp. on Elast.-Plast.Fract.Mech., Freiburg, FRG, Oct. 9.-12.1989, Proc. to be published by MEP, Bury St.Edmunds

The authors like to thank the Deutsche Forschungsgemeinschaft (DFG, German Science Foundation) for funding this research program.

INVESTIGATIONS ON THE CRACK TIP BLUNTING OF MEDIUM STRENGTH STEEL

I. Wagner, M. Schröder

Technische Universität "Otto von Guericke" Magdeburg

The direct observation of crack tip blunting in the in-situ tests is possible with a scanning microscope. A video tape demonstrates the processes of deformation at the crack tip. The different stages from the blunting process to the crack propagation are shown. The investigations were carried out on ferrite-pearlite and austenitic steel.

The blunting process is characterized by a strong stretch-formed area, the stretch zone. This area is represented in Fig. 1 and 2.

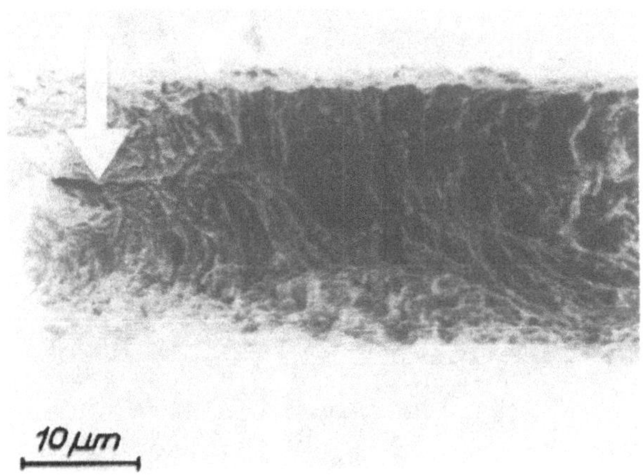

Fig. 1 Crack ground

Fig. 1 shows a view of the crack ground which has already a first crack (see ↓). The side view of this crack blunting stage with local crack initiation (↓) is demonstrated in Fig. 2.

After the exhaustion of deformation ability at the crack tip the stable crack growth starts. At this moment the stretch zone has completely developed on both specimen halves which is shown in Fig. 3. The profile represents the finished stretch zone formation (measured as stretch zone height) and the stable crack propagation along the total specimen width (Fig. 4).

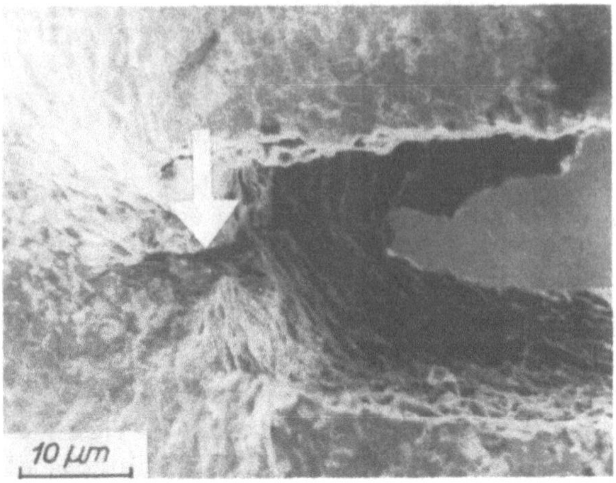

Fig. 2 Profile of the blunted crack tip

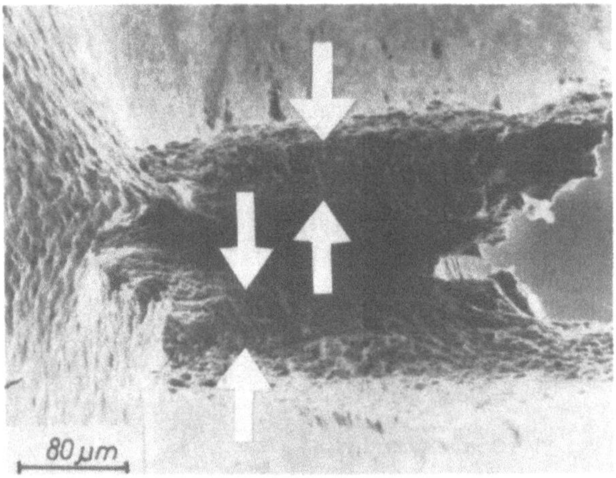

Fig. 3 Stretch zone height on both crack flanks

These observations on the crack tip were made in weldable structural steel St 355 with ferrite-pearlite microstructure. For the purpose of comparison other in-situ tension tests were carried out with austenitic steel X 8 CrNiTi 18.10 . Here essential differences were found in the behaviour of the crack tip blunting compared to the steel St 355. This is due to a different lattice structure.

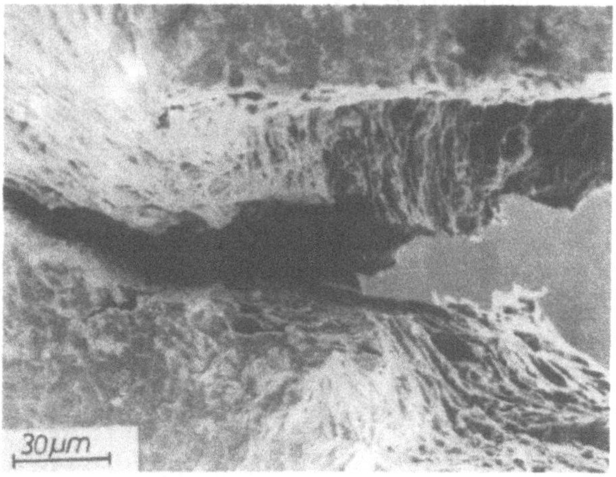

Fig. 4 Crack propagation for finished crack blunting

A typical characteristic of austenitic steel is the formation of a series of stretch zones. In Fig. 5 the completely developed first stretch zone at the crack ground is illustrated. The process of the development of the second stretch zene is continuing.

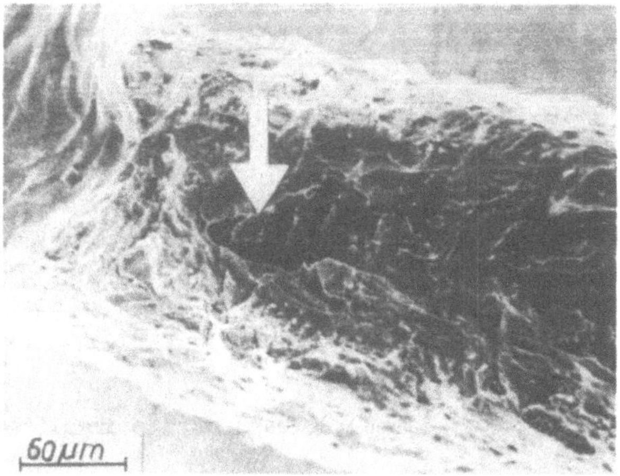

Fig. 5 Crack ground with the beginning formation of the second stretch zone

In Fig. 6 the blunting process of the formation of three stretch zones on both crack flanks has been finished.

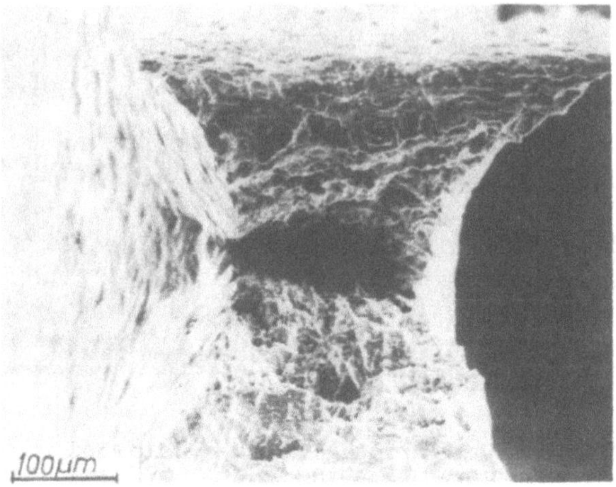

Fig. 6 Crack flanks with three developed stretch zones

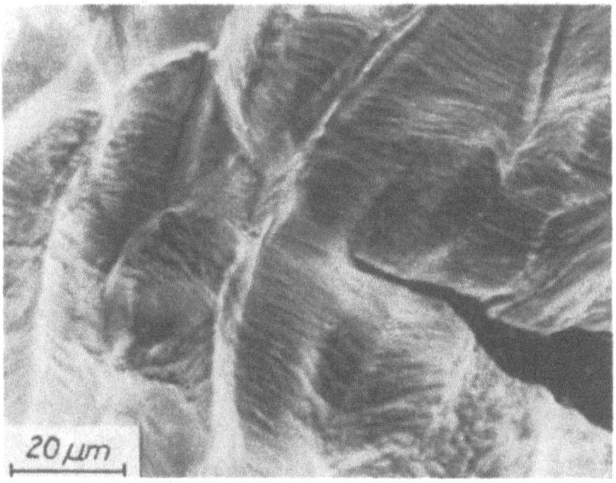

Fig. 7 Slip lines in front of the crack tip

The step-by-step stretching of the crack ground has to be seen in connection with the distinctly visible gliding process going on in front of the crack tip (Fig. 7).

IN SITU EXPERIMENTS IN THE SEM FOR CRACK, DELAMINATION AND FRACTURE RESEARCH OF MATERIALS

K. Wetzig, J. Edelmann, H. Mueller

Central Institute of Solid State Physics and Materials Science, Academy of Sciences of the G.D.R., Helmholtzstr. 20, Dresden 8027 G.D.R.

ABSTRACT

A short review is given about different mechanical in situ deformation techniques inside the SEM. By means of examples these techniques are explained. The laser irradiation of solids is another useful in situ method to study surface reactions and structure changes. The equipment used and some remarkable results are described in detail.

INTRODUCTION

If materials are subjected to mechanical loads usually damage is produced which reduces their life-time. To get continuous information on the damage it is advantageous to submit special samples to in situ treatments in combination with different microscopic methods. These treatments should allow the investigation of

- mechanical properties
- thermally activated processes at surfaces and inside the samples
- surface reactions or
- electrical properties due to applied outer voltages.

In most cases the in situ techniques are used in combination with the scanning electron microscope (SEM). This includes the possibility of

- obtaining high-resolution images
- directly observing the kinematic processes during deformation
- eliminating relaxation and environmental influences during observation under high vacuum conditions
- identifying details, which are visible only under load and
- eliminating the sample transport during the treatment.

For a short summary of the investigated solid state properties and useful recording methods see e.g. [1]. The following remarks will be limited to (1) the investigation of mechanical properties and (2) to thermally activated processes, the latter especially under the aspect of the laser-solid state interaction. The observation of material changes during machining, surface reactions influenced by temperature and chemical attack as well as electrical and stroboscopic measurements are not discussed here.

Static Deformation. In situ investigations of the mechanical properties of solids include the study of deformation, crack formation and fracture processes as well as of structural changes of surface layers. Up to now, several commercially produced deformation attachments for the use inside a SEM are known. Most of them can be used only at room temperature. A detailed review was given formerly [1]. In recent time, some new devices have been designed which allow deformation experiments on ceramic materials [2] or the deformation of larger samples under very high load [3]. As an example, we will describe here a device, developed by WETZIG and coworkers [4], which allows one to realize a more complex program of deformation modes. In the basic state, this device combines a tensile stage with a heater to enable tensile deformations in the temperature range from room temperature up to about 900 °C. Using a cooling attachment instead of the heater, sample temperatures down to -120 °C can be attained.

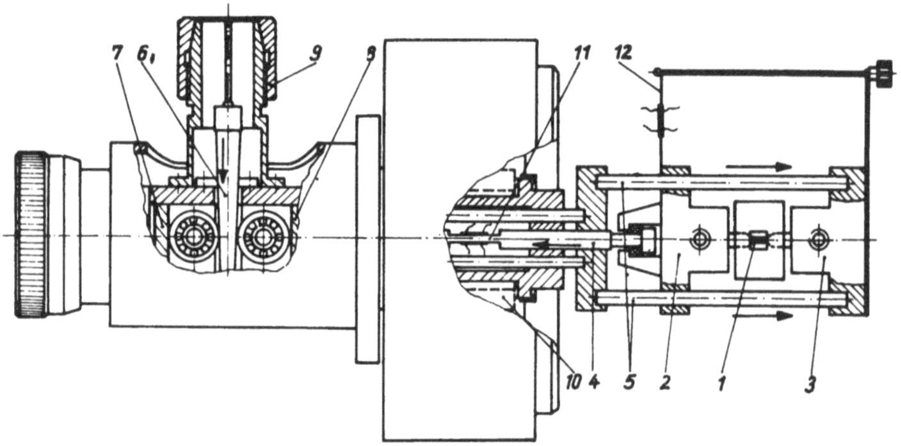

Fig. 1 Cross section of the tensile deformation stage

Fig. 1 gives a schematic view of the cross section through the tensile stage. The sample (1) is fixed in two chucks (2, 3) moving symmetrically against each other by a pair of guiding rods (4, 5). Their movement is forced by a wedge (6) sliding perpendicularly between two ball bearings and driving the pistons (7, 8) of the guiding rods. The tensile load and the sample elongation are measured electrically. For further details see [4]. The technical data of the deformation unit are as follows:

- tensile loading up to 1000 N, either with constant load or constant velocity between 10^{-3} and 0.2 mm/min

- temperature change from -120 °C up to 900 °C
- electrical measurement and plotting of $\sigma(\varepsilon)$ or $\varepsilon(t)$ curves
- registration of acoustic emission rates during deformation.

For deformation experiments at room temperature, the tensile stage can be exchanged for two bending stages, which realize 3 or 4 point bending experiments under different modes of observation. The first stage is used for observing the side face of a bent sample [5]. The second stage, which allows the observation of the sample surface under tension, is shown in Fig. 2 [6]. The main part of this stage is a "scissors-like" arrangement, which presses the sample against a support. The construction uses samples of a length of 25 mm and a cross section of about 4 x 1.5 mm^2. The maximum deformation load is limited to 1000 N, the maximum deflection reaches up to 2 mm. In case of 4 point bending, the distances between inner and outer loading points are l_1 = 4 mm and l_2 = 20 mm, respectively, in the case of 3 point bending l_1 is zero.

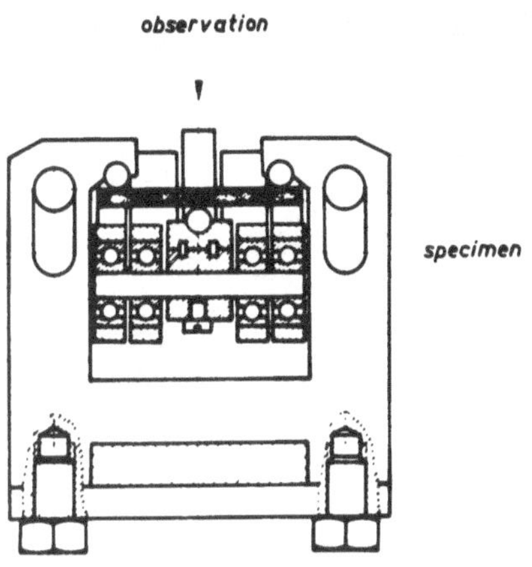

Fig. 2 Cross section of the bending stage

Dynamic Deformation. Using the above-described attachments, it is only possible to get information concerning the static behaviour. To observe samples under dynamic conditions, a cyclic deformation is necessary. In principle, the bending moment can be introduced by electromagnetic or mechanical forces. Due to the main advantages of the mechanical deformation, like higher load and larger amplitudes that can be reached, we used this principle for the construction shown in Fig. 3. This device [7] allows the mechanical variation of the load in the range up to 1000 N and at frequencies from 0 up to 25 Hz. The observation of the side face or of the face under tension can be realized by turning the deformation head by 90°.

The flange 1 contains the cross-table, which allows a sample displacement of about 3 mm in x and y directions. The specimen (6) is moved with the punch (7) by the motor-driven wedge (10). The applied load can be measured by a commercial load-cell (8) in the region up to 2 kN. The gearbox (16) contains the rotating mechanism for the cyclic deformation of the sample. Electronic units are used to control frequency and amplitude of the deformation to synchronize the deformation frequency and the scanning speed of the SEM and to count the number of deformation cycles. Using only the amplitude control, static deformation experiments can also be carried out with this device.

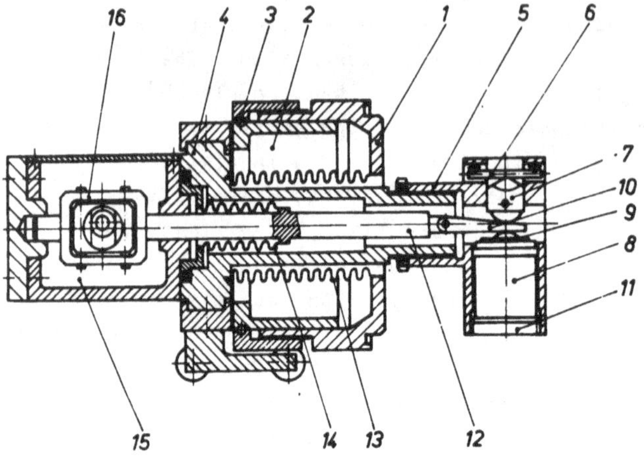

Fig. 3 Cyclic bending device

RESULTS

Using the above-described deformation attachments a large number of investigations were carried out. They include the study of the deformation behaviour of polycrystalline materials [4], crack processes in amorphous metals and the very complex processes in brittle materials and in composites [5]. In this case, the crack processes can precisely be characterized by a registration of the acoustic emission rate [5].

Bending experiments with surface-coated samples demonstrate the formation of nearly parallel cracks (Fig. 4). Later on, a nearly equidistant pattern is formed, showing also delaminations at the interface. As explained by GILLE and WETZIG [6], the fracture behaviour of the coatings can be described theoretically using the mechanical parameters of the substrate and the coating. Measured crack densities and calculated values are in good agreement.

Similar results can be observed during cyclic bending of coated samples. After a great number of cycles, first cracks are formed. Later on the number of cracks increases, delamination processes occur and some cracks propagate into the substrate [7].

In fibre-reinforced plastics, the crack usually propagates discontinuously, stopping at the fibres (Fig. 5). Due to the more complex strain field at the sample surface an additional crack propagates parallel to the fibres.

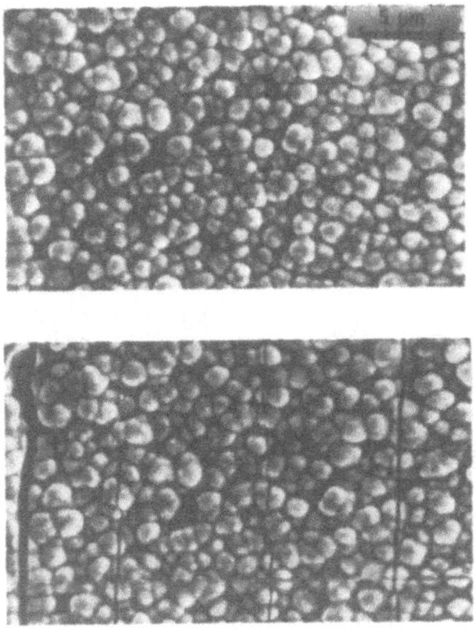

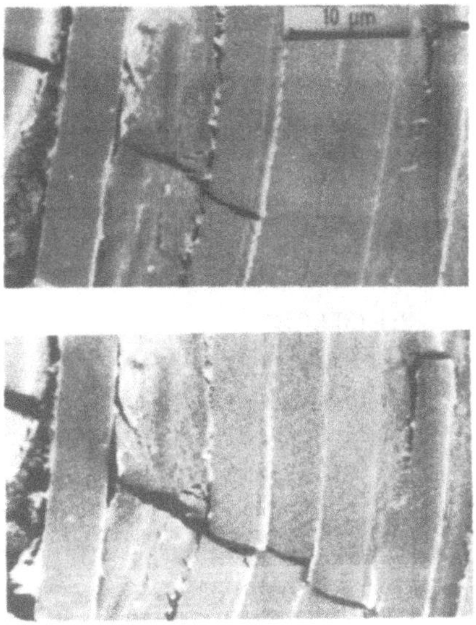

Fig. 4 Static deformation of TiN on C 100 W1 steel

Fig. 5 Cyclic deformation of a carbon fibre-plastic composite

LASER-SOLID INTERACTION

An intense localized heat source inside a SEM can be realized by a laser, giving very short pulses of high power input. This combination, later on called LASEM, was designed by WETZIG et al. [8]. It is used

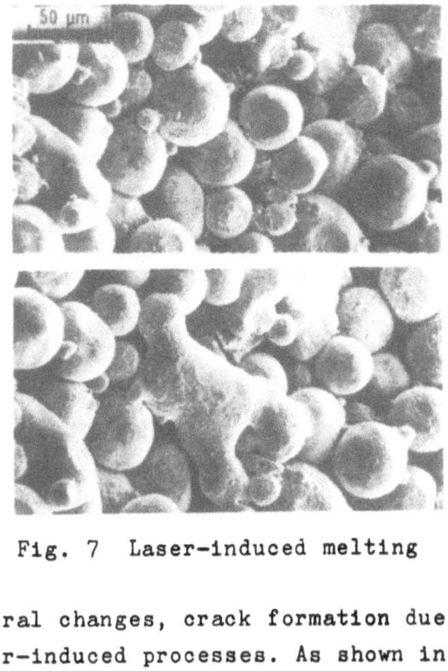

Fig. 6 Crack movement

Fig. 7 Laser-induced melting

e.g. for the investigation of structural changes, crack formation due to thermal shock, melting and other laser-induced processes. As shown in

[8], the laser beam is introduced into the sample chamber by adjusting mirrors and focusing optics.

Fig. 6 shows a crack in a WC-Co hard metal sample. A thermal shock after a second pulse closes the first crack with a nearly parallel crack opening. Melting processes after laser irradiation are shown in Fig. 7. Details will be explained later.

REFERENCES

1 K. Wetzig, 8th Europ. Congr. El.-Micr., Budapest 1984

2 H. Frei and G. Grathwohl, BEDO 22 (1989) 71-8

3 A. Schenk, L. Berka and P. Vasina, to be published

4 K. Wetzig, J. Edelmann and E. Erler, Exp. Techn. Phys. 26 (1978) 589

5 J. Edelmann, G. Fischer, R. Krumphold, K. Wetzig, and G. Zies, Exp. Techn. Phys. 27 (1979) 347-57

6 G. Gille and K. Wetzig, Thin Solid Films 110 (1983) 37-54

7 K. Wetzig, A. Maslov and J. Edelmann, Pract. Met. 21 (1984) 161-72

8 K. Wetzig, J. Edelmann, W. Fischer, and H. Mueller, Scanning 9 (1987) 99

FRACTURE MECHANISM IN DEPENDENCE ON LOADING CONDITION AND MICROSTRUCTURE IN FE - BASE ALLOYS

G. Zouhar

Technical University of Dresden, Department of Mechanical Engineering and Materials Science, 8027 Dresden, Mommsenstraße 13, German Democratic Republic

ABSTRACT

Different loading conditions select different microstructure elements and initiate different fracture mechanisms in structural materials. This is shown by scanning electron microscopy and transmission electron microscopy studies on bcc and fcc Fe- base alloys in connection to fracture models and alloy design.

INTRODUCTION

The present paper shows that scanning electron microscopy and transmission electron microscopy can be effective tools to elucidate microstructure elements and mechanisms controlling the fracture resistance of metallic materials. Results on the fracture behaviour of low tempered ultra- high strength martensite [1] as well as of a metastable austenitic and a two-phase austenitic- martensitic alloy [2] are briefly described.

LOW TEMPERED MARTENSITE

Fig. 1 a shows the true stress- strain curve of a high- temperature thermomechanically treated low tempered martensitic 0.5C-1Cr-0.1 V steel characterized by remarkable mechanical properties during uniaxial tensile deformation : 1830 MPa yield strength, 2230 MPa ultimate tensile strength, 7 % fracture strain and 38 to 45 % reduction of area. Fracture of second phase particles, decohesion of second phase particle/matrix — interfaces, nucleation, growth and coalescence of voids control the fracture mechanism during uniaxial tensile deformation (Fig. 1b, c).

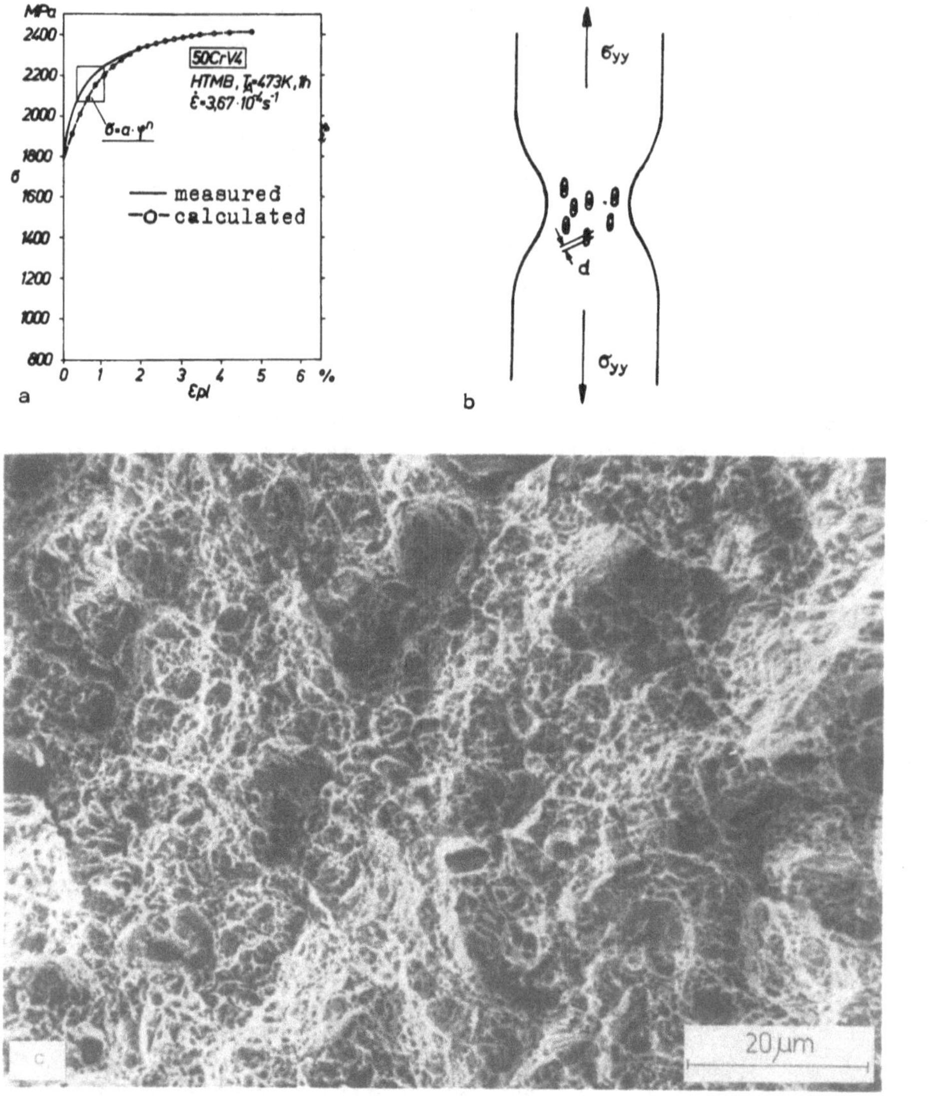

Fig. 1 True stress-strain curve (a) and transcrystalline plastic fracture of the low tempered ultra-high strength martensitic 0.5C-1Cr-0.1V steel, b - schematically, c - scanning electron micrograph.

But, contrary to it intercrystalline brittle fracture is observed during unstable crack propagation in pre-fatigue cracked bend specimen in condition of plane strain (Fig.2 a). The cohesion of the prior austenite grain boundaries is decreased by non-equilibrium segregation of minor impurities and trace elements during hot working and by inheritance of these critical enrichments to the martensite by quenching [3, 4]. The interaction of deformation bands with the prior austenite

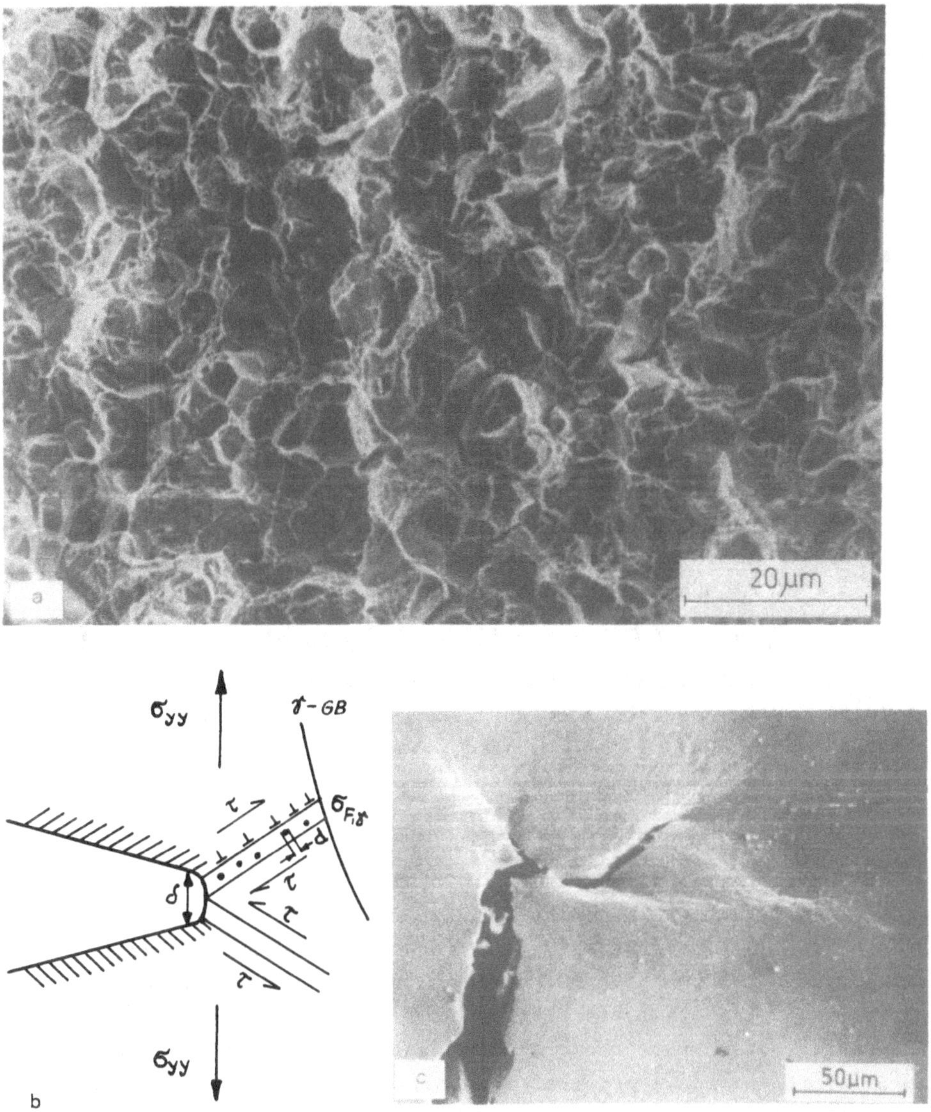

Fig. 2 Intercrystalline brittle fracture of the low tempered ultra-high strength martensitic 0.5C-1Cr-0.1V steel during unstable crack propagation. a - scanning electron micrograph, b - interaction between deformation bands and prior austenite grain boundaries in the near crack tip stress-strain field (schematically), c - deformation bands on the specimen surface, scanning electron micrograph.

grain boundaries in the near crack tip stress-strain field controls the fracture mechanism in this case (Fig. 2b, c). The critical fracture stress σ_{CF} of the prior austenite grain boundaries is reached before developing the decohesion mechanism of the deformation bands.

It should be expected also a brittle fracture at ambient temperature by increasing the deformation rate. Impact loading of notched specimen, notch root radius $\rho = 0.1$ mm, indicates an unexpected tendency to transcrystalline plastic fracture (Fig. 3 a). In order to understand this surprising result the critical stress for intercrystalline cracking σ_{CF} is evaluated by means of the slip-line field solution for small scale yielding [5]. Taking into account the work-hardening of the material the true stress-strain curve (Fig. 1 a) is approximated by an ideal elastic-plastic behaviour with a yield stress of $\sigma_y \approx$ 2250 MPa. The RKR-model [5] yields a critical fracture stress $\sigma_{CF} =$ 3150 MPa for the measured fracture toughness $K_{IC} = 25.5$ MPa$\sqrt{m}$ and for a reasonable critical distance $X_0 \approx 1.5\,\bar{d}_\gamma$ ($\bar{d}_\gamma$ - average diameter of the prior austenite grains). The K^*_{Id}-value for intercrystalline brittle fracture of the investigated charpy specimens (Fig. 3 b) should be in the range of 46 MPa$\sqrt{m}$ according to the equation

$$K^*_{Id} = K_{Id}(\rho) = 2.9 \cdot \sigma_y \left[\exp\left(\sigma_{CF}/\sigma_y - 1\right) - 1\right]^{1/2} \cdot \rho^{1/2}$$

proposed by WILSHAW and TETELMAN [5, 6]. The measured value $K^*_{Id} = 31$ MPa$\sqrt{m}$ is lower than the calculated one. Therefore, transcrystalline plastic fracture should be expected, what is in agreement with the experimental results. The stress-strain state at the notch root appears as more important for the observed fracture mechanism in this case than the high deformation rate in the charpy impact test. The discussed example appears as more dangerous for metallurgists and mechanical engineers than that published by RITCHIE and SERVER [7], because in the present case a fine grained material was selected and the fracture mechanism is not unchanged. The evaluation of the fracture behaviour of ultra- high strength martensitic steels should take into account the influence of notch root radius, notch geometry and stress-strain state in order to elucidate critical combinations of microstructure and loading conditions which are expected to decrease the fracture resistance.

The fatigue crack growth in the near threshold region is controlled by the thickness of the martensite laths (Fig. 4). It is supported by the reasonable agreement between the thickness range of the martensite laths, 0.2 to 1 μm, and the calculated radius of the cyclic plastic zone $r_c \approx 0.2\,\mu m$ according to RICE [8], $r_c = 1/3\pi\,(\Delta K_{th}/2\sigma_y)^2$, for $\Delta K_{th} = 5$ MNm$^{-3/2}$, $\sigma_y \approx 1800$ MPa [1]. The thickness of the martensite laths which appears as upper bound for the tolerable defect size at the fatigue strength level can not be remarkably changed. Therefore, a fatigue strength under rotating bending essentially higher than 10^3 MPa should be scarcely possible for lath martensite [9].

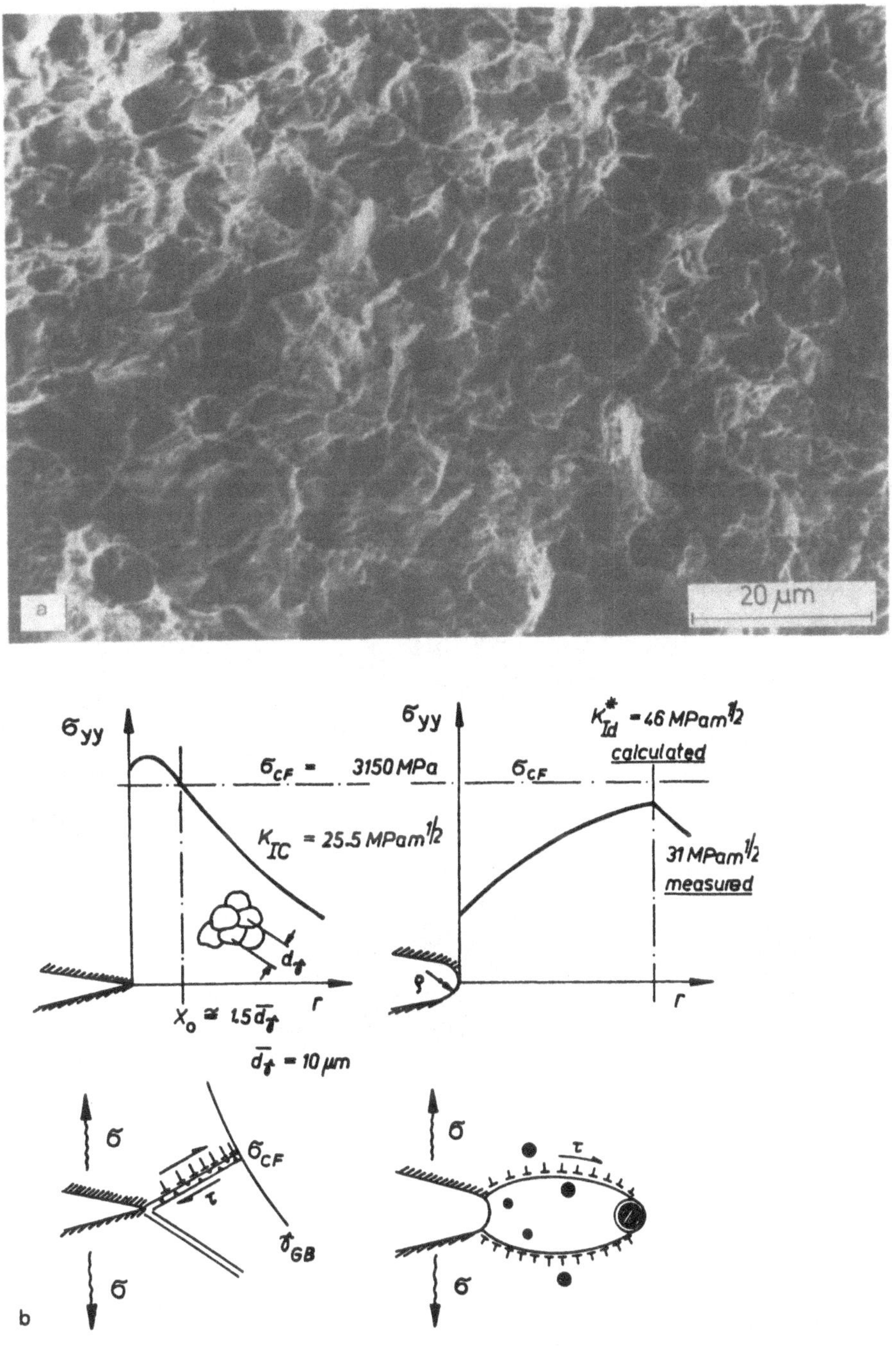

Fig. 3 Typical fracture surface of notched specimens of the ultra-high strength martensitic 0.5C-1Cr-0.1V steel (a, high fraction of transcrystalline plastic fracture, scanning electron micrograph) and fracture models for estimation of the critical fracture stress σ_{CF} (b).

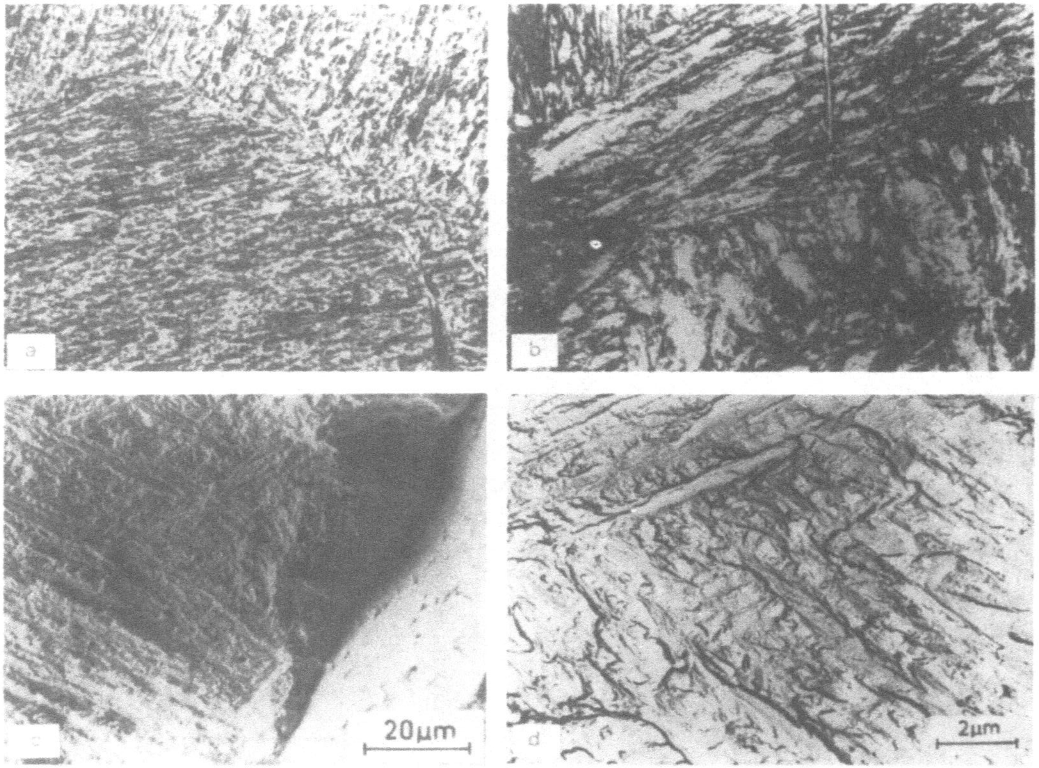

Fig. 4 Correlation between fatigue fracture surface and martensitic microstructure of the ultra- high strength martensitic 0.5C-1Cr-0.1V steel. Martensitic microstructure: a - optical micrograph, b- transmission electron micrograph; fracture surface: c - scanning electron micrograph, d - transmission electron micrograph (replica), bend specimen, load ratio $R = 0$.

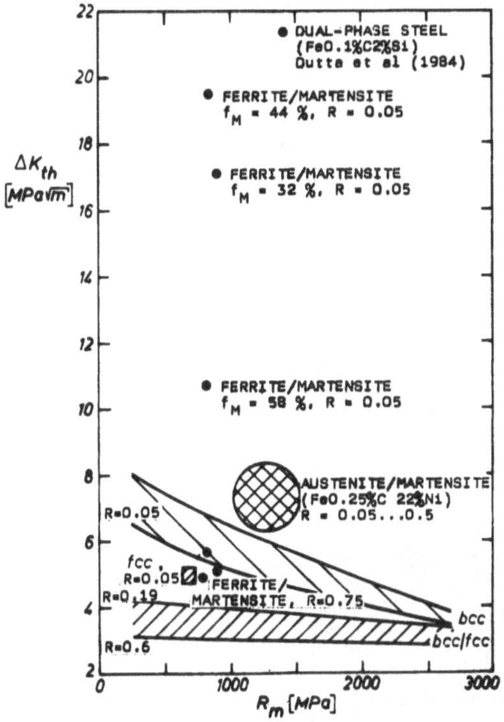

Fig. 5 Fatigue threshold ΔK_{th} in relation to ultimate tensile strength R_m for fcc and bcc materials. f_M- martensite fraction, full circles according to [10].

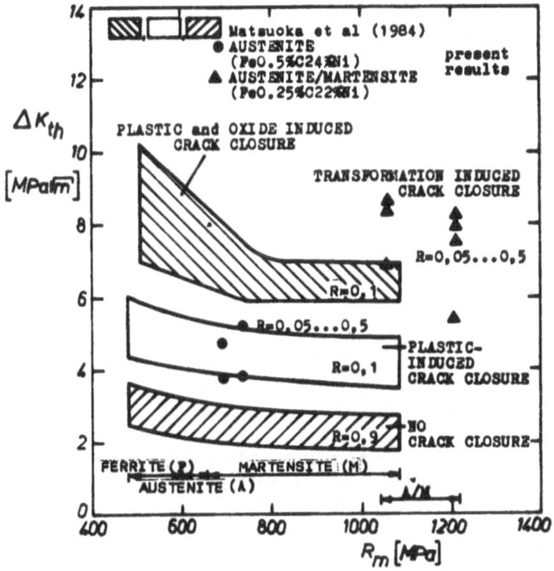

Fig. 6 Fatigue threshold ΔK_{th} in relation to ultimate tensile strength R_m for different crack closure mechanisms and microstruc- tures (according to [11]).

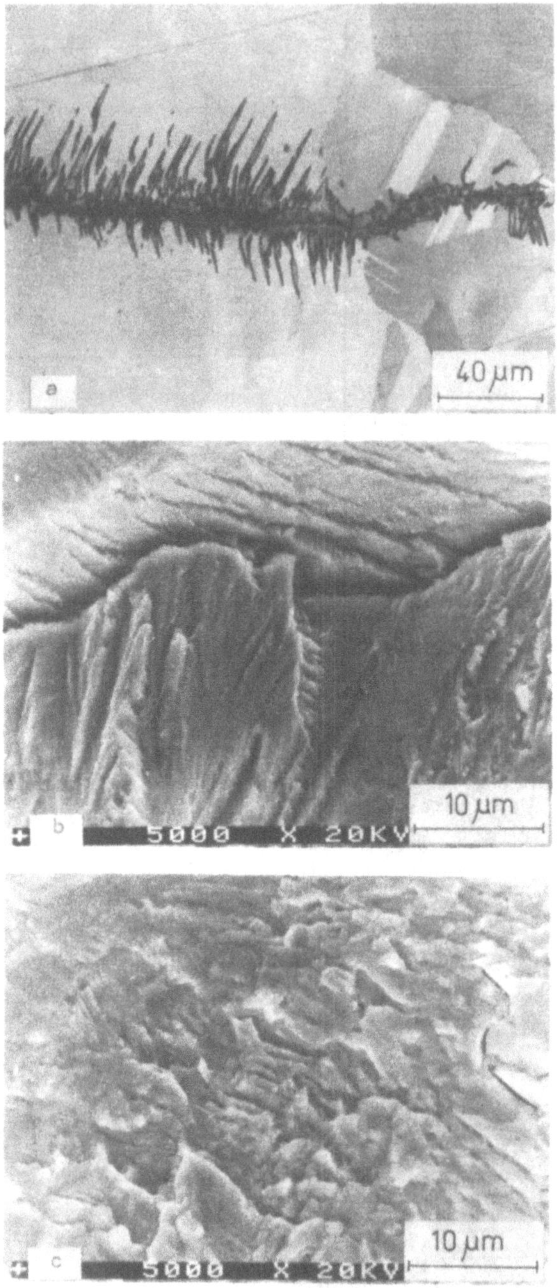

Fig. 7 Deformation induced martensitic transformation (a,c) and shear mode fracture (b) during fatigue crack growth.
a - fatigue crack, specimen centre, Fe-0.5C-24Ni alloy, PARIS-ERDOGAN region, $\Delta K \approx 20$ MPa $\sqrt{m}$, R= 0.2, f = 60...80 Hz, optical micrograph
b - Fe-0.5C-24Ni alloy, ΔK_{th} = 3.8 MPa $\sqrt{m}$
c - Fe-0.25C-22Ni alloy, ΔK_{th}= 5.4 MPa $\sqrt{m}$
b,c - fatigue threshold region, R=0.5, f≈58 Hz, scanning electron micrographs; T= 20°C in all cases.

METASTABLE AUSTENITE AND AUSTENITE-MARTENSITE

Deformation-induced martensitic transformation is well known as effective mechanism for excellent combinations of high ductility and fracture toughness, high strength and fatigue limit of TRIP (Transformation Induced Plasticity) steels. Possible retardations of fatigue crack growth are discussed in terms of crack closure phenomena, near crack tip compression stresses, crack deviation and crack branching [2]. But information on a simultaneous increase of resistance against fatigue crack propagation and strength of metastable austenitic-martensitic alloys by controlling the mechanical stabilization of the austenite phase is only scarce . The ultimate tensile strength of a Fe-0.5 C-24Ni alloy can be increased from 750 to 1500 MPa by increasing the volume content of martensite (deep cooling) and by dynamically polygonized dislocation substructures at nearly constant fatigue threshold values ΔK_{th} = 3.8 MPa$\sqrt{m}$, R = 0.19, and ΔK_{th} = 3.0 MPa$\sqrt{m}$, R = 0.6 (Fig. 5, fcc-range) [2]. Deformation-induced martensite at the fatigue crack tip was observed not until above ΔK = 7 MPa$\sqrt{m}$ but without inflections in the fatigue crack growth curve. Size and orientation of the deformation-induced martensite crystals are controlled by the crack tip stress-strain field and the local orientation of the austenite lattice (Fig. 7 a). A shear mode fracture mechanism is observed in the threshold region (Fig. 7 b) in agreement with the relatively weak crack closure effect (Fig. 6, full circles).

The decreased mechanical stabilization of the austenite phase in the two -phase martensitic-austenitic Fe-0.25C-22Ni alloy yields deformation induced martensitic transformation already in the fatigue threshold region (Fig. 7 c) which is believed to cause the higher threshold values at simultaneously increased ultimate tensile strength values R_m = 1040 to 1200 MPa (Fig. 5 and 6). In general, the investigated two-phase Fe-0.25C -22Ni alloy reveals a markedly decreased load ratio- (R)- sensitivity compared with that of bcc- materials (Fig. 5)[2].

ACKNOWLEDGMENT

The author would like to acknowledge Dr. P. Finke and Dr. K. Edelmann from Central Institute of Solid State Physics and Materials Research Dresden for their assistance in the analysis of fracture surfaces by means of scanning electron microscopy.

REFERENCES

1 G. Zouhar: Dissertation B, Akademie der Wissenschaften der DDR, 1985

2 G. Zouhar, M. Schaper and G. Jäger, In: Proc. 10th Internat.Colloquium on Mechanical Fatigue of Metals, organized by Department of Physics of the Pädagogische Hochschule "K. F. Wander" Dresden, Gaußig, Dec. 12-14, Vol. 2, 1989, p. 108-12

3 G. Zouhar jr., M. Schaper, J. Eickemeyer and P. Finke, In: Proc. 5th Internat. Conf. on Fracture, ICF5, Cannes, Vol. 2, D. Francois (Ed.), Pergamon Press, Oxford 1981, p. 747-54

4 J. Eickemeyer and G. Zouhar, In: Proc. 6th Internat. Conf. on Fracture, ICF6, New Delhi, Vol. 2, S. R. Valluri, D. M. R. Taplin, P. Rama Rao, J. F. Knott, R. Dubey (Eds.), Pergamon Press, Oxford 1984, p. 1457

5 R. O. Ritchie, J. F. Knott and J. R. Rice, J. Mech. Phys. Solids 21 (1973) 395-410

6 A. S. Tetelman, T. R. Wilshaw and C. A. Rau, Int. J. Fracture Mech. 4 (1968), p. 147

7 R. O. Ritchie, B. Francis, and W. L. Server, Metallurg. Transact. 7A (1976) 831- 38

8 J. R. Rice, In: ASTM STP 415, 1966, p. 247

9 G. Zouhar, In: Vorträge zum Problemseminar "Kleine Risse", Technische Universität Dresden, Sektion Grundlagen des Maschinenwesens, Weiterbildungszentrum Festkörpermechanik, Konstruktion und rationeller Werkstoffeinsatz, Nov. 1988.

10 V. B. Dutta, S. Suresh and R. O. Ritchie, Metallurg. Transact. 15A (1984) 1193

11 S. Matsuoka, S. Nishijima, C. Masuda and S. Ohtsubo, In: Proc. 6th Internat. Conf. on Fracture, ICF6, New Delhi, Vol. 3, S. R.Valluri, D. M. R. Taplin, P. Rama Rao, J. F. Knott, R. Dubey (Eds.), Pergamon Press, Oxford 1984, p. 1561- 70

8. PROPERTIES AND INTERACTION OF LATTICE DEFECTS

HREM OF THE NUCLEATION OF PRECIPITATES AT DISLOCATIONS*

P. Haasen, S.Q. Xiao, P.-J. Wilbrandt

Institut für Metallphysik, Universität Göttingen, Hospitalstr. 3-5, D-3400 Göttingen, FRG

ABSTRACT

It is shown by HREM and thermodynamic calculations that the dilated region near an edge dislocation is a favorite site for nucleation of coherent precipitates with positive size misfit. The nuclei are spherical for relatively large supersaturation, cylindrical for a small one.

INTRODUCTION

We have studied the homogeneous precipitation of the γ' (Ni_3Al)-phase out of supersaturated NiAl solutions for a long time by atom-probe-FIM [1] and by HREM [2]. Typical of a Ni14at%Al alloy homogenized at 1150°C, quenched and annealed at 550°C the nucleation of ≈10Å radius particles in the volume is observed for the first 30 min of ageing. At half of this supersaturation, by aging a Ni12at%Al alloy at 575°C, the precipitates are observed at dislocations only. This will be demonstrated in the following by HREM using the superlattice (100) reflections of the $L1_2$ structure. It will be furthermore shown theoretically that this preferential nucleation is the consequence of the formation of a Cottrell atmosphere, rich in solute, in the dislocation zone of edge dislocations. Previous investigators have assumed either the precipitate to be incoherent with the matrix [3] or that the precipitate is actually coherent (as in our alloys) but forms even without a Cottrell atmosphere [4,5,6]. This case can be tested by an anneal slightly above the equilibrium temperature as will be described below. A preliminary report is given in [7,8].

*Dedicated to Prof.H. Bethge on the occasion of his 70th birthday

EXPERIMENTAL

A Philips EM 420 ST microscope operating at 120 kV was used, having a C_s value of 1.35 mm and a damping envelope of the contrast transfer function extended to 12 nm for the usual working conditions. A condensor aperture of 30 μm diameter was used. The Ni12%Al single crystal (from Dr.R.E. Beddoe, Stuttgart) was annealed at 1300°C for 24 h and water quenched. EM discs in (100) orientation were annealed at 575°C for t_A = 2,11 or 53 hrs or at 700°C for 22 hrs under purified argon. The thickness of the transparent region was between 0.5...1.5ζ_{001}, ζ_{001} = 13.6 nm being the periodicity of the (000) beam in the foil.

RESULTS

HREM photographs of precipitates in foils which were annealed at 575°C for various times are shown in figs. 1-3. The areas with visible superlattice fringes (a_0 = 0.356 nm) outline ordered particles. The dark (or bright) spots are dislocations seen end on, often arranged in pairs as expected for dipoles. The dark or white lines mark dislocations lying flat in the foil.

After 2 h annealing small ordered particles are seen at dislocations. Their radii are about 2 nm and they form always at opposite sides of the dipole partners (as if occurring only in their dilation sides. Unfortunately end-on seen dislocations are not suitable for a Burgers vector determination!). With longer annealing the γ'-precipitates grow in proportion to $t_A^{1/2}$ [9]. There are also smaller ordered regions visible in the matrix (radii < 1.5nm) but they do not grow. Fig. 2 shows a bigger particle at an inclined dislocation in a specimen annealed for 11 h. The precipitate clearly is on one side and tangential to the dislocation. The long inclined dislocation in fig. 3 is lined with a chain of precipitates after 53 h at 575°C. As the particles grow together, antiphase boundaries (APB) are often produced between them. They have been quantitatively analyzed in a separate paper [10] with the result that each of the four possible configurations (including the identity) is observed equally frequent, i.e. that the particles are independently nucleated and have grown together. The nuclei are evidently spherical.

THERMODYNAMIC ANALYSIS

The free energy per atom of a binary A,B alloy can be written in terms of the chemical potentials of the components in the phases $j = \alpha, \beta$ and the concentration (molar fraction of B) as

$$F^{j} = (1-C)\mu_A^j + C\mu_B^j \qquad (1)$$

Near an edge dislocation (at the position ρ, θ in cylindrical coordinates) the free energy is changed by

$$\Delta F^{d} = M_0 + M_1 C , \quad M_1 = \frac{3Gb\delta\Omega}{\pi\rho} \sin\theta \qquad (2)$$

according to Cottrell's linear elastic interaction. Here M_0 is the strain energy density of the dislocation itself, G the shear modulus, Ω the atomic volume, b the Burgers vector. Thus the free energy in the strained position is

$$F^{jd} = (1-C)\mu_A^j + C\mu_B^j + M_0 + M_1 C \qquad (3)$$

$$= (1-C)\mu_A^{jd} + C\,\mu_B^{jd}$$

Using the Gibbs-Duhem relation

$$\mu_A^{jd} = F^{jd} - C\,\frac{\partial F^{jd}}{\partial C}\ ; \ \mu_B^{jd} = F^{jd} - (1-C)\,\frac{\partial F^{jd}}{\partial C} \qquad (4a)$$

i.e. $$\mu_A^{jd} = \mu_A^j + M_0,\ \mu_B^{jd} = \mu_B^j + M_0 + M_1 \qquad (4b)$$

Eq. 4b means that the dislocation changes only the chemical potentials of the pure metals, and these changes differ for A and B.

The equilibrium conditions between α and β phases in the strain field of an edge dislocation are

$$\mu_A^{\alpha d} = \mu_A^{\beta d}\ ; \qquad \mu_B^{\alpha d} = \mu_B^{\beta d} \qquad (5a)$$

i.e. $$\mu_A^{\alpha} = \mu_A^{\beta} \qquad \mu_B^{\alpha} = \mu_B^{\beta} \qquad (5b)$$

as without dislocation if the size misfit of a B atom is the same in the α and β phases. This seems to be the case for γ and γ' in <u>Ni</u>Al. The remarkable result means that the equilibrium B concentrations in the matrix and at the dislocation are identical, i.e. $C_e^d = C_e$, $C_p^d = C_p$ (C_p = concentration in the particle). The situation is illustrated in fig. 4. (For size misfits $\delta_\alpha \neq \delta_\beta$ equations are discussed in [8].)

NUCLEATION THEORY NEAR A DISLOCATION

The classical nucleation theory [11] for the homogeneous formation of a second phase particle will change near a dislocation because the solute concentration may differ there ($C_0 \rightarrow C_0^d$) and because of the misfit strain energy contributions. The work of nucleation for a spherical nucleus of radius r far from a dislocation is

$$\Delta F = \frac{4\pi}{3} r^3 \Delta f_v + 4\pi r^2 \sigma + \frac{4}{3}\pi r^3 \Delta f_\delta \qquad (6)$$

Here, Δf_v is the gain in chemical free energy per unit volume of the second phase formed, σ the interfacial energy (per unit area) and Δf_δ the strain energy per unit volume of the misfitting particle. From the tangent construction in fig. 4

$$-\Delta f_v = \frac{N_L(C_p - C_0)}{V_m(1-C_0)} \left(\mu_B(C_0) - \mu_B(C_e)\right)$$

$$= \frac{RT(C_p - C_0)}{V_m(1-C_0)} \ln\left({a(C_0)}/{a(C_e)}\right) \qquad (7)$$

where a = activity, V_m/N_L = volume per atom.
The interfacial energy is according to Becker [12]

$$\sigma = A(C_p - C_0)^2 \qquad (8)$$

where A is a constant. Finally according to Eshelby [13]

$$\Delta f_\delta = \frac{2(1+\nu)}{1-\nu} G\delta^2 (C_p - C_0)^2 \qquad (9)$$

where $\delta = {d\ln a}/{dC}$ is the size misfit, a = lattice parameter in the matrix, ν = Poisson's ratio.

Near the dislocation, C_0 is to be replaced by C_0^d in the expressions (7-9) as elastic interaction energy per unit volume of the precipitate with the dislocation (over that in the solid solution)

$$\Delta f_p = \frac{M_1}{\Omega} \cdot (C_p - C_0^d) \qquad (10)$$

The critical radius of the nucleus near the dislocation is then

$$r^* = \frac{2\sigma^d}{|\Delta f_d^d| - \Delta f_\delta^d - \Delta f_p^d} \qquad (11)$$

where the superscript d indicates the use of C_0^d.
Finally, when the equilibrium Cottrell atmosphere has formed

$$C_0^d = C_0 \exp\left({-M_1}/{kT}\right) \qquad (12)$$

From eqs. (2,6-12) the critical radii have been calculated [8] assuming some of the ΔF^d contributions to act alone and for all of them together. The result is shown in fig. 5 for Ni12%atAl at 575°C and the dilation side of the dislocation ($\sin\theta = -1$). The activities $a(C_0) = 9.9 \cdot 10^{-6}$, $a(C_e) = 6.6 \cdot 10^{-6}$ are extrapolated from [14]. r^* is first calculated for $\sigma = 0.014$ J/m^2 and no Δf^d_δ, Δf^d_p contributions showing the variation of Δf^d_v (ρ) due to the Cottrell atmosphere. r^* (curve 1) is seen to decrease down to $\rho = 10b$ as Δf^d_v increases as expected from fig. 4 at higher C^d_0: Δf^d_p is also negative in the dilation region and largely parallels and amplifies the effect of Δf^d_v on r^* (curve 2). The particle misfit contribution Δf_δ itself is positive and somewhat smaller at $\rho \to 0$ than in the volume (curve 3). The surface energy produces the largest variation in r^* (ρ), curve 4, as in the Cottrell atmosphere the concentration gradient matrix / particle becomes small. The easy nucleation near the dislocation ($r^* \to 0$ for $\rho \to 0$) seen in the sum curve encompassing all contributions (curve 5) is largely a result of the small σ in our system, but for $\rho > 6b \approx 1.5$ nm also due to the dislocation / particle elastic interaction and the gain in chemical driving force in the Cottrell atmosphere.

At a temperature at or a little above the equilibrium solubility decomposition in the matrix no longer takes place as r^* would be infinite. But near a dislocation with its Cottrell atmosphere the critical radius still is small for $\rho < 20b$ and nucleation is observed (fig. 6). The particle formed at 700°C now appears cylindrical along the dislocation as the few nuclei formed under these conditions do not find close neighbors.

In the compression region of the dislocation ($\sin\theta = 1$) all ΔF contributions are positive or small. Therefore γ' nuclei are unstable there in agreement with the observations.

The authors thank Dr. R. Bormann for thermodynamic discussions.

REFERENCES

1 H. Wendt and P. Haasen, Acta Met. 31 (1983) 1649

2 F. Ernst, S.Q. Xiao, F. Haider and P.-J. Wilbrandt, Scripta Met. 21 (1987) 1189

3 J.W. Cahn,, Acta Met. 5 (1957), 169

4 B.Y. Luybov and V.A. Solov'yev, Sov. Phys. Met. 19 (1964) 13
5 C.C. Collins, Acta Met. 18 (1970) 1209
6 D.M. Barnett, Scripta Met. 5 (1971) 261
7 S.Q. Xiao, P.-J. Wilbrandt and P. Haasen, Scripta Met. 23 (1989) 295
8 S.Q. Xiao and P. Haasen, Scripta Met. 23 (1989) 365
9 S.Q. Xiao, thesis Göttingen (1989)
10 S.Q. Xiao and P. Haasen, Scripta Met. 23 (1989) 931
11 K.C. Russell in: Phase Transformations, H.I. Aaronson (ed.) ASM, Metals Park OH (1970) p. 219
12 R. Becker, Ann. Physik 32 (1938) 128
13 J.D. Eshelby, Progr. Sol. Mech. 2 (1961) 89
14 R. Hultgren et al. Selected Values of the Thermodynamic Properties of Binary Alloys, ASM, Metals Park OH (1973)

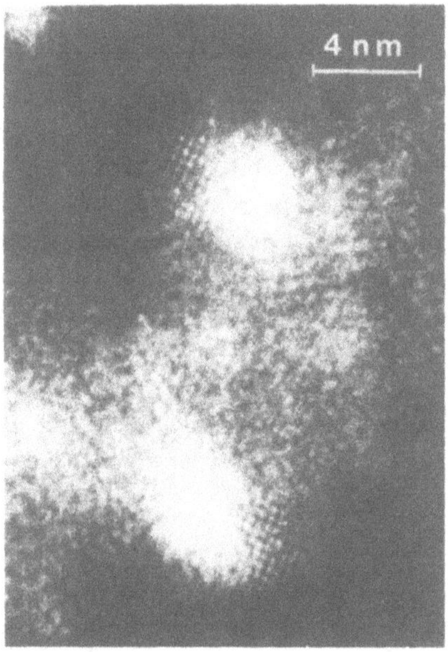

Fig. 1 Two Ni_3Al-precipitates (lattice fringes) at a dislocation dipole seen end on (white points) in a specimen annealed for 2h at 848K.

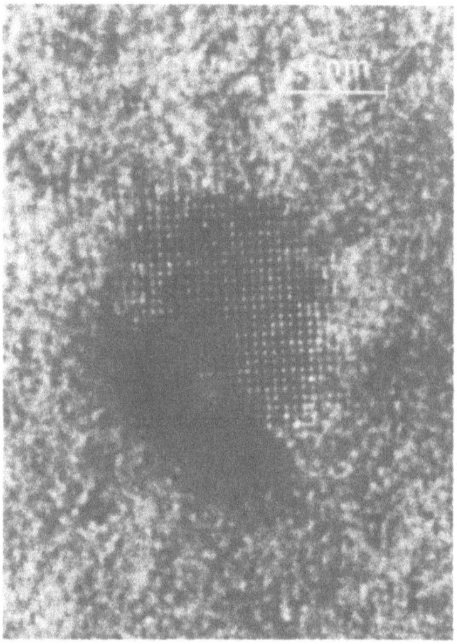

Fig. 2 One Ni_3Al-precipitate (lattice fringes) at an inclined dislocation (dark line) in a specimen annealed for 11h at 848K

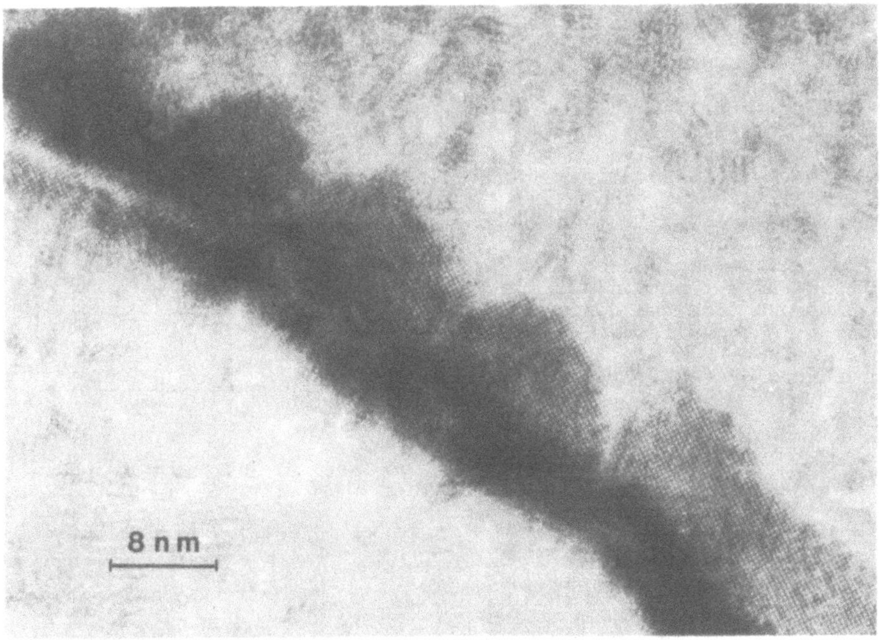

Fig. 3 Chain of Ni_3Al-precipitates at a flat dislocation in a specimen annealed for 53h at 848K.

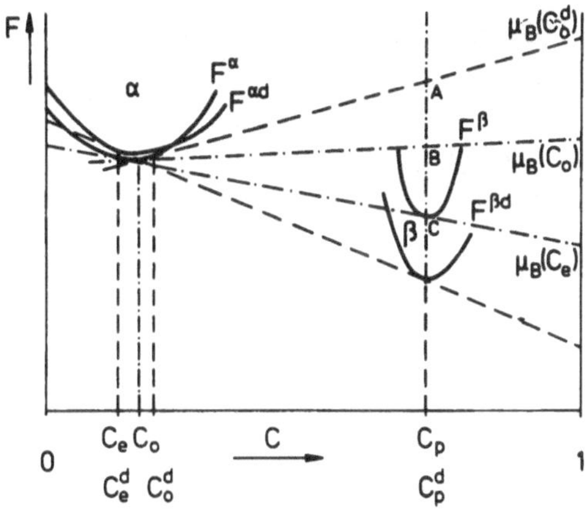

Fig. 4 F^{α}, F^{β} are the free energies and C_e, C_p the equilibrium composition of α, β phases in the absence of a dislocation, $F^{\alpha d}$, $F^{\beta d}$ and C_e^d, C_p^d near an edge dislocation. For $\delta_{\alpha} = \delta_{\beta}$, $C_e^d = C_e$, $C^d_p = C_p$. C_o is the actual composition in the α phase in the absence of a dislocation and C_o^d near an edge dislocation. $\overline{BC} = \Delta f_v$ is the chemical driving force in the absence of a dislocation; $\overline{AC} = \Delta f_v^d$ near an edge dislocation.

Fig. 5 Influences of the elastic strain energy of the precipitate (3), the chemical driving force (1) the elastic interaction energy between dislocation and precipitate (2), the interfacial energy (4) and all of them together (5) on the critical radius of a nucleus as a function of the distance from the dislocation in the dilated region (sinθ=-1) at a temperature of 848K.

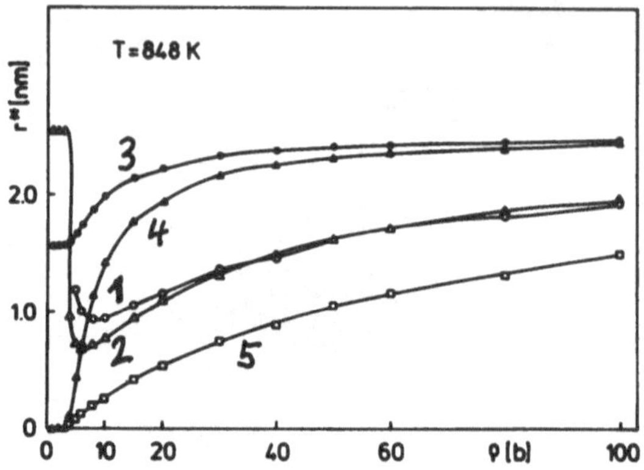

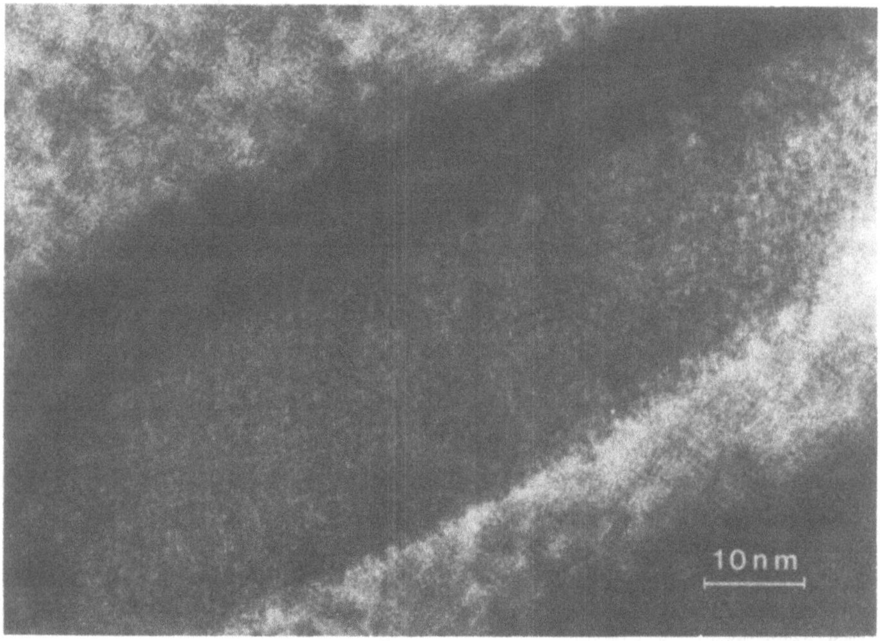

Fig. 6 A cylindrical Ni_3Al-precipitate at a flat dislocation in a specimen annealed for 22h at 973K.

TEMPERATURE DEPEDENCE OF THE STACKING-FAULT ENERGY AND SEGREGATION OF THE SOLUTE ATOMS IN Cu AND Ag BASE ALLOYS

H. Saka

Department of Materials Science and Engineering
Faculty of Engineering, Nagoya University,
Furo-cho, Nagoya 464-01, Japan

ABSTRACT

The variation of the stacking-fault energy with temperature or due to the segregation of solute atoms in noble metal base alloys can be determined accurately by measuring the width of dissociated dislocations using in situ thermal cycling experiments in an electron microspe. The results obtained so far are reviewed, here.

INTRODUCTION

The stacking-fault energy (γ) of an f.c.c. crystal is one of the most important factors, which determine its mechanical and physical properties. Many studies have been carried out to determine the composition and the temperature dependence of γ in f.c.c. alloys [1, 2, 3]. It is well-established that γ decreases monotonously with increasing the electron-to-atom ratio (e/a) in noble metal base alloys such as Cu and Ag base alloys. However, the temperature dependence of γ has been controversially discussed for a long time [2, 3].

In f.c.c. solid solution alloys, the chemical potential of solute atoms in the stacking fault is expected to be different from that in the f.c.c. matrix, and at thermodynamic equilibrium the concentration of the solute atoms in the stacking fault will be different from that in the matrix. This leads to a change in γ. This phenomenon, originally pointed out by Suzuki [4, 5, 6], is called Suzuki effect or Suzuki segregation.

It has been shown that this variation in γ can precisely be measured by in situ thermal cycling experiments in an electron microscope [7 to 13] and the results obtained so far on Cu base and Ag base alloys will be reviewed in this paper.

EXPERIMENTAL TECHNIQUES

In order to measure the temperature dependence of γ, the in situ thermal cycling experiment in an electron microscope is the most powerful method. In this technique, a foil specimen is heated or cooled under observation inside an electron microscope using a specially designed stage and the dissociation width Δ of dissociated dislocations at any temperature (in a certain temperature range) is measured at that temperature. The value γ can be determined from Δ by

$$\gamma = \frac{c_{44}}{2\pi} H_{ij} b_i b_j \frac{1}{\Delta}, \quad \text{(summation convention)} \qquad (1)$$

where C_{ij} is a shear modulus and b_i, b_j are the Burgers vectors of the partial dislocations [14]. H_{ij} is a factor depending on the dislocation line direction, i.e., the angle Ø between the total Burgers vector and the dislocation line.

The advantage of this technique is that the same dislocations can be examined over a wide range of temperature. This is very effective in minimizing the scatter of the data. However, there are a few disadvantages in this technique.

First of all, it is not free from the effects caused by the surfaces of the specimen. Fig. 1 shows an example of the surface effects [15]. A pile-up of dissociated dislocations of the same nature is running from the edge of a wedge-shaped foil to a thicker region. Two types of surface effects are evident. One is that the dissociation width immediately beneath the foil surface is either larger or smaller than in the middle of the foil specimen [16]. The other is that even the dissociation width

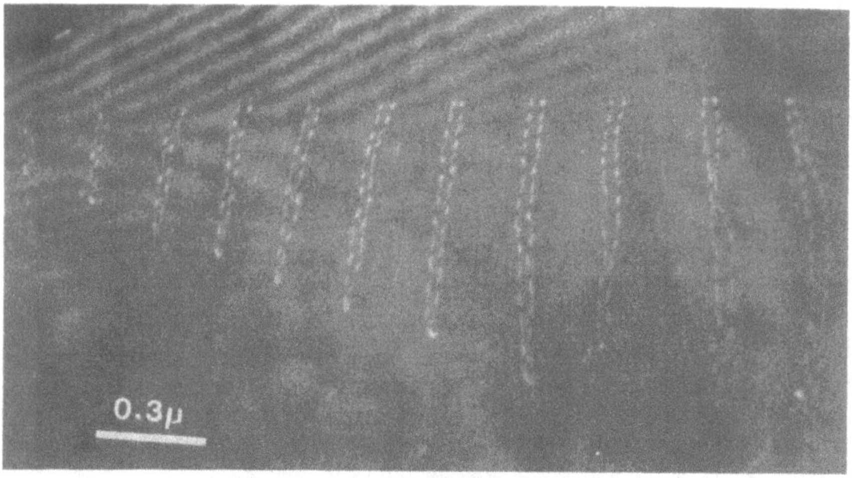

Fig. 1: Surface effects on the dissociation width of dislocations

in the middle of the foils also depends on the thickness of the specimen. In order to avoid such surface effects, two precautions should be fulfilled; (1) dislocations lying in the middle of the foil parallel to the surfaces should be examined, and (2) the specimen should be thicker than a certain value. The former can be attained by using foil specimens with a (111) surface orientation, and the latter by using a high-voltage electron microscope (HVEM). In this respect the weak-beam imaging technique [16], especially when combined with HVEM is most useful, because its resolution is mostly high enough to resolve the dissociation width of a dissociated dislocation in the plane of the foil, and the specimens which can be observed with this technique are reasonably thick.

Another difficulty arises in connection with the use of the HVEM. A specimen is damaged due to the bombardment of electrons during examination if the accelerating voltage exceeds a certain (threshold) value, which depends on the materials [18, 19]. Although this may to some extent be suppressed by using an image intensifier, the safest way is to use an accelerating voltage below the threshold value for the materials to be examined. Even if such precautions are fulfilled, the heating of a thin specimen in the vacuum of an electron microscope may give rise to various undesired phenomena. In addition, in some cases, the value of the dissociation width observed in the in-situ heating experiment does not represent the true value of the dissociation width at that temperature. This is especially true of concentrated alloys, where the interaction of partial dislocations with the high concentration of solute atoms prevents the partials from assuming the equilibrium positions. This is the reason why for reliable conclusions it is essential to carry out the examination of foil specimens which are prepared from crystals annealed in bulk form as a complementary experiment.

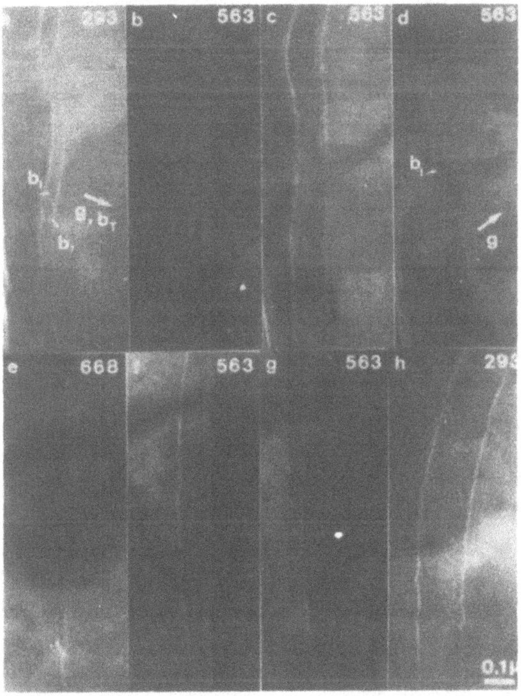

Fig. 2: Typical sequence showing the variation of the dissociation in a dislocation observed during a thermal cycling experiment carried out on Cu-Si [13]

RESULTS

In situ Experiments

Figs. 2 and 3 show the in situ sequence of the weak-beam micrographs and the variation of the dissociation width during thermal cycling experiments carried out on Cu-8.01 at.% Si [13] and Ag-9.6 at.% Al alloys [10], respectively. The dissociation width Δ was plotted against the orientation of a dislocation line, and the apparent stacking-fault energy γ_a was determined by fitting the experimental points to theoretical curves predicted by Eq. (1) (some examples are shown in Fig. 4). The results of the variation of Δ during thermal cycling experiments are summarized in Fig. 5 [12]. It can be seen that the alloys studied can be divided into two groups.

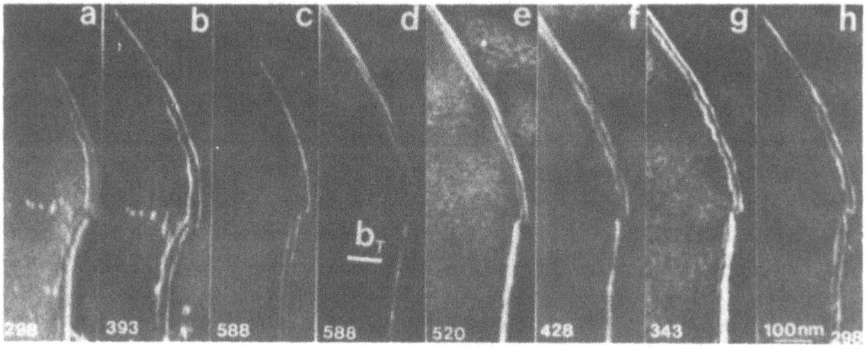

Fig. 3: Typical sequence showing the variation of the dissociation in a dislocation observed during a thermal cycling experiment carried out on Ag-Al [13]

For group I (Ag-Al, Ag-Sn, Ag-In, Cu-Al) the following features are indicated:

1) No variation of Δ occurs if specimens deformed at a low temperature (77 K or 300 K) are heated to temperatures lower than 400 K.

2) At first, Δ drastically decreases between 400 and 450 K and weakly decreases above 450 K.

3) The variation of Δ above 450 K is reversible.

4) Δ increases gradually during cooling, but the room-temperature value of Δ after a complete thermal cycle is lower than the as-deformed value.

For group II(Cu-Si, Cu-Ge, and Ag-Sb) Δ changes as follows:

1) There is no variation of Δ if deformed specimens are heated to temperatures lower than 500 K.

2) On heating deformed specimens to about 500 K, an increase in Δ was observed.

3) On heating to temperatures higher than 600 K, Δ decreases again.

4) The variation of Δ above 560 K is reversible.

5) On cooling, Δ increases, and its room-temperature value after such a complete thermal cycle is much higher than the as-deformed value.

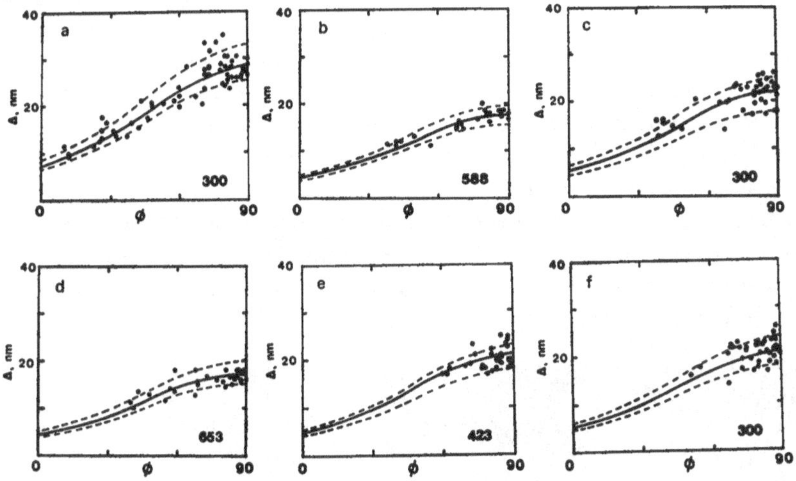

Fig. 4: Examples of the variation of Δ with Ø, where Ø denotes the angle between the total Burgers vector and the dislocation line

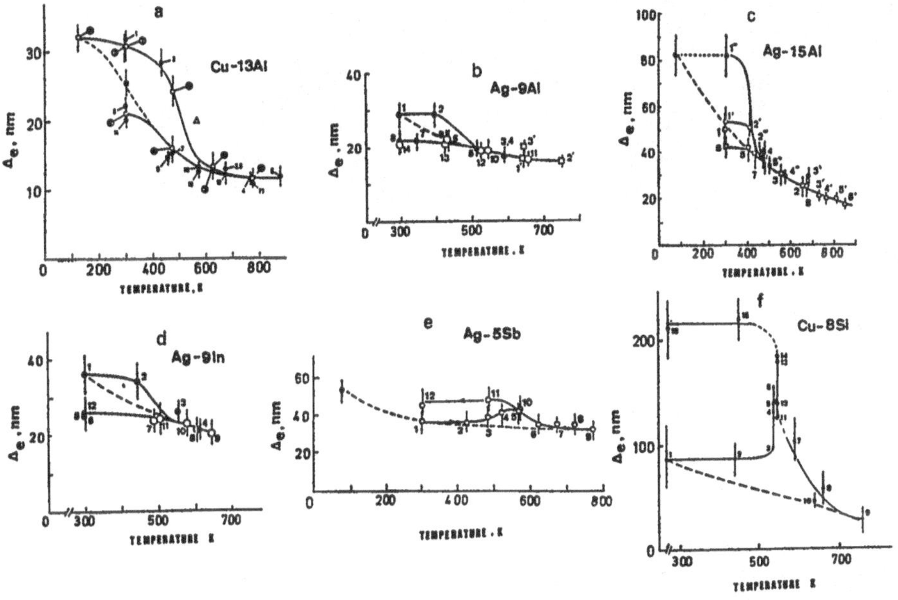

Fig. 5: Temperature variation of Δ during thermal cycling experiments carried out on various Ag and Cu base alloys

Bulk Annealing Experiments [13]

The dissociation width was measured for those specimens which were heat-treated in bulk form. Fig. 6 shows weak-beam micrographs of dissociated dislocations in Cu-13.43 at.% Al alloy cooled at three different rates after annealing for 30 min at 773 K, together with those in an as-deformed specimen. There is no significant difference in the dissociation width Δ. Fig. 7 shows dislocations in three differently treated specimens of a Cu-8.01 at % Si alloy: (a) in the as-deformed specimen, (b) in the furnace-cooled specimen after annealing at 653 K for 1 hr, and (c) in a very slowly cooled specimen (10 K/day). It is clear that the Δ-value in the very slowly cooled specimen is much larger than that in the furnace-cooled or the as-deformed ones.

Fig. 8 shows the values of Δ as a function of the cooling rate after a high-temperature annealing of the alloys studied. Again, they can be divided into two groups: in group II (Ag-Sb, Cu-Si and Cu-Ge) an increase in Δ was observed as a result of the annealing, whereas in group I (Cu-Al, Ag-Al, and Ag-In) such an increase did not occur or was very small, if it existed at all.

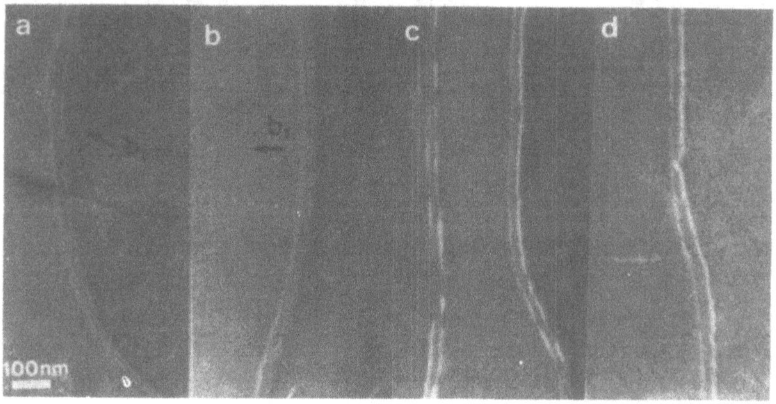

Fig. 6: The effect of the cooling rate on Δ in Cu-14.43 at.% Al. (a) as-deformed, (b) water-quenched, (c) furnace-cooled, and (d) very slowly cooled

EXPLANATIONS OF EXPERIMENTAL RESULTS

In situ Experiments

At low temperatures the Δ-variation determined by in situ thermal cycling experiments is quite similar for the group I and group II alloys. This behaviour can be attributed to the solute impedance effect [1]. That means below a certain temperature T_c, the individual partials are locked at their original sites due to their interaction with solute atoms. Only above T_c the partials can overcome the solute impedance effect by thermally activated processes. Thus, the true value of Δ measured on a specimen deformed at a temperature below T_c must be obtained from the room-temperature value of Δ measured on a specimen deformed at the respective temperature for both groups I and II.

The difference in the behaviour of the Δ-variation above T_c between groups I and II can be explained by postulating that the Suzuki segregation occurs in group II and that it is absent in group I. Fig. 9 sche-

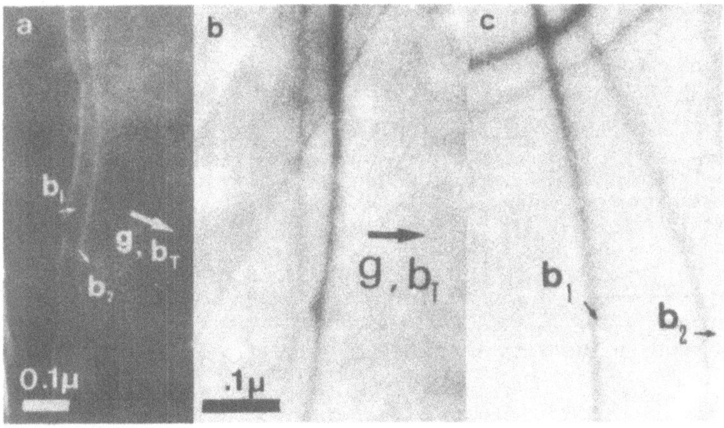

Fig. 7: The effect of the cooling rate on Δ in Cu-8.01 at.% Si. a) as-deformed. (b) furnace cooled, and (c) very slowly cooled

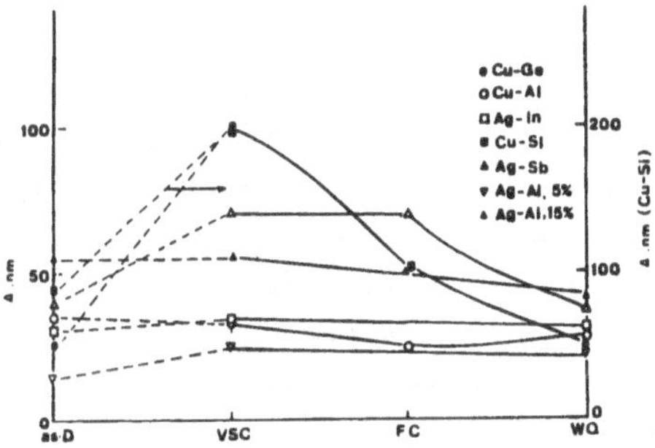

Fig. 8: The effect of the cooling rate on Δ in various of Ag and Cu base alloys, together with the value in as-deformed specimens (as-D)

matically shows the temperature variation of the dissociation width of unsegregated dislocations Δ_u and that of segregated dislocations Δ_s. In as-deformed specimens there is no segregation of the solute atoms to the stacking-fault ribbon of a dissociated dislocation because of a low atomic diffusivity at room temperature. When the specimens are heated to a certain temperature T_D above which the diffusion of the solute atoms becomes substantial, the segregation starts. If T_D is higher than T_c, then Δ increases above T_D as the solute atoms segregate to the stacking fault ribbon, because γ decreases (and equivalently Δ increases) with the concentration of the solute atoms. On further heating a desegregation of the solute atoms away from the stacking fault ribbon takes place because of a reduced amount of the equilibrium segregation with increasing temperature, and Δ decreases again. We can define a temperature T_s above which the amount of segregation is negligibly small. If T_D is higher than T_s (Fig. 9(a)) there is no segregation, being true of group

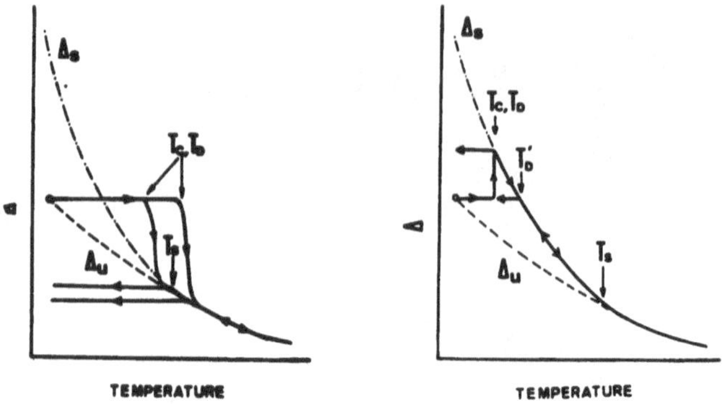

Fig. 9: Schematic illustration of Δ_i, Δ_e, Δ_c, T_c, T_D and T_s

I. On the other hand, if T_s is much higher than both T_D and T_c (Fig. 9(b)), segregation takes place, resulting in an increase of Δ, which holds good for group II.

Bulk Annealing Experiments

The difference between groups I and II after high-temperature annealing in the bulk form can be rationalized by the postulation described above. During high-temperature annealing the equilibrium between a dissociated dislocation and the solute atoms is established. If T_s is much higher than both T_D and T_c (Fig. 9(b)), Δ increases substantially on cooling below T_s because of an increased amount of segregation with decreasing temperature, applying to group II. On the other hand, if T_D is higher than T_s (Fig. 9(a)), there is no segregation during a practical time scale. Similarly, if T_c is higher than T_s (Fig. 9(a)), we cannot detect the segregation by measuring Δ, since Δ cannot assume the equilibrium value below T_c, even if segregation occurs, which holds good for group I. Only if T_s is much higher than both T_c and T_D, the Suzuki segregation can definitely be evidenced as a widening of a dissociated dislocation after a high-temperature annealing.

The dependence of Δ on the cooling rate in the alloys of group II can be explained as follows. During cooling the actual value of Δ_a starts to deviate from the equilibrium value of Δ_e as the temperature approaches T_D even for a very slow cooling so that the room temperature value of Δ_a may correspond to the equilibrium value of Δ_e at a temperature very close to T_D. On the other hand, for a rapid cooling, this deviation starts at T_D', i.e. a temperature still higher than T_D. As a result, the room temperature value of Δ_a after such a rapid cooling corresponds to the equilibrium value of Δ_e at T_D'. It is much smaller than the room temperature value of Δ_a measured after a very slow cooling, which essentially corresponds to the equilibrium value of Δ_e at T_D.

DISCUSSION

Intrinsic Temperature Dependence of the Stacking-Fault Energy

For group I with the Suzuki segregation practically not occurring, the equilibrium value of Δ_e observed above T_D, represents the true SFE γ_i of the unsegregated dislocations. On the other hand, for group II with the Suzuki segregation occurring between T_D and T_s, the equilibrium value of Δ_e observed between T_D and T_s corresponds to the SFE γ_e of the segregated dislocations and not to the true value of SFE γ_i of the unsegregated dislocations. γ_i can be obtained from Δ measured above T_s. An alternative way to determine the temperature dependence of γ_i for group II is only to use the values of γ_i determined below T_c.

The broken lines in Fig. 5 show the temperature dependence of γ_i obtained in that way. In Fig. 10 $d\gamma/dT$ is plotted against e/a for the alloys studied, together with that of pure silver. The following features are revealed.

1) The values of $d\gamma/dT$ for Ag base alloys lie on a single curve. The same is true of Cu base alloys, although the values of $d\gamma/dT$ of Cu-base alloys are slightly higher than those of Ag base alloys.

2) $d\gamma/dT$ increases with e/a, with the steepest slope occurring at $1.10<(e/a)<1.20$.

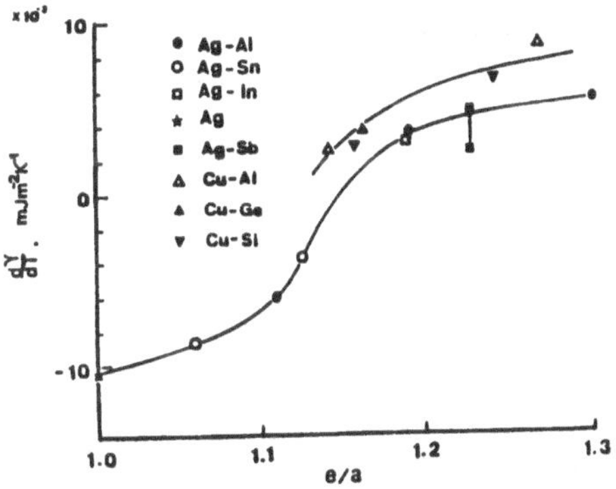

Fig. 10: Effect of the nature of the alloying elements on $d\gamma/dT$ in Ag- and Cu-base alloys

There seem to be inconsistencies between the experimental results of the temperature dependence of γ in noble metal base alloys obtained so far. Apart from inaccuracies involved by TEM techniques used for these previous studies, the inconsistencies may result from ignoring two factors: one is the dependence of $d\gamma/dT$ on the concentration (e/a), and the other is that the Suzuki segregation occurs in some alloys but not in others. If both of them are taken into account, the temperature dependence of γ in noble metal base alloys seems to obey a simple rule.

A theoretical derivation of the temperature dependence of γ in noble metal base alloys has been treated by Tisone, who considered the effect of the temperature on γ in a phenomenological way by involving the relaxation time of conduction electrons [20]. The most important point of Tisone's theory related to the present study is that there is a critical point at $(e/a)_c=1.14$ in a plot of $d\gamma/dT$ versus e/a because of the singularity of the screening of a free electron gas. For $e/a<1.14$, $d\gamma/dT$ is similar to that of pure metals, whereas for $e/a>1.14$, $d\gamma/dT$ is considerably higher. The results obtained in the present study (Fig. 10) agree well with Tisone's theory, and the present experimental results can be said to support the theoretical estimation of the temperature dependence of γ based on the electron theory.

Suzuki Segregation

As pointed out above, the fact whether or not the Suzuki segregation can be detected as a variation in Δ depends on relative positions of T_s, T_D and T_c. T_c is governed by interactions between solute atoms and partial dislocations. Among the interactions possible the elastic interaction is likely to be most important. If so, the difference in the atomic size of solute atoms and solvent atoms is an important factor in determining T_c. Fig. 11 shows the variation of the lattice constants of the Ag base, Cu base, and Au-base solid solutions under consideration [21]. Except for Cu-Si, the variation of the lattice constant $|da/dc|$ of the alloys in group II (which evidence the Suzuki segregation) is larger than that of the alloys of group I (which do not evidence the Suzuki segregation). This does not agree with the aforementioned discussion.

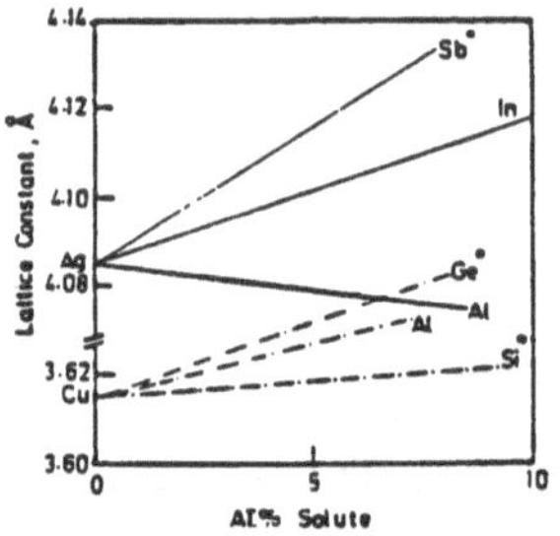

Fig. 11: Variation of the lattice parameter

Recently, it has been shown that in some Au base alloys with solute atoms belonging to the IVb and Vb groups of the periodic table the diffusion of solute atoms is remarkably enhanced [22]. If we assume that this enhanced diffusion of solute atoms is generally valid for other noble metal base alloys such as Cu and Ag base alloys, T_D is likely to be very low in Cu and Ag base alloys with solute atoms belonging to the IVb and Vb groups. All the alloys which show the Suzuki segregation (group II) have solute atoms belonging to either the IVb group or the Vb group. This is exactly what is expected from the consideration of T_D. Thus, it is suggested that the enhanced diffusion of those alloys which have solute atoms belonging to the IVb or Vb groups is responsible for the observed Suzuki segregation in these alloys, and the apparent lack of the Suzuki segregation in group I should be attributed to the low atomic diffusivity of solute atoms in these alloys.

REFERENCES

1 J.W. Christian and P.R. Swann, In: Alloying Behaviour and Effect in Concentrated Solid Solutions, T.B. Massalski (Ed.), Gordon & Breach, New York 1965, p. 105

2 P.C.J. Gallagher, Metall. Trans., 1 (1970) 2429

3 L. Remy, A. Pineau and B. Thomas, Mater. Sci. Engng. 36 (1978) 47

4 H. Suzuki, Sci. Rep. Res. Inst. Tohoku Univ. A4 (1952) 452

5 H. Suzuki, In: Dislocations and Mechanical Properties of Crystals, J.C. Fisher, W.G. Johnston, R. Thomson & T. Vreeland, Jr. (Eds.), John Wiley, New York, 1957, p. 161

6 H. Suzuki, J. Phys. Soc. Japan 17 (1962) 322

7 H. Saka, Y. Sueki and T. Imura, Phil. Mag. A37(1978) 273

8 H. Saka, T. Iwata and T. Imura, Phil. Mag. A37(1978) 291

9 H. Saka, T. Iwata and T. Imura, In: Electron Microscopy 1980, P. Brederoo and J. Landuyt (Eds.), Leiden 1980, p. 326

10 H. Saka, T. Kondo and T. Imura, Phil. Mag. A47(1983) 859

11 H. Saka, In: Dislocations in Solids, H. Suzuki, T. Ninbomiya, K. Sumino & S. Takeuchi (Eds.), Univ. of Tokyo Press, Tokyo 1985, p. 251

12 H. Saka, T. Kondo and T. Imura, Ibid., p. 255

13 H. Saka, Phil. Mag. 47A (1983) 131

14 A.N. Stroh, Phil. Mag. 3 (1958) 625

15 H. Saka, Res Mechanica 11 (1984) 211

16 P.M. Hazzledine, H.P. Karnthaler and E. Winter, Phil. Mag. 32 (1975) 81

17 D.J.H. Cockayne, I.L.F. Ray and M.J. Whelan, Phil. Mag. 20 (1969) 1265

18 M.J. Makin, Phil. Mag. 18 (1968) 637

19 K. Urban, Phys. Stat. Sol. (a), 56 (1975) 157

20 T.C.. Tisone, Met. Trans. 3 (1972) 427

21 W.B. Pearson, Lattice Spacings and Structures of Metals and Alloys, Pergamon Press, London 1958

22 J. Takamura, In: Point Defects and Defect Interactions in Metals, J. Takamura, M. Doyama & M. Kiritani (Eds.), Univ. of Tokyo Press, Tokyo 1982, p. 431

HISTORY AND APPEARANCE FORMS OF DISLOCATIONS+)

Dietrich Schulze

c/o: Akademie der Wissenschaften der DDR,
Zentralinstitut für Festkörperphysik und Werkstofforschung,
Helmholtzstr. 20, DDR 8027 Dresden

HISTORY

The origin of dislocations goes back to the end of the last century. At first, such a configuration was introduced into the physical theory of a mechanical ether by BURTON as early as in 1892. He had the idea of a "strain-figure", a local modification in the structure. Owing to suitable forces it is able to propagate through a granular medium which violates HOOKE's Law for large strains, thus corresponding to the dislocation theory in a crystal lattice. - The model of a wire welded into a ring after it had been twisted - in such a manner also constituting an "intrinsic strain-form - should be mentioned here, too. It was treated theoretically by LARMOR in 1897. LARMOR's concept may be considered a first step to the theory of elastic dislocations.

As early as in 1867, however, REUSCH observed slip steps on the surfaces of rock salt and calcite. Moreover, he used terms such as "slip plane", "direction of easiest translation" (i.e. SCHMID's shear stress law) and he mentioned "weak positions" in crystals. The f i r s t observations of slip lines on metals of MÜGGE in 1899 as well as his denotation "translation" for the glide process have been forgotten for a long time. Up to now, the slip line observations on lead in 1900 by EWING and ROSENHAIN have enjoyed priority. Both the last mentioned scientific events were once again called to mind by the too early deceased G.E.R. SCHULZE in 1967.

The Italian school, WEINGARTEN (1901), VOLTERRA (1907), SOMIGLIANA (1914), developed the dislocation-induced theory of self-strain states in an elastic continuum. LOVE (1920), too, proposed the term "dislocation"; the German translation "Versetzung" was introduced by OROWAN. Further essential stages in the evolution of the understanding of that phenomenon - supposing a lattice structure - were: the deformation on account of "Biegegleitung" (MASING and POLANYI 1923), the idea of "Verhakung" (DEHLINGER 1929), OROWAN's model of local slip (1934), POLANYI's concept of the edge dislocation, and TAYLOR's model of its movement (1934), the introduction of the screw dislocation by BURGERS (1939), and - last not least - the "invention" of a dislocation multiplication model (FRANK-

+)To be published in detail in "Praktische Metallographie" 1989

READ-source, 1950). (For further information see the tables.)

APPEARANCE FORMS

Dislocation-like configurations occur within manifold physical, chemical, botanic, zoological and biological structures, e.g. in BRAGG's soap bubble model, in the flux line lattice of type II superconductors, ferromagnetic domains, liquid crystals, drifting sandhills, zebra stripes, butterfly wings (only "crysiridia madegascariensis"), in the grain pattern of a corn-cob, etc. They play also a dominant role in the replication of the double helix of DNA as well as in that of layer silicates but also in the self-organization of matter.

Obviously, in a generalized sense, the dislocations are universal elements of creation, structure and movement, of multiplication, stabilization and destruction in organic as well as in inorganic states of matter.

HISTORY OF DISLOCATIONS ("First Period")

EVENT	YEAR	AUTHORS
"strain-figure" in a granular medium as a modification in the structure of mechanical ether	1892	G.V. Burton
"intrinsic strain-form" in a medium showing strictly linear elasticity - constituting an electron - moving in the stagnant ether	1897	J. Larmor
Observation of "slip steps" on Pb; PRIORITY?	1900	J.A. Ewing W. Rosenhain
First observation of slip lines (NaCl, $CaCO_3$); origin of terms "slip plane", "direction of easiest translation" and "weak positions" in crystals	1867	F.E. Reusch (Poggendorfs Ann.) once again called to mind by G.E.R. Schulze (1967)
. Very first observation of slip lines on metals (Au, Cu...); . Determination of slip systems; . Originator of the term "translation"	1899	O. Mügge
Theory of self-strain states in an elastic continuum ("distorsioni")	1901 1907 1912/15	G. Weingarten V. Volterra C. Somigliana

EVENT	YEAR	AUTHORS
Theory of hysteresis in a lattice of rotating dipoles with glideable regions of abnormal alignment	1907	J.A. Ewing
"I have ventured to call them DISLOCATIONS"	1920	A.E.H. Love

("Second Period")

EVENT	YEAR	AUTHORS
Proposed notch effect by internal cracks in order to overcome the contradiction between theoretical and measured critical stresses	1921	A.A. Griffith
Systematic measurements of stress-strain curves with Polanyi-apparatus using single and polycrystalline metals	20ies	M. Polanyi, E. Schmid, W. Boas, G. Sachs et al.
shear stress law	1926?	E. Schmid
Model of flexural slip ("Biegegleitung")	1923	G. Masing, M. Polanyi
Interpretation of mechanical hysteresis on the basis of translation periodicity of crystal lattice	1928	L. Prandtl
Model of "Verhakung"	1929	U. Dehlinger
Observation of decorated dislocations; not interpreted as such	1905 1932 1932	H. Siedentopf E. Rexer A. Edner
"Lockerstellen"	about 1930	A. Smekal
Model of local slip	1934	E. Orowan

("Third Period")

EVENT	YEAR	AUTHORS
Models of edge dislocation	1934	E. Orowan
- -		M. Polanyi
First estimation of critical shear stress		G.I. Taylor
Model of screw dislocation	1939	J.M. Burgers
Idea of partial dislocations	1948	W. Shockley
Influence of screw dislocations on crystal growth	1949	W.K. Burton N. Cabrera F.C. Frank
Pictures and energy estimations of small-angle boundaries	1949 to 1953	W. Shockley, W.T. Read
Model of dislocation multiplication (Frank-Read-Source)	1950	F.C. Frank, W.T. Read
Imaging of an edge dislocation in lattice fringes of platinum phthalocyanine	1956	J.W. Menter
Evidence of the actual existence of dislocations in metals by electron diffraction contrast	1956	P.B. Hirsch, R.W. Horne M.J. Whelan
<u>Work hardening theories</u>		
(i) Glide-zone model	1957/58	A. Seeger
(ii) Forest-theories	1958/60	P.B. Hirsch Z.S. Basinski
(iii) Jog-dragging theory	1960	P.B. Hirsch
(iv) Mushroom theory	1962	D. Kuhlmann-Wilsdorf
"The dislocation is the elemental self-stress source"	1958	E. Kröner

(New and "exotic" ideas)

EVENT	YEAR	AUTHORS
Simultaneous slip of <u>all atoms</u> of an ideal lattice plane by <u>tunneling</u>	1929	M. Polanyi, E. Schmid
Dislocations shall overcome barriers by quantum mechanical <u>tunneling</u>	1956	N.F. Mott

EVENT	YEAR	AUTHORS
Experimental indications of the quantum behaviour of moving dislocations at extremely low temperatures	70ies	V.I. Startsev and his co-workers
Existence of dislocations or similar configurations in amorphous materials ? ? ?	1975	R.M. Cotterill et al., J.J. Gilman

APPEARANCE FORMS OF DISLOCATIONS (Selection)

Dislocation-like configurations in	YEAR	AUTHORS
Physics, Chemistry etc.		
Soap-bubble raft	1947	W.L. Bragg, J.F. Nye
Flux lines of superconductors	1967/68	U. Essmann, H. Träuble
Ferromagnetic domains	1968	
(i) High-Voltage Lorentz microscopy of Permalloy		e.g. R.P. Ferrier, I.B. Puchalska
(ii) SEM study of barium ferrite		e.g. D.C. Joy, J.P. Jacubovics
Liquid-crystalline swallow-tailed compounds	1988	W. Weissflog et al.
Liesegang's Rings (self-organization of matter)	since 1898	R.E. Liesegang, see H.K. Henisch
Drifting sandhills	-	see e.g. O. Patzelt; V. Butyrin
Intercalated layer (i.e. inserted half plane) in montmorillionite	1981	A. Weiss
Zoology, Botany, Biology		
Stripes on the neck of a zebra	1942	D'Arcy W. Thompson
Wing-patterns ("lattice fringes") of the butterfly "crysiridia madagascariensis"	1982	D. Schulze
Pattern of grains on a corn-cob	1967	see F.R.N. Nabarro
Partly replicated double helix of DNA	1981	see A. Weiss

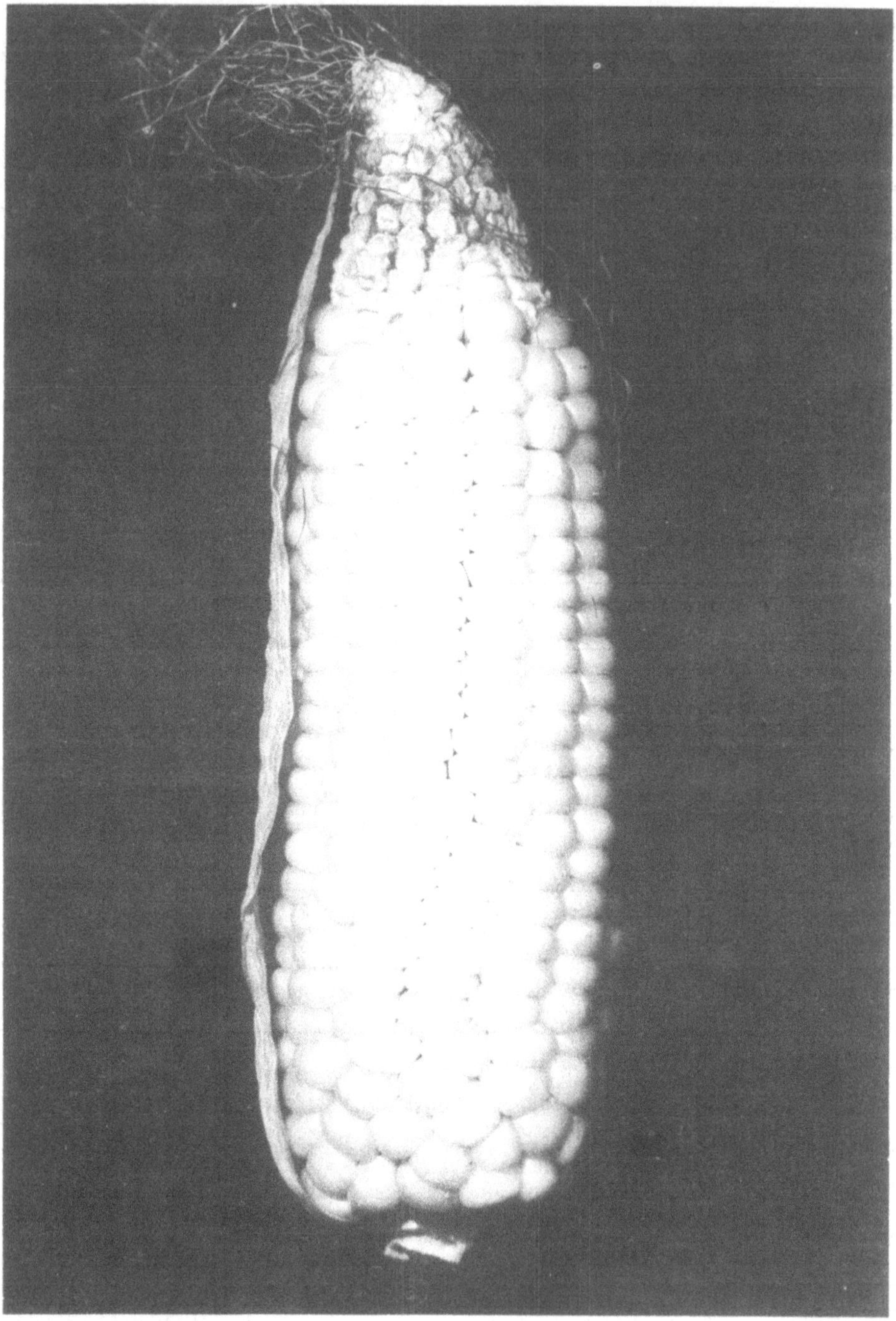

HIGH VOLTAGE / HIGH RESOLUTION STUDIES OF METAL AND SEMICONDUCTOR INTERFACES

K.H. Westmacott, U. Dahmen
National Center for Electron Microscopy, Materials and Chemical Sciences Division, Lawrence Berkeley Laboratory, Berkeley, Ca 94720, USA

ABSTRACT

The application of high resolution transmission electron microscopy to the study of homo- or hetero-phase interface structures requires specimens that meet stringent criteria. In some systems the necessary geometric imaging conditions are established naturally, thus greatly simplifying the analysis. This is illustrated for a diamond-hexagonal/diamond-cubic interface in deformed silicon, a $\Sigma 99$ tilt boundary in a pure aluminum bicrystal, and a germanium precipitate in an aluminum matrix.

INTRODUCTION

The strength of materials is strongly influenced by the presence of interfaces such as grain or interphase boundaries. Consequently, fundamental studies of the structure of interfaces will ultimately lead to a better understanding of the mechanical behavior of materials.

In recent years the availability of transmission electron microscopes with sub-2Å point resolution has allowed atomic resolution imaging of interfaces to be extended to the close-packed metals. The purpose of this short contribution is 1) to emphasize the importance of optimizing the specimen geometry and quality to ensure attainment of the necessary contrast and resolution of interfacial structure and 2) to present brief summaries of three recent high resolution studies of interfaces in metals and semiconductors. Although not all the examples were specifically designed as an interface study, the work provides a clear illustration of three different types of interface, a heterophase interface in a semiconductor (Si), a homophase interface (grain boundary) in a metal (Al), and a heterophase metal/semiconductor interface (Al-Ge).

EXPERIMENTAL DETAILS

An ideal specimen for high resolution imaging is one that is thin (<100Å) flat (not bent) and clean (with a minimum covering of contamination). Most unsuccessful attempts to obtain an interpretable image are attributable to failure to fulfill one or more of these conditions. For the successful imaging of an interface at atomic resolution further stringent geometrical conditions must also be satisfied [1]. Each crystal forming the interface must have a low index zone axis (but not necessarily the same zone in both crystals) parallel to the beam and parallel to a vector in the interface plane. These conditions are illustrated in Figure 1. To achieve atomic resolution throughout the specimen and at the interface the sample is oriented using the biaxial tilt-stage so that these directions are accurately aligned with both the optic axis of the microscope and the electron beam.

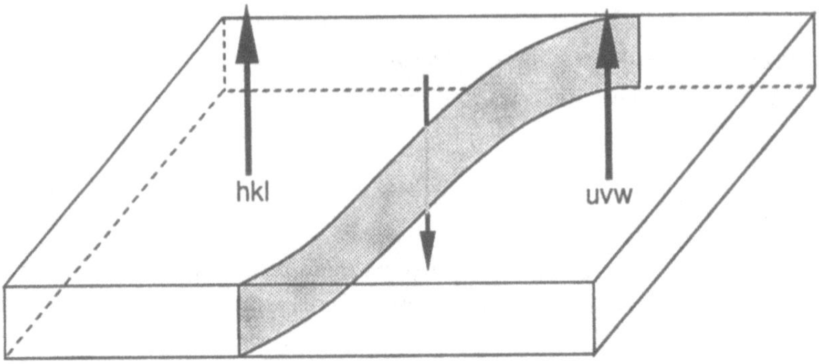

Fig. 1 Schematic illustration of the geometric conditions that must be satisfied to image interface structures at high resolution.

Suitable specimens were prepared by chemical or electropolishing techniques or by dimpling and ion beam thinning and examined in the NCEM JEOL-1000 Atomic Resolution Microscope at 800 kV and other high resolution microscopes.

RESULTS

Hexagonal Silicon (Heterophase Interfaces in a Semiconductor)

The observation of thin ribbons of a diamond-hexagonal (dh) form of silicon in hot-indented specimens was first reported by Eremenko and Nikitenko [2,3]. The ribbons were observed to emanate from the highly deformed regions near the indent and propagate on {511} planes of the matrix. This interesting phenomenon was subsequently studied in detail using a variety of techniques, including high resolution imaging, by Pirouz et al. [4,5] and a comprehensive analysis will shortly appear in print [6]. It is shown that formation of the dh

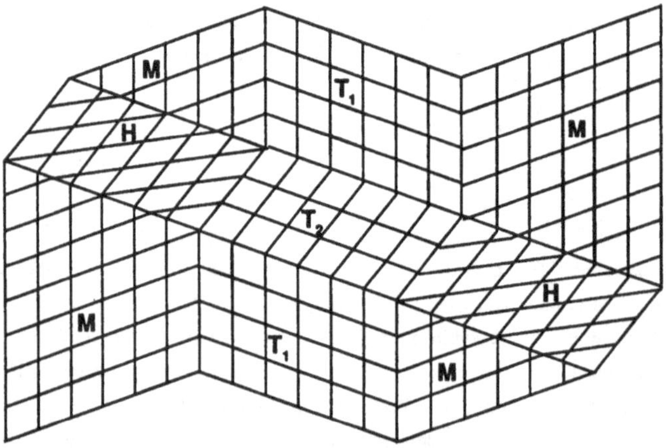

Fig. 2 Schematic diagram showing, in a <110> projection, the formation of a hexagonal ribbon H when a secondary twin T_2 nucleates inside a primary twin T_1 and propagates into the matrix M.

metastable phase can be treated as a martensitic transformation having its origin at the interfaces of deformation twin interactions. A schematic illustration of the formation of hexagonal phase in an fcc lattice in <110> projection is shown in Figure 2. Figure 2 illustrates the condition when a secondary twin T_2 nucleates in a primary twin band T_1 and subsequently propagates into the matrix M. This is the mechanism that gives rise to the long ribbons on {511} habit planes. An example of an hexagonal ribbon imaged at high resolution is given in Figure 3.

Fig. 3 Ribbon of diamond-hexagonal material with {511} habit plane in hot-indented silicon (micrograph courtesy P. Pirouz)

Grain Boundaries in ICB Aluminum Bicrystals (Homophase Interfaces in a Metal)

Unique continuous bicrystal films are formed when aluminum is deposited on a {100} silicon substrate by the Ionized Cluster Beam technique (see Ref. 7 for details). Two different {110} orientation variants of the aluminum form which are rotated relative to each other by 90°. The resulting microstructure consists of islands of one variant in a sea of the other bounded by 90° <110> tilt boundaries all of which close on themselves (i.e. the film contains no grain boundary triple points) [8,9]. If an annealing treatment at 400° C is followed by removal of the silicon substrate a specimen ideal for high resolution TEM observation of the boundaries is obtained. This is because during the annealing the grain boundaries in the aluminum minimize their area by rotating normal to the foil surface. The result is a foil containing the ideal structure depicted in Figure 1. An example of an image of a Σ99 symmetrical boundary.obtained under these conditions is given in Fig. 4. In the center of the image an asymmetrical section is seen to connect two symmetrical {557} boundary segments which exhibit clear atomic relaxations into structural units [10].

Needle Precipitates in Al-Ge (Heterophase Metal/Semiconductor Interfaces.

When a dilute Al-Ge alloy is given a suitable quench/age treatment, long needle-shaped Ge precipitates form lying parallel to <100> directions in the aluminum [11]. Consequently, if thin foils of this alloy are prepared and viewed along a <100> zone axis, observation of the end-on Ge needles allows a detailed examination to be made of the interface between the germanium

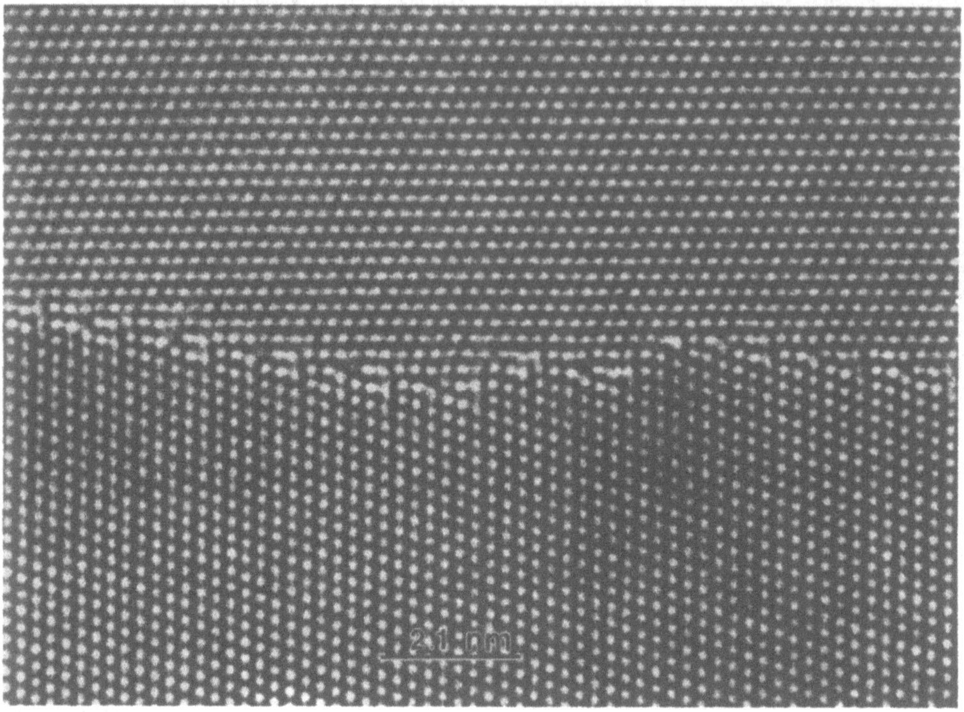

Fig. 4 Faceted segments of a Σ99 89.4° <110> tilt boundary in aluminum

and aluminum. It is found [12,13] that a <110> direction in the Ge is accurately parallel to <100> Al, thus again satisfying the conditions in Figure 1. The precise alignment is attributable to the $\sqrt{2}$ relationship between the lattice parameters of Al and Ge whereby [100] Al almost exactly matches [110] Ge. An example illustrating the image detail of the interface structure as well as the internal precipitate structure is given in Figure 5.

CONCLUSIONS

High resolution imaging is a powerful technique for observing the atomic structure of interfaces. The specimen geometry must be optimized to obtain maximum information on the interface structure.

ACKNOWLEDGEMENTS

This work was supported by the Director, Office of Energy Research, Office of Basic Energy Sciences, Materials Sciences Division of the U.S. Department of Energy under contract No. DE-AC03-76SF00098.

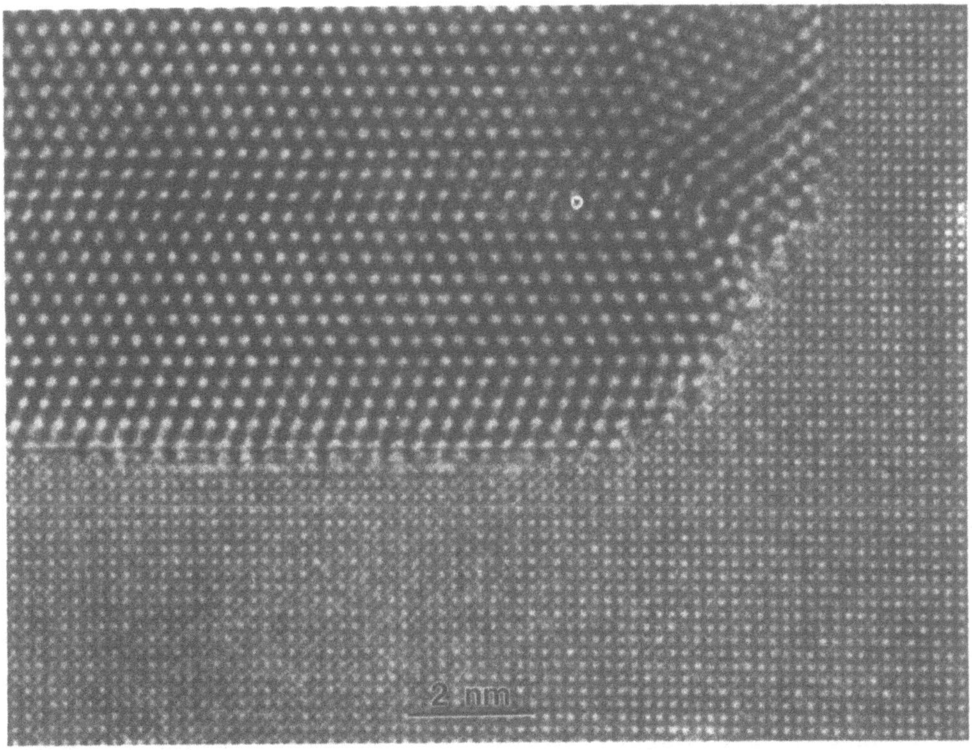

Fig. 5 Interface structure between aluminum matrix and a germanium needle formed by a precipitation reaction. Note the alignment of <100> Al with <110> Ge. (Micrograph by J. Douin)

REFERENCES

1 U. Dahmen, J. Douin, C.J. Hetherington and K.H. Westmacott, MRS Symp. Proc. 139, 87 (1989)

2 V.G. Nikitenko, Sov. Phys. Sol. St.13, 230 (1971)

3 V.G. Eremenko and Nikitenko, Phys. Stat. Sol. (a) 14, 317 (1972)

4 P. Pirouz, R. Chaim, J. Samuels, Proc. 5th Int. Congress. on "The Structure and Properties of Dislocations in Semiconductors", Moscow, March 1986, in press

5 P. Pirouz, R. Chaim and U. Dahmen, MRS Symposium Proc. 104, 133 (1987)

6 P. Pirouz, R. Chaim, U. Dahmen and K.H. Westmacott, Acta Met. in press

7 Proc. ISIAT Conference, Tokyo, Japan (1986-89)

8 U. Dahmen and K.H. Westmacott, Scr. Met. 22,1673 (1988)

9 M.C. Madden, Appl. Phys. Lett. 55, 1077 (1989)

10 A.P. Sutton and V.Vitek, Phil. Trans. Roy. Soc. London A 309, 1 (1983)

11 U. Dahmen, C. Nelson and K.H. Westmacott, 44th EMSA Proc., G.W. Bailey (ed.), p. 538 (1986)

12 U. Dahmen and K.H. Westmacott, Science 233, 875 (1986)

13 J. Douin, U. Dahmen and K.H. Westmacott, to be submitted to Phil. Mag.

MICROSCOPIC INVESTIGATIONS OF LASER DAMAGE IN ALKALI HALIDES

A.V. Gorbunov*, N.V. Klassen*, and K. Schumann**

* Institute of Solid State Physics, USSR Academy of Sciences, Chernogolovka, Moscow District, 142432, USSR
** Institute of Solid State Physics and Electron Microscopy, Academy of Sciences of the G.D.R., P.O.B. 250, Halle/S., 4050, G.D.R.

ABSTRACT

Optical and scanning electron microscopy (SEM) were used to investigate the processes of deformation and fracture in NaCl and KCl single crystals induced by CO_2-laser irradiation. Both pulsed laser optical breakdown and local melting inside the crystals were studied under continuous wave (CW) laser action. X-ray microanalysis was applied to determine the chemical composition of the largest inclusions, which were found to give rise to both the above-mentioned kinds of alkali halide laser damage.

INTRODUCTION

In spite of the evident topicality of the microscopic methods applied to study the optical damage of highly transparent solids, their possibilities are far from being completely realized. Meanwhile, the application of optical microscopy to the in situ observation of the optical breakdown induced by CO_2-laser in bulk crystals of NaCl yields new information [1]. Many more details may be obtained on the crystal damage for irradiation intensities almost one order of magnitude lower than the average statistical breakdown threshold.

This paper presents the experimental results of microscopic investigations of the local damage in the bulk of NaCl and KCl single crystals induced by CO_2-laser irradiation. The optical breakdown at absorbing microinclusions under laser pulses of microseconds and the local melting under CW-laser action are compared with respect to the induced fracture and deformation processes.

EXPERIMENTAL RESULTS

The main features of AHC bulk laser damage at room temperature are generally known (cf.[2, 3], e.g.). They may briefly be described as follows.

For intensities near the bulk breakdown threshold, at the site of the observed light spark, a void is formed of almost spherical shape and a few microns in size (Fig. 1a). It is usually surrounded by mirocracks oriented along {100} cleavage planes. Selective etching reveals that around the void a dense dislocation tangle is located [3], with a dislocation density not lower than 10^8 cm^{-2}. Outside the spots characteristic rosettes are formed by slip bands of screw and edge dislocations. The rosettes are similar to those generated by indentation. The diameter of the spot is typically 200 to 300 μm and the length of the rosettes is about 1mm.

The shape of the voids is distorted mainly by microcracks. We found out that an increase of the surrounding temperature during irradiation reduces the number of cracks and their size. Microcracks are no longer formed at temperatures higher than 600 °C. In the latter case the voids are ideal spheres. (Fig. 1b).

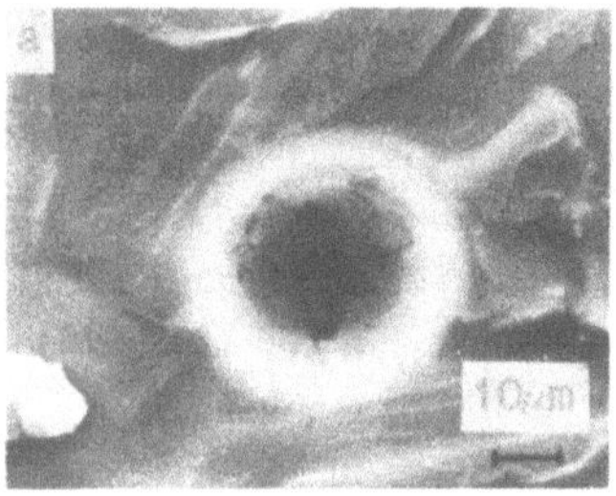

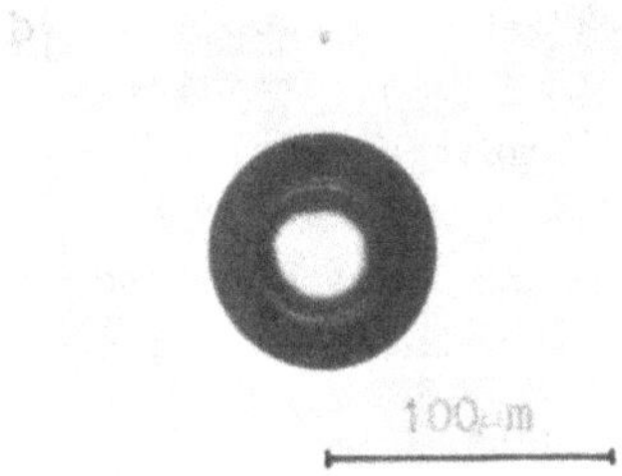

Fig. 1: A void generated by AHC bulk laser damage: a) SEM pattern of room temperature damage; b) optical microscope pattern of the damage at T=600 °C

Recently we realized the melting in the bulk of NaCl and KCl by irradiation with a CW CO_2-laser [4]. In the present study results are described of more careful microscopic in situ observations.

The melting processes are highly localized in the bulk resulting in little liquid drops, which keep a stable size of $d \leq d_{cr} \approx 3$ to 5 μm for a relatively long time during the laser melting. The nucleated droplets are visible as dark dots in Figs 2a to c. During the laser beam heating of the sample some of the droplets exceed the critical size d_{cr} and start to grow quickly. In several seconds the "dark dots" transform into rapidly increasing spheres. At d≤10 μm, there are first indications of crystallographic facetting (Fig. 2a). If the dimensions reach 20 to 30 μm, growth processes in <110> directions are clearly noticeable on the surface of the spherical drops. During the next step of melting a symmetrical 3-dimensional "melting rosette" is formed, with the wings being directed along <110>, Fig. 2c. The shape of the wings is very complicated, since new sub-wings start growing from them along other <110> directions. From these sub-wings again new ones start continuing the growth process. Since the melted structures look like dendrites occurring at crystal growth, they may be designated as liquid dendrites. The similarity is even more pronounced at advanced stages of laser melting (see Fig. 2d). The velocity of the moving tip of a "dendrite" amounts to 100 μm/s.

The most striking feature of the laser melting in the bulk crystal is probably the high mobility of the small liquid droplets (dark dots in Figs. 2a to c) in the solid phase. Their velocity may reach 100 µm/s, which is nearly the same as that of the front of the melting zone during its expansion or its shrinking if the laser beam is switched off. Unlike the dentrites, which grow strictly in crystallographic <110> directions, the motion of the droplets does not show a distinct anisotropy but is generally normal to the melting front.

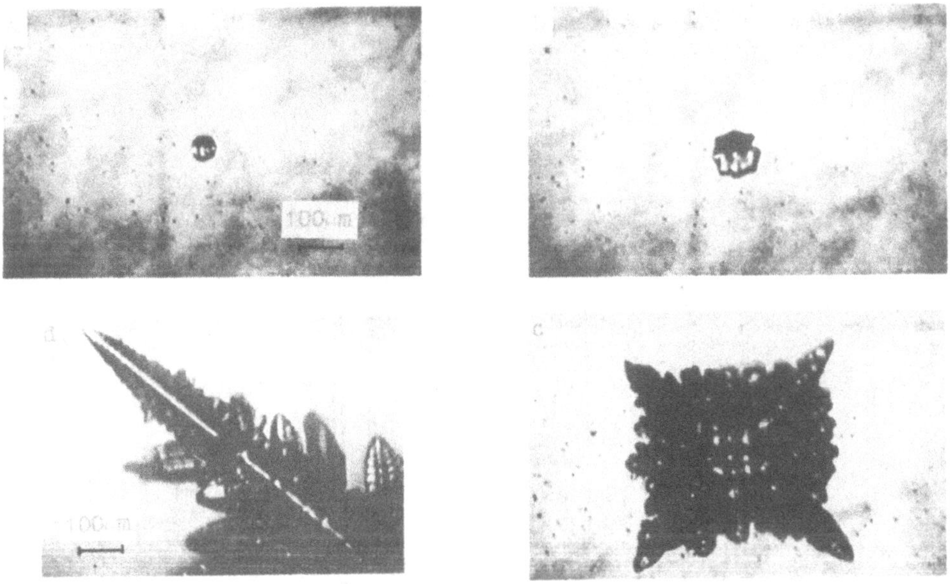

Fig. 2: Laser melting inside NaCl: a to c) basic stages of the melting process of a single nucleated droplet; d) typical view of a moving "liquid dendrite"

The high localization of the optical damage in the volume of transparent crystals is of fundamental interest. It is generally known that the pulsed laser optical breakdown is mostly initiated by absorbing inhomogenities, the nature of which is difficult to identify, however. The problems, on the one hand, arise from the extremely small integral quantity of impurities in the mass of the crystal and, on the other hand, from the small size of the defects such as impurity agglomerates, precipitates, and inclusions.

The largest inclusions which can be identified by optical microscopy in melt-grown commercial grade NaCl and KCl were studied in SEM and analyzed by X-ray microanalysis (Fig. 3). They look like crystallites of various shapes such as genuine spheres or more complicated particles. They are mostly composed of the elements Fe, Ni, Ti, Zn, and Si.

The problem of the localization of the laser melting was investigated by ultramicroscopy (optical microscopy in scattered light), similar to [1]. It turned out that the light scattering centres present in the AHC volume may act as nucleation centres to the melting quite

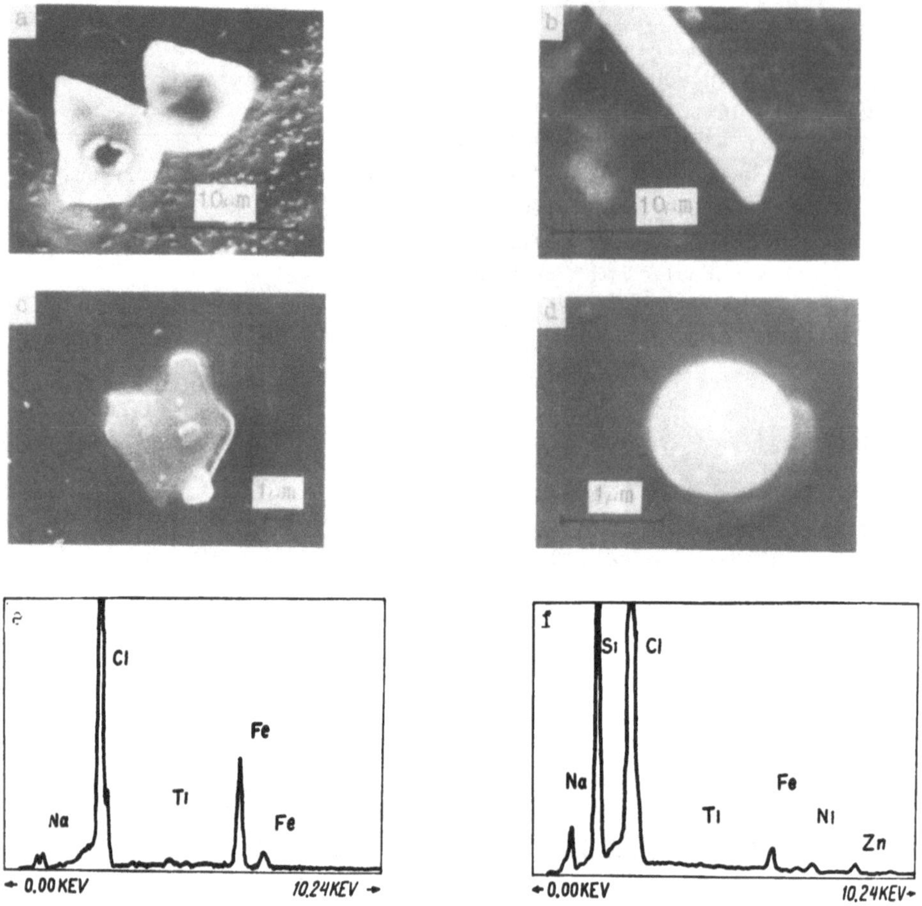

Fig. 3: SEM image of inclusions in NaCl and KCl containing mainly: a)Ni, b) Pt, c) Fe, d) Si. e) and f) electron beam excited X-ray spectra of c) and d)

similar to the optical damage by a pulsed laser. In some cases, however, the melting is nucleated without such centres. In the focus region many dark dots have formed with diameters not larger than 3 μm. They remain constant for a long time. The concentration of the dots strongly increases. At the moment when one or often several of these dots begin to rapidly grow to anisotropic rosettes, the mean distance between them amounts to tens and even hundreds of microns. At this stage there is no correlation to the distribution of the melting centres. However, somewhat later, when the mean distances decrease down to about 10 μm, the melting drops are arranged in <110> directions.

DISCUSSION

The observed large differences in laser damage induced by pulsed or CW-laser irradiation obviously arises from the different times of the mechanical loading of the samples. As to the deformation geometry the two cases are quite similar. In the pulsed mode the mechanical stress arises from the pressure of the dense plasma with temperatures of about 10^4 K, and for the CW-laser the mechanical stress is caused by the pressure of the melt, with its specific volume being 25% higher than that of the solid. At the beginning of both processes the loading geometry corresponds to that of a point source of pressure. At the pulsed breakdown, however, the stress pulse lasts for microseconds, since the time of the light spark does not exceed 5 to 6 μs [1]. Using the experimental results of shock wave loading in LiF [5], of pulsed electron beam deformation of LiF and NaCl [6], and an estimation on the basis of an energy balance made in [3], the magnitude of the pressure in the void may be assumed to be about 1 kbar. Thus, the acting shear stress is about 100 times that of the critical shear stress of the materials investigated. Since the stress is proportional to r^{-3}, the dislocation velocity will rapidly decrease with the distance. The distances the dislocations move during the laser pulse are certainly not much larger than several radii of the void, which corresponds to the dimensions of the dislocation cloud around the void [3]. The dislocation density necessary to move the material of the void volume into the region of the dislocation cloud should be at least 10^8 cm^{-2}. The reason for the isotropic shape of the void is that the interaction between the dislocations prevents them from moving in crystallographic directions. The observed wings of dislocation arrays along <110> outside the dense dislocation tangle [3] probably propagate after a long time after the laser pulse and do therefore hardly influence the shape of the void. In [7] the irradiation of LiF surfaces by the same CO_2-laser was studied to investigate the time evolution of the dipole moment induced by charging dislocations in the wings of similar rosettes. Accordingly, the moment mainly increases during the first 200 to 500 μs after the laser pulse and becomes constant afterwards. The result has been explained by the elastic relaxations, which occur inside the dislocation arrays and result in their relatively slow propagation.

The loading time available during the local melting in the volume is sufficient to relax the mechanical stresses by material transport via diffusion and plastic flow. Near the melting point, in NaCl three types of glide systems are activated with {100}, {110}, and {111} slip planes, respectively, and <110> slip directions in each case. This slip geometry may explain the observed formation of melted channels along <110> in NaCl.

The observed melting structure may alternatively be explained by a change of the melting temperature and (or) of the heat of melting as a consequence of the dislocation structure generated in the volume before melting. The thermal gradients induced by the focused laser beam may be in the order of a few hundreds of $^{\circ}C/cm$ [4], which yield a strong deformation of the crystal. Investigations of the samples in polarized light revealed a quasi-periodical 3-dimensional structure of glide bands with a mean distance of 10 to 30 μm between them. The slip bands mostly intersect along <110> directions. Because of the high disorder and the large internal energy in the regions of intersection melting may start earlier.

The dendritic shape of the melt inside a solid crystal is supposed to result from an instability of the form of the phase front occurring in a solid heated above the melting point. The phenomeneon studied is similar to the growing of "water flowers" or "Tyndall stars" formed inside ice by light or microwave irradiation (cf. [8], e.g., where an overheating of about 0.3 °C was measured). Alkali halide crystals differ from ice so far as the specific volume grows during melting. Furthermore, the processes of plastic deformation are more important, which affect both the development and the nucleation of the melted phase. For the nucleation conditions the work done during the deformation of the surrounding crystal should be taken into account. On the basis of this argument the melting should start in regions where the work necessary for deformation is lower (e.g. at microcracks, microvoids etc.) and not only where the temperature is higher (e.g. at absorbing inclusions). The plastic deformation of a sample, which precedes the melting of the volume (see above) may influence the nucleation of melting, too. The intersection of glide bands results in the formation of compressed or expanded cells of about 10 to 30 μm in size.

The cells underlying the highest expanding stresses are supposed to be most favourable for the nucleation of the melting. According to [9] the melting temperature of NaCl changes with the pressure by about 2 °C/kbar. The generated liquid drop stops growing since the pressure inside the cell increases. Comparing the stresses arising from the volume change during melting with the characteristic stresses of the dislocation rows, on the basis of a force balance, it was possible to estimate the size of the drops to be a few microns. The result agrees well with the value of d_{cr} observed experimentally.

Because of the rigid bonds between the drop and the previously generated dislocations, the high mobility and isotropic motion of the drops can hardly be understood in terms of the traditional mechanisms of diffusion and dislocation glide. A more realistic mechanism is the correlated melting and solidification at the opposite sides of a drop. The reason of this correlation is not yet known. We assume that the defect mechanisms occurring in overheated solids essentially differ from those in usual ones.

REFERENCES

1 A.V. Gorbunov, V.Ya. Emelin, and N.V. Klassen, ZH. Tech. Fiz. 54 (1984) 1383-6

2 V.E. Rogalin, T.I. Samoilova, I.A. Tishchenko, and M.P. Shaskolskaya, Fiz. Tverd. Tela 22 (1980) 3549-54

3 A.V. Gorbunov, E.M. Nadgornyi, and S.N. Valkovskii, phys. stat. sol. (a) 66 (1981) 53-63

4 A.V. Gorbunov, N.V. Klassen, and Yu.I. Koval, Izv. Akad. Nauk SSSR (ser. fiz.) 53 (1984) 586-90

5 Y.M. Gupta, G.E. Duvall and G.R. Fowles, J. Appl. Phys. 46 (1975) 532-46

6 E.V. Darinskaya, I.P. Makarevitch, Yu.I. Meshcheryakov, V.A. Morozov, and A.A. Urusovskaya, Fiz. Tverd. Tela 24 (1982) 1564-66

7 Yu.I. Golovin, A.V. Gorbunov, and A.A. Shibkov, Fiz. Tverd. Tela 30 (1988) 1931-37

8 M. Kass and S. Magun, Zeitschr. f. Krist. 116 (1961) 354-70

9 C.W.F.T. Pistorius, J. Chem. Phys. 45 (1966) 3513-20

TEM INVESTIGATION OF DISLOCATIONS SUBSTRUCTURES AND MICRO-TWINS IN MANGANESE-DOPED TiAl.

G. Hug and P. Veyssière

Laboratoire d'Etude des Microstructures,
Office National d'Etudes et de Recherches Aérospatiales,
BP72, 92322 Châtillon CEDEX, France

ABSTRACT

Dislocation substructures and twins have been studied in γTiAl doped with manganese and deformed at room temperature, 650 and 750°C. At room temperature, whereas the $^1/_2$<110] dislocations dominate, $^1/_2$<112] and <011] superdislocations are also present. Their core structures have been determined and and it is found that the addition of manganese decreases the stacking fault energy. At higher temperature, <011] superdislocations disappear and deformation takes place by propagation of $^1/_2$<110] dislocations and micro-twins; when present $^1/_2$<112] superdislocations are related to micro-twins. The absence of <011] superdislocations is interpreted as the result of the composition-decomposition equilibrium : <011] $\leftrightarrow$ $^1/_2$<110] + $^1/_2$<112] in which the <011] would be unstable. Small $^1/_2$<110] loops, whose size is temperature dependent, have been observed to nucleate homogeneously in the matrix with a high density in the vicinity of the brittle-ductile transition. Finally, a nucleation mechanism of micro-twins from $^1/_2$<112] superdislocations has been identified. The relations between these dislocation mechanisms and the ductile-brittle transition is discussed.

INTRODUCTION

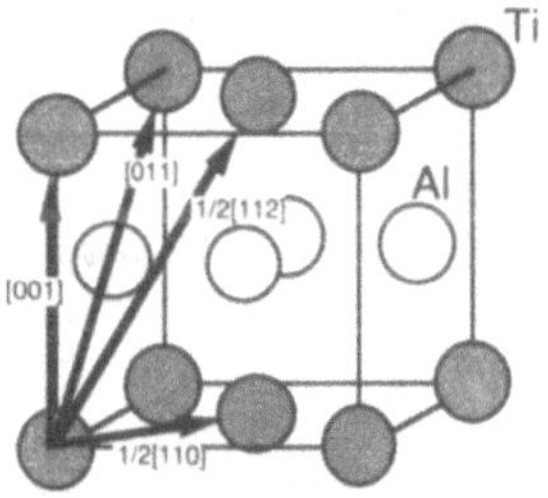

Figure 1 : The $L1_0$ structure and possible Burgers vectors.

Like many other intermetallic ordered alloys with the $L1_2$ structure, TiAl exhibits an anomalous yield stress behavior [1] which increases with test temperature. Such yield-stress peak has been partially attributed to a Kear-Wilsdorf mechanism that produces locks on <011] superdislocations [2].

On the other hand, TiAl, very brittle at room temperature, undergoes a sharp transition toward a highly ductile state near 700°C [3]. Because of its importance in alloy development, the ductility of TiAl has recently received a large attention. It has been shown that adding a little amount of a third element (Mn, V, Cr, Ga...) enhances the room temperature ductility [4]. In the same time several arguments have been put forward in order to explain the brittleness of TiAl (number of glide systems, grain size, grain boundary cohesion...). Amongst them, some are already established to be pertinent as for example the finesse of the microstructure. The impurities which sit in the octahedral cavities (C, N, O) are also thought to reduce the ductility [5, 6]. Since the octahedral cavities are very different in these alloys, solubilities are very low and high in TiAl and in Ti_3Al, respectively, and the effect has been evidenced by Kawabata et al. in the instance of oxygen [7] : γ single-phased samples containing little amount of oxygen are more brittle than the high purity ones. Similar effects may arise from carbon and nitrogen, which are not soluble in the $L1_0$ structure. The adverse effect of impurities may be overcome either by a long annealing at high temperature which produces precipitation of H-phase in which the interstitial atoms segregate or by moving to a composition where the material contains a small amount of Ti_3Al. In the latter case, the γ phase will be free from interstitial atoms since they are widely soluble in α_2, where they find suitable octahedral cavities [8].

However, an impurity free TiAl alloy is still brittle and there is no understanding of the sharpness of the brittle-ductile transition. Recently, a combination of theoretical [9] and experimental [10] studies suggested that the $\frac{1}{2}\langle 110]$ dislocations are locked in deep Peierls valleys created by the "*covalent*" bonding of titanium atoms. It is argued that the covalent bonding is responsible for the low mobility of $\frac{1}{2}\langle 110]$, which in turn would give rise to the brittleness of TiAl, and that covalency is increased by interstitials. So far no investigation has been carried out in order to obtain informations at the core level of dislocations between doped and undoped TiAl or near the brittle–ductile transition. It is the aim of this study to correlate these dislocation substructures with the brittle to ductile transition.

Since the work of Greenberg and co-workers [11] it is known that several complex core configurations are possible within the $L1_0$ structure of TiAl. We will here only focus on and describe configurations which have received experimental evidences. Amongst the four different Burgers vectors (fig. 1) which have been identified in TiAl, three ($\frac{1}{2}\langle 110]$, $\frac{1}{2}\langle 112]$ and $\langle 011]$) are liable to glide in the close–packed planes and produce deformation at low and intermediate temperature. The $\langle 001]\{110)$ glide system is only observed at very high temperature (900-1000°C) [8] and is not in the scope of the present work.

Superdislocations dissociate in two superpartials of the fcc structure with an anti-phase boundary (APB) in between. Due to the tetragonal symmetry, the subsequent dissociation of the superpartials in the octahedral plane gives rise to either a complex stacking fault (CSF) or an intrinsic stacking fault (ISF – a SISF would be bordered by a $\frac{1}{3}\langle 112]$ superpartial) depending on which side of the APB the fault sits. Hence $\langle 011]$ and $\frac{1}{2}\langle 112]$ superdislocations would have a similar core structure (fig. 2.a & 2.b) with a difference on the character of the $\frac{1}{2}\langle 011]$ superpartials. The dissociations are :

$$[011] \rightarrow \tfrac{1}{2}[011] + \text{APB} + \tfrac{1}{6}[\bar{1}21] + \tfrac{1}{6}[112] \quad (1)$$

and :

$$\tfrac{1}{2}[112] \rightarrow \tfrac{1}{2}[101] + \text{APB} + \tfrac{1}{6}[\bar{1}21] + \text{ISF} + \tfrac{1}{6}[112] \quad (2)$$

for the $\langle 011]$ and $\frac{1}{2}\langle 112]$ superdislocations, respectively. When fully dissociated in the octahedral plane these dislocations are considered as *glissile*. The total line energy of superdislocations may be reduced by core transformations. In the Kear–Wilsdorf lock (fig. 2.b) the APB of the $\langle 011]$ superdislocation is located in the low energy (100) cube plane [2]. The $\frac{1}{2}\langle 112]$ undergoes a more complex core transformation leading to an intrinsic/extrinsic (ISF/ESF) pair. In this case three parallel Shockley partials bound two ISFs on two adjacent planes (forming the extrinsic fault), according to :

$$\tfrac{1}{2}\langle 112] \rightarrow \tfrac{1}{6}\langle 112] + \text{ESF} + \tfrac{1}{6}\langle 112] + \text{ISF} + \tfrac{1}{6}\langle 112] \quad (3)$$

The outer Shockley bounding the extrinsic fault is actually formed by reaction of two attractive Shockleys [12, 13, 14] :

$$\tfrac{1}{6}[\bar{1}21] + \tfrac{1}{6}[2\bar{1}1] \rightarrow \tfrac{1}{6}[112] \quad (4)$$

For that reason we name it a **composite Shockley** (partials labelled 2-4 on fig. 2.d). The ISF/ESF can be viewed as two dislocations of the fcc lattice ($\frac{1}{2}[011]$ and $\frac{1}{2}[101]$) gliding on two adjacent planes. However, since they are not perfect dislocations of the $L1_0$ structure, the dislocations form an APB on each plane as they are gliding. Since such a defect is likely to have a high energy, the $\frac{1}{6}\langle 112]$ composite Shockley is sessile (at least at low temperature) and gives rise to long faulted loops observed in TiAl [8]. It has to be emphasized that the transformation between these two different cores requires the crossing of two Shockleys : partials in order 1-2-3-4 in fig. 2.c are in order 1-2-4-3 in fig. 2.d. Finally, the $\frac{1}{2}\langle 110]$ dislocations may dissociate in the standard fcc mode where the stacking fault is a CSF since it involves wrong first neighbors (fig 2.e).

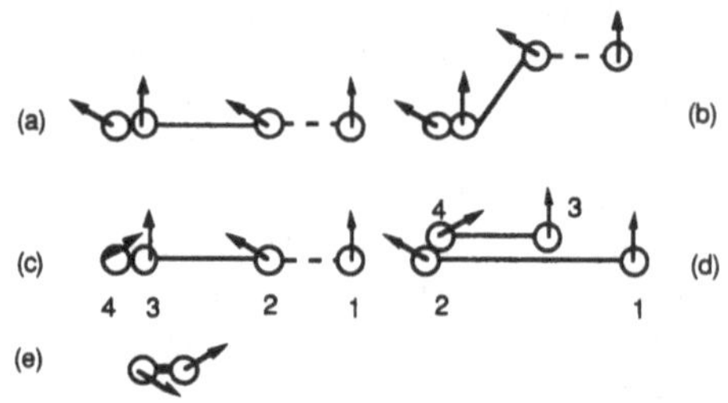

Figure 2 : Dislocation core structures in $L1_0$. The arrows indicate the Burgers vectors of Shockleys. The ISFs are represented with a plain line and the APBs with dashed lines. (a) : $\langle 011]$ APB in the glide plane according to reaction (1). (b) Kear–Wilsdorf lock : APB in the cube plane. (c) $\frac{1}{2}\langle 112]$ glissile configuration in the glide plane according to reaction (2). (d) $\frac{1}{2}\langle 112]$ ISF/ESF pair. (e) $\frac{1}{2}\langle 110]$ dislocation dissociated with a CSF.

EXPERIMENTAL PROCEDURE

Two compositions $(Ti_{46}Al_{54})_{1-x}Mn_x$ with x=1, 1.5 have been chosen in order to make a γ single phased sample. Ingots of TiAl–based alloy are made by arc melting in an argon atmosphere using high purity metals. Compression samples of 8x3x3 mm are cut by spark machining after removal of a 1 mm thick layer on the ingot skin. The samples are subsequently cleaned in a "soft" mixture of HF and HNO_3 at 65°C during a few minutes. They are homogenized at 1300°C during 48 hours and the microstructure is stabilized at 1000°C during 120 hours. The sample are deformed under vacuum at a low strain rate ($1.7 \cdot 10^{-5}$ s^{-1}) at room temperature (RT), 650°C and 750°C and they are rapidly air–cooled (about 100 seconds) after the test in order to minimize a possible evolution of the dislocation microstructure during the cooling. Samples suitable for transmission electron microscopy are then cut at 45° from the compression axis and are thinned using the twin jet polishing method. Transmission electron microscopy (TEM) observations are done using a Jeol 200CX or a Jeol 4000FX microscope operating at 200 and 400 keV, respectively.

Contrary to what was expected from a previous metallography determination of Tsujimoto et al. [4], the samples are not single–phased but constituted of fine γ grains of approximately 5 μm containing some small α_2 isolated precipitates of Ti_3Al (size ≈ 100 nm) which, in fact, can be only detected by TEM. These precipitates exhibit smooth shapes and have orientation relationships with the γ matrix (not determined in this paper). The α_2 precipitates form obstacles to the motion of dislocations and were in no case found to be sheared. This confirms that within a two-phased material γ is the soft phase as noted by several authors. Additionally, grains of lamellar structure (twin related lamella of γ and α_2 phases) also are present with a low volume fraction (probably 5 to 10%). They eventually contain a significant density of dislocations. Weak-beam images are produced with a set of different diffracting conditions to identify each Burgers vector involved in the dissociation reactions and various projections are used to determine stereographically the actual three dimensional core configurations. In the following, $\mathbf{A_z}$ will represent the projection direction that is opposite to the beam direction.

EXPERIMENTAL RESULTS

Deformation Microstructure at Room Temperature and Changes in Surface Energies

After deformation at room temperature the $\frac{1}{2}$<110] dislocations dominate within the micro-structures. They are roughly aligned to the screw direction ; however, at a weak–beam scale, no clear straight and rigid lines that would unambiguously attest to deep covalent Peierls valleys have been found. Since the sample contains a small fraction of second phase (α_2) the covalent bonding of Ti atoms may be smaller than in a single phase material [10]. Other glide systems are also operating, namely super-dislocations with $\frac{1}{2}$<112] and <011] Burgers vectors in order of decreasing density.

The $\frac{1}{2}$<112] superdislocations are gliding in the {111} close–packed planes and they are dissociated according to reaction (3). Along the dislocation line local events take place : the composite Shockley tends to trail giving rise to an elongated stacking fault. This phenomenon is well established in undoped TiAl where long faulted loops with straight bounding partials along the <011] direction have been repeatedly reported [8, 2, 3, 14]. Contrary to the undoped samples, the composite partials adopt a wavy shape in the Mn–doped material. Their mobility is probably higher since few long faulted loops are observed.

Figure 3 is an example of two [011] screw superdislocations (labelled A and B in fig. 3.a). The $\mathbf{A_z}$=[110] projection allows for a complete determination of Burgers vectors of the different partials (see figure caption for details). They have been found to be mainly dissociated according to reaction (1) in the $(11\bar{1})$ plane but cross–slip events onto the $(\bar{1}1\bar{1})$ plane which lead to the corresponding structure on that plane are frequent. In the $\mathbf{g}=\bar{1}1\bar{1}$ picture the extinction is only partial since the ISF is visible as fringes ($\mathbf{R}=\frac{1}{6}[112]$; $\alpha=2\pi\mathbf{g}.\mathbf{R}=-2\pi/3$). It can be seen that the stacking fault vanishes in the bottom part of the superdislocation A where the dislocation has cross–slipped onto the $(\bar{1}1\bar{1})$ plane (X in fig. 3.b) and appears undissociated as a straight line. Since the displacement vector of the stacking fault in the cross–slip plane is $\mathbf{R}=\frac{1}{6}[\bar{1}12]$ the extinction is complete. The $\mathbf{A_z}=[11\bar{1}]$ projection (fig. 3.e) shows the three–fold dissociation and the actual width of planar defects on the B dislocation ; the cross–slip event can be clearly seen on dislocation A. Its ability to cross–slip onto two octahedral planes at room temperature was already noticed in pure TiAl [2] and

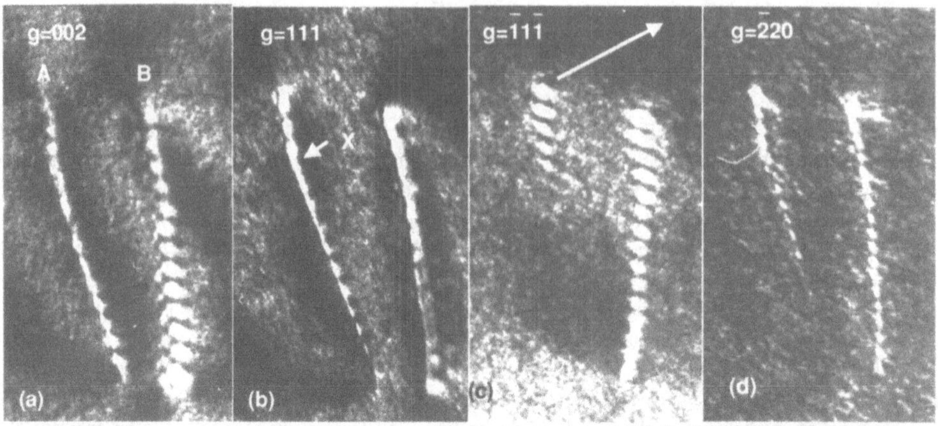

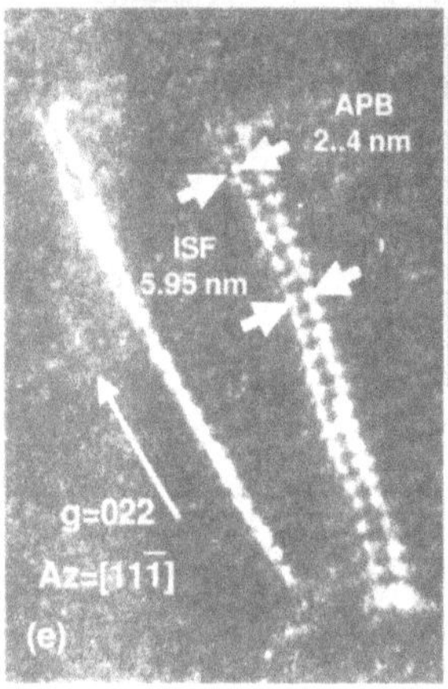

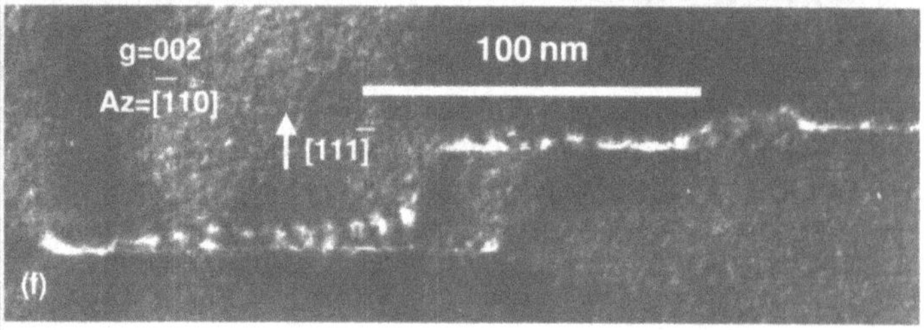

Figure 3 : Screw [011] superdislocations cross–slipping from $(11\bar{1})$ to $(\bar{1}1\bar{1})$. Depending on the octahedral the ISF–superpartial is dissociated according to reaction (1) or its corresponding form in the cross–slip plane. (a) to (d) A_z=[110] projection. (a) **g**=002, CSF–superpartial, ISF and outer Shockley are visible. (b) **g**=111, CSF–superpartial, ISF and inner Shockley are visible. (c) **g**=111, all partials are out of contrast, the ISF in $(11\bar{1})$ is visible but not in $(\bar{1}1\bar{1})$. (d) **g**=2$\bar{2}$0 all planar defects are out of contrast ; only the partials bounding the APB are visible. (e) Projection in the $(11\bar{1})$ glide plane : the three partials are visible and all planar defects are out of contrast (**g** along the Burgers vector). (f) The $(11\bar{1})$ glide plane is edge–on : cross–slip in the$(\bar{1}1\bar{1})$ is visible. (pictures at 400kV).

thus the doped samples exhibit the same behavior. Finally, the glide plane has been imaged edge-on on projection A_z=[110] (fig. 3.f) confirming that the dislocation is located in one single octahedral plane at a time. It excludes the "roof–type dissociation" proposed by Greenberg [11, 15, 16] which involves two ISFs on two cozonal (111) planes.

Measurements of the dissociation distances carried out from the above observations are reported in the table together with the surface energies of the planar defects (APB, ISF) which can be computed using anisotropic elasticity (the same set of elastic constants has been used for both TiAl(Mn) and TiAl). The more noticeable effect is the drop of the ISF energy from 70–85 mJ.m^{-2} in pure TiAl to 50 mJ.m^{-2} in Mn–added specimens while the APB energy tends to increase. It must be pointed out the ISF energy appears to be very sensitive to Mn addition (i.e. 30 mJ.m^{-2} per atomic percent).

Table : ISF and APB energies

		SISF	APB
TiAl(Mn)	Dissociation (nm)	5.95	2.4
TiAl(Mn)	Energy (mJ.m^{-2})	52	180
TiAl	Energy (mJ.m^{-2})	70-85	130-160

Deformation at 650°C (below the Brittle-Ductile transition)

After deformation at 650°C, the microstructure is dramatically different from that in the room temperature deformed specimens :

i. the <011] superdislocations have disappeared ;

ii. the density of $^{1}/_{2}$<112](111) micro–twins has increased ;

iii. only a few isolated $^{1}/_{2}$<112] are still visible, mostly in conjunction with micro–twins ;

iv. $^{1}/_{2}$<110] dislocations are present in a higher density.

A close look at $^{1}/_{2}$<110] dislocations (fig. 4.a) indicates that they exhibit an irregular shape due to numerous pinning points which appears uniformly in the samples strained at this temperature. From place to place the dots are a little wider and it is possible to associate them with small closed loops of 1/2<110] dislocations. The figure 4.b is an extinction condition for the 1/2<110] dislocations but a contrast from the stacking fault (CSF) is still visible. Small fringes can be seen where the loops nucleate suggesting that they might start as a stacking fault.

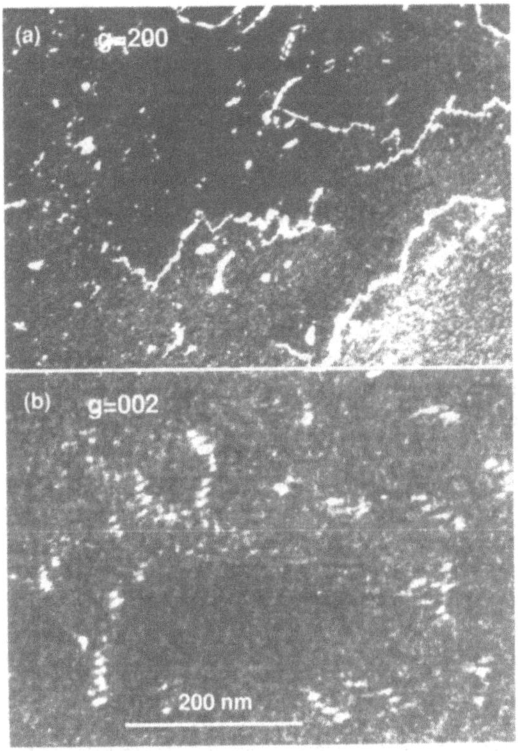

Figure 4: Example of homogeneous nucleation of $^{1}/_{2}$<110] loops after deformation at 650°C. (a) **g**=200 ; the dislocations appear as small dots. (b) When an extinction condition is used (**g**=002) the stacking faults are visible. (200 kV).

Deformation at 750°C (above the Brittle-Ductile transition)

After deformation at 750°C, the substructure differs from that observed in samples strained at 650°C in the fact that the density of $^{1}/_{2}$<110] dislocations is very homogeneous and that no pinning points (dots) are visible any more but many well defined closed loops of $^{1}/_{2}$<110] dislocations are observed to be distributed throughout the specimen. An example of such closed loop projected onto the (100) and the (111) plane, is presented in fig. 5.a and 5.b, respectively. As it can be seen on the first projection a part of the loop exhibit a reverse curvature. When projected along [111], this part consist of a small edge–on segment perpendicular to the $[01\bar{1}]$ direction indicating that this segment is bowing in the $(01\bar{1})$ plane ; since this plane is not a glide plane for a $^{1}/_{2}$<110] dislocation, bowing can only be achieved by climb.

The other particularities of deformation at 750°C are that no <011] superdislocations are present and only a few $^{1}/_{2}$<112] superdislocations could be identified which where found to be correlated with

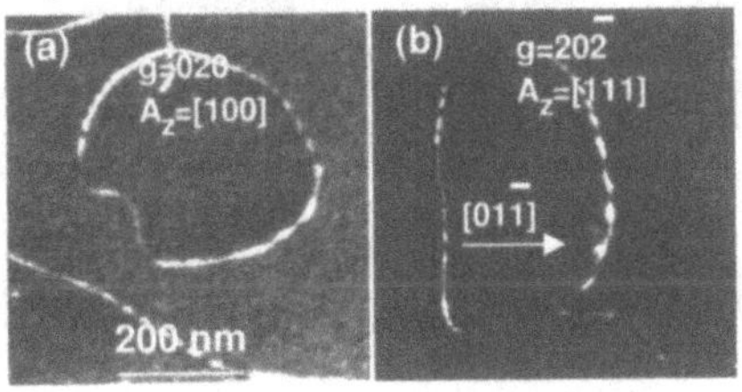

Figure 5 : Example of a closed loop of a $^{1}/_{2}$<110] dislocation after deformation at 750°C. (200 kV).

micro–twins. Three examples of micro–twins are shown in fig. 6 ; each twin is preceded by one single dislocation with $^{1}/_{2}<112]$ Burgers vector. It is important to stress on the fact that **no** other $^{1}/_{2}<112]$ dislocations are present in the specimen and that the leading dislocation has been found in almost all the cases where the tip of the micro–twin was visible. In one example (fig. 6.c) the spreading within the twin of a $^{1}/_{2}<112]$ superdislocation is visible. As a first remark the $^{1}/_{2}[112]$ superdislocation which is dissociated in the ISF/ESF mode can be regarded as a small twin, two atomic planes thick. Two assumptions can be made to explain the nucleation of the twin from the $^{1}/_{2}[112]$ superdislocation : (i) the composite Shockley is pinned and the two $^{1}/_{6}[112]$ Shockleys move away or (ii) the composite Shockley moves on an upper plane (climb or cross–slip) and glides in the same direction than the two other partials. The twin would then be formed by one source of $^{1}/_{2}<112]$ superdislocations (like a Frank-Read source), which changes its glide plane at each turn. The part boxed in fig. 6.c is enlarged in fig. 7 and projected onto the glide plane (fig. 7.b to d). The different conditions of diffraction in this projection allow the determination of Burgers vectors of Shockley partials since each one is perpendicular to one Shockley (for example $\mathbf{g}=2\bar{2}0$ is perpendicular to $\mathbf{b}={}^{1}/_{6}[11\bar{2}]$ etc…). The analysis indicates that the two first partials of the twin have a standard $^{1}/_{6}[11\bar{2}]$ Burgers vector whreas the third one is a composite Shockley, formed from the reaction of a $^{1}/_{6}[\bar{1}2\bar{1}]$ and a $^{1}/_{6}[2\bar{1}\bar{1}]$ partial dislocation (see figure 7 for details). In order to produce a micro–twin without stacking faults, the two Shockleys, forming the composite one have to move on *the same* atomic plane, *different* from the one of the $^{1}/_{6}[11\bar{2}]$ preceding and following partials (fig. 8). This implies that the composite Shockley which is part of the $^{1}/_{2}<112]$ superdislocation has changed its atomic glide plane in order to generate the micro–twin. This is possible by various mechanisms such as cross–slip, climb or interaction with an other glide system. A nucleation mechanism could be the following : a source produces a $^{1}/_{2}<112]$ superdislocation, which then changes its slip plane at each turn

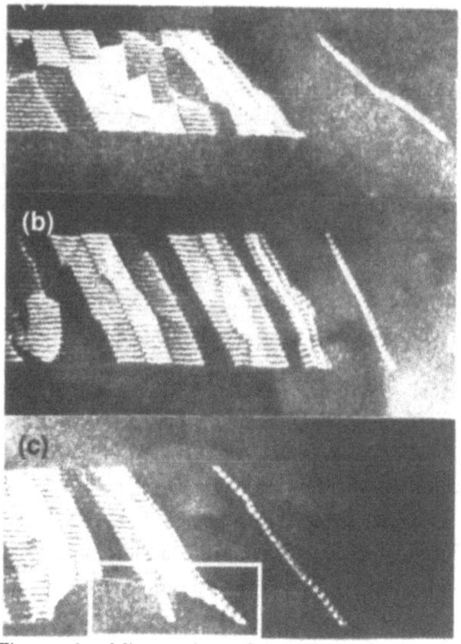

Figure 6 : Micro–twins after deformation at 750°C. Each micro–twin is preceded by a single $^{1}/_{2}<112]$ superdislocation. No other super-dislocations have been found in the sample. In (c), a superdislocation is nucleating the twin. The area boxed is enlarged in next figure. (200 kV).

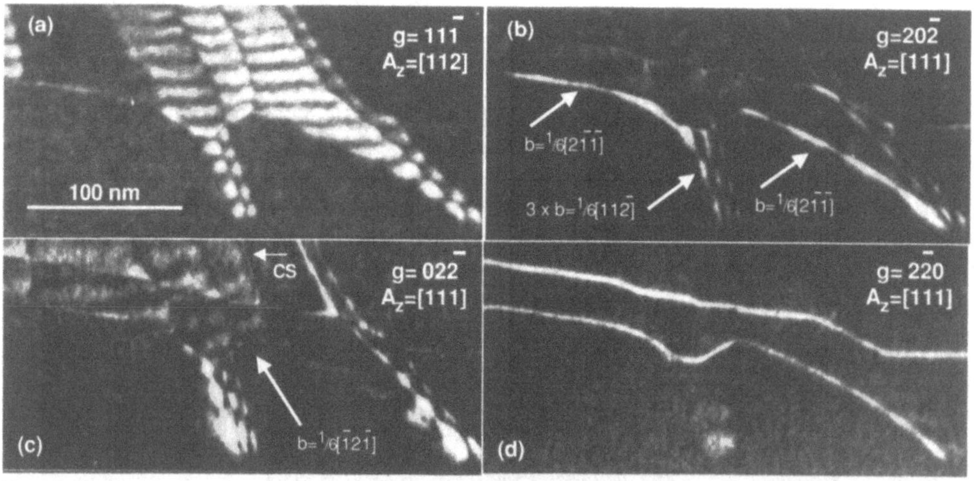

Figure 7 : Details of the nucleation of a micro–twin by a $^{1}/_{2}<112]$ superdislocation. It is possible to identify the different Shockleys from the analysis of extinction conditions. It turns out that the partial labelled CS in (c) (the third one from the twin head) is a composite Shockley since it is formed by the reaction of two Shockley partials with 120° Burgers vectors. (200 kV)

(around a screw dislocation). During the change of slip plane the dislocation transforms into a micro–twin as described above. This explains how the twin is correlated with a single dislocation. This process allows an important part of deformation to be achieved by twinning contributing to the good ductility of TiAl at high temperature.

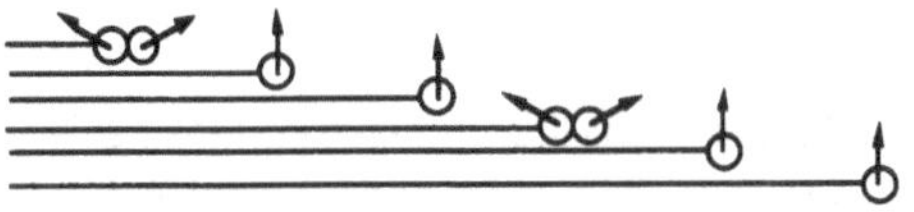

Figure 8 : Structure of a twin nucleated by a super-dislocation. One over three Shockley is a composite.

DISCUSSION

One major result is that a ductile state of TiAl is always correlated with a very low density of <011] superdislocations (high temperature behavior) and that alloying with manganese tends to lower the densities of these dislocations. This result is also consistent with observations in pure TiAl, where only ½<110] and ½<112] dislocations are observed above the brittle–ductile transition temperature [8]. In the present case the absence of <011] superdislocations can be interpreted on the basis of the equilibrium reaction :

$$\langle 011] \leftrightarrow \tfrac{1}{2}\langle 110] + \tfrac{1}{2}\langle 112] \qquad (5)$$

where the <011] and the ½<112] superdislocation are dissociated according to (1) and (2), respectively. Decreasing the ISF energy and increasing the APB energy will most likely favour the right hand member of equation (5) since then only stacking faults are involved. It is therefore concluded that TiAl is subjected to such a decomposition reaction either by alloying or raising the temperature. The way the temperature controls the reaction is not known but one may infer a decrease of the stacking fault energy with temperature.

The observations show that the ½<112] superdislocations are activated at high temperature and the question arises how they can move. As it has been pointed out, the motion of the ISF/ESF pair would produce a double APB to be formed. This high energy double APB can be erased if atoms of titanium and aluminium interchange in the plane of the defect since two APBs on two consecutive planes are involved. This mechanism which does not require any long–range diffusion may start as soon as the jump frequencies of atoms are large enough. Therefore, the ½<112] dislocation, sessile at low temperature, becomes glissile as soon as the temperature is increased to a point where atoms can jump significantly. So far it is the only mechanism that may behave with a threshold and that can be correlated with the sharp brittle to ductile transition in TiAl. The second point correlated with the brittle–ductile transition is the homogeneous nucleation of closed loops of ½<110] ordinary dislocations which provides a high density of such dislocations. The nucleation could be either due to closed loops formed by glide dislocations or to a diffusion–assisted mechanism. The latter one is suggested by the observations at 750°C since climb has there been found on the dislocation loops.

CONCLUSIONS

The ductile state of TiAl is associated with glide of ½<110] and ½<112] dislocations or micro–twinning. The absence of <011] superdislocations is due to their relative instability with respect to the decomposition reaction that yields both, ½<110] and ½<112] dislocations. The instability is achieved either at high temperature or by addition of a third element whose main effect is to reduce the stacking fault energy. The ½<112] dislocations dissociated in a low energy ESF/ISF configuration are sessile at low temperature and can move in the high temperature ductile state provided sufficient thermal activation allows an exchange of first nearest neighbors, a process which does not require any long–distance diffusion. At high temperature TiAl undergoes a slip transition where ½<112] dislocations disappear as they nucleate twinning. All of the above mechanisms are related to diffusion processes, therefore it is concluded that the sharp brittle–ductile transition may be a diffusion assisted dislocation processes.

REFERENCES

[1] T. Kawabata, T. Kanai and O. Izumi, Acta Metall. **33** (1985) 1355-66.

[2] G. Hug, A. Loiseau and P. Veyssière, Philos. Mag. A **57** (1988) 499-523.

[3] H. A. Lipsitt, D. Shechtman and R. E. Schafrick, Met. Trans. **6** A (1975) 1991-96.

[4] T. Tsujimoto and K. Hashimoto, In : "Proceedings of MRS Fall Meeting", C. T. Liu, A. I. Taub, N. S. Stoloff and C. C. Koch (Eds.), Boston, **133** (1988), p. 391-6.

[5] V. K. Vasudevan, S. A. Court, P. Kurath and H. L. Fraser, Scr. Metall. **23** (1989) 907.

[6] E. L. Hall and S. C. Huang, In : "Proceedings of MRS Fall Meeting", C. T. Liu, A. I. Taub, N. S. Stoloff and C. C. Koch (Eds.), Boston, **133** (1988), p. 373-83.

[7] T. Kawabata, M. Tadano and O. Izumi, Scr. Metall. **22** (1988) 1725-30.

[8] G. Hug, 1988, PhD thesis, University of Orsay (France).

[9] B. A. Greenberg, V. I. Anisimov, Yu. N. Gornostirev and G. G. Taluts, Scr. Metall. **22** (1988) 859-64.

[10] S. A. Court, V. K. Vasudevan and H. L. Fraser, Phil. Mag. A (1989).

[11] V. N. Indenbaum, B. A. Greenberg, Yu. N. Gornostirev and L. Ye. Kar'kina, Phys. Met. Metall. **59** (1985) 52-63.

[12] B. C. Carter, Phys. Stat. Sol. (a) **61** (1980) 579-89.

[13] G. Vanderschaeve and B. Escaig, J. Physique Lettres **39** (1978) L74.

[14] G. Hug, A. Loiseau and A. Lasalmonie, Phil. Mag. A **54** (1986) 47-65.

[15] B. A. Greenberg, Phys. Stat. Sol. **42** (1970) 459-68.

[16] B. A. Greenberg, Phys. Stat. Sol. (b) **55** (1973) 59-67.

APPLICATION OF STOCHASTIC GEOMETRY IN TEM STUDIES OF DISLOCATIONS

P. Klimanek[+], D. Stoyan[*]

+Department of Materials Science and *Department of Mathematics,
Academy of Mining, DDR - 9200 Freiberg, P.O. Box 47

ABSTRACT

Interrelations between the characteristics of dislocation structures in crystalline solids and basic models of stochastic geometry are considered, which should be useful in the stereological analysis of TEM micrographs. In this connection the application of pair-correlation functions for random fibre fields is discussed in some detail.

INTRODUCTION

A characterization of the dislocation content of crystalline solids is obviously one of the most important subjects of the application of TEM in materials science. In this connection the following basic problems are of interest:

(i) identification and analysis of individual dislocations or small dislocation groups,

(ii) determination of dislocation densities, and

(iii) qualitative and quantitative description of planar or spatial dislocation arrangements.

From a methodological point of view the problems (i) and (ii) are solved on a level which is suitable in many cases, but a satisfying quantitative treatment of (iii) seems to be still outstanding. As demonstrated in previous work [1,2], ideas of stochastic geometry [3,4] can be helpful in order to improve this situation. In the present paper particularly the characterization of planar or spatial dislocation structures by means of pair-correlation functions as defined in the theory of random fibre fields is discussed.

THEORETICAL CONSIDERATIONS

Dislocation Structures

In order to describe the dislocation content of crystals in terms of stochastic geometry the defect distributions may roughly be divided into the following basic types:

(i) spatial arrangement of point-like defects (e.g. small dislocation loops, stacking-fault tetrahedra),

(ii) planar or spatial configurations (e.g. misfit dislocations in phase boundaries, subboundaries due to polygonization, diffusion-induced dislocation networks) of individual dislocations or dislocation groups (dipoles, multipoles, pile ups), and

(iii) dislocation cell structures with "thick" boundaries or subgrain structures due to plastic deformation and/or annealing processes.

Structures of the kind (i) can be treated as point processes (or fields), the theory and application of which is well developed [3,4]. Arrangements of the type (iii) may be evaluated in terms of spatial tesselations as, for example, the Poisson-Voronoi mosaic [1,3,4]. Finally, adequate models for microstructures of the type (ii) are random fibre fields (or processes) (see [3,4]) which will be considered here.

Basic Properties of Fibre Fields

A mathematical theory of random fibre fields was developed by Mecke (see [3,4]). Such a field is represented by a system of sufficiently smooth curves statistically distributed in a plane or in space. In this connection randomness does not mean complete disorder; there may be short-range (e.g. clustering, multipole formation) or even long-range order (e.g. network formation) in the fibre arrangement. Subsequently the fibre processes are assumed to be homogeneous (i.e. translation invariant) and also isotropic (i.e. rotation invariant), but it should be noted here that it is possible to extend the theory to certain cases of anisotropic fibre fields, too.

The fundamental quantities of a spatial random fibre process are the mean fibre length (dislocation density) L_V per unit volume, and the pair-correlation function (PCF) $g_V(r)$, which completely describes the second-order behaviour of the fibre field. Analogously the mean fibre length L_A per unit area and the function $g_A(r)$ can be defined for planar fibre processes.

The PCF $g_V(r)$ characterizes the correlation between the fibre lengths $L(dV_1)$ and $L(dV_2)$ of volume elements dV_1, dV_2 with a distance r. Accordingly it follows

$$< L(dV_1)\ L(dV_2) > = L_V^2\ g_V(r)\ dV_1\ dV_2$$

with the values $< L(dV) >^2 = L_V^2$, $g_V(r) = 1$ for $r > r_o$ and

$< L_r^2(dV) > \neq L_V^2$, $g_V(r) \neq 1$ for $r \leq r_o$. r_o is an interaction radius, depending on the lengths and the forms of the fibres, above which the dislocation contents of the volume elements dV_1 and dV_2 are independent of each other. r_o cannot become zero, because at small values of r, correlation between fibres is inevitable even in systems with a high degree of randomness [4]. This is due to the fact that every single fiber is a connected geometrical object. Correspondly, $g_V(r)$ always satisfies $g_V(r) \to \infty$ for $r \to 0$, i.e. there is a pole at $r = 0$. On the other hand, the convergence rate of $g_V(r) \to 1$ for $r \to \infty$ is a measure of the randomness of the fibre field, and significant order (i.e. preferred interfibre distances) is indicated by peaks of the PCF.

Further information concerning the properties of PCFs may be taken from [3,4,5]. Particularly important is the fact that operations can be defined (superposition or mixing, see [5]) which allow the construction of the PCFs of complicated fibre fields from those of simpler or different fibre systems.

Models of Dislocation Structures

Dislocation arrangements, which can easily be described in terms of random fibre fields are schematically illustrated in Fig. 1. One limiting case is obviously a defect configuration without any order (Fig. 1a), another a dislocation lattice (Fig. 1d), and defect distributions between these limits may be characterized by the formation of dislocation groups (Fig. 1b) or by local defect clustering (Fig. 1c).

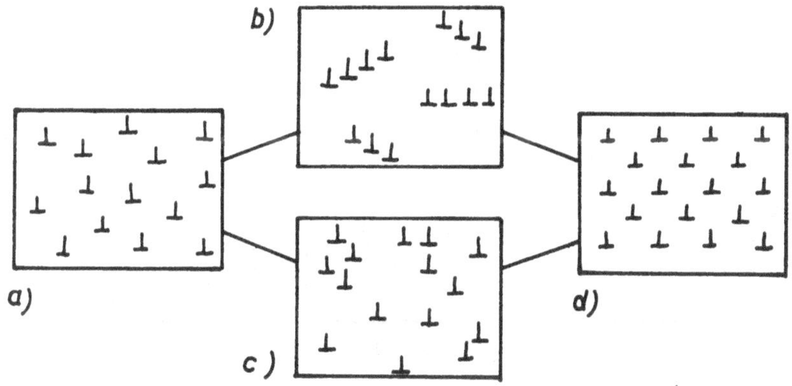

Fig. 1 Models of dislocation distributions in crystals

A mathematical treatment of the corresponding stochastic models was given in [5] (see also [1,4]). Here only two simple examples are briefly considered.

- A suitable model for the dislocation structure of Fig. 1a is a random distribution of straight lines. The PCF of such a fibre process is

  $g_A(r) = 1 + 1 / \pi L_A r$ (planar distribution)

  or (1)

  $g_V(r) = 1 + 1 / 2\pi L_V r^2$ (spatial distribution).

  The PCF $g_V(r)$ is shown in Fig. 2a.

  The simplest model for a dislocation distribution of Fig. 2b is a random arrangement of straight line pairs (e.g. dislocation dipoles or superlattice dislocations) with distance d. For d = const. the PCF $g_V(r)$ of such a configuration is given by

$$g_V(r) = 1 + (1 + r / \sqrt{r^2 \quad d^2}) / 2\pi L_V r^2 \qquad (2)$$

This function has an additional pole at r = d.

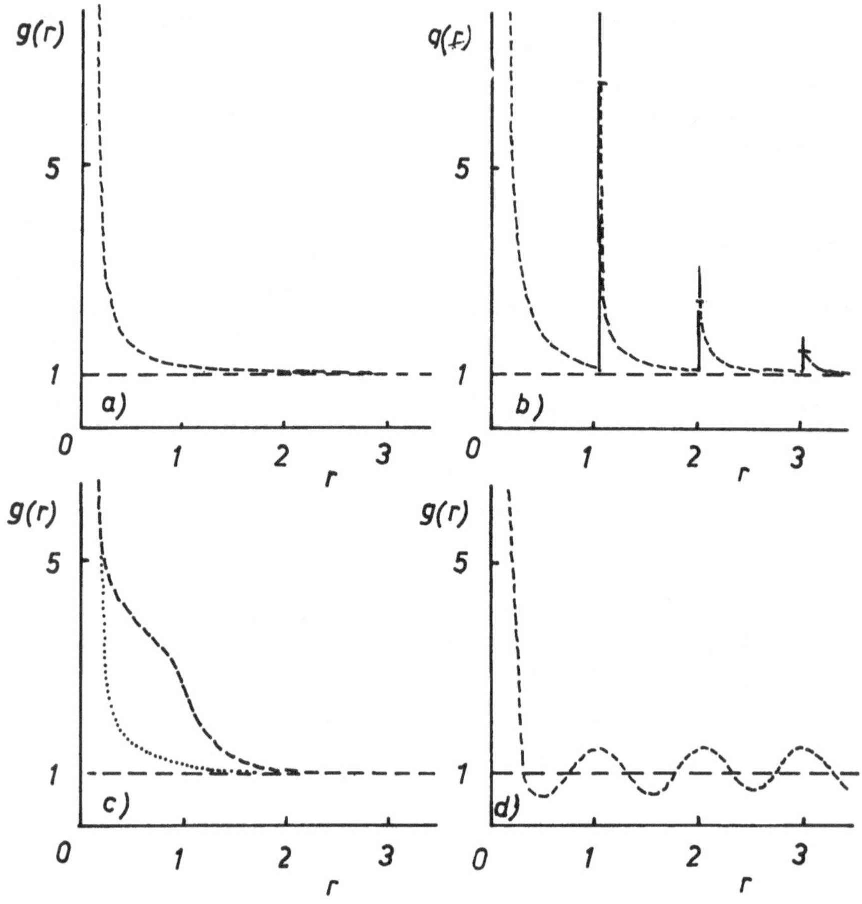

Fig. 2 Spatial PCFs $g_V(r)$ of different dislocation arrangements

If a dislocation group consists of N equidistant defects, a series of poles (maxima) at r = n d (n = 1,2,3...) appears in the PCF $g_V(r)$. The hight of the peaks increase with increasing N. This is shown in Fig. 2b for a random spatial distribution of planar fibre groups with N = 4 (dashed line) and N = 16 (full line) under the assumption of a relatively large, constant interline distance d. If the interline distances become significantly different within the defect groups and small in comparison with the distances between the groups, the peaks of $g_V(r)$ are shifted to small values of r and smeared out over a certain region near r = Ø (Fig. 2c). A similar behaviour of the PCF should also be typical for dislocation clustering. Finally a quasi-periodic PCF appears in a dislocation arrangement which has to be described in terms of a defect lattice or a network (Fig. 2d, see [1]).

EXPERIMENTAL DETERMINATION OF PCFs

A TEM micrograph is a two-dimensional projection topograph of the lattice defects present within the imaged crystal volume. The determination of the PCF of a dislocation arrangement is therefore a problem of stereology, whose solution requires

(i) a numerical determination of the PCF $g_A(r)$ of the dislocation-projection lines onto the plane of the TEM micrograph, and

(ii) a transformation of $g_A(r)$ into a spatial PCF $g_V(r)$ of the dislocation structure by means of the integral equation [6]

$$g_A(r) = (2/t^2) \int_0^t (t - z)\, g_V(\sqrt{r^2 + z^2})\, dz, \tag{3}$$

where t is the foil thickness and z represents a coordinate perpendicular to the foil plane.

However, it can easily be shown by applying of equation (3) to functions as presented in Fig. 2 that there do not appear significant differences between the basic characteristics (type, peak positions due to preferred interfibre distances) of the PCFs $g_V(r)$ and $g_A(r)$. That means, in the stereological analysis of isotropic fibre fields, $g_A(r)$ should often be a good approximation for the spatial PCF $g_V(r)$.

In order to obtain the function $g_A(r)$, two basic techniques of image analysis can in principle be used:

- Fourier analysis of the TEM image by means of optical diffraction and calculation of $g_A(r)$ from the diffraction pattern (e.g. [7])

  or

- direct determination of $g_A(r)$ from a quantitative evaluation of the dislocation-projection line pattern given by the TEM micrograph.

Here only the second way is briefly discussed. Two possible procedures of the evaluation proposed in [1], which can easily be realized by applying an image analyzer (e.g. by the MINI-BVS of Robotron, Dresden), are illustrated in Fig. 3. In the first method (Fig. 3a) the test area of the projected fibre system is intersected by a line system and one has to find the distances $l_{m,i}$ ($i = 1,2,\dots\nu_m$) along a given test line. From the distances $l_{m,i}$ the PCF $g_A(r)$ is obtained via the function

$$k_A(r) = (\pi^2/n) \sum_{m=1}^{n} \sum_{i=1}^{\nu_m} l_{m,i} = 2\pi L_A \int_0^r x\, g_A(x)\, dx \tag{4}$$

The second method (Fig. 3b) is based on pure point counting. The test set is now a strip (or a system of strips) of width δ and the numbers x_n of fibres cutting the strip at distances $r = n$ are counted. In this manner a stationary random process is generated, which is characterized by the function

$$c_L(r) = \langle x_o x_n \rangle / \langle x_o \rangle^2 \quad ; \; r = n \tag{5}$$

$c_L(r)$ is connected with the PCF $g_A(r)$ by an integral equation similar to (3) (substitute t by B and z by a coordinate x in this case; [8]).

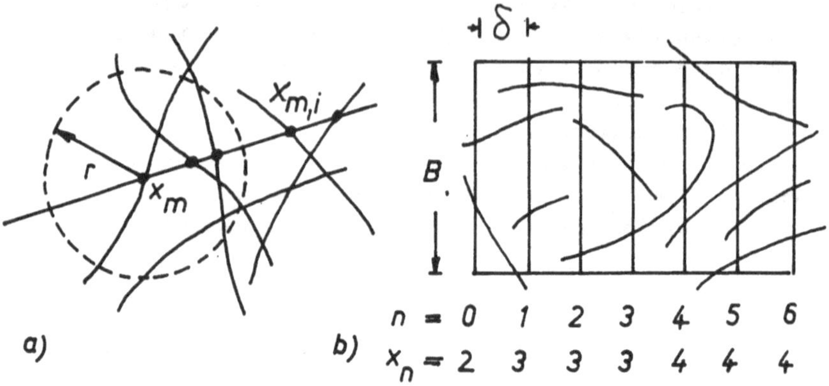

Fig. 3 Procedures for determining of planar PCFs $g_A(r)$ from TEM images

Experimentally determined PCFs $g_V(r)$ of different dislocation arrangements can be found in [1, 9]. The paper [9] also demonstrates some practical application: PCFs could successfully be used in order to separate grain fractions due to different restoration mechanisms (dynamic and postdynamic recrystallization) in high-speed hot rolling of stainless steels.

REFERENCES

1 K.-H. Hanisch, P. Klimanek and D. Stoyan, Cryst. Res. Technol. 20 (1985) 921

2 M. Barthel, P. Klimanek and D. Stoyan, Czech. J. Phys. B35 (1985) 265

3 D. Stoyan and J. Mecke, Stochastische Geometrie, Wiss. Taschenbücher Vol. 275, Akademie-Verl., Berlin 1983

4 D. Stoyan, W.S. Kendall and J. Mecke, Stochastic Geometry and its Applications, Akademie-Verl., Berlin 1984

5 H. Stoyan and D. Stoyan, Phys. Status Solidi A 97 (1986) 163

6 D. Stoyan, Math. Operationsf. Statist., Ser. Statist. 14 (1983) 409

7 K.-H. Hanisch, Math. Nachr. 117 (1984) 235

8 D. Stoyan, Math. Nachr. 102 (1981) 189

9 M. Barthel, P. Klimanek and K.-E. Hensger, Czech. J. Phys. B38 (1988) 377-379

CHARACTERIZATION OF THE INTERACTION BETWEEN DISLOCATIONS AND LOCALIZED OBSTACLES BY HVEM in situ STRAINING EXPERIMENTS

U. Messerschmidt

Institute of Solid State Physics and.Electron Microscopy,
Academy of Sciences of the G.D.R., Halle/Saale, 4010, G.D.R.

ABSTRACT

The interaction between dislocations and strong localized obstacles can well be observed by in situ straining experiments in the high-voltage electron microscope (HVEM). The paper describes which parameters of the obstacle structure can be measured relatively easily from the micrographs of dislocations bowed out under stress. Examples are given for an MgO single crystal and an Al-Zn-Mg alloy. While many features of the dislocation motion are similar in both materials, the relative level of the effective stress and the dislocation kinematics differ remarkably.

INTRODUCTION

In situ straining experiments in the HVEM of precipitation hardened materials show the dislocations bowing out under stress between the obstacles. The bowing is a measure of the stress acting locally. The cusps in the dislocation line mark the places of the obstacles. Thus, several of the parameters describing the obstacle structure as well as the interaction with the moving dislocations can be obtained from evaluations of the geometrical shape of the bowed dislocations. Respective experiments have been done on MgO single crystals and on an Al-4.5wt%Zn-1.2wt%Mg alloy [1 to 6]. It is the aim of the present paper to outline a scheme to determine the relevant parameters and to discuss the question of the proper energy factor of the dislocations necessary to relate the geometric parameters to the acting stresses or forces.

THEORETICAL BACKGROUND

An individual obstacle impedes the dislocation motion by exerting a force f on the dislocation which depends on an activation distance coordinate d in forward direction of the dislocation motion. Such a force-distance curve is sketched in Fig. 1. If the acting force f_e is greater than the obstacle strength f_m, the obstacle is overcome spon-

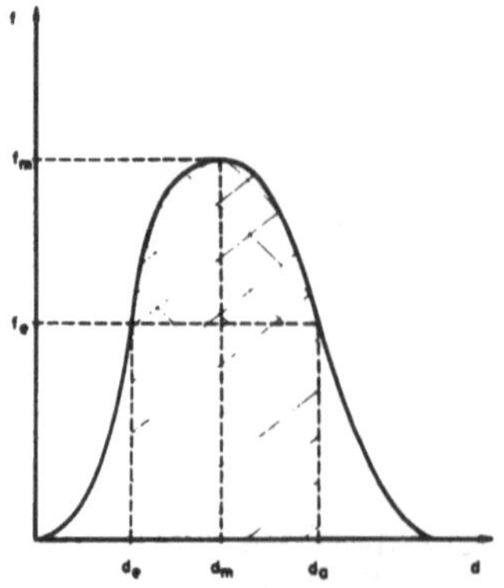

Fig. 1 Force-distance curve for a localized obstacle

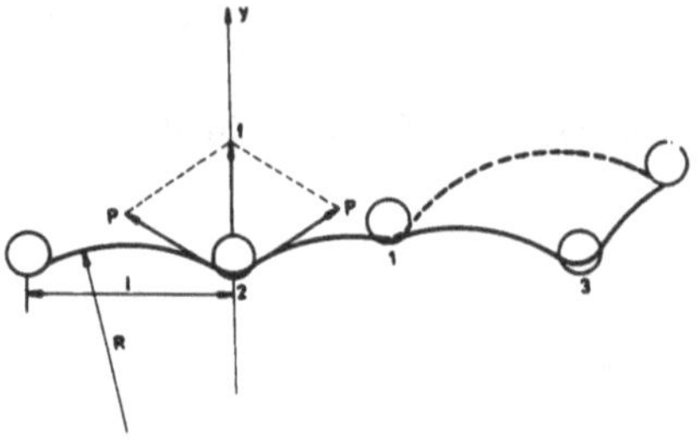

Fig. 2 A dislocation bowing out between localized obstacles

taneously. If $f_e < f_m$, the dislocation forms a stable configuration at the obstacle. The obstacle can then only be overcome by thermal activation after a waiting time t_w, with

$$1/t_w = \nu \exp[-\Delta G(f_e)/kT] = \nu \exp\{-[\Delta F - f_e(d_a-d_e)]/kT\} \qquad (1)$$

ν is an atomic vibration frequency, ΔG the Gibbs free energy of activation (the double-hatched area in Fig. 1), ΔF the free energy, d_a-d_e the activation distance, and k and T have their usual meaning. $f_e(d_a-d_e)$ is the work done by external forces during the activation event (hatched area).

Usually the obstacles are not distributed regularly as pictured in Fig. 2. As a consequence, different forces are applied on the individual obstacles. For strong localized obstacles the so-called Friedel statistics applies (e.g. [7,8], for reviews see [9,10]). Most theories assume that the dislocations bow out between the obstacles to circles of constant radius R and that there exists only one type of obstacles. The density of obstacles may be described by the square lattice distance l_{sq}. l_{sq}^2 is the slip plane area per obstacle.

As shown in Fig. 2, the geometric parameters which can be obtained from the electron micrographs are the individual obstalce distances l the forces f_e and the radii of curvature R. In materials with small anisotropy, the shape of the bowed dislocation segments can well be approximated by arcs of ellipses as calculated by the line tension model in isotropic cystals. The curvature of the segments may be represented by the reciprocal value S of the major half-axis.

A main result of the statistical theories is that the obstacle array can athermally be surmounted if the applied stress exceeds a critical stress

$$\tau_o = 0.95\, f_m^{3/2} / [b\, l_{sq}\, (2\Gamma)^{1/2}], \tag{2a}$$

where b is the absolute value of the Burgers vector and Γ the line tension. In dimensionless form

$$\tau_o' = 0.95\, F_m^{3/2}, \tag{2b}$$

with $F_m = f_m/2\Gamma$. If $\tau < \tau_o$, the dislocation forms stable configurations and exerts individual forces f_e on the obstacles. Further motion is only possible by thermal activation. There does not exist a similar relation between the average value of f_e and the applied stress τ. In the thermally activated case, the average value $\bar{l}$ is larger than l_{sq} and is given by [11]

$$\bar{l} = 0.812\, l_{sq}^{2/3}\, (S/2)^{-1/3}\ . \tag{3}$$

In the line tension model, mentioned above, the curvature S is related to the effective stress τ by

$$\tau = (S\, E_o / b)\, [\ln(l/r_o) + C]\ . \tag{4}$$

Here, E_o is the energy factor of the edge dislocation, r_o the inner cut-off radius and C a constant depending on the dislocation configuration.

EXPERIMENTAL RESULTS

As shown in [12], the shape of bowed-out dislocation segments in materials with low anisotropy corresponds on an average quite well to arcs of ellipses as in isotropic elasticity. A relatively simple method of evaluating dislocation curvatures consists therefore in visually comparing the shape of the segments with a set of ellipses in the proper orientation and projection calculated for different stresses. In materials with more than one direction of the Burgers vector in the respective slip plane, as in the Al-Zn-Mg alloy, the appropriate Burgers vector has to be selected on the basis of the shape of the segments and the magnitude of the orientation factor. A method for simultaneously determining l, S and f_e has been described in [2], but it is restricted to cases where the obstacle distances are much larger than the width of the dislocation images.

Generally, the dislocation images show dark or bright contrast dots at the places of the obstacles so that l can well be measured while the determination of S has to be restricted to sufficiently long segments. Typical images of dislocations under stress, taken from in situ straining experiments, are shown in Fig. 3 for the MgO single crystal and in Fig. 4 for the Al-Zn-Mg alloy. In the following, a procedure is described for determining the geometric parameters of bowed-out dislocation segments which should be applicable quite generally.

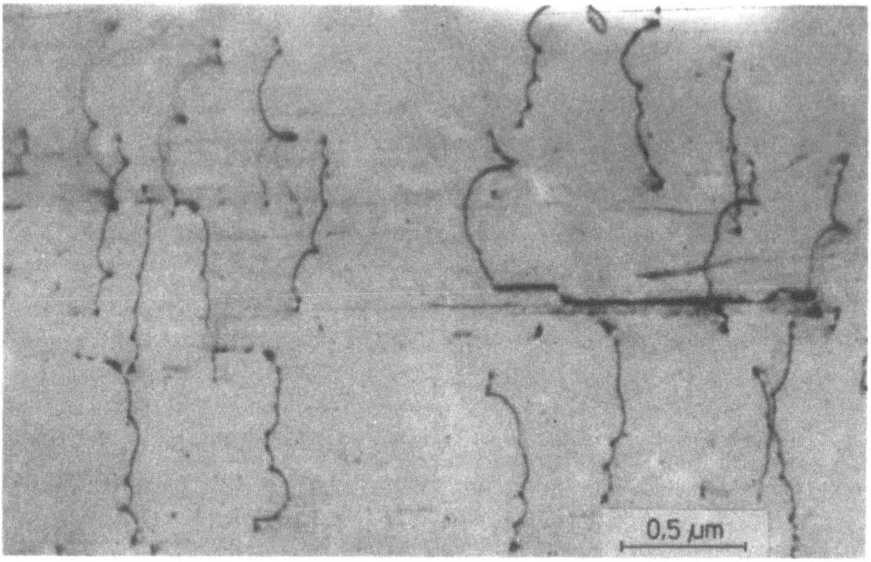

Fig. 3 Screw dislocations in MgO during in situ deformation in the HVEM. Some dislocation segments moved during the exposure of the micrograph

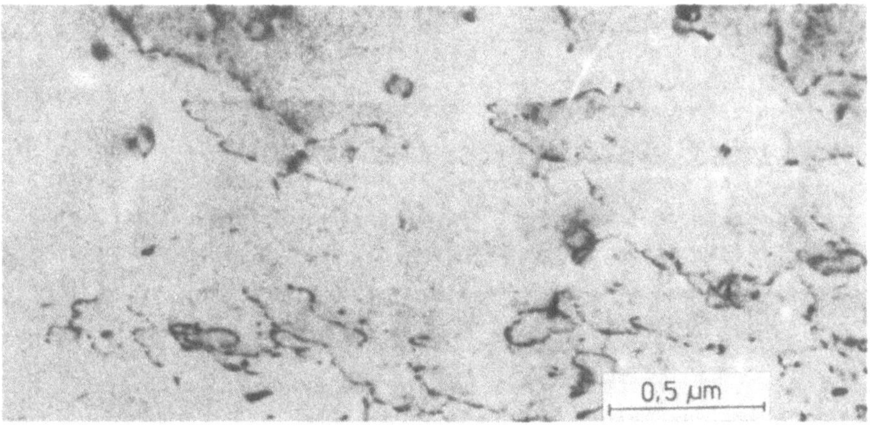

Fig. 4 Dislocations in Al-Zn-Mg under load. Slip trails indicate the glide plane

1) Measure all obstalce distances l (i.e. also the small ones) and calculate their frequency distributions as it is done in Fig. 5 for MgO and Al-Zn-Mg. These histograms turn out to be very similar but they deviate in a characteristic way from the theoretical distribution for a single type of obstacles (for MgO see [11]). Calculate the average value $\bar{l}$ as listed in Tab. 1, Column 3.

2) Deterimine S values on sufficiently long dislocation segments by comparing their shape with calculated loops as outlined above and described in [2] and [13]. Average values are contained in Tab.1, Col. 4. The distribution of S is usually symmetric (for MgO [1]) with a standard deviation of 30 to 35 per cent of the average value [3, 13].

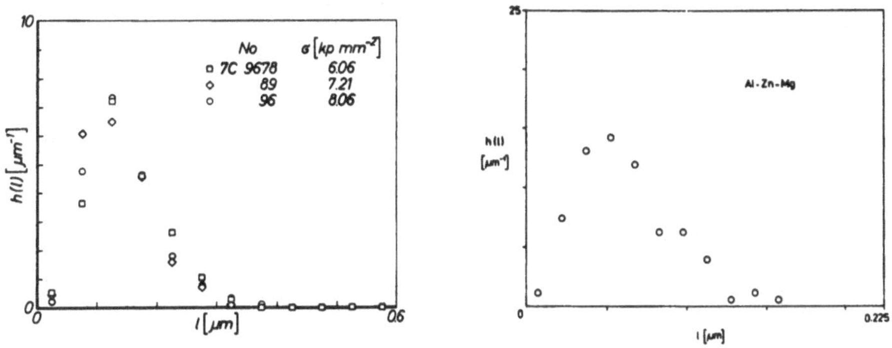

Fig. 5 Frequency distributions of obstacle distances. Left diagram: MgO. Right diagram: Al-Zn-Mg

<u>Table 1</u> Parameters characterizing the interaction between dislocations and localized obstacles in MgO and Al-Zn-Mg

	1 E_o [N]	2 r_o/b	3 $\bar{l}$ [μm]	4 S [μm^{-1}]	5 l_{sq} [μm]	6 τ'
MgO	$1.03*10^{-9}$	0.4	0.137	4.5	0.104	0.23
Al-Zn-Mg	$2.63*10^{-10}$	1	0.062	16	0.060	0.48

	7 τ [MPa] from C=-1.61	8 C	9 τ [MPa] from M	10 τ_o'
MgO	91.6	-5.19	53.5	>0.5
Al-Zn-Mg	59	-4.57	14.5	0.5

3) Calculate l_{sq} by eq. (3) as a measure of the concentration of obstacles in the slip plane (Tab. 1, Col. 5). Both materials exhibit different ratios between $\bar{l}$ and l_{sq} due to different values of the dimensionless stress $\tau' = l_{sq}S/2$ (Col. 6).

4) Calculate individual values of τ from pairs of S and l by eq. (4). This, however, imposes a problem. While the elastic energy factor E_o is well known, the constant C in the logarithmic term is it not. It has been discussed in [12] on the basis of theoretical approaches (e.g. [14]) that the best theoretical value is probably C = -1.61. The respective values of τ are listed in Tab. 1, Col. 7.

If S values are available in a relatively wide range of corresponding l values, C can be determined experimentally from plots of 1/S vs. ln $1/r_0$ with a slope $M = E_0/b\tau$ and an intercept MC, in accordance with eq. (4) (see Fig. 6). The experimental values of C are contained in Tab. 1, Col. 8. τ values based on the experimental C will be much lower than those in Tab. 1, Col. 6. Similar low values follow directly from the slope M (Tab. 1, Col. 9).

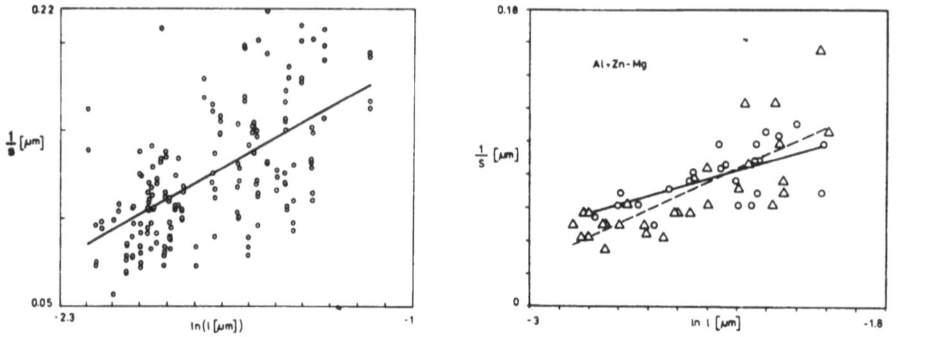

Fig. 6 Dependence of the dislocation curvature on the segment length
Left diagram: MgO (from [12]). Right diagram: Al-Zn-Mg

The evaluations described so far yield information on the density of the obstacles and on the actual level of the effective stress but they say nothing about the obstacle strength or the athermal stress τ_0, respectively. A rough estimate of τ_0 can be obtained from the dislocation kinematics as is briefly described in [6]. According to [15], the motion of the dislocations becomes more and more jerky if τ'/τ_0' increases. It is obvious from the dislocations moved during exposure in Fig. 3 as well as from histograms of slip distances [4] that in MgO the area swept after thermal activation is small. This implies according to [15] that $\tau' < 0.3\ \tau_0'$. In contrast to this, the dislocations move in the Al-Zn-Mg alloy usually over distances of more than a micrometer. Correspondingly, τ' should only be slightly smaller than τ_0'. Thus, the dimensionless athermal stress τ_0' is probably very similar for both materials (Tab. 1, Col. 10).

DISCUSSION

The outlined measurement of obstacle distances and curvatures of bowed-out dislocation segments is a good means to characterize the density of obstacles and, in connection with information on the dislocation kinematics, also their normalized strength. No data, however, can be obtained of the activation energies. All in situ straining experiments should therefore be supplemented with macroscopic measurements of the activation parameters.

A serious problem consists in relating the dislocation curvature to the effective stress. The experimental value of the constant C in eq. (4) leads to a line tension which amounts to only half the usual theoretical value. The low value of the line tension in MgO has been suggested to be due to an electrical charging of the dislocations as a consequence of radiation damage in the HVEM [16], in analogy to a model for charged dislocations in semiconductors [17]. However, the charge density necessary to explain the observed reduction would be unreasonably high. The present new data on Al-Zn-Mg indicate that the effect is probably more general.

At present, there are three possible explanations.

1) In these materials the line tension is really lower than generally assumed. This would affect the interpretation of other experiments, too.

2) The present measurements are not correctly interpreted.

3) It is an artefact of the experimental method. In principle, this is possible since the segments to be evaluated have to be selected appropriately. For both materials, however, the imaging conditions and the parameters (l_{sq}, τ') are quite different. In addition, the curvatures in MgO have been measured by fitting ellipses [1] as well as by a photometric technique [12], but the results are quite independent of these details. Thus, care should be taken in interpreting the curvature of bowed-out dislocation segments. There is hope that further studies help clarify this problem.

REFERENCES

1 U. Messerschmidt and F. Appel, In: Strength of Metals and Alloys, P. Haasen, V. Gerold and G. Kostorz (Eds.), Pergamon, Oxford 1979, p. 1305-10

2 U. Messerschmidt and F. Appel, Krist. u. Techn. 14 (1979) 1331-7

3 F. Appel, M. Bartsch, U. Messerschmidt, E.M. Nadgornyi, and S.N. Valkovskii, Phys. Status Solidi A 83 (1984) 179-94

4 U. Messerschmidt and F. Appel, In: Strength of Metals and Alloys, P.O. Kettunen, T.K. Lepisto and M.E. Lehtonen (Eds.), Pergamon, Oxford 1988, p. 139-44

5 U. Messerschmidt, F.Appel, M. Bartsch, and R. Gerlach, Phys. Status Solidi A 78 (1983) 93-102

6 U. Messerschmidt and F. Appel, Czech. J. Phys. B 35 (1985) 217-20

7 R. Labusch, J. Appl. Phys. 48 (1977) 4550

8 J.W. Morris and D.H. Klahn, J. Appl. Phys. 45 (1974) 2027

9 U.F. Kocks, A.S. Argon and M.F. Ashby, Thermodynamics and kinetics of slip, In: Progress in Materials Science, B. Chalmers, J.W. Christian and T.B. Massalski (Eds.), Vol. 19, Pergamon, Oxford 1975
10 P. Haasen, In: Dislocations in Solids, Vol. 4, F.R.N. Nabarro (Ed.), North-Holland Publ., Amsterdam 1979, p.153
11 F. Appel, U. Messerschmidt, E.M. Nadgornyi, and S.I. Zaitsev, Mater. Sci. Eng. 52 (1982) 69-74
12 U. Messerschmidt, F. Appel and H. Schmid, Philos. Mag. A 51 (1985) 781-96
13 U. Messerschmidt, Thesis Diss. B, Univ. Halle 1986
14 R.O. Scattergood and D.J. Bacon, Philos. Mag. 31 (1975) 179
15 A.I. Landau, Phys. Status Solidi A 61 (1980) 555
16 J.P. Hirth, private communication, 1986
17 J.P. Hirth and H. Ehrenreich, J. Vac. Sci. Technol. A3 (1985) 367

INTERFACE MIGRATION DURING MARTENSITIC TRANSFORMATION IN A CU-ZN-AL ALLOY

J.Stoiber, R.Gotthardt

Institut de Génie Atomique, Ecole Polytechnique Fédérale de Lausanne, 1024 Lausanne, Switzerland

ABSTRACT

Interface migration during thermoelastic martensitic transformation has been investigated by transmission electron microscopy (TEM) observations during in situ transformation. Since the martensitic phase is composed of a high density of stacking faults and interface dislocations, special emphasis has been placed on the effect of these defects on interface mobility. The observations have been made under different conditions variing temperature and applied stress. They are discussed in connection with the results of internal friction (IF) and differential scanning calorimetry (DSC) measurements, which have been performed especially during partial transformation cycling.

INTRODUCTION

Cu-base alloys showing a martensitic phase transformation have received technological interest because of their unusual mechanical properties as pseudo-elasticity and shape memory effect. It is generally accepted that a thermoelastic character of the martensite in these alloys is responsible for their capability of reversible shape changes during transformation.

However one of the most important features of martensitic transformations, namely their hysteretic transformation behaviour, is not yet fully understood. There are many observations during practical applications of Cu-base alloys that the transformation hysteresis becomes unstable after a certain number of transformation cycles. Especially small temperature and stress fluctuations during transformation lead to hysteresis instabilities and make the use of the shape memory effect rather unfavourable.

In a thermodynamical treatment of thermoelastic martensite it is assumed that there exists a local equilibrium of the chemical driving force and non-chemical forces, which controls the growth of a martensite plate (1). The non-chemical forces are the stored elastic energy and frictional forces due to interface movements. The elastic energy opposes the further martensitic transformation and promotes the reverse transformation. The friction energy necessary for the interfaces of the transformation front to overcome certain obstacles is dissipated in the specimen and represents therefore the irreversible contribution of energy involved in the thermoelastic martensitic transformation. Obstacles could be either defects in the parent phase, such as precipitates, or those created in the martensite during growth. Therefore the nature of martensite defects as well as their role during transformation have to be considered in order to understand the hysteresis effect. TEM observations turned out to be suitable in order to get informations about the microstructure of martensite and its relationship to the macroscopic hysteresis behaviour.

The substructure of Cu-Zn-Al martensites has been studied by several authors employing conventional TEM (2) and different types of high resolution electron microscope (HREM) technics (3-8). The monoclinic crystal structure (M18R type) consists of close packed layers with regular stacking faults on every third plane. Moreover random planar defects have been found on the basal planes as well as on planes close to the habit plane. On the planes nearly parallel to the habit planes the faults have a very complex structure, which is not yet identified (3,4). On the basal planes three types of stacking faults have been identified, whereby the displacement vector of the corresponding partial dislocations is not of the shear type, i.e. there exists a component perpendicular to the basal plane (2). Therefore these stacking faults should not be able to move on the basal planes.

Concerning the influence of stacking disorders on the motion of intervariant interfaces two different mechanisms have been proposed. On the one hand it is assumed that these interfaces keep their character while moving and that their motion is retarded by the stacking disorders (5). On the other hand the interfaces are supposed to change their character during motion, either by leaving a defect parallel to the interface behind (6) or by emitting a new stacking disorder (7). Recent in-situ observations of moving intervariant interfaces in a martensitic specimen under the action of an applied stress revealed that no change in stacking disorder during interface motion occures (8). In contradiction to the last statement, conventional TEM observations of in situ transformation have shown frequent changes of stacking fault contrasts during interface migration (the transformation has been induced either by cooling or by stressing the specimens) (9). However the exact nature of the observed defects could not be identified due to the fact that a detailed crystallographic analysis was not possible since this specimen stage, equipped with a cooling and a deformation device, has only one tilt axis.

In the present study the same kind of experiments as in (9) have been done in a more detailed scale in order to get more informations about the interface motion during a thermoelastic martensitic transformation. The role of defects for the interface mobility will be discussed as well as possible consequences for the transformation kinetics and the hysteresis effect in Cu-Zn-Al alloys. Furthermore the obtained results will be compared with results of IF and DSC measurements of the same alloy system.

EXPERIMENTAL

In-situ transformation has been carried out in a Jeol 200CX microscope operating at 200 kV with specimens of a Cu-Zn-Al single crystal (nominal composition 66.3 at.% Cu, 19.3 at.% Zn and 14.4 at.% Al), which was grown by the Bridgeman technic. This composition was chosen in order to have a low transformation temperature (M_s=160K) so that the transformation can take place without thermal diffusion of vacancies. The employed specimen stage allows to induce the martensitic transformation either by cooling or by deforming the specimen. Stress is applied to the specimen by imposing a certain amount of strain, which is then followed by relaxation. Three different types of experiments have been performed:

(i) Observation of martensite-austenite interfaces during cooling.

(ii) Observation of martensite-martensite interfaces during applied stress.

(iii) Observation of martensite-martensite interfaces during heating.

RESULTS AND DISCUSSION

Martensite-Austenite Interfaces

The appearance and growth of a martensitic plate has already been observed and analysed in former studies (9,10). The typical features have also been found in the present work, namely:

- The plates appear suddenly by cooling or stressing the specimen and immediately they have a certain thickness.
- The basal plane stacking faults are immediately built in during growing.
- The tip of the plate creates a local elastic stress field in the matrix, which controls further growing.
- When the specimen is heated or when the stress is released an individual plate retracts immediately due to the elastic stored energy, which promotes the reverse transformation.
- When two individual plates with different orientations are growing together and forming a common interface, reverse transformation can not take place immediately, because elastic strain fields are relaxed due to the accomodation process of the variants.

Therefore one can say that the effect of the transformation hysteresis is certainly dependent on the amount of transformed volume and on the mobility especially of the martensite-martensite interfaces. These interfaces have then to be considered as an important parameter for the hysteresis effect.

Martensite-Martensite Interfaces

Martensite-martensite interfaces were observed during migration under an applied stress as well as under heating. Preliminary as well as recent results have shown that in both cases the interface migrations are accompanied by a change in the stacking fault arrangement (9,11). This can be detected by a change of the contrast of existing stacking faults and by moving limiting partial dislocations. Fig.1 shows three pictures taken during heating of a martensitic part of the specimen with a time difference of 1sec. respectively. The film was stopped at the shown moments, the images recorded by a video-printer and then photographed in the shown arrangement. An intensive stacking fault migration can be noticed by the change of their characteristic superimposed contrasts, even while the position of the interface between these two variants does not change significantly.

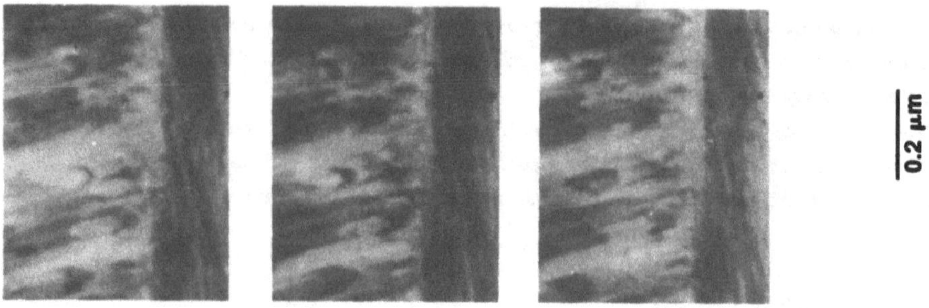

Fig.1
TEM micrographs taken from a video recording during heating a martensite plate. Two variants are shown, separated by a horizontal interface. Growing or shrinking stacking faults can be seen by the change in contrast.

The main result of the present observations is therefore that basal plane stacking faults can propagate through martensite plates, changing their stacking fault sequence, even if no transformation and no martensite-martensite interface displacement occures.

However as already mentioned in the introduction these in-situ observations are in contradiction to the TEM and HREM stacking fault analysis under static condition, which predict that shear type stacking faults do not exist. This discrepancy can be explained by the different conditions during the respective observations. In static cases internal stresses, which must appear during the transformation, are relaxed. The shear type stacking faults, which are seen during in-situ transformation when temperature and existing internal stresses change simultaneously, convert before observation in the static cases. In other words, the internal stresses, which act during in-situ transformation as a driving force for stacking fault migration, are responsible for the observed changes in the martensitic microstructure. Therefore internal stresses get an important parameter for the transformation kinetics of a thermoelastic martensite, as we will also see in the next paragraph.

<u>Effects of Partial Transformation Cycles Measured by IF and DSC</u>

Small temperature cycles in the hysteresis domain have been performed with polycrystaline specimens of a Cu-Zn-Al alloy (M_s=290K) measuring the internal friction (Q^{-1}) and the resonance frequency (f) in an inversed torsion pendelum (for the results in detail see ref.(12)). In addition the same type of temperature cycles has been measured in the DSC.

The results show that the evolution of all parameters (Q^{-1}, f and differential heat ΔH) during one cycle is not independent of the number of cycles. Furthermore a "micromemory" is found when leaving the cycling program in order to proceed the transformation. This "micromemory" is indicated by anomalies of Q^{-1}, f and ΔH when passing the cycling limit temperature, see fig.2 and 3, whereby a local maximum of Q^{-1} and ΔH and a local minimum of f appear.

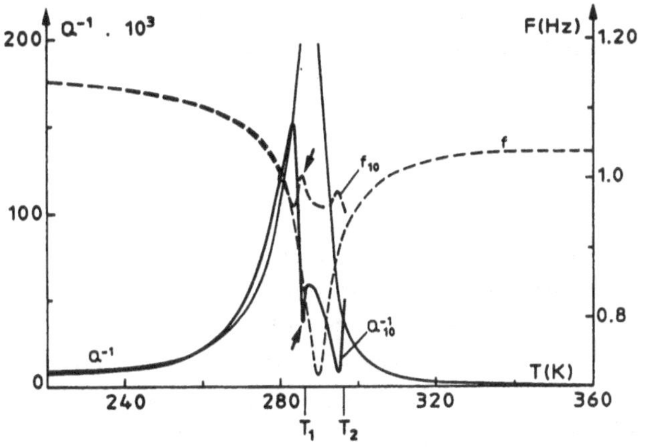

Fig.2

f and Q^{-1} during cooling of a complete cycle and after partial temperature cycling between T_1 and T_2 (f_{10} and $Q^{-1}{}_{10}$). Anomalies are indicated by flashes.

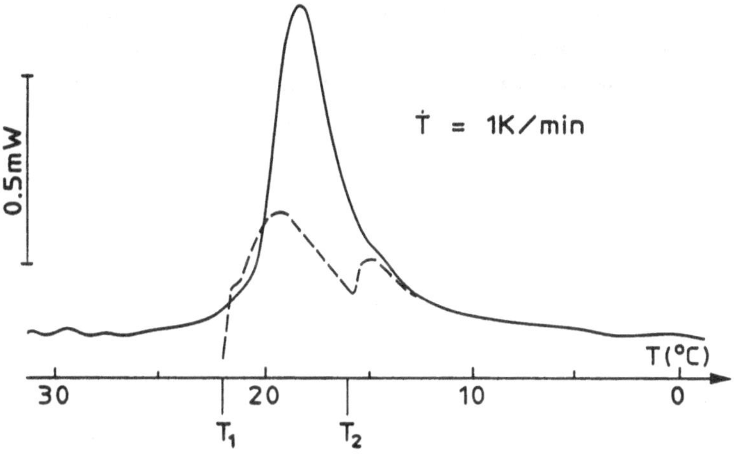

Fig.3

Differential heat (ΔH) during cooling of a complete cycle (full line) and after partial temperature cycling (dashed line) between T_1 and T_2.

From the theory of internal friction a maximum of Q^{-1} and a minimum of f (which is related to the elastic modulus) indicate that the amount of mobile defects is lowered on a local scale. The sudden decrease of ΔH at T_2 (see fig.3) can then be explained by a lowered transformation rate due to less mobile interfaces. Considering the important influence of internal stresses for the stacking fault activity inside a martensitic variant, as we have seen during in situ transformation in the TEM, the following interpretations of the observed "micromemory" after partial temperature cycling can be given:

In a continuously transformed specimen the level of internal stresses is rather high due to contraction/dilatation of the specimen and the formation of the new phase in the material. During small temperature cycles internal stresses can be reduced due to stacking fault and interface migration. When the transformation is continued after partial temperature cycling the reduced level of internal stresses lowers the driving force for further transformation and additional thermal energy is required.

CONCLUSION

TEM in situ observations of the martensitic phase transformation in a Cu-Zn-Al alloy have been performed. The results show that even in a fully transformed specimen, the martensitic microstructure changes when internal stresses are modified, e.g. during heating. In contradiction to most of the HREM investigations of the literature the stacking fault sequencies on the basal planes are frequently changing and limiting partial dislocations are migrating. This discrepancy can be understood by considering the important influence of internal stresses for the microstructure of a thermoelastic martensite.

Furthermore the results of IF and DSC measurements obtained during partial transformation cycling confirm that internal stresses and the mobility of lattice defects (stacking faults, interfaces) are strongly related.

From these results it seems to be clear that internal stresses have an important influence on the transformation kinetics and therefore on the hysteresis effect. Especially when partial transformation cycles inside the hysteresis are performed the distribution of internal stresses can change drastically compared to that of a complete transformation cycle.

REFERENCES

1 J.Ortin, A.Planes, Acta met., 36 (1988), 1873.

2 M.Andrade, M.Chandrasekaran, L.Delaey, Acta met., 32 (1984), 1809.

3 F.C.Lovey, G.Van Tendeloo, J.Van Landuyt, L.Delaey, S.Amelinckx, phys. stat.sol.(a), 86 (1984), 553.

4 M.Andrade, Ph.D.Thesis, Katholieke Universiteit Leuven, Belgium, 1983.

5 M.Fukamachi, S.Kajiwara, Jap.J.appl.Phys., 19 (1980), 479.

6 F.C.Lovey, G.Van Tendeloo, J.Van Landuyt, S.Amelinckx, Scripta Met., 19 (1985), 1223.

7 K.Adachi, J.Perkins, Met.Trans.A, 16 (1985), 1551.

8 T.Katata, H.Saka, Phil.Mag.A, 59 (1989),677.

9 R.Gotthardt, Proc.ICOMAT'82, J.de Physique, 43 (1982), C4-667.

10 R.Gotthardt, O.Mercier, J.de Physique, 42 (1981), C5-995.

11 R.Gotthardt, J.Stoiber, Proceedings of "The Martensitic Transformation in Science and Technology", Bochum 1989, to be published.

12 J.Stoiber, J.Van Humbeeck, R.Gotthardt, Proceedings of ICOMAT'89, Sydney 1989, to be published.

Acknowledgement

The authors would like to thank D.Caillard and A.Couret at CNRS Toulouse for their collaboration in observing in-situ transformation at low temperature by TEM.

AUTHOR INDEX

FSC
www.fsc.org
MIX
Papier | Fördert
gute Waldnutzung
FSC® C083411

Kolibri 360 GmbH
Ferdinand-Jühlke-Straße 7
99095 Erfurt, Deutschland
produktsicherheit@kolibri360.de